Geothermal Heating and Cooling

This publication was supported by ASHRAE Research Project RP-1674
under the auspices of ASHRAE Technical Committee 6.8,
Geothermal Heat Pump and Energy Recovery Applications.

Results of Cooperative Research between ASHRAE and Energy Information Services.

CONTRIBUTORS

Steve Kavanaugh
University of Alabama
Northport, AL
(Chapters 1–6, 9)

Kevin Rafferty
Consulting Engineer
Klamath Falls, OR
(Chapters 7–8)

PROJECT MONITORING SUBCOMMITTEE

Bill Murphy, PhD, PE, *Chair*
University of Kentucky, Paducah Campus
Paducah, KY

Jeremy Fauber, PE
Heapy Engineering
West Chester, OH

Steve Hamstra, PE
Greensleeves LLC
Zeeland, MI

Michael Kuk
CERx Solutions
Oswego, IL

Lisa Meline, PE
Meline Engineering
Sacramento, CA

Gary Phetteplace, PhD, PE
GWA Research
Lyme, NH

Updates/errata for this publication will be posted on the ASHRAE website at www.ashrae.org/publicationupdates.

RP-1674

Geothermal Heating and Cooling

Design of Ground-Source Heat Pump Systems

Steve Kavanaugh
Kevin Rafferty

Atlanta

ISBN 978-1-936504-85-5

1791 Tullie Circle, NE
Atlanta, GA 30329
www.ashrae.org

Cover Design by Tracy Becker

Library of Congress Cataloging-in-Publication Data

Kavanaugh, Stephen P., author.

Geothermal heating and cooling : design of ground-source heat pump systems / Stephen P. Kavanaugh, Kevin D. Rafferty.

pages cm.

"RP-1674."

Includes bibliographical references and index.

Summary: "Best practices for designing nonresidential geothermal systems (ground-source heat pump, closed-loop ground, groundwater, and surface-water systems) for HVAC design engineers, design-build contractors, GSHP subcontractors, and energy/construction managers; includes supplemental Microsoft Excel macro-enabled spreadsheets for a variety of GSHP calculations"-- Provided by publisher.

ISBN 978-1-936504-85-5 (hardcover : alk. paper)

1. Ground source heat pump systems. 2. Heat pumps--Design and construction. I. Rafferty, Kevin D., author. II. American Society of Heating, Refrigerating and Air-Conditioning Engineers. III. Title.

TH7417.5.K38 2014

697--dc23

2014037451

This book is dedicated to our friend Ralph Cadwallader, a tall Texan whose company installed hundreds of miles of vertical ground loops and countless water wells. He was one of the early pioneers of high-production closed-loop ground-source heat pump installations for commercial and institutional buildings. Ralph also contributed immeasurably to the industry through his participation in such organizations as the National Ground Water Association (past president), the Geothermal Heat Pump Consortium, and the International Ground Source Heat Pump Association. May he rest in peace!

Steve Kavanaugh

Dr. Steve Kavanaugh, Fellow ASHRAE, Fellow ASME, served as a professor of mechanical engineering at the University of Alabama from 1984 to 2007 and is now Professor Emeritus. He was the owner of Energy Information Services from 1993 to 2012 and currently maintains the website www.geokiss.com, a resource of HVAC and GSHP information and design tools.

Kavanaugh is the author of the ASHRAE publication *HVAC Simplified* (2006) as well as numerous other articles, and he has presented more than 140 GSHP and HVAC seminars to more than 4500 attendees on the topics of ground-source heat pumps, energy efficiency, and HVAC. These include ASHRAE professional development seminars (PDSs), short courses, and several local chapter-sponsored sessions. In 2001, he was the recipient of ASHRAE's Crosby Field Award for the highest-rated paper presented at an ASHRAE Technical Session, Symposium, or Poster Session for the year.

Kavanaugh is the Handbook Subcommittee chair of ASHRAE Technical Committee (TC) 6.8, Geothermal Energy, and has served as chair of both TC 6.8 and the now merged TC 9.4, Applied Heat Pumps and Heat Recovery. He was also an ASHRAE Scholarship Trustee in 2013–14. He served as the chair of the Board of Directors of Habitat for Humanity of Tuscaloosa from 2001–2003 and 2010–2011, and he was the construction supervisor for five homes of Habitat for Humanity of Tuscaloosa.

He has lived in a home heated and cooled by a GSHP for 30 years.

Kevin Rafferty

Kevin Rafferty, PE, is a consulting engineer and former Associate Director of the Oregon Institute of Technology Geo-Heat Center. He is the coauthor of the original GSHP book and served as co-editor of the ASHRAE special publication *Commercial Ground Source Heat Pump Systems* (1992–1995). He is also the principal author of *Geothermal Direct Use Engineering and Design Guidebook* (1998, Oregon Institute of Technology).

Rafferty has served as Handbook subcommittee chair of TC 6.8 for 16 years and as TC 6.8 chair. He was co-presenter of both the ASHRAE short course and the professional development seminar covering GSHP systems.

He has served as chair of the National Ground Water Association Geothermal Interest Group and has presented seminars on GSHPs for such clients as utilities, universities, professional associations, the U.S. Army Corps of Engineers, Geothermal Resources Council, and ASHRAE.

He has been involved the HVAC industry since 1972, rising from service technician through engineering and research roles to retirement in 2012.

Contents

4 · Applied Ground-Coupled Heat Pump System Design

5 · Surface-Water Heat Pumps

6 · Piping and Pumps for Closed-Loop Ground-Source Heat Pumps

7 · Hydrology, Water Wells, and Site Evaluation

Preface

Geothermal Heating and Cooling is a complete revision of the 1997 ASHRAE publication *Ground Source Heat Pumps: Design of Geothermal Systems for Commercial and Institutional Buildings*. The primary audience includes HVAC design engineers, design-build contractors, GSHP subcontractors, and energy/construction managers of building owners. A unique feature of interest for building owners and architects is that the book provides characteristics of quality engineering firms and information that should be provided by design firms competing for GSHP projects.

This new work takes advantage of the many lessons learned since the time of the original publication, when GSHPs were primarily residential applications. Many improvements have evolved, and performance data, both positive and negative, is available to guide the development of best practices. Information was gathered from ASHRAE and GSHP-industry research and development projects, measured data from long-term installations, and optimized installation practices used by high-production GSHP contractors. As part of the revision, new research was conducted in critical areas not adequately addressed in previous projects.

Seven of the original eight chapters and appendices were completely rewritten and include coverage of closed-loop ground (ground-coupled), groundwater, and surface-water systems, as well as GSHP equipment and piping. Additional information on site characterization has been added, including a new hydrogeological chapter. The final chapter was replaced and contains results of recent field studies, energy and demand characteristics, and updated information to optimize GSHP system cost.

Substantial effort was taken to develop tables, graphs, and equations in both Inch-Pound (I-P) and International System (SI) units, though there are a few instances where content is supplied in I-P units only. Appendix A provides a screenshot of UnitsConverter.xlsx that is useful for manual conversion of units from I-P to SI and vice versa, and Appendix B offers a list of references to publications and standards with information on procedures and specifications that are specific to the GSHP industry.

In addition, this book is accompanied by Microsoft® Excel® macro-enabled spreadsheets, which can be found at www.ashrae.org/GSHP. The spreadsheet tools include *UnitsConverter.xlsx*, *HVACSystemEff.xlsx*, *BoreResistance.xlsm*, *E-PipeAlator14.xlsm*, *WAHPCorrector14.xlsm*, *GroundTemp&Resist.xlsm*, and *Heat Exchanger Temperature Prediction*. These files can be used for a variety of GSHP calculations. If the files or information at the link are not accessible, please contact the publisher.

Acknowledgments

From Steve Kavanaugh

Gratitude is extended to the members of the Project Monitoring Subcommittee who reviewed this text and provided many very useful suggestions for improvement. The reviewers included Bill Murphy (PMS Chair), Jeremy Fauber, Steve Hamstra, Gary Phetteplace and Lisa Meline. Kirk Mescher, Roxanne Scott, Dan Pettway, and Lisa Meline provided the advocacy and support to ensure the project was undertaken.

I feel especially fortunate to have had Dr. Jerald Parker as my advisor at Oklahoma State University. He is a model educator not only in terms of technical knowledge but also in his lifelong joyful commitment to students. I have tried to treat my students as well as he treated me. Thus, a great deal of the information contained in this book resulted from the hard work of many students at the University of Alabama (see listing that follows). In addition to coauthor Kevin Rafferty, this work has also benefitted from association with many colleagues, especially Joey Parker, Allan Skouby, Chuck Remund, Daniel Morris, Barry Johnson, Mike Green, David Dinse, Lonnie Ball, Charles Davis, Charles Smith, Harold Olsen, and, of course my dad, Joe Kavanaugh, who started my interest in GSHPs by installing one in our home in 1959.

From Kevin Rafferty

I'm especially indebted to Steve Kavanaugh for inviting me to join him in the original edition of this book in 1994. In any writing project, and particularly one encompassing as broad a scope as this, the authors, and hence the content, are influenced by a great many individuals. Though only two names appear on the cover, the following have contributed directly or indirectly to its production. Thanks to Earl Baumgartner and Joe Panczak for giving me a start in the HVAC business over 40 years ago. To Gene Culver, Associate Director (retired), OIT Geo-Heat Center, for sharing his geothermal expertise over the past 35 years and for his careful review of Chapters 7 and 8; Darryl Anderson of Anderson Engineering, Lakeview, OR, for his review of Chapters 7 and 8 and sharing his extensive collection of drilling photos; Quinn Dellinger of Cal State Sacramento for the review of Chapters 7 and 8; John Harms of Anderson Engineering for assistance with figures; the hundreds of GSHP seminar attendees from across the United States and Canada whose questions, comments, and arguments have molded the format and content of the information included here. Thanks also to Mike Schnieders of Water Systems Engineering, Ottawa, KS, for permission to reprint his water analysis report (Appendix N).

University of Alabama Students Who Contributed to GSHP Research and Development

Evelyn Baskin
Timothy (Hugh) Calvert
Roman Carter
Kevin Cash
James (David) Deerman
Nickless Devin
Keith Dorsey
Keith Duncan
Bob Falls
Xingshun Gao
Chris Gilbreath
Chris Hill
James Hogland
Joe Hoggle
Kevin Johnson
Errol Jones
Joshua Kavanaugh
Kevin Kavanaugh
Kristofor Kavanaugh
Steven Lambert
Barbara (Hattemer) McCrary
Sanjay Mahaptra
Chad Martin
Daphne Messer
Oddis Mitchell
Eric Nason
Marcus Pezent
Rodney Phillips
Mark Pugh
Richard Rayborn
Randy Roberts
Chris Stripling
Wesley Shearer
James Wilson
Lan Xie
Zer Kai Yap
Jing Yu

Symbols, Acronyms, and Abbreviations

α	thermal diffusivity
AHU	air-handling unit
AHRI	Air-Conditioning, Heating, and Refrigeration Institute
ANSI	American National Standards Institute
AWWA	American Water Works Association
BAS	building automation system
BEP	best efficiency point
bhp	brake horsepower
Btu/h	British thermal units per hour (heat rate unit)
c_p	specific heat
C_v	flow coefficient (flow in gpm that results in Δp = 1.0 psi)
CF (C_f)	correction factor
cfm	cubic feet per minute, ft^3/m
CTS	copper tube size
COP	coefficient of performance, W/W
CO_2	carbon dioxide
Δ	delta (difference)
db	dry bulb (temperature)
DD	drawdown
DOAS	dedicated outdoor air system
DR	dimension ratio (outside diameter/wall thickness)
DX	direct expansion (of refrigerant)
e	roughness (pipe wall)
EAT	entering air temperature
EATDB	entering air dry-bulb temperature
EATWB	entering air wet-bulb temperature
ECM	electronically commutated motor
EER	energy efficiency ratio (for cooling), Btu/Wh or kBtu/kWh
EFLH	equivalent full-load hours
EIA	Energy Information Administration (U. S. Department of Energy)
ELT	entering liquid temperature (used instead of entering water temperature, EWT, when fluid is not pure water)
EPA	U.S. Environmental Protection Agency
ERU	energy recovery unit (sensible and latent heat)

ESP	external static pressure
EWT	entering water temperature
g	acceleration of gravity
g_c	constant to relate mass, length, force, and time [= 32.2 $lb_m \cdot ft/lb_f \cdot s^2$ (I-P), = 1.0 (SI)]
GCHP	ground-coupled heat pump (also called closed-loop ground-source heat pump, GSHP)
GLHP	ground-loop heat pump (also called ground-coupled heat pump, GCHP)
gpm	gallons per minute
GSHP	ground-source heat pump
GWHP	groundwater heat pump (also called open-loop ground-source heat pump, GSHP)
η	efficiency
HC	heating capacity
HDPE	high-density polyethylene (piping material)
hp	horsepower (unit of power, = 0.746 kW)
HVAC	heating, ventilating, and air-conditioning
Hz	frequency unit (cycles/second)
ID (d_i)	inside diameter
IPS	iron pipe size
ISO	International Organization for Standardization
IWL	injection water level
k	thermal conductivity
kW	kilowatt (unit of power or heat rate)
kWh	kilowatt-hour (unit of electrical energy)
kW/ton	kilowatt per ton, electrical demand per unit cooling capacity, kW_{refrig}/kW_{elect}
LEED®	Leadership in Energy and Environmental Design®
LLT	leaving liquid temperature (used instead of leaving water temperature, LWT, when fluid is not pure water)
LMTD	log mean temperature difference, °F (°C)
L/min	litres per minute
L/s	litres per second
LSI	Langlier saturation index
LWT	leaving water temperature
kBtu/h	British thermal units per hour × 1000 (heat rate unit)
NBR	nitrile butadiene rubber
NGWA	National Ground Water Association
NPSH	net positive suction head
NWWA	National Water Well Association
OD (d_o)	outside diameter
Pa	pascal (pressure)
PE	polyethylene
PEX	cross-linked polyethylene (tubing)
PLF	part-load factor
ppm	parts per million
psi	pounds per square inch (unit of pressure)
PVC	polyvinyl chloride (piping material)
PWL	pumping water level
q	heat rate, Btu/h or kW

Q	volumetric flow rate
ρ	density
R	thermal resistance
Re	Reynolds number ($= \rho DV/\mu$)
RSI	Ryznar stability index
rpm	revolutions per minute
Sch	Schedule (pipe dimension)
SEER	seasonal energy efficiency ratio (for cooling), Btu/Wh or kBtu/kWh
SC	sensible cooling capacity (thermal) or specific capacity (of water well flow rate)
SDR	standard dimension ratio (outside diameter/wall thickness)
SWHP	surface-water heat pump
SWHE	surface-water heat exchanger
SWL	static water level
τ	time
t	temperature, °F (°C)
TC	total cooling (capacity) or thermal conductivity
ton	cooling capacity (12,000 Btu/h, rate required to freeze 2000 pounds of water in 24 hours)
UFAD	underfloor air distribution
USGS	U.S. Geological Survey
VAV	variable air volume
VFD	variable-frequency drive (also called variable-speed drive, VSD)
VSD	variable-speed drive (also called variable-frequency drive, VFD)
wb	wet bulb (temperature)
WLHP	water-loop heat pump (a.k.a water-source heat pump, WSHP)
WSHP	water-source heat pump (a.k.a water-to-air heat pump; water-to-water heat pump; water-loop heat pump, WLHP)
X	dimensionless number for line heat source equation $\{= r/[2(\alpha\tau)^{0.5}]\}$

1 Introduction to Ground-Source Heat Pumps

1.1 OVERVIEW, NOMENCLATURE, AND GSHP TYPES

Ground-source heat pump (GSHP) is an all-inclusive term for a variety of systems that use the ground, groundwater, or surface water as a heat source and sink. GSHPs are subdivided by the type of exterior heat exchange system. This includes ground-coupled heat pumps (GCHPs) that are closed-loop piping systems buried in the ground, groundwater heat pumps (GWHPs) that are open-loop piping systems with water wells, and surface-water heat pumps (SWHPs) that are closed-loop piping coils or open-loop systems connected to lakes, streams, or other reservoirs. Heat pumps are located in the buildings and cool by removing indoor heat and rejecting it to the exterior GSHP loop. In heating, the process is reversed as heat is removed from the outdoor loop by the heat pumps and is delivered to the building. Many parallel terms exist for GSHPs, such as *geothermal heat pumps* (GHPs), *earth energy systems*, and *GeoExchange®* systems that are used to meet a variety of marketing or institutional needs. However, ASHRAE (2011) has established a standard nomenclature to which this book attempts to conform.

GSHPs initially were more widely applied to residential buildings but are now increasingly being utilized in the commercial and institutional sectors. The economics of GSHPs can be very attractive in larger buildings because elaborate equipment and controls are not required to provide comfort and high efficiency. When simple design approaches are followed, the added cost of ground heat exchangers can be offset to a large extent. Simple designs also have the advantage of reducing maintenance requirements, which can be very attractive to building owners with minimal maintenance resources (e.g., schools). However, simply attaching a ground heat exchanger, groundwater loop, or surface-water coils to conventional water-cooled HVAC systems (e.g., chilled-water variable-air-volume systems) usually results in higher installation costs, poor efficiency, and added maintenance requirements. Typical installation recommendations, design guides, and conventional approaches must be amended in order to take full advantage of these systems. This book provides engineers with GSHP design methods that deal with larger multiple-zone buildings with diverse loads and occupancy patterns. Other sources (Remund 2011; Kavanaugh 1991) provide detailed treatment of the design and installation of residential and light commercial GSHPs.

GSHPs are rarely effective in cooling-only or heating-only applications. Thus, heat pumps of some type are connected to the exterior ground, groundwater, or surface-water loops to provide cooling and heating inside the building. The most widely used unit is a water-to-air heat pump as shown in Figure 1.1. Water or water-antifreeze solution circulates through a liquid-to-refrigerant heat exchanger. Air to be heated or cooled is circulated through a conventional finned-tube air-to-refrigerant heat exchanger and air distribution system. In applications where the heat pumps are located near an area where a water heating load is present (i.e., a kitchen), optional heat recovery heat exchangers can be included. Packaged heat pumps in the range of 0.5 to 50 tons (2 to 175 kW) are available. Note that small and mid-size units typically have higher efficiencies because of the lower fan power requirements compared to larger units that often have fans with much higher total static pressure in order to provide circulation through more extensive air distribution networks.

Water-to-water heat pumps as shown in Figure 1.1 are also commonly used and can be especially effective when the building water-loop temperatures are not extreme. Thus, in-floor heating systems that might only require maximum temperatures near 100°F (38°C) and chilled-beam applications with temperatures near 55°F (13°C) tend to have higher efficiencies. Good efficiencies can also be attained using low-static-pressure fan-coil units (FCUs) and water-to-water heat pumps with supply water-heating temperatures below 115°F (46°C). However, large central air-handing units (AHUs) with high total-static-pressure fans and/or systems that require higher heating-mode supply temperatures (>120°F [49°C]) are not recommended if system efficiency and low operating costs are primary goals.

A third type of GCHP is the direct-expansion (DX) GCHP, which uses a buried copper piping network as one of the heat pump coils through which refrigerant is circulated. These systems normally incorporate a forced-air distribution system, although hydronic systems can also be used. Systems using water-to-air and water-to-water heat pumps are often referred to as *GCHPs with secondary solution loops* to distinguish them from DX GCHPs. This book concentrates on the design of secondary solution systems; DX GCHPs are not covered.

Chapter 2 of this book covers in more detail heat pump equipment, system efficiencies, and accompanying accessories.

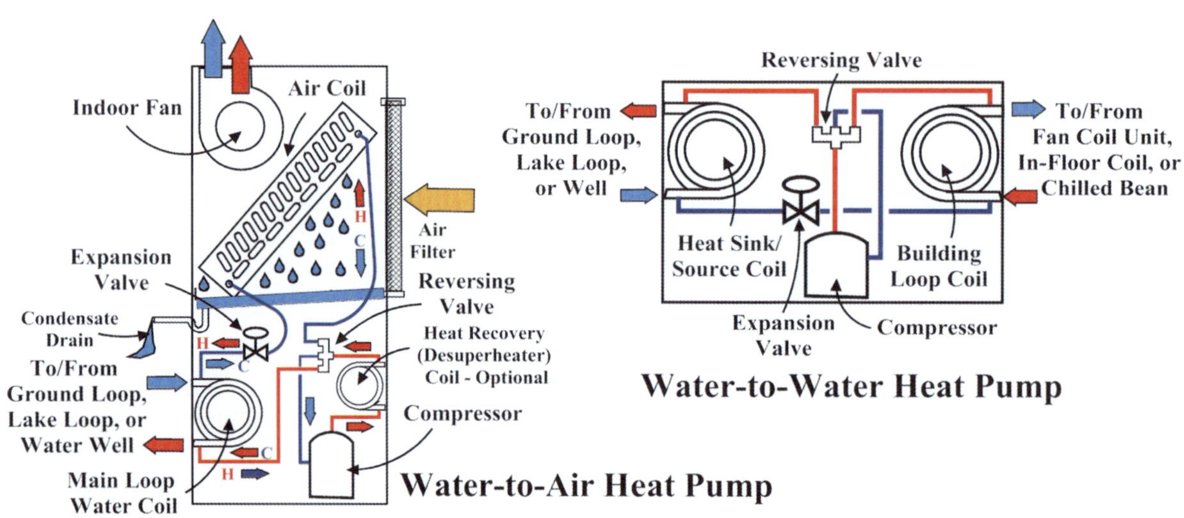

Figure 1.1 Primary GSHP Equipment Options

1.2 GROUND-COUPLED HEAT PUMPS

GCHPs are a subset of GSHPs and are often referred to as *closed-loop ground-source heat pumps*. A GCHP refers to a system that consists of a network of heat pumps that are linked to a closed ground heat exchanger buried in the soil. GCHPs are further subdivided according to ground heat exchanger design.

Vertical GCHPs are by far the most common type. The ground heat exchanger is usually constructed by placing two high-density polyethylene (HDPE) tubes in a vertical borehole as shown in Figure 1.2. The tubes are thermally fused at the bottom of the bore to a close return U-bend. Standard prefabricated vertical tube sizes range from 3/4 to 1 1/4 in. (25 to 40 mm) nominal diameter. Common bore depths range between 200 and 300 ft (60 and 90 m), but local drilling conditions may dictate they be shorter or, in many cases, over 400 ft (150 m) in depth. Deeper bores are not common and caution is required to offset deep-bore hydrostatic conditions and added pipe head losses even when the largest standard-sized U-tubes are applied (see Appendix C).

The advantages of vertical GCHPs are that they require relatively small plots of ground, are in contact with soil that varies very little in temperature and thermal properties, require the smallest amount of pipe and pumping energy, and can yield the most efficient GCHP system performance. The disadvantage is that they are typically higher in cost because of limited availability of appropriate equipment and installation personnel. In some cases, when the cooling requirements exceed the heating needs, installation cost can be reduced by installing a hybrid system with ground loop sized to meet the heating requirement in parallel with a fluid cooler or cooling tower. These systems require added maintenance, added controls, and following ASHRAE (2000) guidelines to minimize the risks associate with cooling towers. The system design of vertical GCHPs is the focus of Chapters 3 and 4 of this book.

Horizontal GCHPs can be divided into three subgroups: single pipe, multiple pipes, and coiled pipe that looks like a Slinky™ toy. Initial designs of single-pipe horizontal

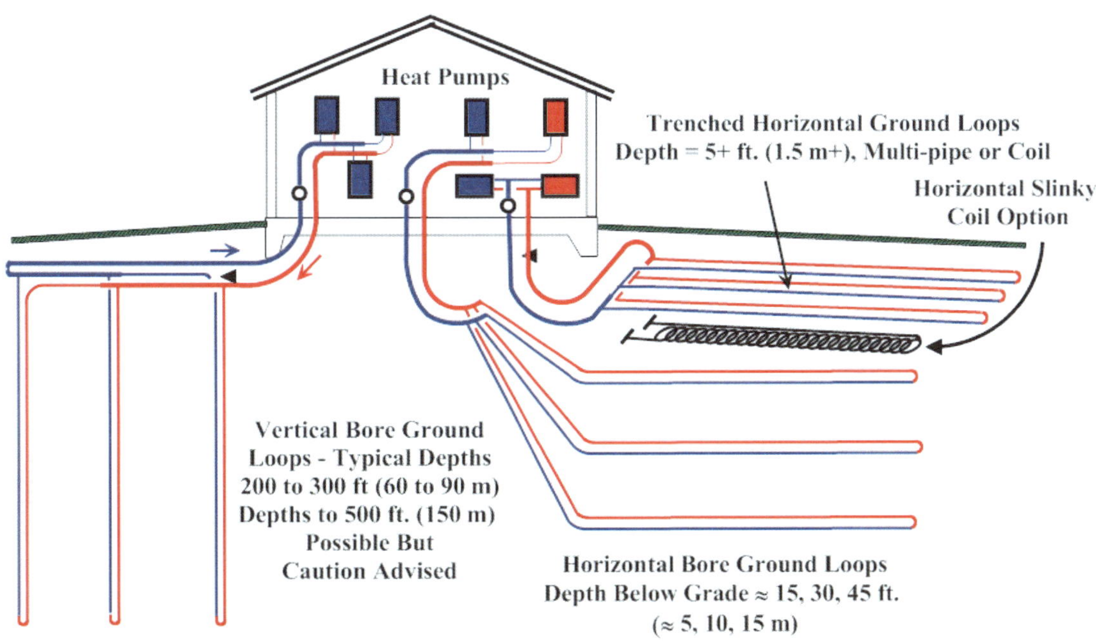

Figure 1.2 Closed-Loop Ground-Coupled Heat Pump with Three Ground-Loop Options

GCHPs had them placed in narrow trenches at least 5 ft (1.5 m) deep. These designs require the greatest amount of ground area. Multiple pipes (usually two or four) placed in a trench at a greater depth than the minimum (5 ft [1.5 m]) can reduce the amount of required ground area. Contractors have used either deep, narrow trenches (dug with a chain-type trencher) or wide trenches (dug with a backhoe) with pipes separated by 12 to 24 in. (30 to 60 cm). Although trench length can be reduced, total pipe length must be increased with multiple-pipe GCHPs in order to overcome thermal interference with adjacent pipes in the same trench. The slinky coil is reported to also reduce required ground area. These horizontal ground heat exchangers are constructed by stretching small-diameter HDPE tubing from the tight coil in which it is shipped into an extended coil that can be placed vertically in a narrow trench or laid flat at the bottom of a wide trench.

Horizontally bored ground loops are a crossover between vertical and horizontal ground loops. Horizontal drilling machines can install heat exchangers deeper and use multilayer placement of U-tubes, which substantially reduces the required land area compared to shallow horizontal loops. As with vertical loops, the surrounding ground temperature and thermal properties vary little with season. Thus, horizontally bored ground loops are well suited to larger building applications. (See Appendix D for information on vertical-loop installation equipment and procedures.)

The advantages of horizontal GCHPs are that they are typically less expensive than vertical GCHPs in residential and small (< 20 ton [70 kW]) commercial building applications because appropriate installation equipment is often more widely available and many residential applications have adequate ground area. These GCHPs (except for deep horizontally bored loops) are less commonly used in commercial and institutional buildings because of the larger ground area required. Other disadvantages include greater adverse variations in performance because horizontal ground temperatures and thermal properties fluctuate with season, rainfall, and burial depth; slightly higher pumping energy requirements; and lower system efficiencies. Remund (2011) covers the design and installation of horizontal GCHPs in greater detail.

1.3 GROUNDWATER HEAT PUMPS

The second subset of GSHPs is groundwater heat pumps (GWHPs). Until the recent development of GCHPs, GWHPs were the most widely used type of GSHP. GCHP systems were developed in part in response to the widespread water quality problems experienced by residential GWHP systems in the 1960s and 1970s. In the commercial sector, plate heat exchangers are used to isolate the building loop from exposure to groundwater, eliminating water quality problems in the building. While the cost of the ground heat exchanger per ton of capacity is relatively constant for a GCHP, the cost of a well-water system (on a per-ton [per-kW] basis) is much lower for a large system (Rafferty 1995), as discussed in Chapter 8. A single high-volume well can serve an entire building. Properly designed GWHP systems require more maintenance than GCHP or closed-loop SWHP systems, but this cost is small in the context of the potential capital cost savings (see Chapter 8).

Various systems are possible. A widely used system places a central water-to-water heat exchanger between the groundwater and a closed water loop that is connected to water-to-air heat pumps located in the building (Figure 1.3). In smaller buildings (<20 tons [70 kW]), it is possible to circulate the groundwater directly through each heat pump at the risk of corrosion and fouling of heat exchangers and control valves that may result when untreated water is circulated through a distributed system. A third possibility is to circulate groundwater through a central chiller (or heat pump) and to heat and cool

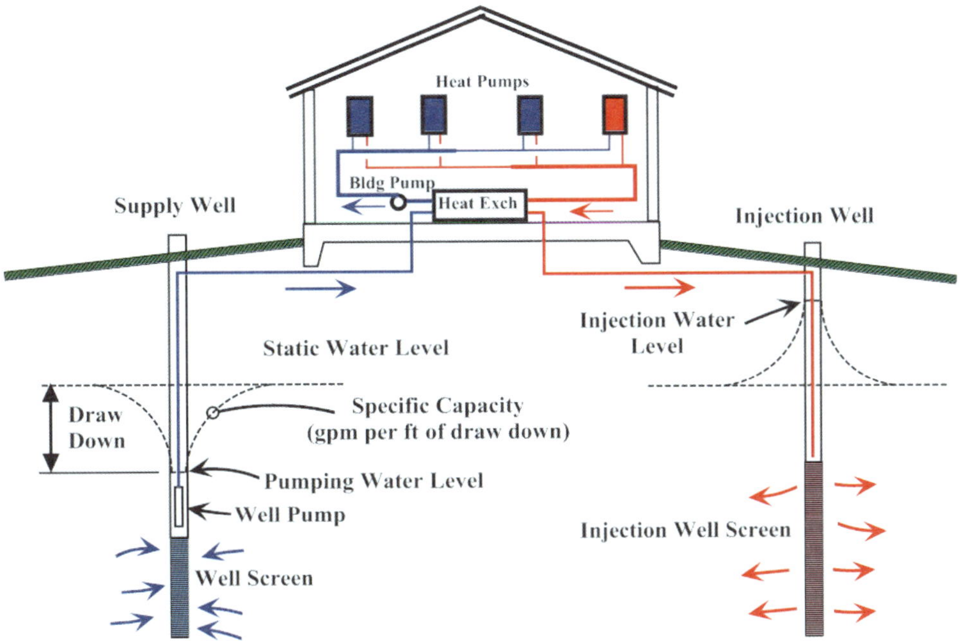

Figure 1.3 Open-Loop Groundwater Heat Pump with Isolation Heat Exchanger

the building with a conventional chilled- and hot-water distribution system, though central chiller systems tend not be as energy efficient as unitary designs.

All three types of systems (and other variations) lend themselves to the possibility of direct precooling or cooling in much of the United States. Low-temperature groundwater (<58°F [15°C]) can be circulated through hydronic coils in conjunction with heat pumps. This can displace a large amount of energy required for cooling, especially when precooling outdoor ventilation air. Direct cooling is possible with colder water found in the northern portion of the US.

The advantages of GWHPs are that they are lower in cost compared to GCHP systems, the water well is very compact, water well contractors are widely available, and the technology has been used for decades. Disadvantages are that local environmental regulations may preclude use or injection of groundwater, water availability may be limited, fouling precautions may be necessary if the well is not properly developed or water quality is poor, and pumping energy may be excessive if the pump is oversized or poorly controlled.

1.4 SURFACE-WATER HEAT PUMPS

Surface-water heat pumps (SWHPs) have been included as a subset of GSHPs because of the similarities in applications and installation methods. SWHPs can be either closed-loop systems similar to GCHPs or open-loop systems similar to GWHPs. However, thermal characteristics of surface water bodies are quite different from those of the ground. Some unique applications are possible and special precautions are warranted.

Closed-loop SWHPs consist of water-to-air or water-to-water heat pumps located in a building and connected to a piping network placed in a lake, river, or other open body of water (Figure 1.4). A pump circulates a water-antifreeze solution through the heat pump's water-to-refrigerant coils and the submerged piping loop that transfers heat to or from the

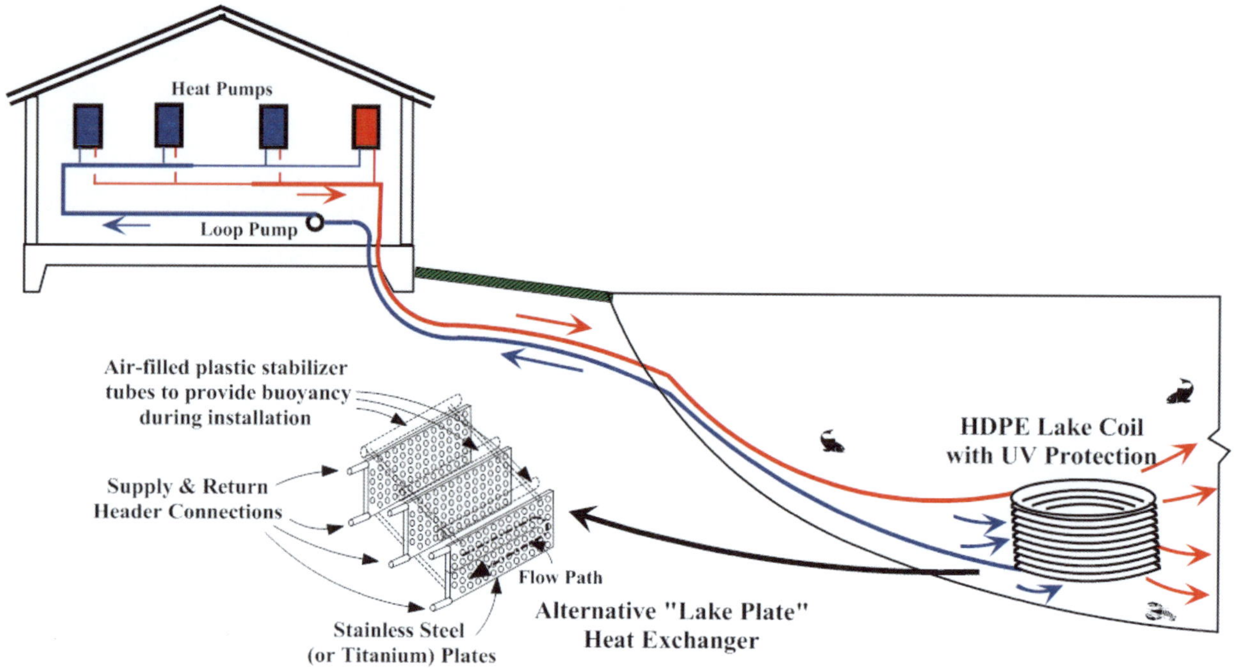

Figure 1.4 Closed-Loop Surface-Water Heat Pump with Two Lake Coil Options

lake. The recommended piping material is thermally fused HDPE with some type of ultraviolet radiation protection. Copper and other types of plastic tubing have also been used, but polyvinyl chloride (PVC) should be avoided. Many installations have used 3/4 in. or 1 in. (25 or 32 mm) HDPE tubing for the primary heat exchanger coils. Larger-diameter, thicker-wall tubing is recommended for areas in which damage from boats is a possibility. Coils are normally arranged in multiple parallel piping patterns to minimize pressure losses. Plate heat exchangers as shown in Figure 1.4 are also available with stainless steel or titanium materials. The main header pipes connecting the primary heat exchanger coils are sized to minimize losses, and they are normally of larger diameter than the individual coil tubing. Additional ASHRAE research is in progress to develop design tools for SWHPs systems (ASHRAE 2009), but results are not yet available.

The advantages of closed-loop SWHPs are relatively low cost (compared to GCHPs), low pumping energy requirements, high reliability, low maintenance requirements, and low operating costs. Disadvantages are the possibility of coil damage in public lakes and wide temperature variations with outdoor conditions if lakes are small and/or shallow. This would result in some undesirable variations in efficiency and capacity, but they would not be as severe as with air-source heat pumps.

Open-loop SWHPs can use surface water bodies in a manner similar to cooling towers, without the need for fan energy or frequent maintenance. In warm climates, lakes can also serve as heat sources during the winter heating mode. However, closed-loop systems are the only viable option for heating in moderate and colder climates.

Surface water can be pumped directly to water-to-air or water-to-water heat pumps or through an intermediate heat exchanger that is connected to the units with a closed piping loop. Direct systems tend to be smaller, with only a few heat pumps. In deep lakes (≥40 ft [≥12 m]), thermal stratification often exists throughout the year to the extent that direct cooling or precooling is possible. Water can be pumped from the bottom of deep lakes through heat exchangers in the return air duct. Total cooling is a possibility if water is

50°F (10°C) or less. Precooling is possible with slightly warmer water that can then be circulated through the heat pump units. Section 5.8 in Chapter 5 provides recommendations for direct cooling and precooling system design.

Water pump options fall into three categories: above surface, vertical pumps with submerged impellers and above-surface motors, and submersible. Above-surface pumps must have low net positive suction head (NPSH) requirements, and precautions must be taken to ensure water remains in the pump during off cycles. Vertical pumps with submerged impellers connected to above-surface motors are often an alternative if precautions are taken for lake level fluctuations. Submersible pumps can serve as a flexible alternative. Low-head single-stage types can be used if the building is located near the lake. Multistage units can provide water for greater elevations and distances. Filtration of coarse particles and objects can be accomplished on the suction side of any of the above pumps. This is often sufficient if heat exchangers are equipped to be periodically flushed. A thorough feasibility study for a large central New York chilled-water system presents detailed design, environmental, and economic information on existing direct cooling systems (SUNY 2011). Although somewhat dated, the information by Kavanaugh (1991) provides some additional details regarding residential SWHP systems and design recommendations for direct cooling and precooling with surface water or groundwater.

1.5 EXTERIOR AND BUILDING LOOP PIPING OPTIONS

Conventional wisdom assumes the best practice for large piping loops is to incorporate a central loop with large variable-speed pumps and two-way control valves on the HVAC equipment. As discussed in the following sections, field studies have shown this assumption is often incorrect for closed-loop GCHP systems (Kavanaugh and Kavanaugh 2012). While this practice has some economic advantages in conventional chilled-water systems, GWHPs, and SWHPs, the economy of scale is not present to the same degree with GCHPs, especially in large-footprint one- and two-story buildings. Although building diversity often results in reduced length for central ground loops, the total cost of the system (especially in large-footprint buildings) will be greater because of the added cost of extensive runs of large-diameter interior piping. Multiple interior loops also afford the possibility of using HDPE and fiber-core polypropylene, thus eliminating the need for corrosion inhibitors. This could be an important factor in locations where certain chemicals are prohibited from being circulated in deep underground piping that is in contact with sensitive aquifers. Engineers should carefully consider other options, some of which are shown in Figure 1.5. Figures 1.6 to 1.11 demonstrate other common options, which are discussed in greater detail with additional variations in later chapters.

1.6 FIELD STUDY RESULTS

Results of a field study of long-term performance of 40 commercial and institutional buildings with GCHP systems have appeared in a series of seven articles in *ASHRAE Journal*. Energy performance in terms of ENERGY STAR® rating (EPA 2012) was categorized by the loop types shown in Figures 1.6, 1.7, 1.8, and 1.9 (Kavanaugh and Kavanaugh 2012). An explanation of the ENERGY STAR rating method and additional details of the long-term GSHP performance monitoring project appear in Chapter 9. Additional buildings were monitored to supplement background information for this book. Appendix E presents results for one of the monitored buildings.

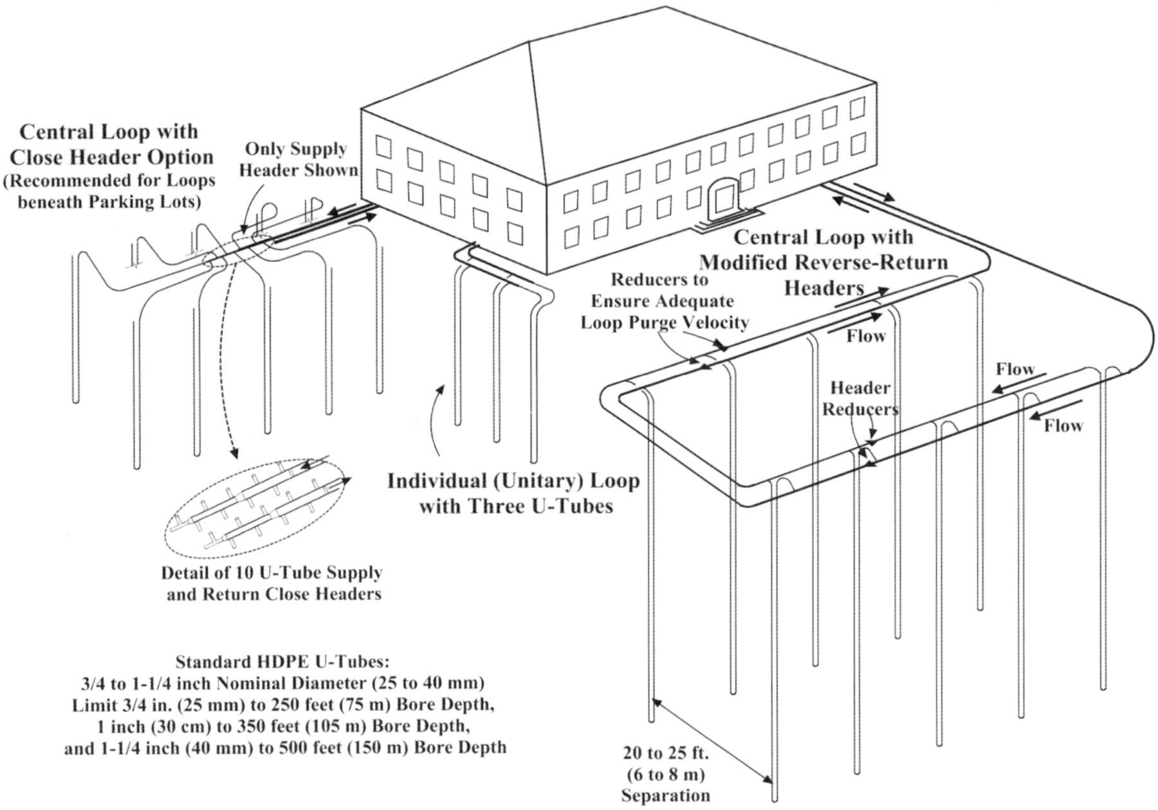

Figure 1.5 Three Options for Closed-Loop Heat Pump Vertical Ground-Loop Circuits

Four of the monitored buildings in the field study have unitary loop systems as shown in Figure 1.6. Each unit is connected to an individual ground loop consisting of two, three, or four vertical U-tubes. Water-loop circulation is provided by small on-off pumps. Larger, less frequently occupied spaces such as cafeterias and gyms are conditioned by air-cooled equipment. All four buildings are schools (two elementary, one middle, and one high school) located in a hot climate and were built between 1996 and 2001. The classrooms, offices, and libraries are heated and cooled by water-to-air heat pumps. ENERGY STAR ratings ranged from 93 to 100 with an average of 97. (An ENERGY STAR rating of 97 indicates the building uses less source energy than 97% of buildings of this type when corrections for climate, occupancy, schedule, and internal loads are applied. EPA [2012] provides details.)

Six of the monitored buildings in the study are served by multiple water-to-air heat pumps connected to a one-pipe building loop as shown in Figure 1.7. When a unit is activated, liquid is removed from the loop by a low-head circulator pump on each unit and discharged a short distance downstream. Main pumps, controlled by loop temperature, provide continuous circulation to ensure no recirculation occurs. As shown in Figure 1.7, the ground loop is a conventional two-pipe reverse-return network. All six sites are schools (five elementary and one middle school) located in Illinois. One school was built in 1938 and the others were built in the 1950s. The buildings were retrofitted with the GSHPs between 2006 and 2008. Each school is heated and cooled by water-to-air heat pumps connected to the central one-pipe loop. ENERGY STAR ratings ranged from 82 (1938 school) to 99 with an average of 94. When the older building is not considered the average rating of the five 1950 vintage schools was 96.

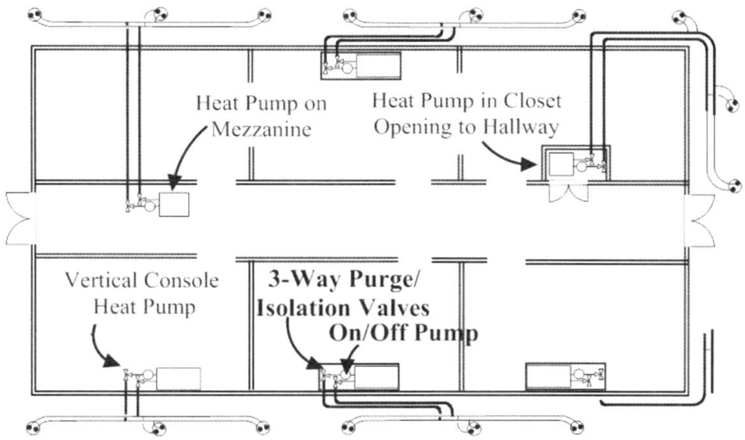

Figure 1.6 Unitary-Loop GCHP with Each Heat Pump Connected to Individual Loops

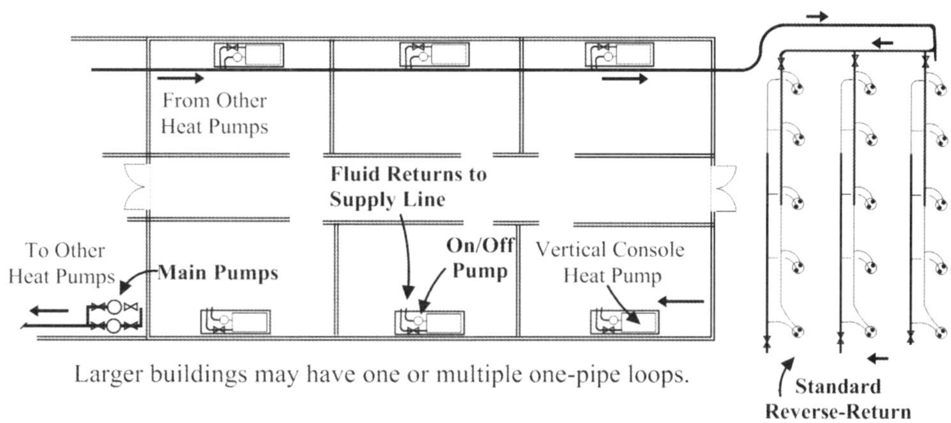

Figure 1.7 One-Pipe Loop GCHP with Reverse-Return Header Ground Loop

Five of the monitored buildings had common-loop systems as shown in Figure 1.8. Multiple water-to-air heat pumps are connected to a common two-pipe loop. Each unit has its own on-off circulator pump that circulates water through the entire common building and ground loop. Check valves are installed on the pump discharge to prevent reverse circulation from other units when the pump and heat pump are not operating. Four of the buildings have multiple common loops (thus the alternative term *subcentral* for *common*) with 2 to 15 heat pumps on each loop. One building has a single common loop for the entire building with flow provided by small circulator pumps on each heat pump. Four of the sites are schools (three elementary and one middle school) and one is an office. Four buildings are located in Alabama and one elementary school is in Kentucky. The Kentucky school was built in 2007 and the Alabama office was built in 1993. The Alabama middle school was built in 1929 and the elementary schools in the 1950s. Portions of all three schools were retrofitted with the GSHPs in 2002. ENERGY STAR ratings ranged from 97 for the Kentucky school down to 21 for the Alabama office. The low score for the office resulted from the use of multiple large pumps that operated continuously. Only 29% of the middle school was conditioned with a GCHP, and it received an ENERGY STAR rating of 56. The Alabama elementary schools had ENERGY STAR ratings of 82 and 85 with 45% and 69% of the floor areas being conditioned with GCHPs.

Eighteen of the monitored buildings are served by multiple water-to-air heat pumps connected to a central building loop as shown in Figure 1.9. Two of buildings are served by the setup of a central chiller connected to a central ground loop with some portions being served by water-to-air heat pumps. Fourteen of the buildings have variable-speed pumps controlled primarily by differential pressure on the building supply and return headers. Four systems have constant-speed continuously operating pumps. Fourteen of the sites are schools (seven elementary, three middle, and four high schools), four are offices, one is a hotel, and one is an active senior living facility. The sites are located in Florida, Georgia, Mississippi, Tennessee, and Kentucky. At one site a fluid cooler was installed after the first year of operation due to high loop temperatures. Two additional sites (in the same school district) were equipped with coolers at installation, but because

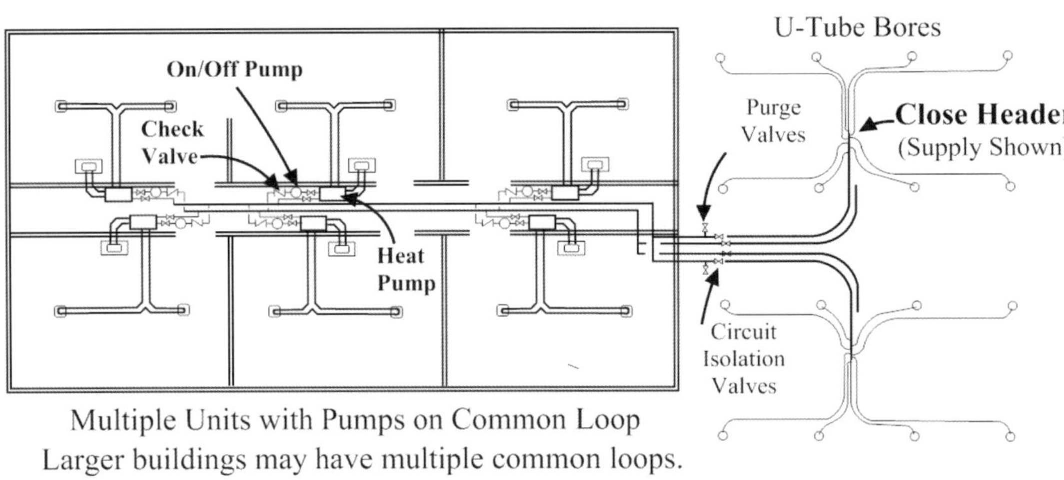

Figure 1.8 Common (Subcentral) Loop GCHP with Close Header Ground Loop

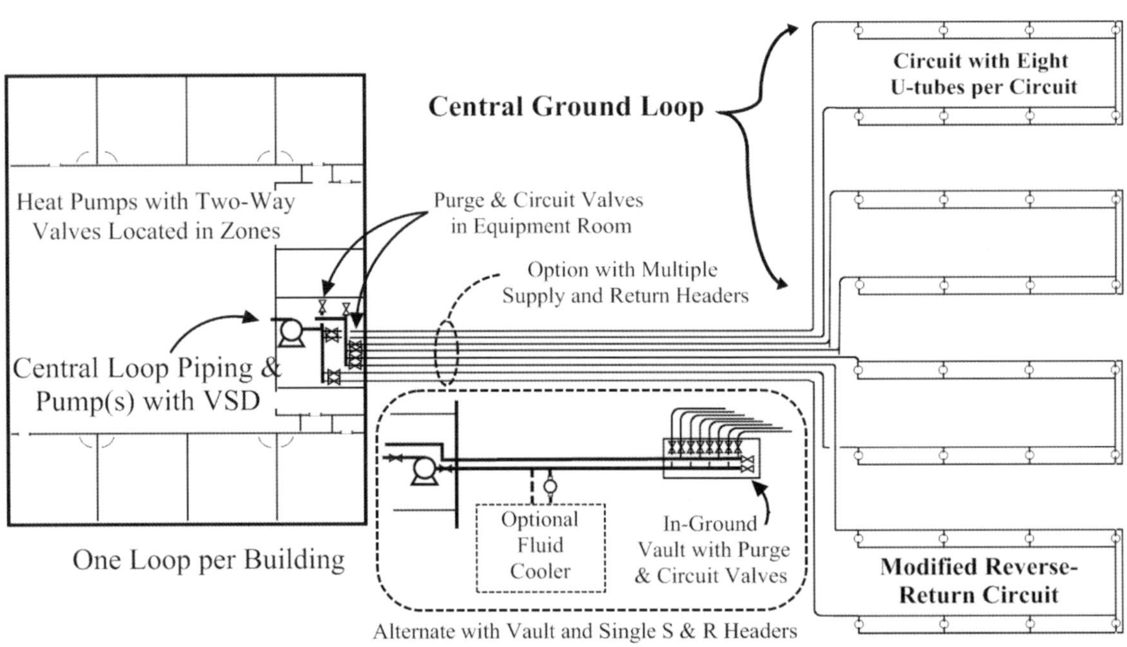

Figure 1.9 Central Loop GCHP with Modified Reverse-Return Header Ground Loop

the ground loops were 50% larger than those of the first school, the coolers did not need to operate. Five of the GCHPs were retrofits and the remaining systems were installed when the buildings were constructed. Dates of GCHP installations range from 1988 to 2008. ENERGY STAR ratings ranged from 1 (hotel) to 93 (retrofit school). If the rating of 1 were not considered, the average ENERGY STAR rating would be 60. The systems with variable-speed drive pumps had an average rating of 57, the constant-speed pump systems had an average of 72, and the systems with chillers had an average rating of 21. The hybrid (fluid cooler equipped) system with the smaller loop had an ENERGY STAR rating of 79, while the systems with larger loops and unused fluid coolers had ratings of 93 and 87.

None of the monitored buildings were GWHP systems as shown in Figure 1.10 or SWHP systems as shown in Figure 1.11.

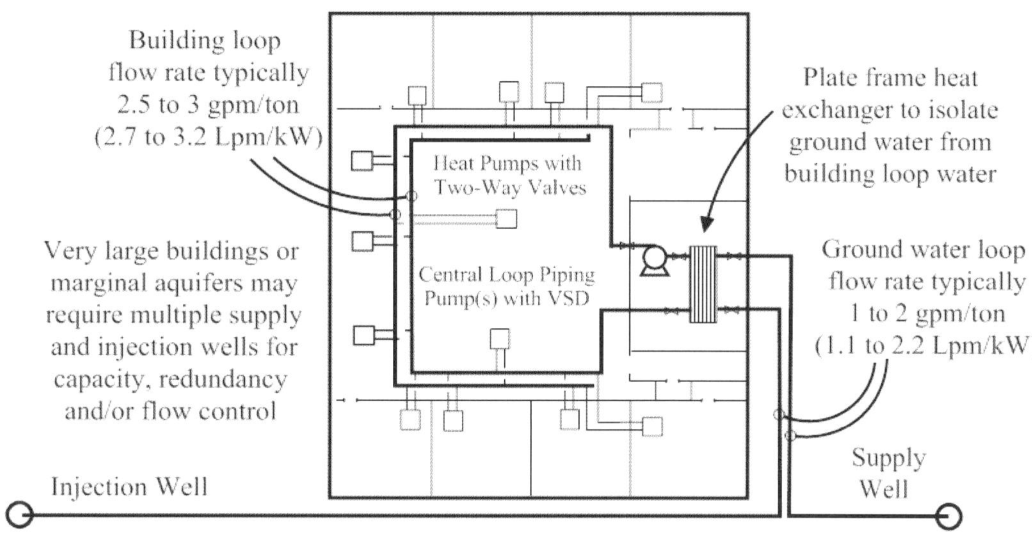

Figure 1.10 Central-Loop GWHP with Plate-Frame Isolation Heat Exchanger

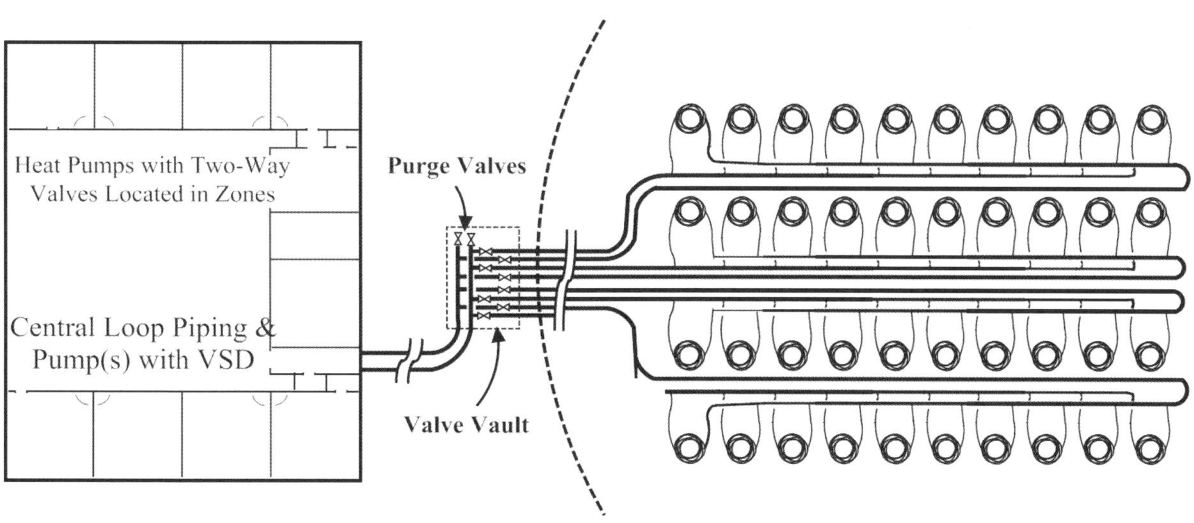

Figure 1.11 Central-Loop SWHP with Reverse-Return Header Lake Coils

1.7 PRELIMINARY ASSESSMENT, DESIGN STEPS, AND DELIVERABLES

During the preliminary stages of any GSHP project, three considerations must be evaluated to determine what type of system (ground-coupled, groundwater, or surface water) is optimal for the building and the site:

- Hydrogeological characteristics and land availability of the site
- Local, state, and federal regulations and cost of permitting
- Building cooling/heating requirements and layout, which dictate the most appropriate HVAC system that is affordable and maintainable by the owner

The characteristics of the site should be considered before the type of GSHP is chosen. A great amount of state and U.S geological survey information is well documented to assist in determining drilling and formation conditions. A book is available from ASHRAE (Sachs 2002) that helps HVAC engineers familiarize themselves with hydrogeological concepts. Local, state, and federal regulations vary significantly and must be identified. A comprehensive GSHP regulation study was conducted in the 1990s (Den Braven and Jensen 1996; Den Braven 1998), but it has not been updated recently. Highly regulated locations may have permitting fees that can be a considerable percentage of total ground heat exchanger costs. Equally important, a preliminary evaluation of the *system* efficiency and equipment costs for the HVAC system is critical to the success of a project, as the HVAC cost has been found to be approximately three-fourths of the total GSHP system cost (Kavanaugh et al. 2012).

GCHPs seem to be the most common GSHP type in both commercial and residential buildings. The lack of exposure of the "outdoor" unit, which eliminates weather-related and environmental damage, theft, and maintenance requirements, is an especially attractive characteristic to building owners with limited operation resources (schools, small building owners, etc.). However, the land area requirement can eliminate GCHPs from consideration, especially in urban, high-density applications. Consider that a single vertical bore can typically support one to two cooling tons (3.5 to 7 kW), which requires approximately 400 ft^2 (40 m^2) of land area. In buildings where the cooling load is much greater than the heating requirement, the required land area can be reduced significantly with hybrid GCHPs. Also, designers are attempting to drill to greater depths to reduce the required land area. Caution is advised with deeper drilling because pump requirements will likely be greater, bore separation should be increased to reduce the possibility of cross-drilling during installation, and the potential for pipe failure for depths beyond 500 ft (150 m) is not yet well established (see Appendix C). Additional details of GCHP site selection can be found in Sections 3.5 and 3.6 of Chapter 3.

The presence of a nearby reservoir or the site requirement of a water retention pond would sway the decision toward using a SWHP. SWHPs tend to be less expensive than GCHPs and can be more efficient in cooling if the summer water temperatures are lower than ground temperatures, as may be the case in deep reservoirs or large open bodies of water. Reservoir size and depth requirements are discussed in Section 5.10 and temperature profiles are found in Section 5.3.

The availability of plentiful groundwater would sway the choice toward a GWHP. This is especially true for larger buildings and where the groundwater is shallow, because the economics of GWHPs compared to GCHPs and SWHPs improves with larger building size and shallow water wells. The required separation distance between the supply and the injection well in some cases may impact the site requirement. These issues are addressed in Section 7.5. General groundwater availability information can be obtained from state and federal geological surveys, but the level of detail needed for system design typically requires a well flow test, as discussed in Chapter 7.

Too often designers attempt to attach traditional HVAC systems to groundwater or surface-water heat exchangers. In some cases two different design teams are separated at the building wall, one responsible for the HVAC and the other responsible for the outdoor heat exchanger. These decisions almost always drive down system efficiency and elevate system installation costs.

Section 2.4 outlines the recommended procedure for evaluating GSHP system efficiency that considers the input of all primary HVAC components, including heat pumps or chillers, supply fans, terminal fans, return fans, indoor pumps, outdoor pumps, and fluid cooler/cooling tower fans (for hybrid systems). This procedure is critical but it is rarely performed unless there is a complaint of high energy use by the owner. Section 9.2 provides such an analysis for a LEED Platinum building with an underperforming GSHP system. Section 9.4 provides a recommended procedure for estimating the cost of the (inside-the-building) HVAC system. Performing this procedure before the final design is initiated may prevent time-consuming, painful, and ill-advised redesign to bring the GSHP system cost to within an allowable budget.

The recommended design steps for GCHP systems provided below are an update of previous versions provided in an *ASHRAE Transactions* paper (Kavanaugh 2008) and the Geothermal Energy chapter of *ASHRAE Handbook—HVAC Applications* (2011). While several of the steps are also common to GWHPs and SWHPs, steps in which the procedures are different are subdivided into three substeps, one for each type of system.

1. Calculate peak zone cooling and heating loads and estimate off-peak loads.
2. Provide suggestions to reduce building envelope, lighting, and ancillary loads with estimates of reduction in HVAC and ground-loop costs.
3. Estimate the annual heat rejection into and absorption from the loop field to account for potential ground, groundwater, or reservoir-water temperature change.
4. Select the preliminary loop operating temperatures and flow rate to begin optimization of first cost and efficiency (selecting temperatures near the normal source temperature will result in high efficiencies but larger and more costly ground loops).
5. Correct heat pump performance at rated conditions to actual design conditions (Chapter 2). *Note that some designers prefer to reverse the order of Steps 5 and 6.*
6. Select heat pumps to meet cooling and heating loads and locate units to minimize duct cost, fan power, and noise.
7. Arrange heat pumps into ground-loop circuits to minimize system cost, pump energy, and electrical demand (Chapters 4 and 6).
8. Conduct a site survey.
    a. For closed-loop GCHPs, conduct a thermal property test to determine ground thermal properties and drilling conditions (Chapter 3). For small projects a survey of geological reports can be used to conservatively estimate these values.
    b. For open-loop GWHPs, conduct a well flow test (Chapter 7).
    c. For closed-loop SWHPs, determine or conduct a survey of the surface-water reservoir depth and, if time permits, water temperature in late winter (February, early March) and late summer (late August, September). If temperature surveys are not possible, consult references (such as EIS 2014) for temperature profiles for lakes of similar dimensions and locality. Additional

information may be available in the final report of the SWHP heat pump investigation (ASHRAE 2009) when it becomes available.

9. Assess outdoor heat exchanger options.
   a. For closed-loop GCHPs, determine and evaluate possible loop field arrangements that are likely to be optimum for the building and site (bore depth, separation distance, completion methods, annulus grout/fill, and header arrangements). Include subheader circuits (typically 5 to 15 U-tubes on each) with isolation valves to permit air and debris flushing of sections of the loop field through a set of full-port purge valves.
   b. For open-loop GWHPs, site the production well(s) and injection well(s) to provide adequate separation and access to the wellhead for maintenance.
   c. For closed-loop SWHPs, estimate the number of coils or plates necessary and locate them in a deeper portion of the reservoir that is in reasonable proximity (i.e., the required pump power is less than 10% of total heat pump power).
10. Determine the optimum ground, groundwater, or surface-water heat exchanger dimensions with calculations provided in this book or by commercial software. Recognize one or more alternatives that provide equivalent performance and that may yield more competitive bids.
11. Evaluate alternative designs: loop field arrangements, operating temperatures, flow rates, heat exchanger dimensions and materials, grout/fill materials, etc.
12. Lay out interior piping and compute head loss through the critical path, and select pumps and control method.
13. Determine system efficiency and consider modifying the water distribution system if pump demand exceeds 10% of the system total demand, modify the air distribution system if fan demand exceeds 15% of the system total, select more efficient pumps, or redesign ground/groundwater/surface-water loop.

ASHRAE Handbook—HVAC Applications (2011) lists the minimum deliverables necessary to adequately specify a closed-loop GCHP installation; items are added here for GWHPs and SWHPs:

- Heat pump specifications at rated conditions.
- Pump specifications, expansion tank size, and air separator.
- Fluid specifications (system volume, inhibitors, antifreeze concentration if required, water quality, etc.).
- Design operating conditions (entering and leaving ground-loop temperatures, return-air temperatures [including wet bulb in cooling], airflow rates, and liquid flow rates.
- Pipe header details with ground-loop layout, including pipe diameters, spacing, and clearance from building and utilities.
- Specifications for outdoor heat exchanger.
  - For closed-loop GCHPs: bore depth, approximate bore diameter, bore separation, and grout/fill specifications (thermal conductivity, acceptable placement methods to eliminate any voids).
  - For open-loop GWHPs: well depth, casing material and diameter, well screen specifications, filters, injection-well specifications, and precautions to avoid air entrainment.
  - For closed-loop SWHPs: surface-water heat exchanger materials, length of tubing (or size of plates), number of loops, numbers of circuits, header size, and burial method.

- Piping material specifications and visual inspection and pressure testing requirements.
- Purge provisions and flow requirements to ensure removal of air and debris without reinjection of air when switching to adjacent subheader circuits.
- Instructions on connections to building loop(s) and coordination of building and ground-loop flushing.
- Sequence of operation for controls.

1.8 REFERENCES

ASHRAE. 2000. Guideline 12-2000, *Minimizing the Risk of Legionellosis Associated with Building Water Systems*. Atlanta: ASHRAE.

ASHRAE. 2009. *Development of design tools for surface water heat pump systems*. ASHRAE RP-1385. Final Report in Progress. Atlanta: ASHRAE.

ASHRAE. 2011. *ASHRAE Handbook—HVAC Applications*, Chapter 34, Geothermal Energy, pp. 34.9–34.34. Atlanta: ASHRAE.

Den Braven, K.R. 1998. Survey of Geothermal Heat Pump Regulations in the United States. *Proceedings of the Second Stockton International Geothermal Conference*. Galloway, NJ: The Richard Stockton College.

Den Braven, K.R., and J. Jensen. 1996. State and federal vertical borehole grouting regulations. Final report to the Electric Power Research Institute on Project RP 33881-01, July.

EIS. 2014. Surface Water Temps. Ground-Source Heat Pump Design—Keep it Simple and Solid. Northport, AL: Energy Information Services. www.geokiss.com/surwatertemps.htm

EPA. 2012. How the Rating System Works. www.energystar.gov/index.cfm?c=evaluate_performance.pt_neprs_learn

Kavanaugh, S.P. 1991. Ground and water source heat pumps. Northport, AL: Energy Information Services.

Kavanaugh, S.P. 2008. A 12-step method for closed-loop ground-source heat pump design. *ASHRAE Transactions* 114(2).

Kavanaugh, S.P., and J.S. Kavanaugh. 2012. Long-term commercial GSHP performance, part 1: Project overview and loop circuit types. *ASHRAE Journal* 54(6).

Kavanaugh, S.P., M. Green, and K. Mescher. 2012. Long-term commercial GSHP performance, part 4: Installation costs. *ASHRAE Journal* 54(10).

Rafferty, K. 1995. A capital cost comparison of commercial ground-source heat pump systems. *ASHRAE Transactions* 101(2).

Remund, C. 2011. *Ground Source Heat Pump Residential and Light Commercial Design and Installation Guide*. Stillwater, OK: International Ground Source Heat Pump Association.

Sachs, H. 2002. *Geology and Drilling Methods for Ground Source Heat Pump System Installation: An Introduction for Engineers*. Atlanta: ASHRAE.

SUNY. 2011. Assessing the feasibility of a central New York naturally chilled water project. Final Report, USEPA Award XA-97264106-01. Albany, NY: The Research Foundation, The State University of New York.

2 Equipment for Ground-Source Applications

2.1 HEAT PUMP TYPES

The most common type of heat pump used with ground-source applications is the water-to-air unit as shown in Figure 2.1. The water-to-refrigerant coil is linked to the external (source) water loop and serves as the condenser in cooling and the evaporator in heating. The air-to-refrigerant coil is usually linked to a forced-air system. However, there is increasing use of water-to-water heat pumps (and dedicated cooling or heating units). Water-to-water units are used for hydronic floor heating, dedicated domestic water heating, outdoor air preconditioning, and hydronic heating and cooling.

Water-to-air cooling-only units have also been used in refrigeration applications, while heating-only units have been used to heat water. Caution is advised against cooling-only and heating-only GSHP systems in order to minimize the long-term heat imbalance within the ground, groundwater, or surface-water source. Thus, cooling-only or heating-only equipment should be integrated into systems that also have heat pumps that provide both heating and cooling to more closely balance the amount of heat delivered to or removed from the source. In some cases prudent combinations of heating-only and cooling only equipment can reduce the size of a shared ground loop. Examples are a convenience store with a car wash, as shown in Figure 2.2, or a food-service kitchen that has refrigeration equipment always adding heat to the ground loop and water heater units always removing heat. In the convenience store example, the heat rejection of the cooler and freezer is coupled to a loop that also has heat pump water heaters for the car wash. In a kitchen, the refrigeration equipment and cooling units could be connected to the same loop with heat pump water heaters for the dish washers.

Development of water-source heat pumps has been primarily directed toward satisfying the needs of the residential sector. Advances can be applied to the commercial sector with little or no modifications in units with capacities of less than 65,000 Btu/h (19 kW). Development of larger high-efficiency units has been slower, which means systems with multiple small heat pumps will typically consume less energy than those with fewer large units.

GSHP systems installed before 1980 often used heat pumps that were intended for water-loop heat pump applications in which a cooling tower is used to reject heat and a boiler is used to provide heat. System efficiencies suffered because this equipment was not optimized for heating with water below 60°F (16°C). Also, the cooling efficiency was

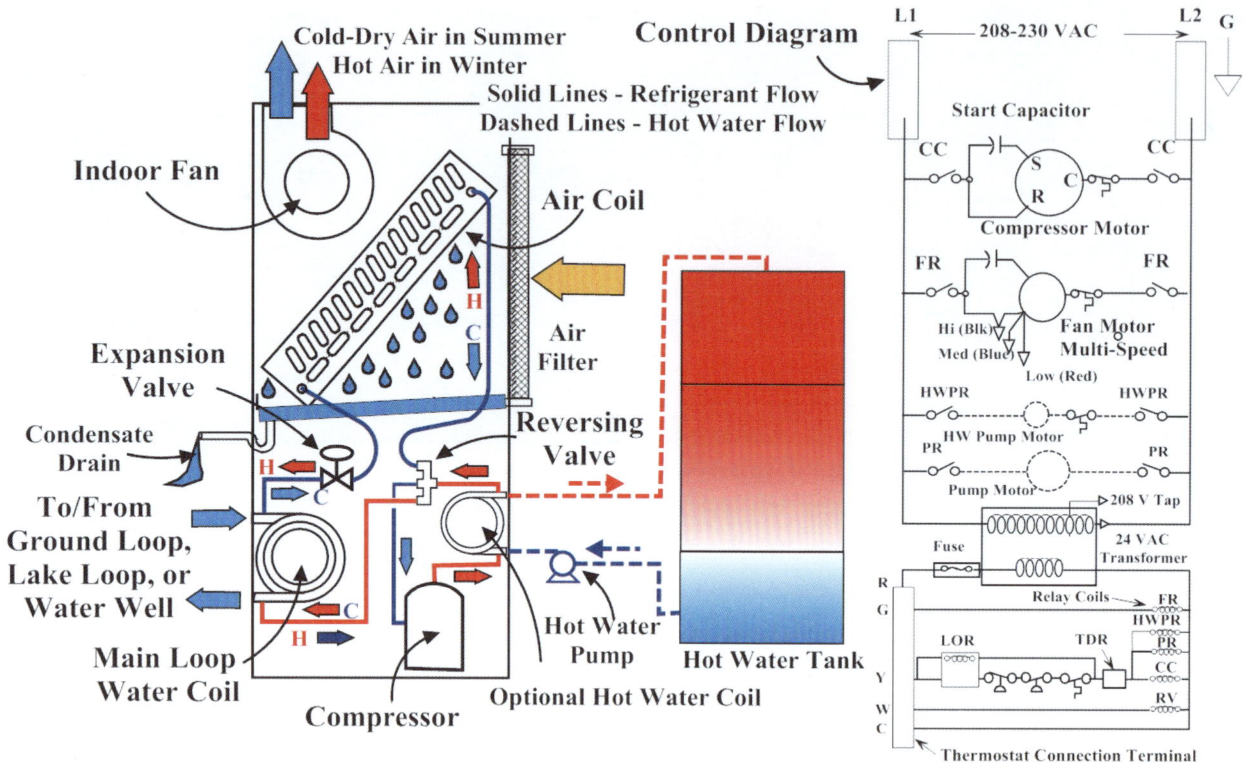

Figure 2.1 Vertical Water-to-Air Heat Pump for Ground-Source Applications

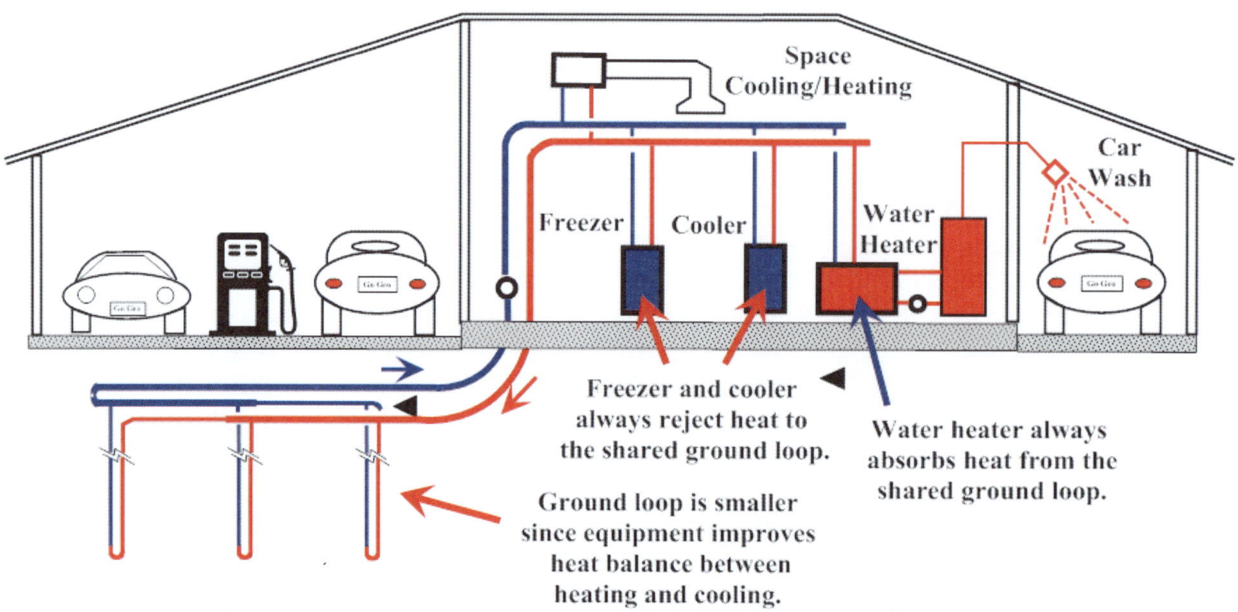

Figure 2.2 Convenience Store Application with Heating and Cooling Requirements

often low in these systems, and little attempt was made to minimize head loss through the water coil. After 1980 several manufacturers introduced extended-range equipment with refrigerant control that allowed operation at a wide range of liquid temperatures. In the late 1980s equipment was introduced that used high-efficiency compressors, large water and air coils, and high-efficiency fan motors. This equipment is well suited to commercial applications. More recently, manufacturers have introduced multispeed, multistage, and variable-speed water-to-air and water-to-water heat pumps.

The equipment is often compact, and in many cases cabinets are similar in size to indoor units of split-system heat pumps and air handlers of equivalent capacity. However, this equipment requires more room for service because the compressor, water coil, and controls must be accessed. Figure 2.3 shows three water-to-air heat pumps located in an equipment room with adequate spacing for duct installation and service; they are elevated off the floor to minimize cabinet corrosion from condensation. Figure 2.4 shows the location of a unit on a mezzanine above a hallway in a school. The supply and return ducts are routed over to the ceiling and into an adjacent classroom. Service is possible without disrupting the occupants or using a ladder. Figure 2.5 shows a unit with a factory-installed circulator pump. Figure 2.6 shows a large horizontal water-to-air heat pump hung from a gymnasium ceiling. Figure 2.7 displays a vertical classroom unit with an internal energy recovery unit (ERU) (note the two additional air registers). Smaller spaces can be served by console units with capacities as low as 6000 Btu/h (1.8 kW), as exhibited in Figure 2.8. Figure 2.9 shows a bank of eight water-to-air heat pumps located in a basement equipment room. In this application the units serve the building outdoor air coils but could also be used for heating and cooling spaces.

Service technicians are especially sensitive to equipment that is installed with little consideration for serviceability. Access for routine maintenance, such as filter changes, is

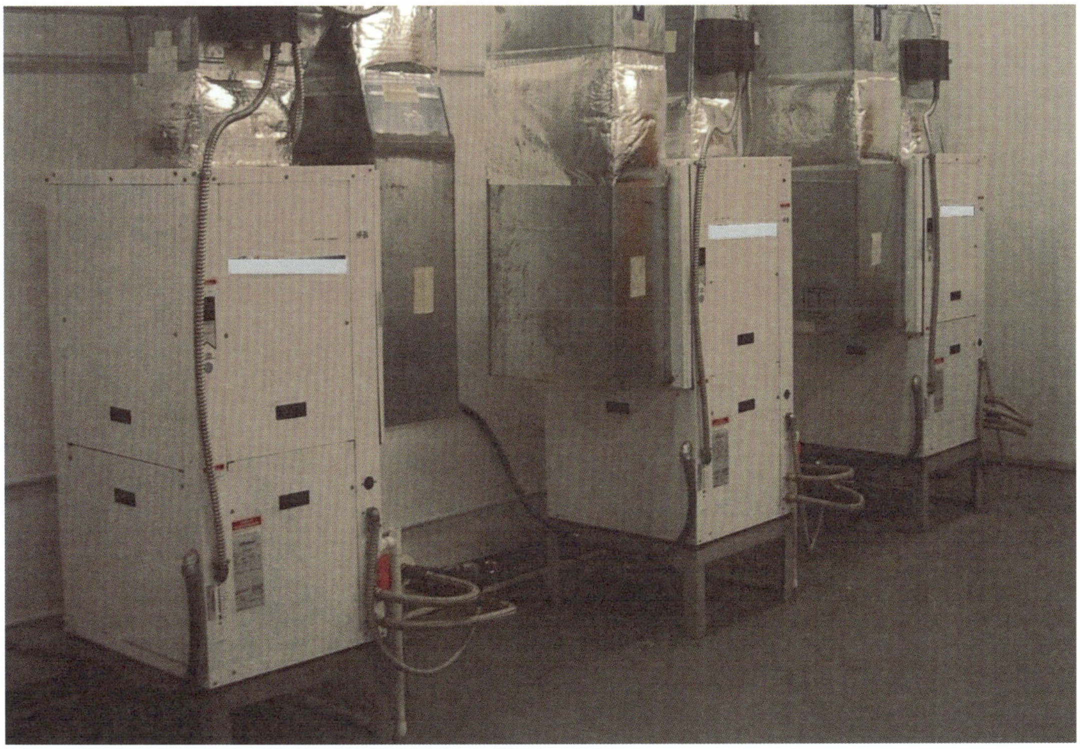

Figure 2.3 Accessible Water-to-Air Heat Pump Equipment Room Installation

Figure 2.4 Water-to-Air Heat Pump on Mezzanine above School Hallway

Figure 2.5 Water-to-Air Heat Pump with Internal Pump

Figure 2.6 Horizontal Water-to-Air Heat Pump in Gymnasium

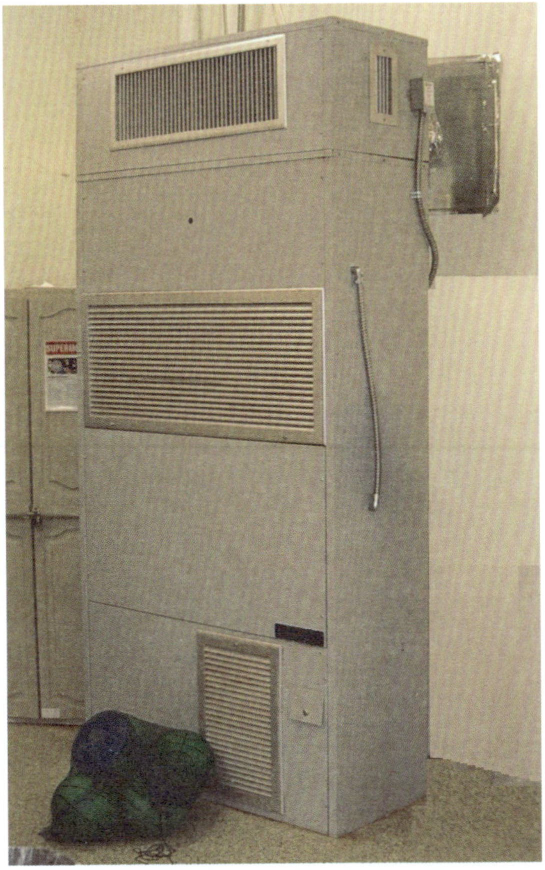

Figure 2.7 Classroom Water-to-Air Heat Pump with Internal Energy Recovery Unit

Figure 2.8 Water-to-Air Heat Pump Classroom Console Unit

Figure 2.9 Bank of Eight Water-to-Water Heat Pumps

important because tasks that are difficult to perform are more likely to be neglected. The required time to complete difficult repair and component replacement is especially troubling when equipment is poorly located. Figure 2.10 shows a classroom heat pump that replaced a unit ventilator. Although the unit's height is much greater, the footprint is the same as that of the unit ventilator. The left portion of the figure shows the location of the unit with the return air grille at desktop height, the overhead supply air register, and the programmable thermostat. The right portion of the figure demonstrates the accessibility of the components in the lower cabinet.

Figure 2.11 is an example of a nonconventional approach to problem solving that resulted when poor attention is given to serviceability. A horizontal water-to-air heat pump was installed in a ceiling space above a light fixture and water sprinkler head. The fan motor failed and replacement without removing the heat pump was impossible. Fortunately, an enterprising but time-constrained service technician noted that access could be gained by removing a portion of the gypsum board covering the access panel. A picture was placed over the newly created access path to avoid an additional maintenance task.

Figure 2.12 shows a similar situation with a unit installed above the ceiling in a closet. In order to perform service, storage items had to be moved from the space closet and service was performed by the technician while standing on a ladder.

Figure 2.13 displays the complexity of controls that accompany modern water-source heat pumps with multispeed and variable-speed capacities. Designers should carefully weigh the potential added maintenance cost to owners with the limited benefits of complex equipment. This is especially true for applications such as schools that have very limited maintenance personnel and budgets. The circuit boards are proprietary equipment, and some manufacturers require specialized factory training for installation and service technicians. This could be a serious financial burden to owners with multiple buildings, heat pumps from multiple manufacturers, and multiple proprietary control networks that have limited periods of product support as a result of frequent product "upgrades."

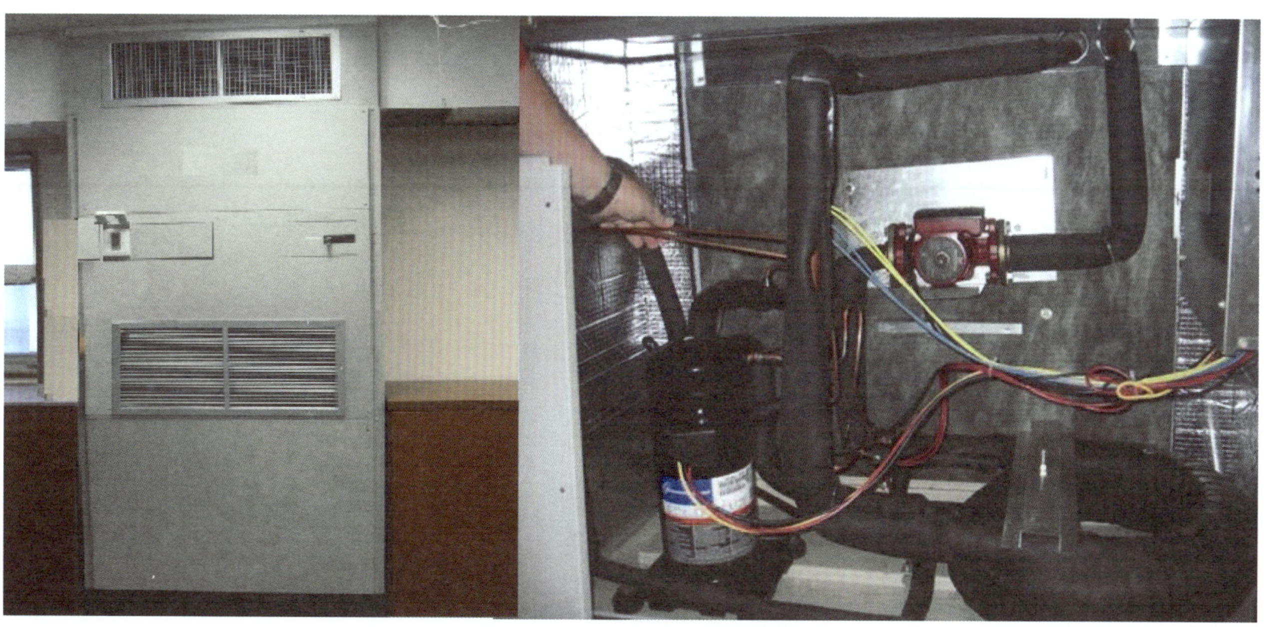

Figure 2.10 Classroom Unit (left) and with Panel Removed (right)

Figure 2.11 Technician Solution to Servicing Heat Pump with Limited Access

Figure 2.12 Difficult-to-Service Heat Pump Location

Figure 2.13 Controls for Multiple-Capacity Water-to-Air Heat Pump

2.2 WATER-SOURCE HEAT PUMP STANDARDS

AHRI/ASHRAE ISO Standard 13256-1 (ASHRAE 2012a) dictates the testing and performance rating for water-to-air heat pumps, and AHRI/ASHRAE ISO Standard 13256-2 (ASHRAE 2012b) covers water-to-water heat pumps. Table 2.1 summarizes the air and water temperatures dictated by these standards to rate performance. Reported values are total cooling (TC) in Btu/h (kW), energy efficiency ratio (EER) in Btu/Wh (COP_c in $W_{cooling}/W_{electrical}$), heating capacity (HC) in Btu/h (kW), and coefficient of performance (COP_h) in $W_{heating}/W_{electrical}$. Four sets of rating conditions are used to represent approximations of conditions occurring for various applications, as shown in the table. The water-loop heat pump[1] (WLHP) rating uses entering liquid temperatures[2] (ELTs) to the heat pumps and assumes the units are connected to a cooling tower and boiler. However, the cooling-mode ELT is in most cases appropriate for well-designed ground-coupled heat pumps (GCHPs). Groundwater heat pump (GWHP) ELTs are based on groundwater being pumped directly to the units and are appropriate for residential applications in moderate climates. The full-load and part-load ELTs for ground-loop heat pumps[3] (GLHPs) are appropriate for cold-climate residential applications but not optimal for most commercial systems or moderate- or warm-climate residential systems (Kavanaugh 2008).

---

1 The term used by the International Organization for Standardization (ISO), *water-loop heat pump (WLHP)*, is equivalent to the ASHRAE term *water-source heat pump (WSHP)*.

2 The term *entering liquid temperature (ELT)* is used because liquids are often a combination of water and other liquids, creating solutions with lower freeze points. Some publications may use ELT and entering water temperature (EWT) interchangeably. The ISO also uses the term *brine* rather than *antifreeze*; antifreeze implies the solutions will never freeze at lower temperatures, which is not the case.

Table 2.1 AHRI/ASHRAE ISO Standard 13256-1 Rating Conditions for Water-to-Air Heat Pumps (ASHRAE 2012a)

Entering Liquid and Air Temperatures	WLHP	GWHP	GLHP	GLHP-PL
ELT—Cooling Exterior Loop	86°F (30°C)	59°F (15°C)	77°F (25°C)	68°F (20°C)
ELT—Heating Exterior Loop	68°F (20°C)	50°F (10°C)	32°F (0°C)	41°F (5°C)
EAT—Cooling Dry Bulb/Wet Bulb	80.6°F / 66.2°F (27°C / 19°C)			
EAT—Heating	68°F (20°C)			

Notes: PL = part-load. Values for TC, EER, HC, and COP do not include fan or pump power required to circulate air and water through the air distribution system and piping loop. Values for TC do not include the loss of capacity due to the heat of the fan. The power to circulate air and water through the unit itself is included in the calculation.

Table 2.2 AHRI/ASHRAE ISO Standard 13256-2 Rating Conditions for Water-to-Water Heat Pumps (ASHRAE 2012b)

Entering Liquid Temperatures	WLHP	GWHP	GLHP	GLHP-PL
ELT—Cooling Exterior Loop	86°F (30°C)	59°F (15°C)	77°F (25°C)	68°F (20°C)
ELT—Heating Exterior Loop	68°F (20°C)	50°F (10°C)	32°F (0°C)	41°F (5°C)
ELT—Cooling Interior Loop	53.6°F (12°C)			
ELT—Heating Interior Loop	104°F (40°C)			

Notes: PL = part-load. Values for TC, EER, HC, and COP do not include pump power required to circulate water through the exterior and interior piping loops. Likewise, the fan power of terminal units (fan coil units, air handling units) is not included. Values for TC do not include the loss of capacity due to the interior piping loop pump heat or air terminal unit fan heat.

The footnote to Table 2.1 is significant in that the power used to determine the rated capacity and efficiency assumes the external static pressure (ESP) to overcome air distribution losses is zero. The logic is that the designer is aware of this limitation and has access to the necessary tools to make the corrections to actual capacity and efficiency once the pressure losses of the air distribution system and filters are known. The pump pressure required for water circulation through the building and external loop system is also assumed to be zero.

Note also the entering air dry-bulb (EATDB) and entering air wet-bulb (EATWB) temperatures in cooling (80.6°F/66.2°F [27°C/19°C]) do not reflect typical operating conditions. Values assume the return air is mixed with raw outdoor air, a practice that is becoming less common with the increase in preconditioning of ventilation air.

Procedures for correcting performance for fan power, water and air temperatures, airflow rates, and water flow rates are presented in the following section. The spreadsheet performance correction tool *WAHPCorrector14.xlsm* follows these procedures. It is available with this book at www.ashrae.org/GSHP.

Table 2.2 summarizes the water temperatures used to rate performance of water-to-water heat pumps. Source loop temperatures and efficiency indicators are identical to those for water-to-air heat pumps. The building loop ELT for cooling is 53.6°F (12°C), which results in a supply chilled-water temperature in the 41°F to 48°F (5°C to 9°C) range. These values are reasonable for chilled-water systems with fan-coils. The building loop ELT for heating is 104°F (40°C), which results in a supply hot-water temperature in the 110°F to 115°F (43°C to 46°C) range. These values are slightly lower than the values used in heat pump and condensing boiler applications with fan-coils. Thus, some adjustment is necessary to reduce efficiency and capacity when higher temperatures are

[3] The term used by the ISO, *ground-loop heat pump (GLHP)*, is equivalent to the ASHRAE term *ground-coupled heat pump (GCHP)*.

required. However, in-floor heating applications often operate with lower temperatures, so capacity and efficiency can be slightly higher.

Similar to the water-to-air heat pump standard, the water-to-water heat pump standard assumes zero pump pressure for the ground loop and has no consideration of building loop pump power or fan power.

2.3 PERFORMANCE OF WATER-SOURCE HEAT PUMPS

The performance of water-to-air and water-to-water heat pumps is rated at multiple exterior (source) ELTs. This is perhaps the most significant variable in unit performance, and interpolation to intermediate values is often necessary. Other important variables that must be considered for correction are the following:

- Fan power
- Airflow rate
- Liquid flow rate
- Entering air temperatures (for water-to-air heat pumps)
- Entering building loop liquid temperatures (for water-to-water units)
- Pump power for source loops
- Pump power for building loops (water-to-water units)

The process of correcting rated performance to actual conditions is somewhat cumbersome, but it is critical because conditions vary dramatically. The following section outlines the process of correcting performance.

For water-to-air heat pumps the recommended procedure is as follows:

1. Correct for ELT by interpolating (or extrapolating) the heat pump TC and EER using rated values for nearest ELTs. Repeat for HC and COP.
2. Compute the input power by dividing the TC (Btu/h [W]) by the EER (Btu/Wh) or COP_c.
3. Correct for entering air temperatures (EATs) using correction factors for TC, input power in cooling, HC, and input power in heating.
4. Correct for airflow rate using correction factors for TC, input power in cooling, HC, and input power in heating.
5. Correct for liquid flow rate using correction factors for TC, input power in cooling, HC, and input power in heating.
6. Compute the added fan power required to overcome air distribution network and filter losses. Convert heat pump gross capacities to net capacities by reducing TC and increasing HC by the added fan heat.
7. Compute the added pump power required to overcome ground-loop head losses. Add the pump power to the heat pump power and fan power.
8. Correct EER and COP using the corrected net capacity divided by the corrected input power (heat pump, fan, and pump).

This procedure requires a large amount of effort. To assist in the process, the spreadsheet tool *WAHPCorrector14.xlsm*, which is based on the eight-step manual heat pump performance calculation procedure, has been used to develop a time-saving (but less accurate) alternative. (*WAHPCorrector.xlsm* is available with this book at www.ashrae.org/GSHP.) In the spreadsheet, multipliers are applied to the rated TC, EER, HC, and COP values to correct performance to conditions and constraints likely to occur in actual applications. These conditions are as follows:

- Cooling indoor air temperatures of 75°F db/63°F wb (24°C/17°C) (from 80.6°F/66.2°F [27°C/19C°])
- Heating indoor air temperatures of 70°F db (from 68°F [20°C])
- Fan power/heat required to distribute air through average duct/filter systems

The correction factors from AHRI/ASHRAE ISO Standard 13256-1 (ASHRAE 2012a) rating conditions are as follows:

- **Multiply rated TC by 0.93**
- **Multiply rated EER by 0.80**
- **Multiply rated HC by 1.03**
- **Multiply rated COP by 0.89**

These factors apply to rated TC and EER for ELTs at 86°F, 77°F, and 59°F (30°C, 25°C, and 15°C) but not to part-load values at 68°F (20°C) and to rated HC and COP for ELTs at 68°F, 50°F, and 32°F (20°C, 10°C, and 0°C) but not for part-load values at 41°F (5°C). These corrections do not account for added pump power, which also must be applied.

Systems with water-to-water heat pumps typically contain multiple units and additional auxiliary equipment and are even more challenging to correct. To assist in this process, the spreadsheet *HVACsystemEff.xlsx* is available with this book at www.ashrae.org/GSHP. This program can also be used to determine the system efficiency of a wide variety of non-GSHP HVAC options.

For individual water-to-water heat pumps the recommended correction procedure is as follows:

1. Correct for ELT by interpolating (or extrapolating) the heat pump TC and EER using rated values for nearest ELTs. Repeat for HC and COP.
2. Compute the input power by dividing the TC (Btu/h [W]) by the EER (Btu/Wh) or COP_c.
3. Correct for building liquid flow rate using correction factors for TC, input power in cooling, HC, and input power in heating.
4. Correct for source (exterior loop) liquid flow rate using correction factors for TC, input power in cooling, HC, and input power in heating.
5. Compute the added pump power required to overcome building head losses. Add the pump power to the heat pump power and fan power. Deduct building pump heat from TC and add building pump heat to HC.
6. For systems with fan-coil terminals, compute the added fan power required to overcome air distribution network and filter losses. Convert heat pump gross capacities to net capacities by reducing TC and increasing HC by the added fan heat. Add the fan power to the rated heat pump power.
7. Compute the added pump power required to overcome ground-loop head losses. Add the pump power to the heat pump power and fan power.
8. Correct EER and COP using the corrected net capacity divided by the corrected input power (heat pump, fan, and pump).

Tables 2.3a and 2.3b provide the ratings for one manufacturer's product line of high-efficiency water-to-air heat pumps, including nine single-speed models and three variable-speed units, in I-P and SI units, respectively. Cooling and heating capacity and efficiency values are provided for the previously mentioned WLHP, GWHP, and GLHP operating conditions. This includes part-load values for the variable-speed models. Note that the part-load ELTs for the GLHP applications differ from the full-load ratings. How-

Table 2.3a Rated Capacity and Efficiency Values for Water-to-Air Heat Pumps—I-P

				Single-Speed Water-to-Air Heat Pumps											
				Water-Loop Heat Pump				Groundwater Heat Pump				Ground-Loop Heat Pump			
				Clg—86°F ELT		Htg—68°F ELT		Clg—59°F ELT		Htg—50°F ELT		Clg—77°F (FL)		Htg—32°F (FL)	
Model	**Load**	**cfm**	**gpm**	**TC**	**EER**	**HC**	**COP**	**TC**	**EER**	**HC**	**COP**	**TC**	**EER**	**HC**	**COP**
15	Full	500	4	14.4	16.5	18.5	5.3	16.7	27.0	15.5	4.7	15.0	18.1	12.0	4.0
18	Full	600	5	18.0	16.5	23.0	5.3	21.0	26.8	19.0	4.7	18.5	19.0	14.7	4.1
22	Full	850	8	20.7	17.5	25.3	6.2	23.5	30.0	19.8	5.3	21.7	21.0	15.0	4.0
30	Full	900	8	28.3	19.2	32.7	5.8	31.3	28.8	25.8	5.0	29.4	21.9	20.0	4.0
36	Full	1200	9	34.5	19.6	38.0	6.1	37.2	30.1	30.3	5.2	35.0	22.0	24.1	4.4
42	Full	1300	11	40.6	19.2	44.1	5.9	45.2	29.5	34.9	5.2	42.0	21.4	27.5	4.2
48	Full	1500	12	47.0	17.5	55.4	5.5	52.0	26.1	45.1	4.8	49.3	19.7	35.3	4.0
60	Full	1800	15	64.3	17.2	69.8	5.4	72.0	26.1	55.1	4.7	66.8	19.5	43.3	3.9
70	Full	2000	18	70.6	16.0	84.3	5.1	79.1	23.8	66.1	4.4	73.2	18.2	52.0	3.7

					Variable-Speed Water-to-Air Heat Pumps											
Model	**Load**	**cfm Clg**	**cfm Htg**	**gpm**	WLHP and GWHP Part-Load (PL) ELTs = Full-Load (FL) ELTs								Clg—68°F (PL)		Htg—41°F (PL)	
36	Full	1300	1500	9	32.0	18.0	50.0	5.3	38.0	31.5	41.0	4.6	36.0	22.0	32.0	3.5
36	Part	1300	1500	9	11.0	21.0	17.0	7.5	13.0	47.2	14.0	5.9	14.0	37.0	13.0	5.3
48	Full	1500	1800	12	41.0	17.6	67.0	5.0	49.0	31.7	55.0	4.3	46.0	21.7	43.0	3.6
48	Part	1500	1800	12	16.0	22.5	24.0	7.6	19.2	53.2	19.0	5.9	19.0	41.0	16.0	5.3
60	Full	1800	2200	15	50.0	16.3	78.0	4.8	60.0	28.6	65.0	4.3	56.0	19.4	51.0	3.5
60	Part	1800	2200	15	20.0	21.7	29.0	7.5	23.2	45.8	23.0	6.0	23.0	36.0	20.0	5.1

Cooling EAT = 80.6°F db/66.2°F wb, Heating EAT= 68°F db, TC and HC in Btu/h × 1000, EER in Btu/Wh, COP in W/W

Table 2.3b Rated Capacity and Efficiency Values for Water-to-Air Heat Pumps—SI

				Single Speed Water-to-Air Heat Pumps											
				Water-Loop Heat Pump				Groundwater Heat Pump				Ground-Loop Heat Pump			
				Clg—30°C ELT		Htg—20°C ELT		Clg—15°C ELT		Htg—10°C ELT		Clg—25°C (FL)		Htg—0°C (FL)	
Model	**Load**	**L/s**	**L/min**	**TC**	**COP_c**	**HC**	**COP_h**	**TC**	**COP_c**	**HC**	**COP_h**	**TC**	**COP_c**	**HC**	**COP_h**
15	Full	235	15	4.2	4.8	5.4	5.3	4.9	7.9	4.5	4.7	4.4	5.3	3.5	4.0
18	Full	280	19	5.3	4.8	6.7	5.3	6.2	7.9	5.6	4.7	5.4	5.6	4.3	4.1
22	Full	400	30	6.1	5.1	7.4	6.2	6.9	8.8	5.8	5.3	6.4	6.2	4.4	4.0
30	Full	425	30	8.3	5.6	9.6	5.8	9.2	8.4	7.6	5.0	8.6	6.4	5.9	4.0
36	Full	579	34	10.1	5.7	11.1	6.1	10.9	8.8	8.9	5.2	10.3	6.4	7.1	4.4
42	Full	610	42	11.9	5.6	12.9	5.9	13.2	8.6	10.2	5.2	12.3	6.3	8.1	4.2
48	Full	710	45	13.8	5.1	16.2	5.5	15.2	7.6	13.2	4.8	14.4	5.8	10.3	4.0
60	Full	850	57	18.8	5.0	20.5	5.4	21.1	7.6	16.1	4.7	19.6	5.7	12.7	3.9
70	Full	940	68	20.7	4.7	24.7	5.1	23.2	7.0	19.4	4.4	21.5	5.3	15.2	3.7

					Variable-Speed Water-to-Air Heat Pumps											
Model	**Load**	**L/s Clg**	**L/s Htg**	**L/min**	WLHP and GWHP Part-Load (PL) ELTs = Full-Load (FL) ELTs								Clg—20°C (PL)		Htg—5°C (PL)	
36	Full	610	710	34	9.4	5.3	14.7	5.3	11.1	9.2	12.0	4.6	10.6	6.4	9.4	3.5
36	Part	610	708	34	3.2	6.2	5.0	7.5	3.8	13.8	4.1	5.9	4.1	10.8	3.8	5.3
48	Full	710	850	45	12.0	5.2	19.6	5.0	14.4	9.3	16.1	4.3	13.5	6.4	12.6	3.6
48	Part	710	850	45	4.7	6.6	7.0	7.6	5.6	15.6	5.6	5.9	5.6	12.0	4.7	5.3
60	Full	850	1040	57	14.7	4.8	22.9	4.8	17.6	8.4	19.1	4.3	16.4	5.7	14.9	3.5
60	Part	850	1040	57	5.9	6.4	8.5	7.5	6.8	13.4	6.7	6.0	6.7	10.6	5.9	5.1

Cooling EAT = 27°C db/19°C wb, Heating EAT= 20°C db, TC and HC in kW, COP_c and COP_h in W/W

ever, the ELTs for the WLHP and GWHP applications are the same for part-load and full-load ratings. Table 2.4 provides similar information for a product line of water-to-water heat pumps.

Table 2.5 is a set of cooling-mode correction factors for entering air conditions in I-P and SI units. Rated capacity and efficiency from Tables 2.3a and 2.3b are multiplied by the factors for the corresponding increase or decrease in EATDB or EATWB. Note that TC and EER are corrected using the EATWB while the sensible cooling capacity (SC) is corrected using both dry-bulb and wet-bulb temperatures. Table 2.6 is a similar set of heating-mode correction factors for HC and COP based on only EATDB. Table 2.7 is a set of correction factors for airflow rate as a percentage of rated flow for both cooling and heating.

The correction for liquid flow rate is complicated by variation in reported performance based on liquid flow rate. Values can range from specific liquid flow rates less than 2 gpm/ton (2.2 L/min·kW) to values greater than 3 gpm/ton (3.2 L/min·kW). Figures 2.14 and 2.15 are used to determine correction factors. Values for rated flows are entered on the horizontal axis and followed vertically to intersect the actual specific flow rate. A horizontal line is followed from this intersection point to find a correction factor on the vertical axis. Note that an example is shown in Figure 2.14 indicating that for a specific rated

Table 2.4 Rated Capacity and Efficiency Values for Water-to-Water Heat Pumps

	Liquid Flows		Water-Loop Heat Pump				Groundwater Heat Pump				Ground-Loop Heat Pump			
	Source	Bldg	Clg—86°F ELT		Htg—68°F ELT		Clg—59°F ELT		Htg—50°F ELT		Clg—77°F (FL)		Htg—32°F (FL)	
Model	gpm	gpm	TC	EER	HC	COP	TC	EER	HC	COP	TC	EER	HC	COP
96	23	23	93	14.6	125	4.0	105	22.0	103	3.3	100	16.8	82	2.8
108	28	28	103	14.0	142	4.0	123	21.6	118	3.3	114	16.2	93	3.0
120	32	32	128	13.8	175	3.8	151	21.0	145	3.2	139	16.0	115	2.8
140	36	36	143	14.5	193	4.2	166	22.5	160	3.8	155	17.0	127	3.1
180	45	45	170	14.0	209	3.9	183	20.0	189	3.5	177	15.8	153	2.8
210	52	52	202	14.8	257	4.2	227	21.8	219	3.8	212	17.0	173	3.1
240	60	60	222	13.3	286	3.9	257	20.0	244	3.5	242	15.5	193	2.8
360	86	86	335	14.3	453	4.3	na	na	na	na	351	16.2	297	3.2
540	135	135	533	15.2	691	4.3	na	na	na	na	559	16.4	486	3.3

Building Loop: Cooling ELT = 53.6°F, Heating ELT = 104°F. TC and HC in Btu/h × 1000, EER in Btu/Wh, COP in W/W

	Liquid Flows		Water-Loop Heat Pump				Groundwater Heat Pump				Ground-Loop Heat Pump			
	Source	Bldg	Clg—30°C ELT		Htg—20°C ELT		Clg—15°C ELT		Htg—10°C ELT		Clg—25°C (FL)		Htg—0°C (FL)	
Model	L/min	L/min	TC	COP_c	HC	COP_h	TC	COP_c	HC	COP_h	TC	COP_c	HC	COP_h
96	87	87	27.3	4.3	36.6	4.0	30.8	6.4	30.2	3.3	29.3	4.9	24.0	2.8
108	106	106	30.2	4.1	41.6	4.0	36.0	6.3	34.6	3.3	33.4	4.7	27.3	3.0
120	121	121	37.5	4.0	51.3	3.8	44.3	6.2	42.5	3.2	40.7	4.7	33.7	2.8
140	136	136	41.9	4.2	56.6	4.2	48.7	6.6	46.9	3.8	45.4	5.0	37.2	3.1
180	170	170	49.8	4.1	61.3	3.9	53.6	5.9	55.4	3.5	51.9	4.6	44.8	2.8
210	197	197	59.2	4.3	75.3	4.2	66.5	6.4	64.2	3.8	62.1	5.0	50.7	3.1
240	227	227	65.1	3.9	83.8	3.9	75.3	5.9	71.5	3.5	70.9	4.5	56.6	2.8
360	326	326	98.2	4.2	132.8	4.3	na	na	na	na	102.9	4.7	87.0	3.2
540	511	511	156.2	4.5	202.5	4.3	na	na	na	na	163.8	4.8	142.4	3.3

Building Loop: Cooling ELT = 12°C, Heating ELT = 40°C. TC and HC in kW, COP_c and COP_h in W/W

Table 2.5 Cooling Capacity and Input Power Correction Factors (CFs) for EATs*

EATWB, °F	Total Capacity	Sensible Cooling Correction Factor					Cooling Power CF
		70°F db	75°F db	80°F db	80.6°F db	85°F db	
55	0.914	0.989	1.118				0.986
60	0.928	0.83	1.017	1.174	1.26		0.995
63	0.962	0.725	0.905	1.018	1.134	1.271	0.997
65	0.984	0.655	0.831	1.018	1.05	1.198	0.998
66.2	1	0.618	0.794	0.981	**1**	1.162	1
67	1.017	0.581	0.76	0.943	0.965	1.125	1.001
70	1.071		0.654	0.829	0.849	1.014	1.004
75	1.188			0.648	0.675	0.825	1.008
EATWB, °C	**Total Capacity**	**Sensible Cooling Correction Factor**					**Cooling Power CF**
		21°C db	**23°C db**	**25°C db**	**27°C db**	**29°C db**	
13	0.915	0.976					0.987
15	0.925	0.862	0.97	1.11	1.302		0.993
17	0.957	0.739	0.85	1	1.151	1.24	0.997
19	1	0.618	0.72	0.87	**1**	1.12	1
21	1.004	0.581	0.6	0.74	0.849	0.99	1.004
23	1.151			0.59	0.768	0.86	1.007
25	1.235				0.559	0.73	1.01

** Bold indicates rated values.*

Table 2.6 Heating Capacity and Input Power Correction Factors (CFs) for EATs

EAT, °F	Heating Capacity CF	Heating Power CF	EAT, °C	Heating Capacity CF	Heating Power CF
50	1.045	0.809	10	1.045	0.809
55	1.032	0.863	12.5	1.033	0.858
60	1.02	0.915	15	1.022	0.905
65	1.007	0.968	17.5	1.011	0.952
68	1	1	20	1	1
70	0.995	1.025	22.5	0.989	1.05
75	0.982	1.074	25	0.977	1.095
80	0.97	1.126	27.5	0.966	1.142

Table 2.7 Capacity and Input Power Correction Factors (CFs) for Airflow Rate

% Rated Flow	Total Capacity	Sensible Cooling CF	Cooling Power CF	Heating Capacity CF	Heating Power CF
70	0.946	0.833	0.926	0.96	1.138
80	0.968	0.888	0.948	0.976	1.057
90	0.985	0.941	0.97	0.988	1.025
100	1	1	1	1	1
110	1.01	1.052	1.033	1.01	0.986
120	1.018	1.097	1.07	1.019	0.98
130	1.022	1.132	1.113	1.026	0.975

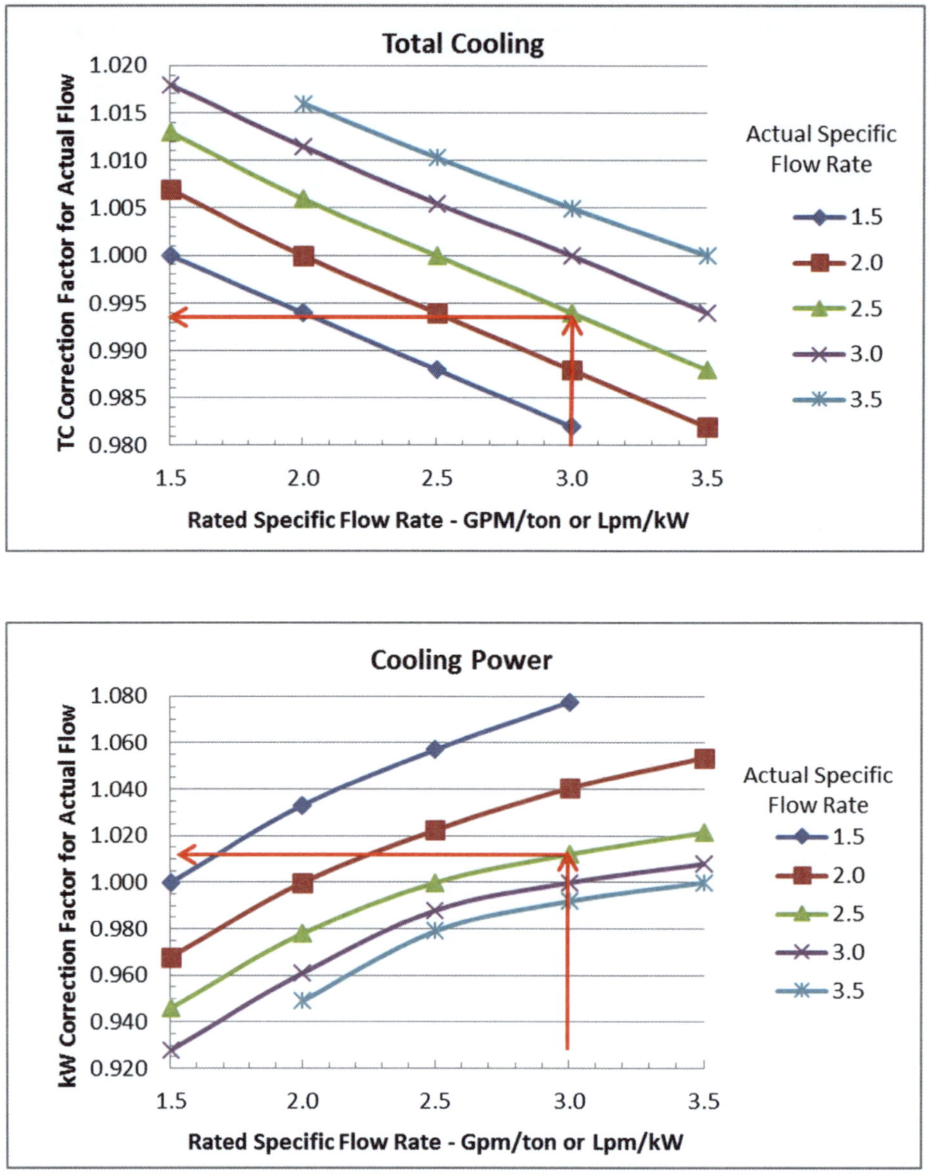

Figure 2.14 Cooling Capacity and Input Power Correction Factors for Liquid Flow Rate

flow of 3.0 (gpm/ton [L/min·kW]) and an actual flow rate of 2.5 (gpm/ton [L/min·kW]) the correction factor for cooling mode power is 1.01.

In addition to ELT, another significant factor affecting heat pump performance is fan power. The standard ratings do not include the power required to deliver the ESP required to distribute air through ducted systems or the pressure required to overcome filter losses. This correction is significant, especially in the cooling mode, because the added fan power is converted to heat and negatively impacts net cooling capacity. In heating this is a benefit in terms of capacity but a penalty in terms of input power.

The power must be corrected to include a reasonable ESP and loss representative of modern filters. Typical ESP requirements for unitary equipment are 0.4 to 0.6 in. H_2O (100 to 150 Pa) (Parker and Proctor 2001). When filters are clean friction losses typically range from 0.2 to 0.5 in. H_2O (50 to 125 Pa) and when dirty can be as high as 1.0 in. H_2O

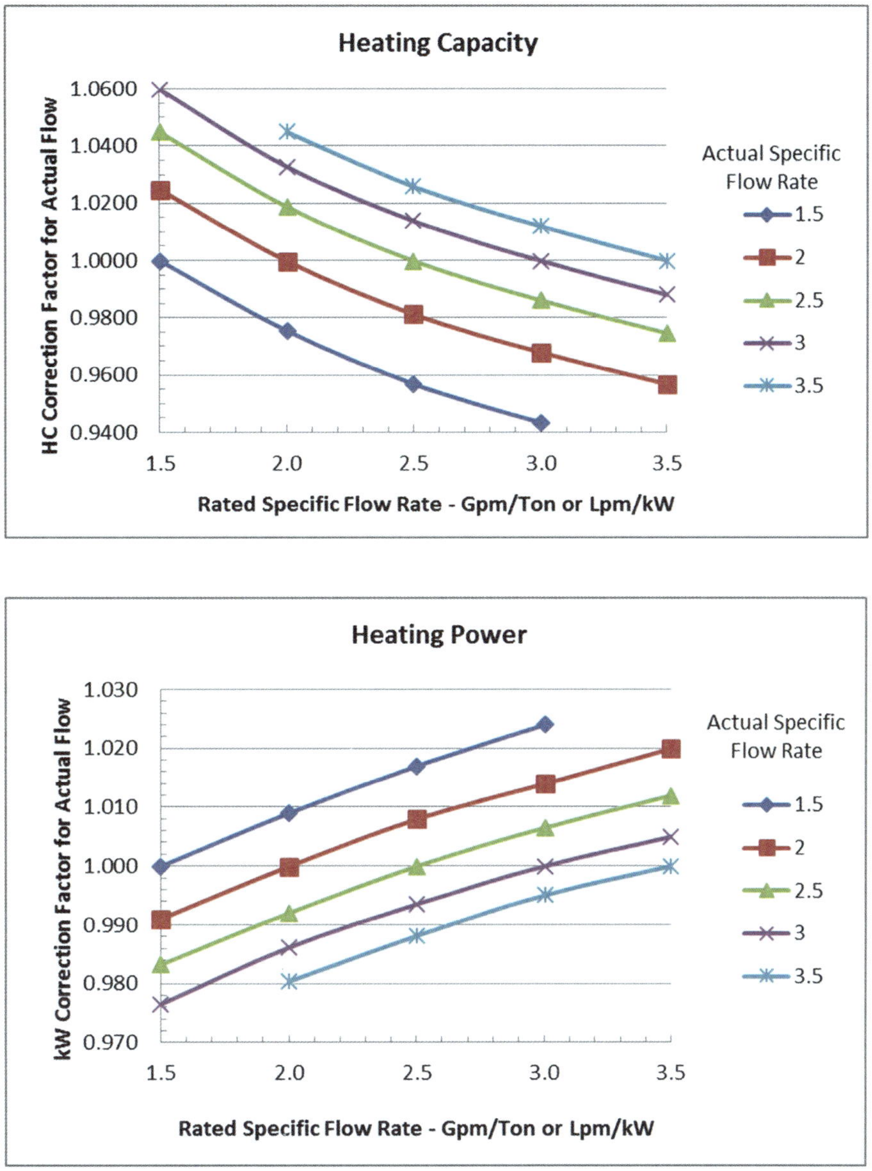

Figure 2.15 Heating Capacity and Input Power Correction Factors for Liquid Flow Rate

(250 Pa) (AAF 2012). The amount of pressure required to be delivered by the fan must be corrected to include the ESP and filter loss:

$$P_{Cor} = \text{ESP} + \textit{filter loss} \tag{2.1}$$

The additional fan power is

$$W_{Fan} = \frac{P_{Cor} \times Q}{\eta_{w-a}} \tag{2.2}$$

where

Q = volumetric airflow rate, cfm (L/s)

$\eta_{w\text{-}a}$ = $\eta_{fan} \times \eta_{motor}$ = fan wire-to-air efficiency, % (AHRI/ASHRAE ISO Standard 13256-1 assumes 30% [ASHRAE 2012a])

When conventional I-P and SI units are applied, Equation 2.2 can be expressed as

$$W_{Cor}\text{ (watts)} = \frac{746\text{ (W/hp)} \times P_{Cor}\text{ (in.)} \times Q\text{ (cfm)}}{6350 \times \eta_{w\text{-}a}} \quad \text{(I-P)} \quad (2.3a)$$

$$W_{Cor}\text{ (watts)} = \frac{P_{cor}\text{ (Pa)} \times Q\text{ (L/s)}}{1000\text{ (L/m}^3) \times \eta_{w\text{-}a}} \quad \text{(SI)} \quad (2.3b)$$

The uncorrected power input for the heat pump for cooling and heating can be determined from the equipment capacity and efficiency.

$$W_{Rated}\text{ (watts)} = TC_{Rated}\text{ (Btu/h)} \div \text{EER (Btu/Wh)} \quad \text{(I-P)} \quad (2.4a)$$

$$W_{Rated}\text{ (watts)} = TC\text{ (watts)} \div COP_c \quad \text{(SI)} \quad (2.4b)$$

$$W_{Rated}\text{ (watts)} = HC_{Rated}\text{ (Btu/h)} \div [3.412\text{ (Btu/Wh)} \times COP_h] \quad \text{(I-P)} \quad (2.5a)$$

$$W_{Rated}\text{ (watts)} = TC\text{ (watts)} \div COP_h \quad \text{(SI)} \quad (2.5b)$$

When the fan power is included the heat pump input power is

$$W_{hp}\text{ (watts)} = W_{Rated} + W_{Cor} \quad (2.6)$$

The entire input power of the fan is converted to heat because the unitary equipment motor losses, fan losses, and air distribution friction are within the conditioned space. The rated cooling capacity (TC_{Rated}) without the effects of the fan heat is often referred to as *gross capacity* and is converted to net total cooling capacity as

$$TC_{net}\text{ (Btu/h)} = TC_{Rated}\text{ (Btu/h)} - 3.412\text{ (Btu/Wh)} \times W_{Cor}\text{ (watts)} \quad \text{(I-P)} \quad (2.7a)$$

$$TC_{net}\text{ (watts)} = TC_{Rated}\text{ (watts)} - W_{Cor}\text{ (watts)} \quad \text{(SI)} \quad (2.7b)$$

and the net efficiencies are

$$EER_{net}\text{ (Btu/Wh)} = TC_{net}\text{ (Btu/h)} \div W_{hp}\text{ (watts)} \quad \text{(I-P)} \quad (2.8a)$$

$$COP_{net} = TC_{net}\text{ (watts)} \div W_{hp}\text{ (watts)} \quad \text{(SI)} \quad (2.8b)$$

In heating, the fan heat is added to the rated heating capacity:

$$HC_{net}\text{ (Btu/h)} = HC_{Rated}\text{ (Btu/h)} + 3.412\text{ (Btu/Wh)} \times W_{Cor}\text{ (watts)} \quad \text{(I-P)} \quad (2.9a)$$

$$HC_{net}\text{ (watts)} = HC_{Rated}\text{ (watts)} + W_{Cor}\text{ (watts)} \quad \text{(SI)} \quad (2.9b)$$

and the coefficient of performance is

$$COP_{net} = HC_{net}\text{ (Btu/h)} \div 3.412\text{ (Btu/Wh)} \times W_{hp}\text{ (watts)} \quad \text{(I-P)} \quad (2.10a)$$

$$COP_{net} = HC_{net}\text{ (watts)} \div W_{hp}\text{ (watts)} \quad \text{(SI)} \quad (2.10b)$$

EXAMPLE 2.1—
HEAT PUMP PERFORMANCE CORRECTION, COOLING MODE (I-P)

The Model 36 water-to-air heat pump (Table 2.3a) is operated with 80°F ELT, 7 gpm liquid flow, 1080 cfm airflow, and 75°F db/63°F EATWB. The system requires an ESP of 0.4 in. H_2O, a filter with a friction loss of 0.3 in. H_2O, and a pump that draws 190 W. Calculate the net total cooling capacity, total input power, and system EER.

Solution

Step 1 is to correct TC and EER for 80°F ELT using TC at 86°F (34,500 Btu/h*), TC at 77°F (35,000 Btu/h*), EER at 86°F (19.6 Btu/Wh), and EER at 77°F (22.0 Btu/Wh). (**TC and HC values in Table 2.3a are expressed in Btu/h × 1000 and are converted to Btu/h for calculations.*)

$$\begin{aligned} TC_{80} &= TC_{77} + (80°F - 77°F) \times \frac{TC_{86} - TC_{77}}{86°F - 77°F} \\ &= 35{,}000 + (80 - 77) \times \frac{34{,}500 - 35{,}000}{86 - 77} \\ &= 34{,}800 \text{ Btu/h} \end{aligned}$$

$$\begin{aligned} EER_{80} &= EER_{77} + (80°F - 77°F) \times \frac{EER_{86} - EER_{77}}{86°F - 77°F} \\ &= 22.0 + (80 - 77) \times \frac{19.6 - 22.0}{86 - 77} \\ &= 21.2 \text{ Btu/Wh} \end{aligned}$$

Step 2 is to compute the input power using TC and EER at 80°F ELT.

$$W_{80} = TC_{80} \div EER_{80} = 34{,}800 \text{ Btu/h} \div 21.2 \text{ Btu/Wh} = 1642 \text{ W}$$

Step 3 is to correct TC and input power from 66.2°F EATWB to 65°F using Table 2.5 correction factors of 0.962 for TC and 0.997 for power input.

$$TC_{63} = Cf_{66.2 \to 63} \times TC_{66.2} = 0.962 \times 34{,}800 \text{ Btu/h} = 33{,}480 \text{ Btu/h}$$

$$W_{63} = Cf_{66.2 \to 63} \times W_{66.2} = 0.997 \times 1642 = 1637 \text{ W}$$

Alternate Step 3 would be to correct the sensible cooling capacity (SC) for 75°F/63°F EAT. If SC is available, the correction factor from Table 2.5 for converting $SC_{80.6/66.2}$ to $SC_{75/63}$ is 0.905.

Step 4 is to correct TC and input power from 1200 cfm to 1080 cfm using Table 2.7. The flow rate of 1080 cfm is 90% of 1200 cfm. The correction factors are 0.985 for TC and 0.990 for power.

$$TC_{1080} = Cf_{1200 \to 1080} \times TC_{1200} = 0.985 \times 33{,}480 \text{ Btu/h} = 32{,}980 \text{ Btu/h}$$

$$W_{1080} = Cf_{1200 \to 1080} \times W_{1200} = 0.990 \times 1637 = 1621 \text{ W}$$

Step 5 is to correct TC and input power from 9 gpm to 7 gpm using Figure 2.14. It is suggested that the specific flow rates in the figure be calculated for rated values at the nearest rated ELT, which would be 77°F. From Table 2.3a, TC is 35,000 Btu/h, which is 2.92 tons (= 35,000 Btu/h ÷12,000 Btu/h·ton). Therefore,

$$\text{Rated specific flow} = 9 \text{ gpm} \div 2.92 = 3.1 \text{ gpm/ton}$$

$$\text{Actual specific flow} = 7 \text{ gpm} \div 2.92 = 2.4 \text{ gpm/ton}$$

Figure 2.14 indicates the correction factor is 0.993 for TC and 1.015 for power. **However, the correction factors are applied to the values in Step 4, not the rated values.** Thus,

$$TC_7 = Cf_{9 \to 7} \times TC_9 = 0.993 \times 32{,}980 \text{ Btu/h} = 32{,}750 \text{ Btu/h}$$

$$W_7 = Cf_{9 \to 7} \times W_{90} = 1.015 \times 1621 = 1645 \text{ W}$$

Step 6 is to correct TC and power for the additional fan power required to overcome friction in the air distribution network and air filter using the AHRI/ASHRAE ISO Standard 13256-1 fan wire-to-air efficiency of 30% (ASHRAE 2012a).

$$P_{Fan} = \text{ESP} + \text{filter loss} = 0.4 + 0.3 = 0.7 \text{ in. } H_2O$$

$$W_{Fan} \text{ (watts)} = \frac{746 \text{ (W/hp)} \times 0.70 \text{ (in.)} \times 1080 \text{ (cfm)}}{6350 \times 30\%} = 296 \text{ W}$$

$$W_{hp+fan} = 1645 + 296 = 1941 \text{ W}$$

$$\mathbf{TC_{net} \text{ (Btu/h)} = 32{,}750 \text{ (Btu/h)} - 3.412 \text{ (Btu/Wh)} \times 296 \text{ (watts)} = 31{,}740 \text{ Btu/h}}$$

Step 7 is to add the pump power to find the total system power.

$$\mathbf{W_{total} = W_{hp+fan+pump} = 1645 + 296 + 190 = 2131 \text{ W}}$$

Step 8 is to compute system EER using the corrected net cooling capacity and the total system power for 80°F ELT, 7 gpm water flow, 1080 cfm airflow, 75°F db/63°F EATWB, an ESP of 0.4 in. H_2O, a filter friction loss of 0.3 in. H_2O, and a 190 W pump.

$$\mathbf{EER_{system} = 31{,}740 \text{ Btu/h} \div 2131 \text{ W} = 14.9 \text{ Btu/Wh}}$$

(This is 32% less than the EER of 22.0 Btu/Wh at GLHP conditions.)

EXAMPLE 2.2— HEAT PUMP PERFORMANCE CORRECTION, HEATING MODE (SI)

The Model 48 water-to-air heat pump (Table 2.3b) is operated with 5°C ELT, 40 L/min, 745 L/s, and 22°C EAT. The system requires an ESP of 125 Pa, a filter with a friction loss of 80 Pa, and a pump that draws 250 W. Calculate the net heating capacity, total input power, and system COP.

Solution

Step 1 is to correct HC and COP for 5°C ELT using HC at 10°C (13,200 W*), HC at 0°C (10 300 W*), COP at 10°C (4.8), and COP at 0°C (4.0). (**TC and HC values in Table 2.3b are expressed in kW and are converted to W for calculations.*)

$$HC_5 = HC_{10} - (5°C - 0°C) \times \frac{HC_{10} - HC_0}{10°C - 0°C}$$

$$= 13\ 200 - (5 - 0) \times \frac{13\ 200 - 10\ 300}{10 - 0}$$

$$= 11\ 750\ W$$

$$COP_5 = COP_{10} - (5°C - 0°C) \times \frac{COP_{10} - COP_0}{10°C - 0°C}$$

$$= 4.8 - (5 - 0) \times \frac{4.8 - 4.0}{10 - 0}$$

$$= 4.4$$

Step 2 is to compute the input power using HC and COP at 5°C ELT.

$$W_5 = HC_5 \div COP_5 = 11\ 750\ W \div 4.4 = 2670\ W$$

Step 3 is to correct HC and input power from 20°C EAT to 22°C using Table 2.6 correction factors (via interpolation) of 0.991 for HC and 1.04 for power input.

$$HC_{22} = Cf_{20 \to 22} \times HC_{20} = 0.991 \times 11\ 750\ W = 11\ 644\ W$$

$$W_{10} = Cf_{20 \to 22} \times W_{20} = 1.04 \times 2670 = 2777\ W$$

Step 4 is to correct HC and input power from 710 L/s to 745 L/s using Table 2.7. The flow rate of 745 L/s is 105% of 710 L/s. The correction factors (via interpolation) are 1.005 for HC and 0.993 for power.

$$HC_{745} = Cf_{710 \to 745} \times HC_{710} = 1.005 \times 11\ 644\ W = 11\ 700\ W$$

$$W_{745} = Cf_{710 \to 745} \times W_{710} = 0.993 \times 2777 = 2758\ W$$

Step 5 is to correct HC and input power from 45 L/min to 40 L/min using Figure 2.15. The specific flow rates in the figure are calculated for rated values at 10°C. From Table 2.3b, HC is 13.2 kW. Therefore,

$$\text{Rated specific flow} = 45\ \text{gpm} \div 13.2\ \text{kW} = 3.4\ \text{L/min·kW}$$

$$\text{Actual specific flow} = 40\ \text{gpm} \div 13.2\ \text{kW} = 3.0\ \text{L/min·kW}$$

Figure 2.15 indicates the correction factor is 0.990 for HC and 1.004 for power. **However, the correction factors are applied to the values in Step 4, not the rated values.** Thus,

$$HC_{40} = Cf_{45 \to 40} \times HC_{45} = 0.990 \times 11\ 700\ W = 11\ 580\ W$$

$$W_{40} = Cf_{45 \to 40} \times W_{45} = 1.004 \times 2758 = 2769\ W$$

Step 6 is to correct HC and power for the additional fan power required to overcome friction in the air distribution network and air filter using the AHRI/ASHRAE ISO Standard 13256-1 fan wire-to-air efficiency of 30% (ASHRAE 2012a).

$$P_{Fan} = \text{ESP} + \text{filter loss} = 125 + 80 = 205 \text{ Pa}$$

$$W_{Fan} \text{ (watts)} = \frac{205 \text{ (Pa)} \times 750 \text{ (L/s)}}{1000 \text{ (L/m}^3\text{)} \times 30\%} = 513 \text{ W}$$

$$W_{hp+fan} = 2769 + 513 = 3282 \text{ W}$$

$$\mathbf{HC_{net} \text{ (watts)} = 11\,580 + 513 = 12\,093 \text{ W}}$$

Step 7 is to add the pump power to find the total system power.

$$\mathbf{W_{total} = W_{hp+fan+pump} = 2769 + 513 + 250 = 3532 \text{ W}}$$

Step 8 is to compute system COP using the corrected net heating capacity and the total system power for 5°C ELT, 40 L/min, 745 L/s, 22°C EAT, an ESP of 125 Pa, a filter loss of 80 Pa, and a 250 W pump.

$$\mathbf{COP_{system} = 12\,093 \text{ W} \div 3532 \text{ W} = 3.4}$$

(This is 20% less than the COP of 4.0 at GLHP conditions.)

2.4 GSHP SYSTEM PERFORMANCE

The correction procedures for individual heat pumps presented in the preceding section are often inadequate for larger, more complex HVAC systems. A procedure for computing system efficiency for more complex systems is discussed in more detail in the ASHRAE publication *HVAC Simplified* (Kavanaugh 2006). In the procedure, the capacities of the primary cooling or heating devices and the power, efficiency, or fuel rate are entered. This is followed by a systematic listing of all auxiliary devices related to distribution of air and water. Cooling capacity deductions for cooled air or chilled water are computed from fan and pump characteristics (flow rate, total pressure, fan/pump efficiency, motor efficiency). Capacity additions are made in the heating mode using similar information. Although power input for condenser equipment is included, capacity corrections are not provided. As mentioned previously, a spreadsheet that follows this procedure (*HVACsystemEff.xlsx*) is available with this book at www.ashrae.org/GSHP.

Proponents of GSHPs have gone to great lengths to extol the great benefit of the moderate temperatures of the earth to enhance the efficiency and performance compared to conventional HVAC systems. An equal, and in some cases even greater, benefit is the minimal amount of auxiliary equipment necessary to cool and heat commercial buildings. Conversely, if designs incorporate a high level of auxiliary equipment, the efficiency benefits of GSHPs can be nullified.

Comparative evaluations using the above-mentioned procedure for larger systems are highly recommended during the early design phase of the HVAC system. A demonstration follows that compares the system efficiency of a conventional GSHP system (that has very few auxiliary devices) with a traditional HVAC system (with multiple auxiliary devices) connected to a vertical ground heat exchanger. The total cooling capacities of both systems are nearly equal.

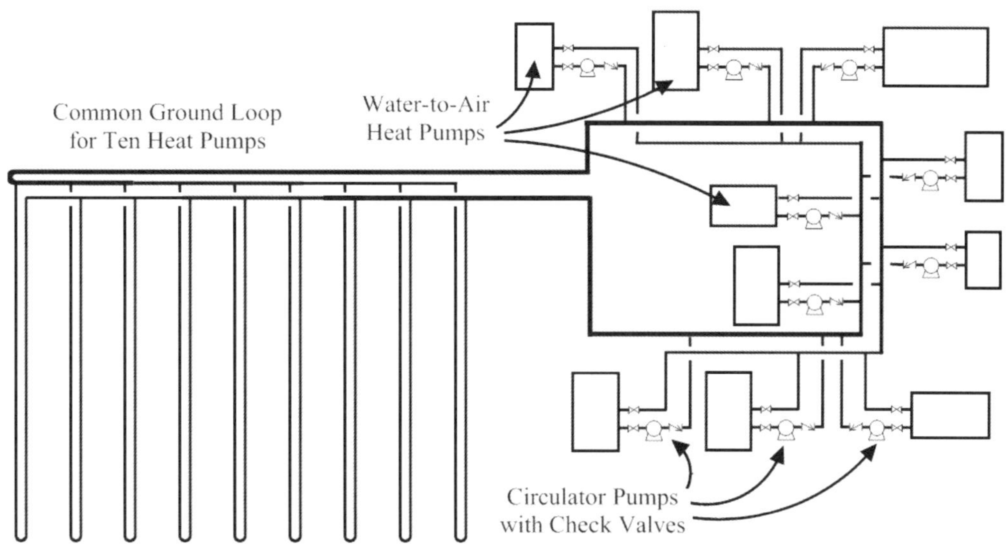

Figure 2.16 Ten-Heat-Pump Common Loop—One of Twenty in Example

The conventional GSHP system is a common-loop system as shown in Figure 2.16. The configuration of multiple small networks results in minimal pump and fan power requirements. The example system consists of 200 heat pumps connected to 20 individual ground-loop circuits with approximately 10 heat pumps on each circuit. Figure 2.17 demonstrates the *HVACsystemEff.xlsx* output.

The quantities, EERs, and capacities for three different heat pump models are entered into the rows for Item 1. Values are taken from Table 2.3a using the recommended WSHP rating point (ELT = 86°F [30°C]). The corrections for fan power are done in Item 2b by including the recommended ESP for filter and air distribution losses and the rated values for airflow rate. Note that the cooling capacity deductions are computed in the rightmost columns using a wire-to-air efficiency of 30%. The power input for the ground-loop pumps is included in the Item 5 rows using the rated liquid flow rate, a relatively low pump head (a result of the multiple small ground loops), a pump efficiency of 50%, and a motor efficiency of 50%. This results in a relatively low wire-to-water efficiency of 25% ($\eta_{w\text{-}w} = \eta_{pump} \times \eta_{motor}$), which is typical for small wet-rotor pumps.

The resulting system EER is 14.6 Btu/Wh (COP_c = 4.27). The net system cooling capacity is 557 tons (1960 kW). An important item to note is that the fan heat penalty is only –17.5 tons (–61 kW) and the ASHRAE/IES Standard 90.1-2010 indicator for fan power is 0.36 hp/1000 cfm (ASHRAE 2010). The total system power input requirement is 458.6 kW.

The system for comparison links a vertical ground loop to a conventional chilled-water variable-air-volume (VAV) air distribution network as shown in Figure 2.18. Two 340 ton (2000 kW), 0.5 kW/ton (COP_c = 7.0) chillers provide water to sixteen air-handling units (AHUs) ranging in capacity from 4000 to 40,000 cfm (1900 to 19,000 L/s). The AHUs are equipped with variable-speed fans and deliver air to a network of 230 fan-powered VAV terminals ranging in capacity from 800 to 1600 cfm (380 to 760 L/s). The system also includes eight 34,000 cfm (16 000 L/s) return air fans. Three sets of pumps provide flow to the ground loop, chilled water to the building, and flow through the chillers. Figure 2.19 demonstrates the *HVACsystemEff.xlsx* output.

Item	HVAC System Cooling Efficiency Calculator (Reference: *HVAC Simplified*, ASHRAE, 2006)											
			Yellow Colored Cells for Input					Do Not Type in Blue (Output) Cells				
Item		Enter Chiller Data in Cells Below			OR	Enter AC/HP Data						
1	Chillers or Heat Pumps Click Here for Explanation	Qty.	kW/ton	Tons		Qty.	EER	kBtu/h	kW-each	kW-total	kBtu/h	tons
	WC chillers based on CoLWT=95 F*, ChWLT=44 F*				---	50	19.2	28.3	1.5	73.7	1415.0	117.9
	AC chillers based on OAT = 95 F*, ChWLT = 44 F*				---	100	19.6	34.5	1.8	176.0	3450.0	287.5
	Enter EER at 95 F* (air source) or EER at 86 F* (WSHP)					50	19.2	40.6	2.1	105.7	2030.0	169.2
	*Enter ratings at non-standard conditions when appropriate.				---				0.0	0.0	0.0	0.0
2a	Air Handling Unit Fans (≥1.0 hp) Click Here		Air Flow (cfm)	TP (in. wtr)	η fan (%)	η mtr. (%)	η VSD (%)	hp-each	---	---	---	---
								0.00	0.00	0.0	0.0	0.0
								0.00	0.00	0.0	0.0	0.0
								0.00	0.00	0.0	0.0	0.0
2b	DX Coil, Water Coil, or VAV Fans (<1.0 hp) Click Here		Air Flow (cfm)	TP (in. wtr)	η wire-air (%)	OR hp-each	η mtr. (%)		---	---	---	---
	Systems with Unitary ACs or Heat Pumps	50	900	0.6	30.0%			0.28	0.21	10.6	-36.1	-3.0
		100	1200	0.7	30.0%			0.44	0.33	32.9	-112.2	-9.4
		50	1300	0.7	30.0%			0.48	0.36	17.8	-60.8	-5.1
3	Return Air Fans - Enter exhaust air % ➔	20%	Air Flow (cfm)	TP (in. wtr)	η fan (%)	η mtr. (%)	η VSD (%)		---	---	---	---
								0.00	0.00	0.0	0.0	0.0
								0.00	0.00	0.0	0.0	0.0
								0.00	0.00	0.0	0.0	0.0
4	Chilled Water Pumps		Wtr. Flow (gpm)	ΔH (ft. wtr)	η pump (%)	η mtr. (%)	η VSD (%)		---	---	---	---
								0.00	0.00	0.0	0.0	0.0
								0.00	0.00	0.0	0.0	0.0
								0.00	0.00	0.0	0.0	0.0
5	Condenser or Ground Loop Water Pumps		Wtr.Flow (gpm)	ΔH (ft. wtr)	η pump (%)	η mtr. (%)	η VSD (%)		---	---	Click Here	
		50	8	30	50.0%	50.0%	100.0%	0.12	0.18	9.0	---	---
		100	9	30	50.0%	50.0%	100.0%	0.14	0.20	20.3	---	---
		50	11	30	50.0%	50.0%	100.0%	0.17	0.25	12.4	---	---
6	Condenser (or Fluid Cooler/Cooling Tower) Fan	0	Air Flow (cfm)	TP (in. wtr)	η fan (%)	η mtr. (%)	η VSD (%)	hp-each	---	---	---	---
		0	---	---	---	75.0%	100.0%		0.00	0.0	---	---
			---	---	---				0.00	0.0	---	---
			---	---	---				0.00	0.0	---	---
										kW	kBtu/h	tons
								System Totals		458.6	6685.9	557.2
				kW/ton :	0.82	EER =	14.6	Btu/W-	COP =	4.27		
		ASHRAE 90.1 check		Q (cfm) =	230000	Fan hp =	82	hp/1000cf	0.36			

Figure 2.17 *HVACsystemEff.xlsx* Output—System Cooling Efficiency for Common-Loop GCHP System with 200 Heat Pumps

Although the chillers in this analysis are very efficient, the resulting system EER is 7.9 Btu/Wh ($COP_c = 2.3$), which is substantially lower than the value of 14.6 Btu/Wh ($COP_c = 4.27$) of the heat pump system. The primary cause of low efficiency is the size and number of fans in the air distribution system. The sum of the fan power is 374 kW, which is larger than the 340 kW input of the chillers. Additionally, the heat generated by the fans reduces the gross capacity of the chillers by 100 tons (350 kW). This amount cannot be considered excessive for these typical VAV systems because the fan power limit of 1.72 hp/1000 cfm is 25% below the limit set by ASHRAE/IES Standard 90.1-2010 (Bolt 2012).

Another item to consider is the added required length of ground heat exchanger because of the low system efficiency and large amount of fan heat. The fan heat delivered to the building at full load is equivalent to 100 tons (352 kW). The ground loop must dissipate this added load in cooling. In the winter, the building can be heated by the fans to a large extent, which reduces chillers operating in heating and the amount of heat removed from the ground. This shifts the annual ground heat balance further toward the cooling mode. To offset the additional heat remaining in the ground, the ground heat exchanger must be further enlarged to meet requirements in the cooling mode. HVAC systems with large auxiliary power requirements, such as a chilled-water VAV system with high fan and pump pressure requirements, are not recommended for GSHP applications.

Water-to-water heat pump and reversible chiller applications can be efficient if the following conditions are met:

- Fan power requirements are minimized with low-static-pressure fan-coil units (FCUs) or chilled beams or are completely eliminated with in-floor (radiant) heat.

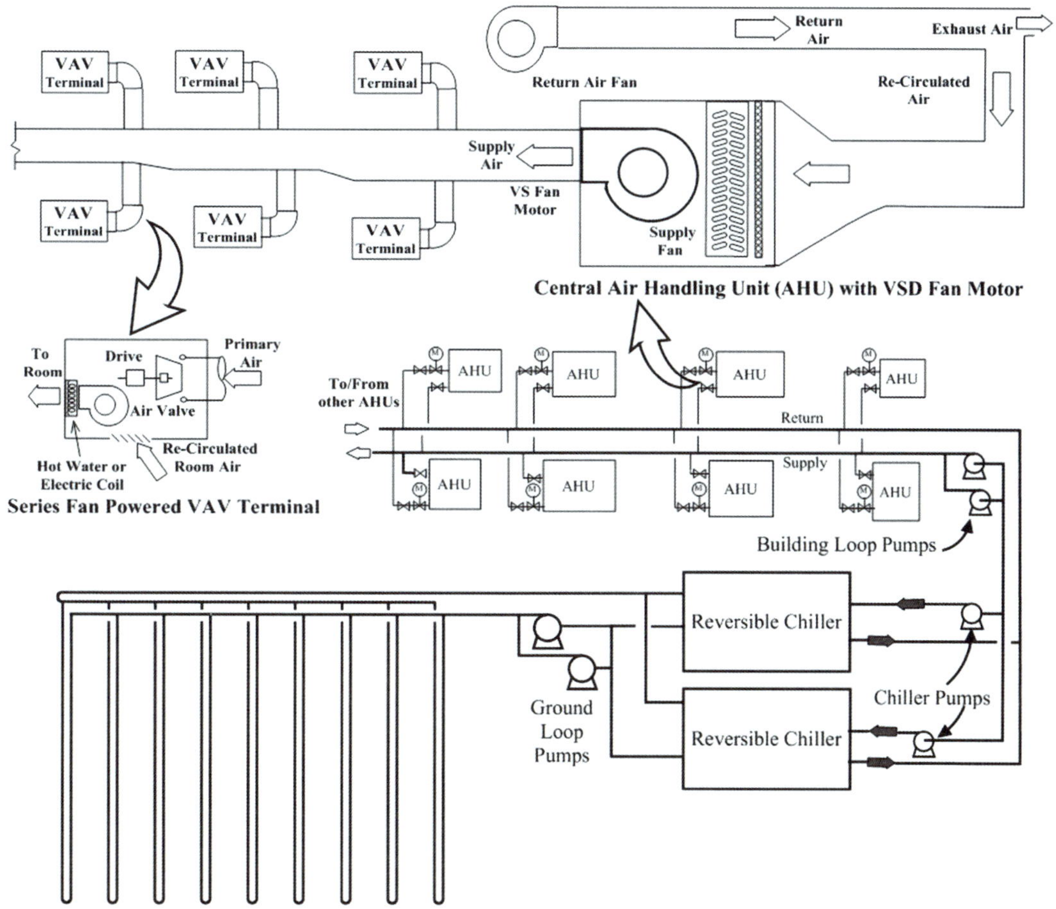

Figure 2.18 Chilled-Water VAV Vertical Ground-Loop System

HVAC System Cooling Efficiency Calculator (Reference: *HVAC Simplified*, ASHRAE, 2006)

Yellow Colored Cells for Input — Do Not Type in Blue (Output) Cells

Enter Chiller Data in Cells Below OR Enter AC/HP Data

Item															
1	Chillers or Heat Pumps Click Here for Explanation	Qty.	kW/ton	Tons		Qty.	EER	kBtu/h	kW-each	kW-total	kBtu/h	tons			
	WC chillers based on CoLWT=95 F*, ChWLT=44 F*	2	0.5	340	---				170.0	340.0	8160.0	680.0			
	AC chillers based on OAT = 95 F*, ChWLT = 44 F*				---				0.0	0.0	0.0	0.0			
	Enter EER at 95 F* (air source) or EER at 86 F* (WSHP)								0.0	0.0	0.0	0.0			
	*Enter ratings at non-standard conditions when appropriate.				---				0.0	0.0	0.0	0.0			
2a	Air Handling Unit Fans (≥1.0 hp) Click Here		Air Flow (cfm)	TP (in. wtr)	η fan (%)	η mtr. (%)	η VSD (%)	hp-each	---	---	---	---			
		4	40000	5	75.0%	93.0%	98.0%	41.99	34.37	137.5	-469.1	-39.1			
		4	20000	4	75.0%	92.0%	98.0%	16.80	13.90	55.6	-189.7	-15.8	SA cfm		
		8	4000	2	70.0%	88.0%	98.0%	1.80	1.56	12.5	-42.5	-3.5	272000		
2b	DX Coil, Water Coil, or VAV Fans (<1.0 hp) Click Here		Air Flow (cfm)	TP (in. wtr)	η wire-air (%)	OR hp-each	η mtr. (%)		---	---	---	---			
	Systems with Unitary ACs or Heat Pumps	50	1600	0.5	30.0%			0.42	0.31	15.7	-53.4	-4.5			
		120	1200	0.5	28.0%			0.34	0.25	30.2	-103.1	-8.6	VAV cfm		
		60	800	0.5	25.0%			0.25	0.19	11.3	-38.5	-3.2	272000		
3	Return Air Fans - Enter exhaust air % →→→	20%	Air Flow (cfm)	TP (in. wtr)	η fan (%)	η mtr. (%)	η VSD (%)		---	---	---	---			
		8	34000	2.5	75.0%	92.0%	97.0%	17.85	14.92	119.4	-325.8	-27.2			
								0.00	0.00	0.0	0.0	0.0	RA cfm		
								0.00	0.00	0.0	0.0	0.0	272000		
4	Chilled Water Pumps		Wtr. Flow (gpm)	ΔH (ft. wtr)	η pump (%)	η mtr. (%)	η VSD (%)		---	---	---	---			
		2	825	100	70.0%	90.0%	97.0%	29.76	25.43	50.9	-173.5	-14.5	CHW gpm		
		2	825	50	70.0%	90.0%	97.0%	14.88	12.72	25.4	-86.8	-7.2	1650		
								0.00	0.00	0.0	0.0	0.0			
5	Condenser or Ground Loop Water Pumps		Wtr.Flow (gpm)	ΔH (ft. wtr)	η pump (%)	η mtr. (%)	η VSD (%)		---	---	Click Here				
		2	1000	100	70.0%	90.0%	97.0%	36.08	30.83	61.7	---	---	CW gpm		
								0.00	0.00	0.0	---	---	2000		
								0.00	0.00	0.0	---	---			
6	Condenser (or Fluid Cooler/Cooling Tower) Fan	0	Air Flow (cfm)	TP (in. wtr)	η fan (%)	η mtr. (%)	η VSD (%)	hp-each	---	---	---	---		Fan kW	Fan tons
		0	---	---	---	75.0%	100.0%		0.00	0.0	---	---	AHUs	205.5	-58.4
			---	---	---				0.00	0.0	---	---	VAV term	57.2	-16.3
			---	---	---				0.00	0.0	---	---	RAF	119.4	-27.2
										kW	kBtu/h	tons		382.1	-101.8
								System Totals		860.0	6677.6	556.5			
				kW/ton :	1.55	EER =	7.8	Btu/W-h	COP =	2.28					
		ASHRAE 90.1 check		Q (cfm) =	272000	Fan hp =	469	hp/1000cfm=	1.72						

Figure 2.19 *HVACsystemEff.xlsx* Output—Component Specifications and System Efficiencies for Chilled-Water VAV GSHP

- Pump power requirements are minimized in both ground and building loops.
- Leaving hot-water temperatures are kept below 110°F (43°C). Lower values are better for applications such as in-floor heat that provide comfort with lower temperatures.
- Leaving chilled-water temperatures are above 44°F (7°C). Higher values are better for applications such as outdoor air coils and chilled beams that can work well with slightly higher temperatures.

The *HVACsystemEff.xlsx* spreadsheet tool can be used to evaluate other alternatives if the heat pump or chiller performance is corrected for any nonrated liquid temperatures.

2.5 SUGGESTED GSHP SPECIFICATIONS

The recommendations of Table 2.8 are intended to serve as *minimum requirements* for water-to-air and water-to-water heat pumps for GSHP applications. The rating points are consistent with values for source and building loop ELTs used in AHRI/ASHRAE ISO Standards 13256-1 (water-to-air heat pumps) and 13256-2 (water-to-water heat pumps) (ASHRAE 2012a, 2012b). It is suggested that the use of part-load efficiency ratings be avoided when developing specifications. Part-load airflow rates can be well outside normal practice and efficiency values will be inflated because the proportionally large air and water distribution losses are not included in the standard ratings.

2.6 OUTDOOR AIR AND GSHPs

The conventional practice of delivering outdoor ventilation air through a multizone central air distribution system is not an option when unitary equipment, such as water-to-

Table 2.8 Recommended Minimum Allowable Heat Pump Efficiencies—
Efficiency Values Based on Ratings According to AHRI/ASHRAE ISO Standards 13256-1 and 13256-2 (ASHRAE 2012a, 2012b)

Source Water ELT Range	23°F to 104°F (–5°C to 40°C)
Minimum Allowable Water-to-Air Heat Pump Cooling-Mode Efficiency	
At ELT = 86°F (30°C)	EER (WLHP) = 14.0 Btu/Wh, (COP_c = 4.1)
At ELT = 77°F (25°C)	EER (WLHP) = 15.5 Btu/Wh, (COP_c = 4.5)
Minimum Allowable Water-to-Air Heat Pump Heating-Mode Efficiency	
At ELT = 50°F (10°C)	COP_h = 4.0
At ELT = 32°F (0°C)	COP_h = 3.2
Minimum Allowable Water-to-Water Heat Pump Cooling-Mode Efficiency	
(Building Loop ELT = 53.6°F [12°C])	
At ELT = 86°F (30°C)	EER (WLHP) = 13.0 Btu/Wh, (COP_c = 3.8)
At ELT = 77°F (25°C)	EER (WLHP) = 14.0 Btu/Wh, (COP_c = 4.1)
Minimum Allowable Water-to-Water Heat Pump Heating-Mode Efficiency	
(Building Loop ELT = 104°F [40°C])	
At ELT = 50°F (10°C)	COP_h = 3.0
At ELT = 32°F (0°C)	COP_h = 2.5
Maximum Allowable Liquid Coil Loss @ 68°F (20°C)	
Source	12 ft water @ 2.8 gpm/ton (35 kPa @ 3.0 L/min·kW)
Building	10 ft water @ 2.4 gpm/ton (30 kPa @ 2.6 L/min·kW)

air heat pumps, serves individual zones. Designers may choose to apply conventional central chilled-water systems, with or without a ground loop, because they are fixed on the multizone method of supplying fresh air. However, dedicated outdoor air systems (DOASs) are increasingly being applied in all types of systems, central and unitary, because of their energy-saving potential and control simplicity.

Figures 2.20 and 2.21 compare the approaches of the air delivery methods. In the multizone approach, the ventilation air is mixed with the recirculated air, which results in every zone receiving the same fraction of ventilation air to primary air (Z_{pz}). An issue arises if one or more zones require a very high fraction compared to the other zones. An example is shown in Figure 2.20 of a conference room with many occupants sharing an air distribution system with multiple single occupant offices. In this case the offices may require 10% to 20% or less outdoor air, but they will receive the same fraction as the conference room, which may be 25% to 50%. In mild seasons it is also possible that the high fraction of ventilation air could result in excess air being delivered to an office, overcooling the occupants.

Figure 2.21 is a schematic of a DOAS for a similar application. The obvious disadvantage is the required additional duct system. In this approach, the ventilation air and the supply air are not mixed prior to entering the zone. Thus, the offices receive only the amount of ventilation air necessary to satisfy the needs of the zone occupants.

Another significant advantage of the DOAS is that supply air fans, which are much larger than ventilation air fans, do not have to operate continuously to supply occupants with fresh air. While it is true that VAV supply fans can reduce speed to save energy, their minimum allowable flow is often much greater than the amount required to meet ventilation air requirements.

The result of this situation is that energy-saving ventilation air delivery systems are the same for both central and unitary systems. Therefore, the decision to use a unitary GSHP, a central GSHP, or a conventional HVAC central system is independent of the ventilation air system. The possible savings in heat pump capacity and energy use with a combination of DOAS and ventilation air energy recovery units (ERUs) is an important tool in GSHP design optimization.

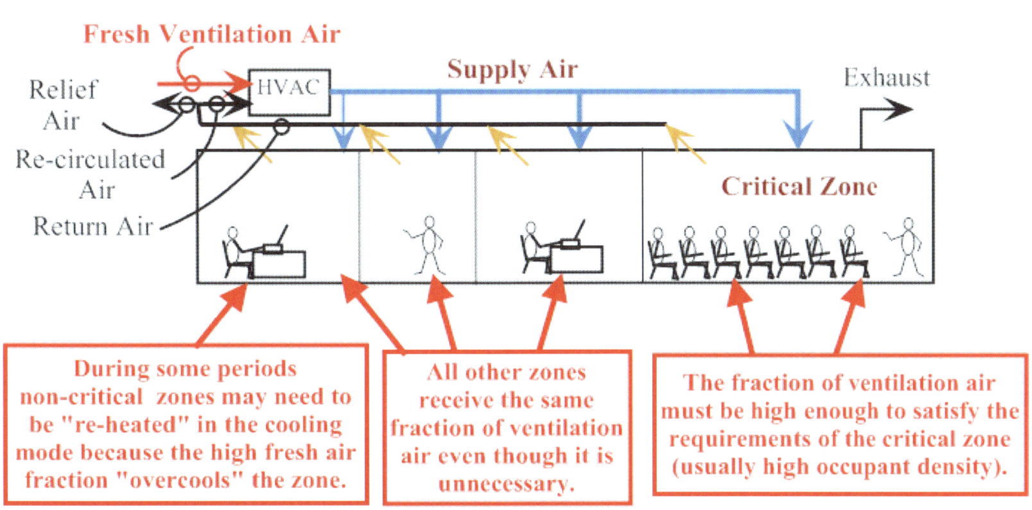

Figure 2.20 Multizone Ventilation Air Delivery

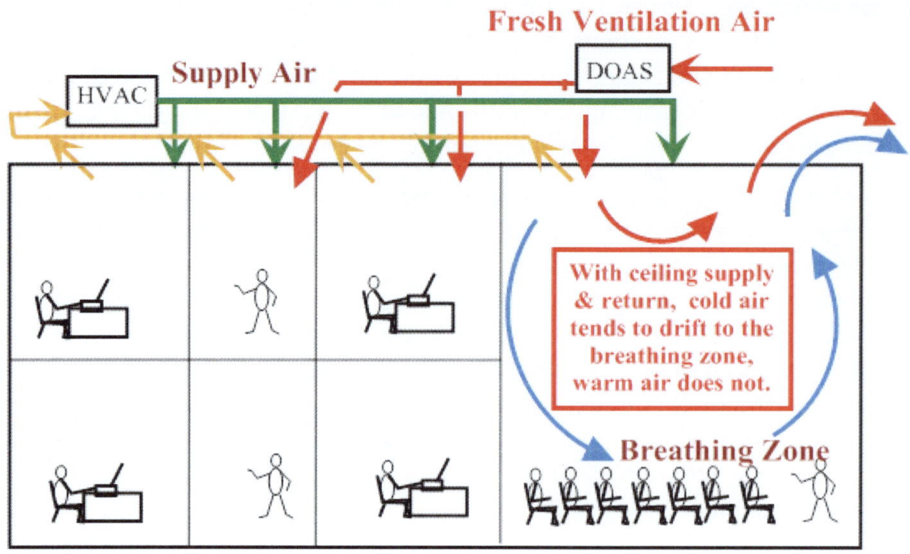

Figure 2.21 DOAS for Ventilation Air Delivery

An overview of ASHRAE Standard 62.1, *Ventilation for Acceptable Indoor Air Quality* (ASHRAE 2013), is presented here to demonstrate the impact of load reduction upon GSHP design. **This standard has experienced frequent changes, and readers are encouraged to use the edition that applies to local codes.** The issues with providing acceptable ventilation in buildings with unitary equipment will likely remain and therefore are presented here using the most recent edition of the standard.

Standard 62.1 dictates the minimum amount of ventilation air provided to the breathing zone (V_{bz})[4] for each zone is based on the number of occupants and the floor area.

$$V_{bz} = R_p P_z + R_a A_z \qquad (2.11)$$

where

R_p = outdoor airflow rate per person from Table 2.9
P_z = zone population (maximum, calculated average, or default if unknown)
R_a = outdoor airflow rate per unit area in cfm/ft^2 (L/s·m^2) from Table 2.9
A_z = zone floor area in ft^2 (m^2)

Table 2.9 also includes default occupancy values per unit area for each type of space. These default values of (P_z/A_z) can be applied to Equation 2.11 to arrive at default values of airflow rate per person:

$$V_{bz}\,(\text{cfm}) = \left(R_p + \frac{R_a}{(P_z/A_z)_{Default}}\right) \times P_z \qquad (2.12)$$

In many cases the outdoor air intake flow (V_{ot}) may not be delivered to the breathing zone if the distribution system is ineffective and correction procedures are not incorporated. The zone air distribution effectiveness (E_z) accounts for how well the ventilation supply air is delivered to the breathing zone, which is prescribed to be 4.5 ft (1.4 m) above

4 Standard 62.1 applies the symbol V (normally the symbol for velocity) for *airflow* rather than the standard ASHRAE practice of using the symbol Q for *volumetric flow rate* (ASHRAE 2013).

Table 2.9 Minimum Ventilation Rates in Breathing Zone—Abbreviated (Complete listing found in Table 6.2.2.1 of ASHRAE Standard 62.1-2013)

Occupancy Category	People Outdoor Air Rate, R_P		Area Outdoor Air Rate, R_a		Default Values		
	cfm/person	L/s·person	cfm/ft^2	L/s·m^2	People per 1000 ft^2 (100 m^2)	cfm/person	L/s·person
Education							
Day Care	10	5	0.18	0.9	25	17	8.6
Classroom (ages 5–8)	10	5	0.12	0.6	25	15	7.4
Classroom (ages 9+)	10	5	0.12	0.6	35	13	6.7
Lecture Classroom	7.5	3.8	0.06	0.3	65	8	4.3
Food and Beverage							
Restaurant Dining	7.5	3.8	0.18	0.9	70	10	5.1
Cafeteria/Fast Food	7.5	3.8	0.18	0.9	100	9	4.7
Hotels, Dorms							
Bed/Living Rooms	5	2.5	0.06	0.3	10	11	5.5
Lobbies/Prefunction	5	2.5	0.06	0.3	20	8	4.0
Assembly	5	2.5	0.06	0.3	120	6	2.8
Office Buildings							
Office Space	5	2.5	0.06	0.3	5	17	8.5
Reception Area	5	2.5	0.06	0.3	30	7	3.5
Telephone/Data Entry	5	2.5	0.06	0.3	60	6	3.0
Public Assembly							
Conference	5	2.5	0.06	0.3	50	6	3.1
Auditorium	5	2.5	0.06	0.3	150	5	2.7
Library	5	2.5	0.12	0.6	10	17	8.5
Museum	7.5	3.8	0.06	0.3	40	9	4.6

the floor. Figure 2.21 shows an example in which the ventilation air is supplied and returned at the ceiling. In cooling, cold, dense air will tend to drift down to the breathing zone and will result in a higher value for E_z compared to warm, less dense air (heating mode) that will tend to stay near the ceiling. The outdoor airflow (V_{oz}) that must be supplied to the zone is (ASHRAE 2013)

$$V_{oz} = V_{bz} / E_z \tag{2.13}$$

where

E_z = 1.2 (floor supply of cool air and ceiling return provided low-velocity displacement ventilation achieves unidirectional flow and thermal stratification)

E_z = 1.0 (ceiling supply of cool air; ceiling supply of warm air with floor return; floor supply of warm air with floor return; ceiling supply of warm air less than 15°F (8°C) above room air temperature with ceiling return provided the diffuser jet velocity of 150 fpm (0.75 m/s) reaches the breathing zone; or floor supply of cool air with ceiling return provided a jet velocity of 150 fpm (0.75 m/s) reaches the breathing zone)

E_z = 0.8 (ceiling supply of warm air 15°F (8°C) greater than room air temperature and ceiling return or makeup supply air drawn in on the opposite side of the room from the exhaust and/or return)

E_z = 0.7 (floor supply of warm air and ceiling return)
E_z = 0.5 (makeup supply air drawn in near the exhaust and/or return location)

For single-zone systems where one air handler supplies outdoor and recirculated air to only one zone, the outdoor air intake flow (V_{ot}) is equal to the zone outdoor airflow (V_{oz}).

$$V_{ot} = V_{oz} \tag{2.14}$$

When one air handler supplies only outdoor air to one or more zones, the outdoor air intake flow (V_{ot}) is equal to the sum of the zone outdoor airflows. This type of system is referred to as a *100% outdoor air system* or, in the case where the ventilation air system is decoupled from the primary air system, a *dedicated outdoor air system* (DOAS).

$$V_{ot} = \Sigma V_{oz} \tag{2.15}$$

A DOAS separates the ventilation air system from the primary HVAC system. Typically, the ventilation air is conditioned (cooled, dehumidified, heated, humidified) to near indoor conditions and is delivered to the space in a separate distribution system or partially integrated into the primary system. This permits simple control (Coad 1996).

Multiple-zone systems that deliver a mixture of outdoor air and recirculated air to several zones can be corrected for the occupant diversity (D) in the zones and system ventilation efficiency (E_v). Occupants may move from normally occupied zones to normally unoccupied zones (e.g., meeting rooms). The occupant diversity is used to compute the uncorrected outdoor intake (V_{ou}).

$$V_{ou} = D\Sigma_{all\ zones}R_pP_z + \Sigma_{all\ zones}R_aA_z \tag{2.16}$$

where

D = $P_s/\Sigma_{all\ zones}P_z$
P_s = total population in the area served by the multizone system

Multiple-zone systems can provide only a single outdoor air to supply air fraction. However, each zone has an individual requirement for this fraction. Thus, zones with higher outdoor air requirements may not receive adequate ventilation air because they are receiving the average fraction. The zone primary outdoor air fraction (Z_{pz}) is computed for every zone from the ratio of the zone ventilation airflow rate (V_{oz}) to the primary airflow rate (V_{pz}).

$$Z_{pz} = V_{oz}/V_{pz} \tag{2.17}$$

The maximum value of Z_{pz} is found and used to determine the system ventilation efficiency (E_v).

$$E_v = 0.9 \text{ if Max } (Z_{pz}) \leq 0.25$$

$$E_v = 0.8 \text{ if Max } (Z_{pz}) \leq 0.35$$

$$E_v = 0.7 \text{ if Max } (Z_{pz}) \leq 0.45$$

$$E_v = 0.6 \text{ if Max } (Z_{pz}) \leq 0.55$$

Use Appendix F of ASHRAE Standard 62.1 if Max (Z_{pz}) > 0.55

The corrected value of outdoor air intake flow (V_{ot}) for a multiple-zone system is

$$V_{ot} = V_{ou}/E_v \tag{2.18}$$

Table 2.10 presents the computation of V_{ot} for a 10-zone office for both a DOAS and a multizone system. Two separate values are provided with the assumption that the air supply and return are in the ceiling for both cases. The greater of the two values represent requirements in heating ($E_z = 0.8$) and cooling ($E_z = 1.0$).

What is not apparent in the results of Table 2.10 is the level of control complexity that is required for multizone VAV systems. In these types of systems, V_{bz} remains constant for a constant occupancy while the V_p is reduced according to load. This increases the value of Z_{pz} in every zone, which in turn increases Max (Z_{pz}), lower ventilation efficiency, and increase required part-load ventilation airflow. With a DOAS the ventilation efficiency remains constant at 100%.

A great many efforts have focused on lowering the cost of GSHPs and almost all of them have concentrated on reducing the cost (and size) of the exterior loop (ground, groundwater, surface water). Few of these efforts have been successful. However, Figure 2.22 shows a most effective device for reducing GSHP loop size when properly installed *and maintained*. With the improvement of building envelopes, the ventilation air load has become a primary, and in some cases, the largest component of the heating and cooling loads. Thus, reducing this load significantly will effectively reduce the size of the ground heat exchanger and the energy consumed of the heat pumps.

Chapter 4 contains an example 10,000 ft^2 (930 m^2) office building in St. Louis, Missouri, with a calculated cooling load of 266 kBtu/h (78 kW) and a heat loss of 191 kBtu/h (56 kW). The addition of a ventilation air ERU reduces the cooling load by 15% to 227 kBtu/h (67 kW) and the heat loss by 37% to 121 kBtu/h (36 kW). In this case the outdoor

Table 2.10 Outdoor Indoor-Air Intake Flow Rates for 10-Zone Office—DOAS and Multizone

			Single Zone and DOAS							
	R_p	**People**	R_a	**Area**	E_z					
	5	17	0.06	2800	0.80	V_o =	**316**	**cfm**	(243 cfm for E_z = 1.0)	
				Multizone Systems						
Zone Type	**Zone #**	R_p	**# People**	R_p**·people**	R_a	A (ft^2)	R_a**·Area**	V_{bz}	V_p	Z_{pz}
Office	1	5	1	5	0.06	300	18	23	300	0.08
Office	2	5	1	5	0.06	200	12	17	200	0.09
Office	3	5	1	5	0.06	200	12	17	200	0.09
Office	4	5	2	10	0.06	200	12	22	200	0.11
Office	5	5	2	10	0.06	200	12	22	200	0.11
Office	6	5	2	10	0.06	200	12	22	200	0.11
Office	7	5	4	20	0.06	400	24	44	400	0.11
Office	8	5	4	20	0.06	400	24	44	400	0.11
Reception	9	5	2	10	0.06	300	18	28	300	0.09
Conference	10	5	18	90	0.06	400	24	114	400	0.29
		Totals	17	95		2800	144	353	Max (Z_{pz}) =	0.29
Max occupants	17	Diversity	1.00	E_z =	0.80					
V_{ou} =	299	cfm	E_v =	0.8		V_o =	**373**	**cfm**	(299 cfm for E_z = 1.0)	

Note: yellow indicates inputs, blue indicates outputs.

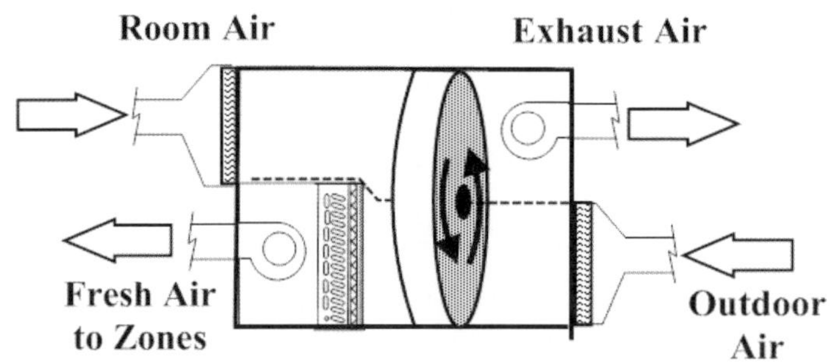

Figure 2.22 Energy Recovery Unit: An Effective GSHP Loop Reduction Device

ventilation airflow was 15% of the primary airflow. In applications with higher outdoor air fractions (e.g., schools), the percent load reduction would be even more dramatic.

It is important to understand the impact ERUs have on GSHP systems to properly apply them. In most applications, ERUs are more effective in reducing the heat loss and heating energy use. Note in the example St. Louis office building that the peak reductions were 15% in cooling and 37% in heating. The reasons for this are as follows:

- The temperature and humidity differences between the entering outdoor air and the exhausted outdoor air are typically larger in the heating season.
- The fan heat in the cooling mode will reduce the capacity of the ERU while it will be useful in the heating mode.
- In well-insulated and sealed buildings the ventilation air will often be the largest load.

The downside of ERUs being more effective in heating is that the hours the heat pumps spend in heating will be reduced relative to the hours spent in cooling. Therefore, the annual ground-loop heat balance will be further shifted toward cooling mode heat rejection. This may result in the percent reduction in GSHP loop size due to the ERU being less than the percent reduction in the cooling load in this case. For example, the ground loop size in the previous example may only be 10% to 12% for the 15% cooling load reduction. The issue of annual heat storage effects is discussed in greater detail in Chapters 3 and 4.

It is also important to be able to deactivate the ERU when outdoor conditions (temperature and/or humidity) are such that free cooling is possible. An active ERU would be counterproductive when the outdoor air temperature is lower than the indoor air and cooling is required. The temperature (and possibly the humidity) of the ventilation air would be increased, thereby adding to the cooling load rather than reducing it. The ERU in Figure 2.22 has a rotating wheel that can be stopped to disable operation when free cooling is possible. Other passive ERU designs may need bypass ductwork and dampers to enable this mode of operation.

Figure 2.23 presents another detail for improving system performance and efficiency regarding ventilation air delivery. When the ventilation air is introduced into the return air plenum, the heat pump fan must operate continuously even when the heat pump is not operating. Minimum speed with a standard permanent split capacitor fan motor will likely result in only a small reduction in energy use. Even with variable-speed motors at 30% speed, airflow will be nearly 50% of full-load flow. Furthermore, moisture that remains on the heat pump coil and in the drain pan will evaporate when the heat pump is off. Figure 2.23 also demonstrates the option of introducing the ventilation air directly into the space in a location not directed upon occupants but opposite from the return grille so that

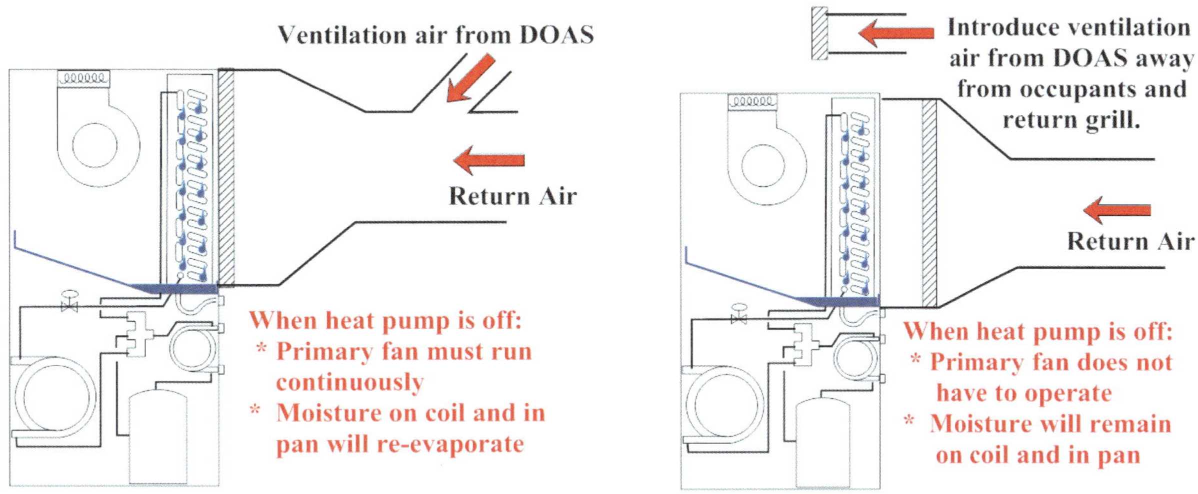

Figure 2.23 Zone Ventilation Air Delivery Options and Issues with Unitary Heat Pumps

the fresh air will reach the breathing zone. The issues resulting from introducing the ventilation air into the heat pump return air duct are avoided with the direct delivery of ventilation air as shown in the right portion of Figure 2.23.

2.7 REFERENCES

AAF. 2012. Perfect Pleat Extended Surface, Pleated Filter, MERV 7. Louisville, KY: American Air Filter International.

ASHRAE. 2010. ANSI/ASHRAE/IES Standard 90.1, *Energy Standard for Buildings Except Low-Rise Residential Buildings*. Atlanta: ASHRAE.

ASHRAE. 2012a. ANSI/AHRI/ASHRAE ISO Standard 13256-1: 1998 (RA 2012), *Water-Source Heat Pumps-Testing and Rating for Performance—Part 1: Water-to-Air and Brine-to-Air Heat Pumps*. Atlanta: ASHRAE.

AHRI. 2012b. ANSI/AHRI/ASHRAE ISO Standard 13256-2: 1998 (RA 2012), *Water-Source Heat Pumps Testing and Rating for Performance—Part 2: Water-to-Water and Brine-to-Water Heat Pumps*. Atlanta: ASHRAE.

ASHRAE. 2013. ANSI/ASHRAE Standard 62.1-2013, *Ventilation for Acceptable Indoor Air Quality*. Atlanta: ASHRAE.

Bolt, J. 2012. How 90.1-2010 Will Affect Health Care Facilities. *ASHRAE Journal* 54(8).

Coad, W.J. 1996. Indoor air quality: A design parameter. *ASHRAE Journal* 38(6).

Kavanaugh, S.P. 2006. *HVAC Simplified*. Atlanta: ASHRAE.

Kavanaugh, S.P. 2008. A 12-step method for closed-loop ground-source heat pump design. *ASHRAE Transactions* 114(2).

Parker, D.S., and J. Proctor. 2001. Hidden power drains: Trends in residential heating and cooling fan watt power demand. FSEC-PF361-01. Cocoa, FL: Florida Solar Energy Center.

3 Fundamentals of Vertical Ground Heat Exchanger Design

3.1 OVERVIEW

The design of vertical ground heat exchangers is complicated by the variety of geological formations and properties that affect thermal performance. Proper identification of materials, moisture content, and water movement is an involved process and cannot be economically justified for every project. Therefore, the necessary information for complex computation and analysis is often unavailable. A more prudent design approach is to apply empirical data to an analytic solution of heated and cooled pipes placed in the ground. Thermal property tests of the ground provide improved accuracy over standard geological surveys and are highly recommended when designing GSHPs for commercial and institutional buildings. For residential and very small commercial projects, in which the cost of a thermal property test is difficult to justify, conservative values can be estimated by using values for soils in a particular group and moisture content when this information is available from state geological surveys (see Chapter 7). Initial designs are usually conservative and can be amended based on the performance results of early installation if they are monitored.

For commercial and institutional buildings, the cost of thermal property tests (see Section 3.6) are justifiable given the sizeable added cost of ground heat exchangers that are designed using conservative thermal property values. Additionally, the drilling and installation of a thermal property test bore provides a wealth of information to drilling contractors who will submit bids for the project. Bid prices will typically be more competitive if the installation contractors have good characterization of the formation. Thus, thermal property tests provide valuable information to both the design engineer and the ground-loop contractor.

Another factor affecting the uncertainty of ground-loop performance is the possibility of some permanent change in the local ground temperature for systems with large annual differences between the amount of heat extracted (heating mode) and the amount rejected (cooling mode). This effect is compounded in larger systems because earth heat exchangers are more likely to be installed in close proximity because of more limited ground area availability. A residence may be located on a 1/2 acre (2000 m^2) lot and have a balanced 3 ton (10.6 kW) heating/cooling load, whereas an office tower may be located on a 1 acre (4000 m^2) plot and require 200 tons (700 kW) of cooling and only 50 tons (175 kW) of heating.

Moisture change and movement produce a cooling effect that can result in a significant mitigation of long-term ground temperature rise. The thermal energy required to lower soil moisture content 1% is equivalent to a 30°F (17°C) rise in ground temperature (EIS 2009). Groundwater movement can provide sufficient moisture recharge during heating mode operation and idle periods to return the ground to its natural moisture content. In very cold climates where the ground loops may operate below 32°F (0°C), the latent heat of the freeze-thaw process of the moisture near the ground heat exchanger provides a thermal capacity that also mitigates undesirable ground temperature change.

Not only do the phenomena of moisture change and freezing complicate the prediction of long-term thermal performance, but the information required to formulate these predictions resides in a complex mixture of sands, clays, rocks, moisture, and other unknowns well below the surface of the earth. Methods to gather sufficient details of this data are not currently available and will likely never be developed at reasonable costs. However, it is possible to better define the range of uncertainty using a combination of well logs (see Chapter 7), thermal property tests, and measured field data of ground-loop temperatures in actual buildings.

There is a limited amount of ground-loop performance data that suggests long-term temperature change is not the dominant reason for hot loops when cooling is the dominant operating mode (Kavanaugh and Kavanaugh 2012). It is essential to trend ground heat exchanger entering liquid temperatures (ELTs) and leaving liquid temperatures (LLTs) in combination with building cooling and heating loads for the first 5 to 10 years of system operation. This helps determine if higher (or lower) than expected temperatures are the result of poor design and installation or long-term change. This additional information is very much needed to improve the accuracy of GSHP design tools and ground-loop models, including those presented in this book.

The method described Section 3.2 is based on the solution of the equation for heat transfer from a cylinder buried in the earth developed and evaluated by Carslaw and Jaeger (1947). The equation and solution were suggested by Ingersoll et al. (1954) as an appropriate method of sizing ground heat exchangers in cases where the line source equation may result in error. The simpler line source equation yields good results for daily average loop temperatures, but errors will result when time periods are less than six hours. Therefore, accurate predictions of hourly loop temperature variations require the cylindrical heat source equation. A procedure to apply the methods of Ingersoll et al. to account for the U-tube arrangement and hourly heat rate variations has been developed (Kavanaugh 1992).

It has been demonstrated that the thermal performance of a ground heat exchanger is a strong function of the amount of heat that has been extracted from or rejected to the ground (Claesson and Eskilson 1987). Minimum and maximum temperatures may take several years to occur. This is especially true if multiple vertical bores are located in close proximity. The worst-case design condition might occur several years after installation. Therefore, the design of the ground loop should consider system performance for an extended period. However, it is suggested that complex and detailed simulation for a great many years (10+) is unnecessary since the data to drive the simulation is not available. Therefore, simple heat transfer models that use empirical data will result in design tools that are likely to be more accurate than sophisticated models that do not consider field-measured ground-loop performance.

3.2 EQUATIONS FOR REQUIRED GROUND HEAT EXCHANGER LENGTH

The method of Ingersoll et al. (1954) can be applied to closed-loop ground heat exchanger design. This method is a modification of a simple steady-state heat transfer equation per unit length:

$$q = \frac{L_{bore}(t_g - t_w)}{R_{ov}} \quad (3.1)$$

where

q = heat transfer rate to/from ground
L_{bore} = ground heat exchanger bore length
t_g = ground temperature
t_w = average water-loop temperature
R_{ov} = overall resistance of ground and bore, ft·h·°F/Btu (m·K/W)

It is critical to understand that the rate of heat rejected to the ground in cooling per ton (kW) of capacity is 60% to 70% greater than the rate of heat absorbed from the ground in heating per ton (kW) of capacity. The electrical energy required to drive the compressor, fans, and pumps is converted to heat that must be rejected into the ground. For a cooling EER of 13.6 Btu/Wh (COP of 4.0), four units of heat are removed from the building, one unit of input power is converted to heat, and these combined five units of heat are transferred to the ground. Thus, the rate of heat delivered to the ground is 125% of the cooling capacity of the heat pump. In heating the compressor, fan and pump power is converted to useful heat, which is delivered to the building. For a heating COP of 4.0, four units of heat are delivered to the building, one unit of input power is converted to heat, and therefore only three units need to be removed from the ground. Thus, the heat taken from the ground is only 75% of the heating capacity of the heat pump.

The adjustment from the heat rate removed from the building to the ground heat is a function of the heat pump system cooling efficiency (EER or COP_c):

$$\frac{q_{cond}}{q_{lc}} = \frac{\text{EER} + 3.412}{\text{EER}} \quad \text{(I-P)} \quad (3.2a)$$

$$\frac{q_{cond}}{q_{lc}} = \frac{\text{COP}_c + 1.0}{\text{COP}_c} \quad \text{(SI)} \quad (3.2b)$$

where

q_{cond} = heat pump condenser heat rate to ground, Btu/h (W)
q_{lc} = building design cooling block load, Btu/h (W)
EER = energy efficiency ratio, Btu/Wh
COP_c = cooling mode coefficient of performance, $W_{cooling}/W_{electrical}$

However, the input heat (electrical) into the heat pump and auxiliary equipment in the heating mode is delivered to the building. Thus, the heat removed from the ground by the evaporator is

$$\frac{q_{evap}}{q_{lh}} = \frac{\text{COP}_h - 1}{\text{COP}_h} \quad (3.3)$$

where

q_{evap} = heat pump evaporator heat rate from ground, Btu/h (W)
q_{lh} = building design heating block load, Btu/h (W)
COP_h = heating mode coefficient of performance, $W_{heating}/W_{electrical}$

The net annual heat transfer rate (q_a) is computed using the equivalent full-load hours in cooling (EFLH$_c$) and heating (EFLH$_h$). Values for EFLH for a variety of locations and building types can be found in Table 4.5 (Carlson 2001).

$$q_a = \frac{q_{cond} \times \text{EFLH}_c + q_{evap} \times \text{EFLH}_h}{8760} \tag{3.4}$$

When the simple Equation 3.1 is rearranged to solve for the vertical heat exchanger bore length, the basis for design optimization is noted:

$$L_{bore} = \frac{q \times R_{ov}}{(t_g - t_w)} \tag{3.1a}$$

The heat rate (q) is fixed by the building heating and cooling requirements and the ground temperature (t_g) is fixed by the earth. The overall resistance (R_{ov}) is constrained by the thermal properties of the ground, the design of the heat exchanger, and the heat rate to and from the ground. The design optimization is between the average water-loop temperature (t_w) and the heat exchanger length (and cost). In cooling mode, a lower value for t_w results in more efficient heat pump operation but a longer and more expensive ground loop. In heating mode, a higher value for t_w results in improved heat pump operation but a longer and more expensive ground loop.

Equation 3.1 is a steady-state equation and can be transformed to represent the variable heat rate of a ground heat exchanger by using a series of constant heat rate pulses as suggested by Ingersoll et al. (1954). The thermal resistance (R) of the ground per unit length is calculated as a function of time, which corresponds to the time over which a particular heat pulse occurs. Equations 3.5 and 3.6 include a minimum of three heat pulses: an average annual pulse, an average monthly pulse preceding the design day, and a short-term pulse that is typically the maximum pulse during the design day of one to six hours in length. A term is also included for the bore resistance (R_b) that accounts for the thermal resistance of the tube wall (R_t), the film resistance between the fluid and tube (R_{film}), and the resistance of the fill or grout material ($R_{annulus}$) in the annual region between the tube(s) and the bore wall, illustrated in Figure 3.1.

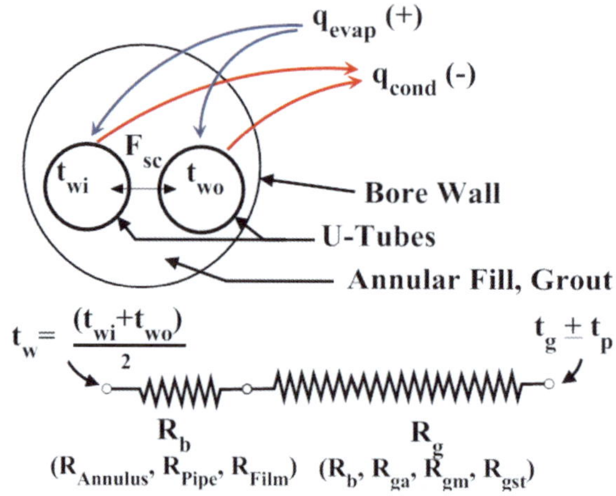

Figure 3.1 Schematic and Thermal Network for U-Tube Ground Heat Exchanger

The resulting equation for ground heat exchanger bore length for cooling takes the form

$$L_c = \frac{q_a R_{ga} + q_{cond}(R_b + \text{PLF}_m R_{gm} + F_{sc} R_{gst})}{t_g - \dfrac{\text{ELT} + \text{LLT}}{2} + t_p} \tag{3.5}$$

The required length for heating is

$$L_h = \frac{q_a R_{ga} + q_{evap}(R_b + \text{PLF}_m R_{gm} + F_{sc} R_{gst})}{t_g - \dfrac{\text{ELT} + \text{LLT}}{2} + t_p} \tag{3.6}$$

where

F_{sc} = short-circuit heat loss factor between supply and return tubes in bore (see Figure 3.7)
L_c = required bore length for cooling, ft (m)
L_h = required bore length for heating, ft (m)
PLF_m = part-load factor during design month
q_a = net annual average heat transfer to the ground, Btu/h (W)
R_{ga} = effective thermal resistance of the ground—annual pulse, h·ft·°F/Btu (m·K/W)
R_{gst} = effective thermal resistance of the ground—short-term pulse, h·ft·°F/Btu (m·K/W)
R_{gm} = effective thermal resistance of the ground—monthly pulse, h·ft·°F/Btu (m·K/W)
R_b = thermal resistance of bore, h·ft·°F/Btu (m·K/W)
t_g = undisturbed ground temperature, °F (°C)
t_p = long-term ground temperature penalty caused by ground heat transfer imbalances, °F (°C)
ELT = heat pump entering liquid temperature, °F (°C)
LLT = heat pump leaving liquid temperature, °F (°C)

The sign convention for Equations 3.5 and 3.6 assumes the energy balance is done on the heat pumps; therefore, q_{evap} is positive, q_{cond} is negative, q_a is positive if the annual amount of heat removed from the ground in heating (q_{evap} × operating time) is greater that the heat added to the ground in cooling (q_{cond} × operating time), and t_p is positive for a long-term rise in ground temperature.

The optimal trade-off between system efficiency and ground-loop length typically occurs when the maximum value for the heat pump ELT in the cooling mode is 20°F to 30°F (11°C to 17°C) greater than the undisturbed ground temperature (t_g). The optimum tends to be on the lower end of this range for warmer climates (t_g > 60°F [15°C]) and toward the upper end of the range for cooler climates. For heating, the optimum value for the ELT is typically 8°F to 15°F (5°C to 8°C) less than the undisturbed ground temperature (t_g). Buildings in warmer climates or those with high internal cooling loads tend to have optimal values on the lower end of this range while buildings in cold climates with high heat losses tend to have optimum values on the higher end of this range.

Optimum liquid flow rates for closed-loop systems are typically in the 2.5 to 3.0 gpm/ton (2.7 to 3.2 L/min·kW) range. The following estimates can be used with good accuracy for the heat pump LLT. These values assume water is the fluid; values will be 3% to 5% higher for typical antifreeze solutions used with GSHPs (see Appendix F for properties of antifreeze solutions).

- For a flow rate of 3.0 gpm/ton (3.2 L/min·kW) the LLT will be approximately 10°F (5.6°C) higher than the ELT in cooling and 6°F (3.3°C) lower than the ELT in heating.
- For a flow rate of 2.5 gpm/ton (2.7 L/min·kW), the LLT will be approximately 12°F (6.7°C) higher than the ELT in cooling and 7.2°F (4°C) lower than the ELT in heating.
- For a flow rate of 2.0 gpm/ton (2.15 L/min·kW), the LLT will be approximately 15°F (6.7°C) higher than the ELT in cooling and 9°F (5°C) lower than the ELT in heating.

The required bore length (L_{bore}) is the larger of the two lengths resulting from Equations 3.5 and 3.6. If the length required for cooling is larger than that for heating, the heating mode t_{wi} can be increased until the resulting value of L_h is similar to that of L_c. This will result in a higher value for system COP_h because the liquid entering the heat pump is higher than the value assumed for the initial heating mode calculation. The inverse is true if the initially calculated heating mode length is greater than the cooling mode length.

In applications where the cooling length (L_c) is much greater than the heating length (L_h), one option is to install the smaller heating length and a fluid cooler or a cooling tower with an isolation heat exchanger typically placed in parallel with the ground loop to compensate for the smaller ground loop. This is referred to as a *hybrid ground-coupled heat pump (GCHP)*. Until recently these systems were primarily used to remedy poorly designed or installed GCHPs that were experiencing high ground heat exchanger temperatures. More frequently now they are being used as either a first or the primary alternative option when the building loads for cooling are greater than those for heating. In some cases the hybrid GCHP option is chosen because of an installation cost advantage, while in some applications the land area for a ground heat exchanger sized for cooling is not available.

While hybrid systems can reduce installation cost, they also lose the primary low-maintenance advantage of GCHPs in terms of both the absence of aboveground outdoor equipment and simplicity of controls. The added auxiliary equipment will also lower system efficiency unless the coolers are sized to provide substantially lower ELTs than those possible with a ground heat exchanger alone. These types of systems should be used with caution in buildings such as schools that have minimal maintenance staffs and occupants susceptible to potential health risks from poorly maintained or located evaporative cooling equipment.

In applications where the heating length (L_h) is much greater than the cooling length (L_c), the option to add supplemental heating capacity in parallel or series with the ground loop is highly problematic. If a boiler is connected to the ground loop, the possibility of high-temperature water entering the ground heat exchanger could result in failure of the high-density polyethylene (HDPE) tubing. This is especially true in installations where internal tube static pressures are high (tall buildings and/or deep bores in formations with low groundwater tables). It is suggested that the heating loads be carefully reviewed so that credit for energy recovery units (ERUs) is considered in reducing heating requirement and therefore design heating length (L_h). It is also recommended that conventional air-side heat pump auxiliary heat be considered, such as electrical resistance in the heat pump or hot-water distribution system. In commercial buildings this added requirement is typically much lower than it is for residential applications. If the supplemental need is modest, the added cost of the equipment and electrical distribution system is likely to be much lower than the added cost of a boiler and piping distribution network.

There is some benefit to heat transfer to and from the horizontal header network connecting the vertical heat exchangers. This heat transfer is not typically considered because the effects are limited. However, if ground loop headers are buried at shallow depths in small systems that may sit idle during winter set back, a brief period of low-temperature fluid entering some heat pumps could cause them to shut down. Details of shallow-earth ground temperatures and heat transfer in horizontal pipes are addressed in Chapter 5. This includes headers for both ground-coupled and surface-water heat pumps.

A difficult but important item to address is the long-term temperature change (t_p) that can occur when the amount of heat rejected annually to the ground in cooling is much different than the amount of heat removed from the ground in heating. Conduction heat transfer equations, such as Equations 3.5 and 3.6, apply to a line or cylinder heat source in a semi-infinite medium with no interference from adjacent heat sources. Modifications are necessary to prevent excessive long-term variations when these heat exchangers are placed in rows or grids. The issue manifests itself more frequently in the cooling mode in the form of ground temperature increase since both the building load and the heat pump system power must be rejected to the ground. In heating mode, the heat pump input power is converted to beneficial building heat, which proportionally reduces the amount of heat required to be extracted from the ground loop. Thus, an imbalance will occur toward heat rejection even if the cooling and heating annual loads are identical. Excessive temperature decline is also possible in colder climates and/or in buildings with modest internal heat gains. The most obvious methods of reducing the negative effects are longer bore lengths, greater separation distances from adjacent bores (S_{bore}), and bore field arrangements that have fewer bores that are surrounded by four other bores (e.g., a 2 × 18 grid rather than a 6 × 6 grid). This could of course result in ground loops that are economically nonviable because of the length and land area required.

However, field measurements from installations that have been in operation for several years indicate the increase in long-term temperature is mitigated by the fact that the ground is not a simple solid whose thermal behavior can be predicted by conduction heat transfer models alone. Phase change (evaporation-condensation and freeze-thaw) and convection heat transfer effect must be included. Figure 3.2 compares the maximum average ground-loop temperature rise above the local ground temperature at twenty cooling-dominant GSHP installations (Kavanaugh and Kavanaugh 2012). These results do not indicate a consistent rise in temperature for systems that have been operating for several years. The warmest loops are those that are relatively short or have bores installed close together and have grouts with poor thermal properties.

Older GSHP systems appear to actually have lower approach temperatures. Results are not adjusted for many important factors such as vertical bore length, ground thermal properties, and vertical bore separation distance. The newer systems tend to have slightly shorter ground loops, but this is offset somewhat because older systems tend to have smaller vertical bore separation distances and lower-conductivity grout and fill. Figure 3.2 does provide some factors that likely influence the loops with the largest approach factors. Three of the newer systems with high approach temperatures have vertical bore lengths less than 120 ft/ton (10.4 m/kW). Two systems with long loops but large approach temperatures have low-thermal-conductivity grout (0.38 Btu/h·ft·°F [0.66 W/m·K]), 15 ft (4.6 m) bore separation, and cooling mode indoor air temperatures below 70°F (21°C).

It is recognized that this data set is small and that the presence of significant long-term temperature change cannot be excluded at this time. Although much more field data is highly desirable, the absence of any significant trend of increased ground temperature (noted by elevated maximum approach temperature) with increased years of GSHP oper-

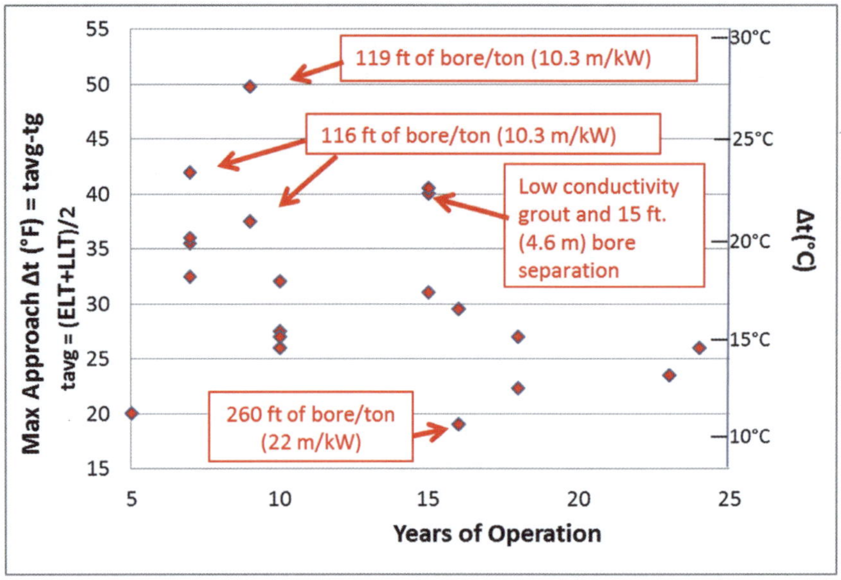

Figure 3.2 Measured Increase in Average Loop Temperature Above Initial Ground Temperature

ation would indicate that long-term ground temperature change is not dramatic. The position that even well-designed and installed ground heat exchangers with imbalanced loads will have to eventually be abandoned does not appear to be true. Elevated temperatures in vertical ground loops are also a result of inadequate heat exchanger length, inadequate bore separation distance, and low-conductivity grout. Improper completion methods and insufficient air purging may also contribute to very warm or cold loops.

In cooling mode, the long-term temperature rise is mitigated by the cooling effect from reductions in moisture content (evaporation), as shown in Figure 3.3. The amount of heat required to reduce the moisture content by 1% in a typical formation is equivalent to the amount of heat necessary to raise the ground temperature by 30°F (17°C) (EIS 2009). When ground temperature increases within the loop field, the saturation pressure of water vapor increases, which also increases the evaporation rate. This drying effect can reduce formation thermal conductivity if the temperature increase is excessive during the cooling season. When heat exchanger lengths and bore separation distances fall within recommended values, moisture from natural groundwater movement and moisture migration toward the cooler pipe during the heating season will recharge the formation.

Results cannot be applied to long-term temperature decline in which the amount of heat removed from the ground in heating far exceeds the heat rejected in cooling. The transfer mechanisms are entirely different. In cold climates the latent heat capacity available at the freeze point of water is significant and mitigates loop temperature decline below the freeze point. Later in this chapter, long-term temperature change is discussed in more detail.

3.3 BOREHOLE THERMAL RESISTANCE

The design equations for ground heat exchanger sizing have four terms for thermal resistance *per unit length of bore* (not unit length of pipe). Three of these involve the resistance of the ground. They have the form of steady-state values but are actually derived from transient heat rates during the most critical periods of building cooling and

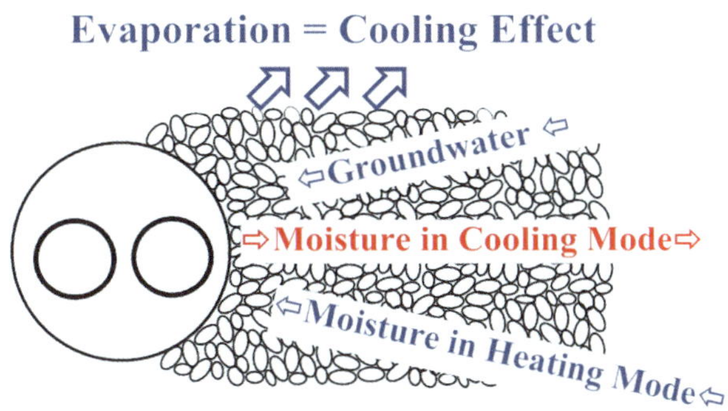

Figure 3.3 Ground Heat Exchanger Moisture Migration and Evaporative Cooling Mechanisms

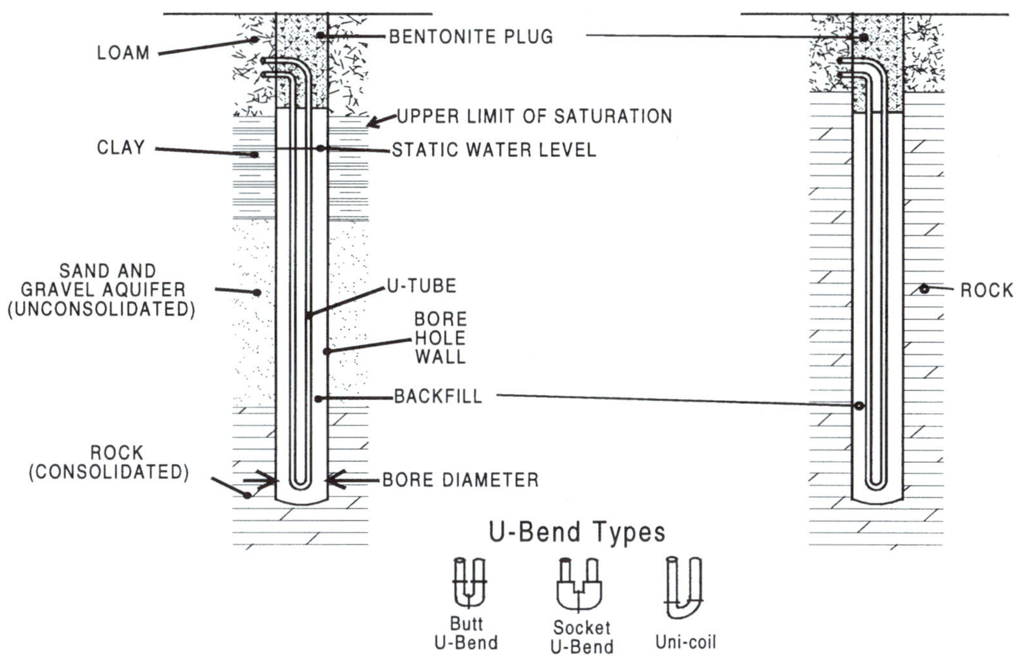

Figure 3.4 Typical U-tube Installations for Unconsolidated and Consolidated Formations

heating requirements. Examples of computation of these three values are presented in Section 3.4. The remaining term is the equivalent thermal resistance of the bore (R_b). Since the liquid inside the loop, the piping, and the backfill material has very little thermal mass compared to the surrounding ground, R_b can be treated as a constant (steady-state) value. Figure 3.1 represents a cross section of a typical bore with a U-tube heat exchanger. Figure 3.4 is a representation of vertical sections of two U-tube installations that supplement the discussion that follows.

The thermal resistance of the ground heat exchanger vertical bore considers the effects of the pipe resistance and bore annulus grout resistance.

$$R_b = R_p + R_{gt} \tag{3.7}$$

The pipe resistance includes the convective film resistance of the fluid and the conductive resistance of the pipe walls. Contact resistances between the pipe walls and fill material are negligible compared to the high resistance of plastic pipe walls and annular grouts. For a single U-tube (two tubes), the pipe resistance is

$$R_p = (R_{film} + R_{tube})/2 = [(1/(\pi d_i h_{conv}) + \ln(d_o/d_i)/2\pi k_p)]/2 \tag{3.8}$$

For two U-tubes (four tubes) in a bore, the pipe resistance is

$$R_p = (R_{film} + R_{tube})/4 = [(1/(\pi d_i h_{conv}) + \ln(d_o/d_i)/2\pi k_p)]/4 \tag{3.8a}$$

A correlation for the thermal resistance of the grout has been developed using shape factor correlations (Remund 1999):

$$R_{grt} = \left[\beta_0 \left(\frac{d_b}{d_o}\right)^{\beta_1} \times k_{grt}\right]^{-1} \tag{3.9}$$

Coefficients for Equation 3.9 (β_0, β_1) have been developed for three locations of the tubes, as shown in Figure 3.5. The positions are (A) centered in the bore and in contact with each other, (B) centered and spaced evenly in the bore, and (C) centered and in contact with the bore wall. However, the most likely location of the U-tubes is BC—but coefficients for this location are unavailable. A similar but slightly more detailed approach was developed by Hellström (1991) and applied to a design procedure (Philippe et al. 2010).

Because the actual installed locations of the U-tubes cannot be determined even when spacers are installed, exact computation of bore resistance values is somewhat uncertain. It is possible to apply the results from thermal property tests to calculate the bore resistance if the U-tube dimensions, grout conductivity, and borehole diameter are known (Kavanaugh 2010). Thermal property tests were conducted at 15 installations where these values were known and the bore resistance was calculated. The bore resistances calculated using thermal property test results best matched the values computed with Equations 3.7, 3.8, and 3.9 when the following U-tube locations were used:

- Location C at 4 (27%) of the sites
- An average of locations B and C at 5 (33%) of the sites
- Location B at 5 (33%) of the sites
- Location A at 1 (7%) of the sites

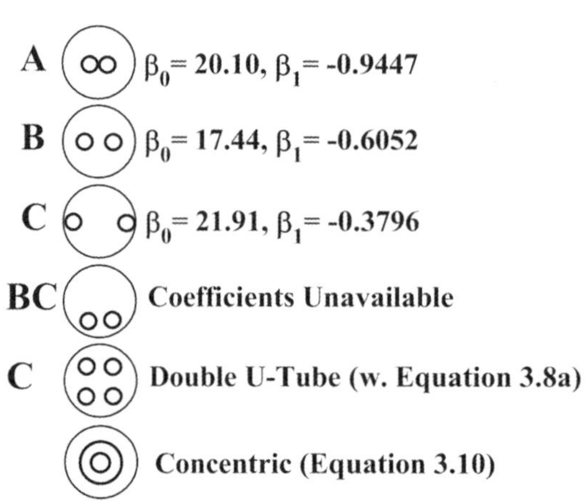

Figure 3.5 Bore Resistance Shape Factors for U-Tube Locations in Vertical Boreholes

Table 3.1 provides the bore resistances computed using Equations 3.7, 3.8, and 3.9 for three different grout conductivities, three different fluid flow regimes (laminar, transition, and fully turbulent), three different U-tubes sizes, and three different bore diameters for locations B and C. The resistance is also computed for a double U-tube in a bore. Designers can use the values of location B (conservative), BC (average), C (risky), or Double. The values in the tables provide only two digits of accuracy, which reflect the uncertainty of being able to determine the locations of the tubes in deep vertical bores. The spreadsheet tool *BoreResistance.xlsm*, available with this book at www.ashrae.org/GSHP, calculates resistances for traditional U-tube vertical heat exchangers for a broader variation of pipe materials, grout conductivities, flow rates, and fluid types.

The borehole thermal resistance for a concentric arrangement can also be calculated with Equation 3.7. The terms for pipe and grout resistance can be combined into a single equation:

$$R_b = R_{film} + R_{tube} + R_{grt} = 1/(\pi d_i h_{conv}) + \ln(d_o/d_i)/2\pi k_p + \ln(d_b/d_o)/2\pi k_{grt} \tag{3.10}$$

Note that the short-circuit heat loss factor (F_{sc}) in the overall heat exchange length could be much higher for concentric arrangements compared to U-tubes if the inner tube is not well insulated. It is highly recommended that novel heat exchanger designs be evaluated using the procedures discussed by Kavanaugh (2010).

Tables 3.2a and 3.2b are provided to assist in the determination of the thermal resistance of the grout (R_{grt}) or borehole fill in the annular region between the heat exchanger tubes and the borehole wall. Note that the term *thermal conductivity* (TC) is the same as k_{grt} in Equations 3.9 and 3.10. An important task for the material in the annulus is to prevent the flow of surface water (or undesirable groundwater) into the ground and groundwater aquifers. Surface-water and some groundwater aquifers may contain pollutants or minerals that could contaminate sensitive drinking or irrigation water sources. Many of the more effective grouts for sealing boreholes, such as high-solids sodium bentonite grout (>20% solids), are poor heat conductors. Conversely, some materials that have effective heat transfer properties are not suitable for preventing water migration in the boreholes. In some locations, regulations permit the use of these porous materials if the upper-section boreholes (typically 20 ft [6 m]) are sealed with a nonporous grout. Cement-based materials that traditionally have been used to seal water-well casings are typically not suitable for closed-loop heat pump boreholes. Unlike bentonite-based grouts, materials that set up solid will not be effective in sealing around HDPE pipe that shrinks with the lower temperatures experienced during heating mode operation. However, special additives can be added to cement-based grout with close tolerances, as listed in Tables 3.2a and 3.2b.

Bentonite-based grouts can be thermally enhanced with the addition of large volumes of silica sand or smaller volumes of sand in combination with graphite. These recipes retain the ability to provide an effective seal. The material handling costs of the sand-only enhancement increases the cost per unit length of the ground heat exchanger, but in most cases the reduction in required bore length offsets the added material cost. The introduction of graphite dramatically reduces the amount of material handled, but the cost of graphite itself is a factor to consider. In some cases contractors do not have pumping equipment that can handle the enhanced grouts with the abrasive sands. One option is to allow alternatives for contractors in a format such as the following:

- Install fifty (50) 1 in. nominal (32 mm) DR 11, HDPE U-tube ground heat exchangers in a 5 × 10 grid at 20 ft (6 m) separation with

Table 3.1 Thermal Resistances of Bores with U-Tubes for Various Conditions

Tube Diameter and Dimension	Tube Location	Bore Diameter, in.	Thermal Resistance of Bore, h·ft·°F/Btu								
			Fluid Reynolds No. = 2000 Grout Conductivity, Btu/h·ft·°F			Fluid Reynolds No. = 4000 Grout Conductivity, Btu/h·ft·°F			Fluid Reynolds No. = 10,000 Grout Conductivity, Btu/h·ft·°F		
			0.40	0.80	1.20	0.40	0.80	1.20	0.40	0.80	1.20
3/4 in. DR 11 HDPE U-Tube	B	4	0.47	0.30	0.25	0.40	0.24	0.19	0.39	0.23	0.18
		5	0.51	0.33	0.27	0.45	0.26	0.20	0.44	0.26	0.20
	C	4	0.33	0.24	0.20	0.27	0.17	0.14	0.26	0.17	0.14
		5	0.35	0.27	0.21	0.29	0.18	0.15	0.28	0.18	0.14
	Double	5	0.28	0.17	0.14	0.25	0.14	0.11	0.24	0.14	0.11
1 in. DR 11 HDPE U-Tube	B	4	0.42	0.28	0.24	0.36	0.22	0.17	0.35	0.21	0.17
		5	0.46	0.30	0.25	0.40	0.24	0.19	0.39	0.23	0.18
		6	0.50	0.32	0.26	0.44	0.26	0.20	0.43	0.25	0.19
	C	4	0.32	0.23	0.20	0.25	0.17	0.14	0.25	0.18	0.13
		5	0.33	0.24	0.21	0.27	0.17	0.14	0.26	0.17	0.14
		6	0.35	0.24	0.21	0.28	0.18	0.15	0.28	0.17	0.14
	Double	5	0.26	0.17	0.13	0.23	0.13	0.10	0.23	0.13	0.10
		6	0.27	0.17	0.14	0.24	0.14	0.11	0.24	0.14	0.10
1 1/4 in. DR 11 HDPE U-Tube	B	5	0.42	0.28	0.23	0.36	0.22	0.18	0.35	0.21	0.17
		6	0.45	0.29	0.24	0.39	0.23	0.18	0.38	0.23	0.18
	C	5	0.31	0.22	0.20	0.26	0.17	0.14	0.25	0.16	0.13
		6	0.32	0.23	0.20	0.26	0.17	0.14	0.26	0.16	0.13
	Double	6	0.25	0.16	0.13	0.23	0.13	0.10	0.22	0.13	0.10

Tube Diameter and Dimension	Tube Location	Bore Diameter, mm	Thermal Resistance of Bore, m·°C/W								
			Fluid Reynolds No. = 2000 Grout Conductivity, W/m·°C			Fluid Reynolds No. = 4000 Grout Conductivity, W/m·°C			Fluid Reynolds No. = 10,000 Grout Conductivity, W/m·°C		
			0.70	1.40	2.10	0.70	1.40	2.10	0.70	1.40	2.10
25 mm DR 11 HDPE U-Tube	B	100	0.26	0.17	0.14	0.24	0.14	0.11	0.23	0.14	0.11
		125	0.29	0.18	0.15	0.26	0.16	0.12	0.26	0.11	0.12
	C	100	0.18	0.13	0.11	0.16	0.10	0.09	0.15	0.10	0.08
		125	0.19	0.13	0.11	0.17	0.11	0.09	0.16	0.10	0.08
	Double	125	0.16	0.10	0.08	0.14	0.08	0.06	0.14	0.08	0.06
32 mm DR 11 HDPE U-Tube	B	100	0.24	0.16	0.13	0.21	0.13	0.10	0.21	0.13	0.10
		125	0.26	0.17	0.14	0.23	0.14	0.11	0.23	0.14	0.11
		150	0.28	0.18	0.14	0.26	0.15	0.12	0.25	0.15	0.11
	C	100	0.17	0.12	0.11	0.15	0.10	0.08	0.14	0.09	0.08
		125	0.18	0.13	0.11	0.16	0.10	0.08	0.15	0.10	0.08
		150	0.19	0.13	0.11	0.17	0.11	0.09	0.16	0.10	0.08
	Double	125	0.15	0.09	0.07	0.13	0.08	0.06	0.13	0.08	0.06
		150	0.15	0.10	0.08	0.14	0.08	0.06	0.14	0.08	0.06
40 mm DR 11 HDPE U-Tube	B	125	0.24	0.16	0.13	0.22	0.13	0.11	0.21	0.13	0.10
		150	0.26	0.17	0.14	0.23	0.14	0.11	0.23	0.14	0.11
	C	125	0.17	0.12	0.11	0.15	0.10	0.09	0.14	0.09	0.08
		150	0.18	0.13	0.11	0.16	0.11	0.09	0.15	0.10	0.08
	Double	150	0.14	0.09	0.07	0.13	0.08	0.06	0.13	0.08	0.06

Table 3.2a Properties of Grouts, Fills, and Pipe Materials (Allan 1996; GPI 2014)—I-P

Sodium Bentonite Recipes

Bentonite, lb	Silica Sand, lb	Graphite, lb	Water, gal	Note	Yield, gal	TC (k_{grt}), Btu/h·ft·°F	Density, lb/gal
50	0	0	33		36	0.38-0.40	9.0
50	0	0	24		27	0.41-0.43	9.3
50	0	0	14		17	0.43-0.45	9.8
50	100	0	15		23	0.65-0.75	12.0
50	200	0	18		32	0.85-0.95	12.5
50	400	0	22		42	1.2-1.3	15.1
50	0	8	16	HPG*	18	0.85-0.95	10.6
50	50	8	18	HPG*	23	0.85-0.95	11.2
50	0	15	16	SPG*	19	0.85-0.95	10.4
50	50	10	24	SPG*	31	0.85-0.95	10.0
50	0	15	18	HPG*	21	1.2-1.3	10.2
50	50	15	20	HPG*	25	1.2-1.3	11.3
50	0	20	15	SPG*	18	1.2-1.3	10.8
50	100	15	16	SPG*	23	1.2-1.3	13.0

Cement Recipes

Cement, lb	Silica Sand, lb	Other, lb	Water, gal	S. Plasticisizer, oz	Yield, gal	TC (k_{grt}), Btu/h·ft·°F	Density, lb/gal
94	200	0	Neat Cement—Not Recommended				
94	200	300-400	Concrete—Not Recommended				
94	200	1	6	21	19	1.2-1.4	18.2

Engineered, High-Yield Cement for GSHP Applications

Cement, lb	Silica Sand, lb	Graphite, lb	Water, gal	Note	Yield, gal	TC (k_{grt}), Btu/h·ft·°F	Density, lb/gal
50	0	0	11		13	0.45-0.50	10.9
50	0	8	11	HPG*	13	0.85-0.95	11.5
50	0	15	11	HPG*	14	1.20-1.40	11.2

Sands—Gravel, Aggregrate, Crushed Limestone, Cuttings, etc.

Dry Density, lb/ft^3	Moisture, %	TC (k_{grt}), Btu/h·ft·°F
80	5	0.6-0.9
80	15	0.7-1.1
100	5	1.0-1.2
100	15	1.3-1.5
120	5	1.3-1.8
120	15	1.5-2.1

Properties unknown: Laboratory and in-situ thermal testing recommended

Caution: Borehole bridging and voids likely; surface grout plug required

Pipe Materials

Material	TC (k_p), Btu/h·ft·°F	Density, lb/ft^3	Material	TC (k_p), Btu/h·ft·°F	Density, lb/ft^3
HDPE—3xxx	0.25	58 - 60	Aluminum	137	170
HDPE—4xxx	0.26	58 - 60	Carbon Steel	30	560
Polypropylene	0.14	56.5	Copper	230	490
Polyvinyl chloride (PVC)	0.08	87	Stainless Steel (304)	10	500
Cross-linked polyethylene (PEX)	0.25	58 - 60			

** HPG = high-performance graphite; SPG = standard-performance graphite.*

Table 3.2b Properties of Borehole Grouts and Fills (Allan 1996; GPI 2014)—SI

Sodium Bentonite Recipes

Bentonite, kg	Silica Sand, kg	Graphite, kg	Water, L	Note	Yield, L	TC (k_{grt}), W/m·K	Density, kg/m^3
23	0	0	125		36	0.68	1077
23	0	0	91		27	0.73	1113
23	0	0	53		17	0.76	1173
23	45	0	57		23	1.2	1436
23	91	0	68		32	1.6	1496
23	181	0	83		42	2.2	1807
23	0	4	61	HPG*	18	1.6	1269
23	23	4	68	HPG*	23	1.6	1340
23	0	7	61	SPG*	19	1.6	1245
23	23	5	91	SPG*	31	1.6	1197
23	0	7	68	HPG*	21	2.2	1221
23	23	7	76	HPG*	25	2.2	1352
23	0	9	57	SPG*	18	2.2	1293
23	45	7	61	SPG*	23	2.2	1556

Cement Recipes

Cement, kg	Silica Sand, kg	Other, kg	Water, L	S. Plasticisizer, oz	Yield, L	TC (k_{grt}), W/m·K	Density, kg/m^3
43	91	0	Neat Cement—Not Recommended				
43	91	135-180	Concrete—Not Recommended				
43	91	0	23	21	72	2.2	2178

Engineered, High-Yield Cement for GSHP Applications

Cement, kg	Silica Sand, kg	Graphite, kg	Water, L	Note	Yield, L	TC (k_{grt}), W/m·K	Density, kg/m^3
23	0	0	42		49	0.8	1305
23	0	4	42	HPG*	49	1.6	1376
23	0	7	42	HPG*	53	2.3	1340

Sands—Gravel, Aggregrate, Crushed Limestone, Cuttings, etc.

Dry Density, kg/m^3	Moisture, %	TC (k_{grt}), W/m·K	
1280	5	1.0	1.6
1280	15	1.2	1.9
1600	5	1.7	2.1
1600	15	2.3	2.6
1920	5	2.3	3.1
1920	15	2.6	3.6

Properties unknown: Laboratory and In-situ thermal testing recommended

Caution: Borehole bridging and voids likely; surface grout plug required

Pipe Materials

Material	TC (k_p), W/m·K	Density, kg/m^3	Material	TC (k_p), W/m·K	Density, kg/m^3
HDPE—3xxx	0.43	940	Aluminum	237	2720
HDPE—4xxx	0.45	940	Carbon Steel	52	8960
Polypropylene	0.24	900	Copper	398	7840
Polyvinyl chloride (PVC)	0.14	1400	Stainless Steel (304)	17	8000
Cross-linked polyethylene (PEX)	0.43	940			

** HPG = high-performance graphite; SPG = standard-performance graphite.*

- Each bore being 240 ft (73 m) in length using a grout with a thermal conductivity of 1.0 Btu/h·ft·°F (1.7 W/m·K)
- Alternate 1: Each bore being 260 ft (79 m) in length using a grout with a thermal conductivity of 0.85 Btu/h·ft·°F (1.5 W/m·K)
- Alternate 2: Each bore being 300 ft (91 m) in length using a grout with a thermal conductivity of 0.4 Btu/h·ft·°F (0.7 W/m·K)

***Note:* Do not infer the added lengths used in the above example are always proportional to the change in grout conductivity. Total borehole length must be calculated for each case to provide equivalent performance.**

The thermal resistance of pipe (R_p) is a combination of the resistance of the tubing wall itself (R_{tube}) and the resistance of the fluid film (R_{film}) inside the pipe wall (Equations 3.8 and 3.10). Calculation of R_{tube} is straightforward and requires knowledge of only the pipe thermal conductivity (k_p), inside diameter (d_i), and outside diameter (d_o). Values for pipe thermal conductivity are provided in the bottom rows of Tables 3.2a and 3.2b.

Calculation of R_{film} is much more difficult because the equations used to determine film heat transfer coefficients (h_{film}) are complex and in some situations highly uncertain. Fortunately, this resistance is typically much smaller than the resistance of the grout, plastic pipe wall, and the ground. Therefore, errors in this calculation typically do not result in large errors in the overall resistance of the vertical ground heat exchanger. (This is not always true in other GSHP applications, such as surface-water heat exchangers in which high values for R_{film} tend to make a larger, but not dominant, contribution to overall thermal resistance.)

Determination of film coefficients begin with the Reynolds number ($Re = \rho DV/\mu$), which provides an indication of the flow regime (laminar, transition, turbulent) inside the pipe. Low flow rates in cold, viscous fluids will result in laminar flow and higher thermal resistance at the fluid-wall interface. It is important to recall the other component materials in the ground heat exchanger are plastic tubing, grout, soil, and rock, none of which have outstanding thermal properties. Thus, the negative effect of laminar flow upon the overall heat exchange rate in this application is not nearly as dramatic as it is in compact heat exchangers having materials with outstanding thermal properties (such as copper). More details of the procedure to determine film coefficients are presented in the surface-water heat pump discussion in Chapter 5 because the inside film resistance plays a more significant role in this application when the flow regime is laminar.

The flow rate through individual U-tubes is determined by dividing the total system flow rate by the number of parallel U-tube flow paths in the bore field. Almost always the number of parallel flow paths is equal to the number of vertical bores, unless two U-tubes are placed in each borehole or U-tubes are placed in series when bore depths are shallow (see Figure 3.7). Table 3.3 provides the Reynolds numbers for a variety of flow rates, tube sizes, fluids, and temperatures that are common in ground heat exchangers. Values can be used in conjunction with Tables 3.1 and 3.2a or Tables 3.1 and 3.2b to estimate bore resistance in lieu of Equations 3.7, 3.8, and 3.9. Furthermore, these equations require a value for the heat transfer coefficient (h), which necessitates a more rigorous computation. It is also important to recognize that equations used to determine fluid heat transfer coefficients were developed for horizontal tubes. The actual values in vertical tubes will likely be much higher because the buoyancy-induced natural convection effects are not significant in horizontal tubes (Kavanaugh 1984). Thus, the thermal resistance values in Table 3.1 are likely to be somewhat conservative.

Table 3.3 Reynolds Numbers in DR 11 HDPE Pipe for Various Pipe Diameters and Flow Rates

Fluid	Temperature, °F	3 gpm			5 gpm			10 gpm		
		3/4 in.	1 in.	1 1/4 in.	3/4 in.	1 in.	1 1/4 in.	1 in.	1 1/4 in.	1 1/2 in.
Water	68	10700	8500	6800	17800	14200	11300	28500	22600	19700
20% Propylene glycol	32	2800	2200	1800	4700	3700	2900	7400	5900	5200
20% Propylene glycol	50	4000	3200	2500	6700	5300	4200	10700	8500	7400
20% Propylene glycol	86	7500	6000	4700	12400	9900	7900	19800	15700	13700
30% Propylene glycol	32	1600	1300	1000	2700	2100	1700	4300	3400	3000
30% Propylene glycol	50	2500	2000	1600	4200	3300	2600	6600	5300	4600
30% Propylene glycol	86	5300	4200	3300	8800	7000	5600	14100	11200	9800
25% Methyl alcohol	32	3300	2600	2100	5500	4400	3500	8800	7000	6100
25% Methyl alcohol	50	4800	3900	3100	8100	6400	5100	12900	10200	8900
25% Methyl alcohol	86	8900	7100	5600	14800	1180	9300	23600	18700	16300

To estimate loop water flow: gpm ≈ q (Btu/h) ÷ [500 × Δt (°F) × No. of Parallel U-Tubes]

Fluid	Temperature, °C	10 L/min			20 L/min			40 L/min		
		25 mm	32 mm	40 mm	25 mm	32 mm	40 mm	32 mm	40 mm	50 mm
Water	20	10030	7769	6293	20129	15657	12616	31342	25165	20080
20% Propylene glycol	0	2625	2011	1666	5315	4080	3238	8138	6570	5300
20% Propylene glycol	10	3750	2925	2314	7577	5844	4689	11767	9465	7543
20% Propylene glycol	30	7030	5484	4350	14022	10916	8820	21774	17482	13964
30% Propylene glycol	0	1500	1188	925	3053	2316	1898	4729	3786	3058
30% Propylene glycol	10	2343	1828	1481	4749	3639	2903	7258	5902	4689
30% Propylene glycol	30	4968	3839	3054	9951	7718	6252	15506	12471	9989
25% Methyl alcohol	0	3093	2376	1944	6220	4852	3908	9678	7795	6218
25% Methyl alcohol	10	4499	3565	2869	9160	7057	5694	14186	11358	9072
25% Methyl alcohol	30	8343	6490	5183	16736	1301	10383	25953	20823	16614

To estimate loop water flow: L/min ≈ q (kW) ÷ [0.0692 × Δt (°C) × No. of Parallel U-Tubes]

EXAMPLE 3.1— CALCULATION OF BORE THERMAL RESISTANCE

Determine the bore thermal resistance for a ground heat exchanger consisting of a 1.0 in. (32 mm) DR 11 HDPE tube placed in a 5 in. (127 mm) diameter bore grouted with thermally enhanced sodium bentonite (one part bentonite/four parts sand) that is flowing at 4 gpm (15 L/min) with 20% propylene glycol at 50°F (10°C).

Solution

Tables 3.2a and 3.2b indicate the midrange grout conductivity is 0.9 Btu/h·ft·°F (1.56 W/m·K). Table 3.3 is used to interpolate the Reynolds number for 4 gpm (15 L/min) for the 20% propylene glycol mixture using values for 3 gpm (11 L/min) (Re = 3200) and 5 gpm (19 L/min) (Re = 5300) to find a value that is slightly above the value when Re = 4000. Table 3.1 contains columns for Re = 4000 and grout conductivities of 0.8 and 1.2 Btu/h·ft·°F (1.39 and 2.08 W/m·K). For a 5 in. (127 mm) diameter bore, these grout conductivities result in values for bore resistance of 0.24 and 0.19 h·ft·°F/Btu (1.39 and 1.73 W/m·K), respectively, if location B (in Figure 3.4) is assumed.

These values are used to find a value of 0.23 h·ft·°F/Btu (0.133 m·K/W) for a grout conductivity of 0.9 Btu/h·ft·°F (1.56 W/m·K). If location C is assumed, the resulting interpolated value is 0.16 h·ft·°F/Btu (0.092 m·K/W). The recommended value for design would be the average of locations B and C, resulting in 0.20 h·ft·°F/Btu (0.116 m·K/W), with the location-B result of 0.23 h·ft·°F/Btu (0.133 m·K/W) suggested for conservative designers.

Alternate Solution

The spreadsheet *BoreResistCalc.xlsm*, which is available with this book at www.ashrae.org/GSHP, generates values of 0.196 h·ft·°F/Btu (0.113 m·K/W) for location BC and 0.226 h·ft·°F/Btu (0.131 m·K/W) for location B.

3.4 GROUND THERMAL RESISTANCE AND BASIC HEAT EXCHANGER DESIGN

In Equations 3.5 and 3.6 the most difficult parameters to evaluate are the equivalent thermal resistance of the ground. The solutions of Carslaw and Jaeger (1947) require that the time of operation, outside pipe diameter, and thermal diffusivity of the ground be related in the dimensionless Fourier number (Fo):

$$\mathrm{Fo} = \frac{4\alpha_g \tau}{d^2} \tag{3.11}$$

The cylindrical heat source solution of Carslaw and Jaeger is modified to permit calculation of equivalent thermal resistances for varying heat pulses. Consider a system that can be modeled by three heat pulses, a 10 year (3650 day) pulse of q_a, a one month (30 day) pulse of q_m, and a 4 hour (0.167 day) pulse of q_d. Three times are defined as

$$\tau_1 = 3650$$

$$\tau_2 = 3650 + 30 = 3680,$$

$$\tau_f = 3650 + 30 + 0.167 = 3680.167 \text{ days}$$

The Fourier number is then computed using the following values:

$$\begin{aligned} \mathrm{Fo}_f &= 4\alpha\tau_f/\mathrm{d}^2 \\ \mathrm{Fo}_1 &= 4\alpha(\tau_f - \tau_1)/\mathrm{d}^2 \\ \mathrm{Fo}_2 &= 4\alpha(\tau_f - \tau_2)/\mathrm{d}^2 \end{aligned} \tag{3.12}$$

The G-factor for each of the Fourier values is then determined from Figure 3.6. The three equivalent thermal resistance values during each heat pulse are found from

$$R_{ga} = \frac{G_f - G_1}{k_g}$$

$$R_{gm} = \frac{G_1 - G_2}{k_g} \tag{3.13}$$

$$R_{gst} = \frac{G_2}{k_g}$$

There is some degradation of performance due to short-circuit heat losses between the upward and downward flowing legs of any type of ground heat exchanger. For conventional U-tubes the loss is approximately 4% when liquid flow rates are 3 gpm/ton (3.2 L/min·kW), which represents a 10°F (6°C) differential (Kavanaugh 1984). Losses can be accounted for by multiplying the equivalent thermal resistance for the short-term heat pulse (R_{gst}) by 1.04. For a 15°F (9°C) differential F_{sc} will be greater at 1.06. The losses are reduced considerably if there are two or three U-tubes in series. The differential temperature between the upward and downward legs will be lower, as shown in Figure 3.7, using a 10°F (6°C) differential temperature on the supply and return headers as an example. This arrangement is not standard practice but may occur in situations

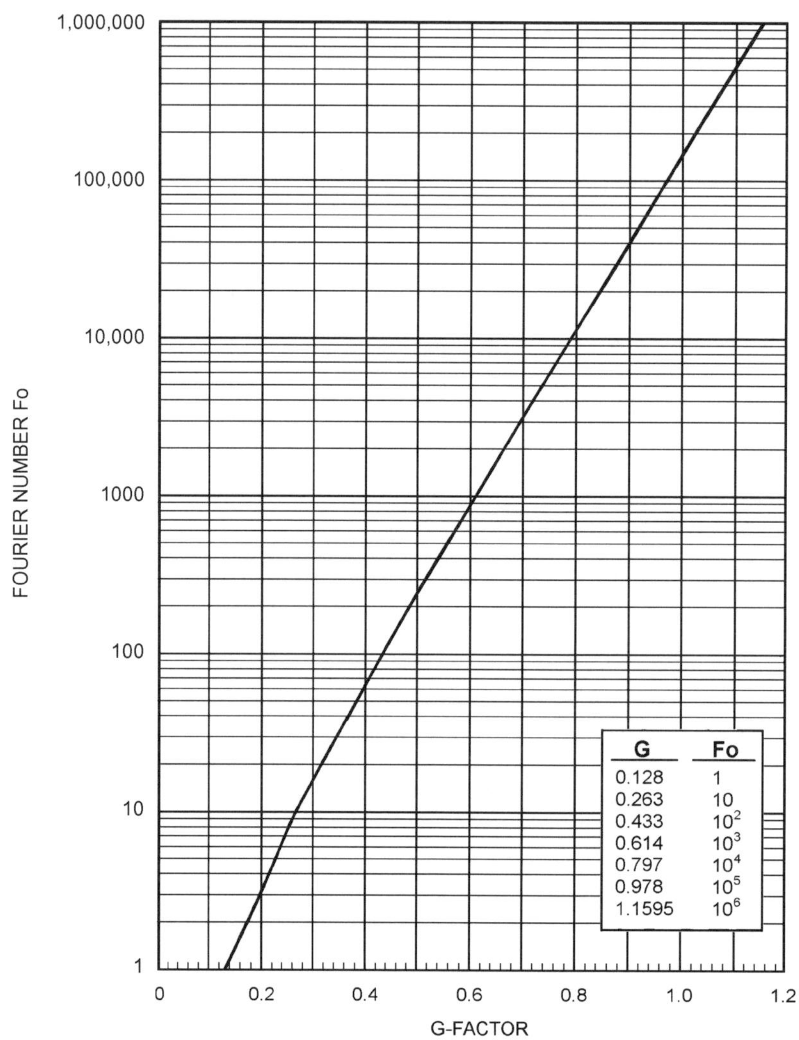

Figure 3.6 Fourier/G-Factor Graph for Ground Thermal Resistance (Ingersoll et al. 1954)

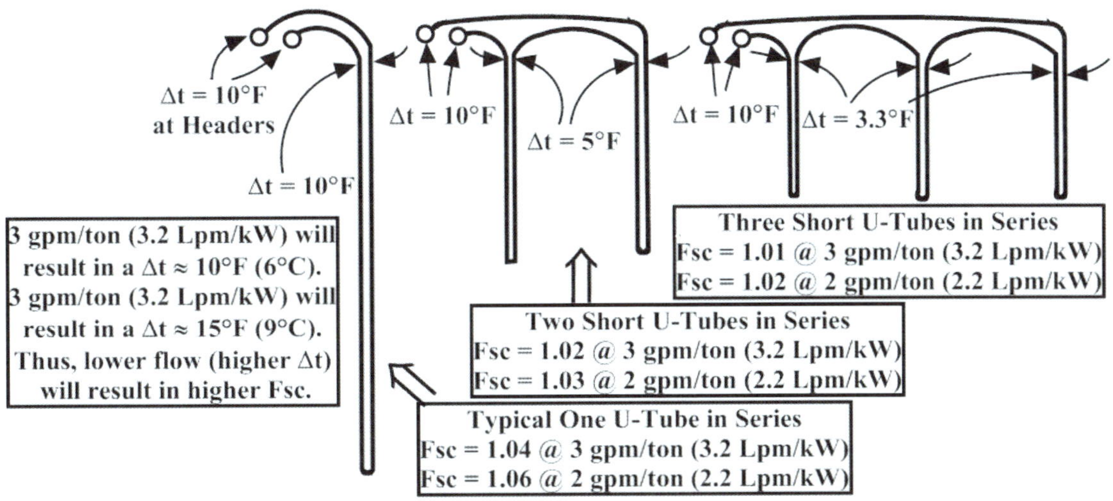

Figure 3.7 Short-Circuit Factor (F_{sc}) for Standard and Shallow Bore U-Tube Applications (Kavanaugh 1984)

where drilling depths are limited because of environmental concerns, difficult formations, or rig limitations.

Consider an application in which the original plan is to install 30 200 ft (60 m) U-tubes. However, a sensitive drinking water aquifer is present at a depth of 150 ft (45 m), so limitations are imposed on the drilling depth. Placing U-tubes only 100 ft (30 m) in depth would result in twice the number of parallel circuits, lower velocity tube flow, a greater challenge when purging air and debris at start-up, and double the number of take-off fittings from the headers. Thus, two U-tubes could be placed in series as shown in Figure 3.7 before returning to the horizontal headers. The result would be 60 U-tubes, 100 ft (30 m) in depth with 30 parallel flow paths rather than 60. In this case the temperature difference and heat loss between the U-tube legs would be less than the difference in the standard single-bore, parallel loop. Thus, the short-circuit heat loss factor (F_{sc}) would be lower, as indicated in Figure 3.7.

EXAMPLE 3.2—
VERTICAL GROUND HEAT EXCHANGER DESIGN—I-P

Find the required vertical ground heat exchanger for the building described.

- Office in Atlanta, Georgia, with eight zones
- Cooling block load (q_{lc}) = 300,000 Btu/h (25 tons)
- Heating block load (q_{lh}) = 180,000 Btu/h
- Design month (August) part-load factor (PLF_m) = 0.28
- Vertical U-tube = 1.0 in. nominal, DR 11, HDPE, 5 in. borehole diameter
- 5 × 5 square grid (25 vertical bores) with 20 ft separation
- Heat pump ELT = 85°F
- Heat pump LLT = 95°F
- Heat pump cooling efficiency (EER) = 14.1 Btu/Wh

- Heat pump heating efficiency (COP) = 4.1
- Ten year (3650 day), one month (30 day), and four hour (0.167 day) heat pulse analysis
- $EFLH_c$ = 1220 h (see Table 4.5)
- $EFLH_h$ = 590 h (see Table 4.5)
- A thermal property test provided the following information:
  - Ground Temperature (t_g) = 65°F
  - Ground conductivity (k_g) = 1.4 Btu/h·ft·°F
  - Ground diffusivity (α_g) = 1.0 ft²/day
  - Bore fill conductivity (k_b) = 1.0 Btu/h·ft·°F
  - Static water table at 50 ft below surface

Solution

Determine the ground heat transfer rates in cooling and heating and net annual heat to and from the ground (Equations 3.2, 3.3, and 3.4):

$$q_{cond} = q_{lc} \times \frac{\text{EER} + 3.412}{\text{EER}} = -300{,}000 \text{ Btu/h} \times \frac{14.1 + 3.412}{14.1} = 372{,}000 \text{ Btu/h}$$

$$q_{evap} = q_{lh} \times \frac{\text{COP} - 1}{\text{COP}} = 180{,}000 \text{ Btu/h} \times \frac{4.1 - 1}{4.1} = 136{,}100 \text{ Btu/h}$$

$$q_a = \frac{q_{cond} \times \text{EFLH}_c + q_{evap} \times \text{EFLH}_h}{8760 \text{ h}}$$

$$= \frac{372{,}000 \text{ Btu/h} \times 1220 \text{ h} + 136{,}100 \text{ Btu/h} \times 590 \text{ h}}{8760 \text{ h}} = -42{,}700 \text{ Btu/h}$$

Determine the thermal resistances of the ground for the three prescribed heat pulses (Equations 3.11, 3.12, and 3.13):

$$\text{Fo}_f = 4 \times 1.0 \text{ ft}^2\text{/day} \times 3680.167 \text{ days} \div (5 \text{ in.} \div 12 \text{ in./ft})^2 = 84{,}800, \text{ from Figure 3.6, } G_f = 0.96$$

$$\text{Fo}_1 = 4 \times 1.0 \text{ ft}^2\text{/day} \times (3680.167 - 3650) \div (5 \text{ in.} \div 12 \text{ in./ft})^2 = 695, \text{ from Figure 3.6, } G_1 = 0.58$$

$$\text{Fo}_2 = 4 \times 1.0 \text{ ft}^2\text{/day} \times (3680.167 - 3680) \div (5 \text{ in.} \div 12 \text{ in./ft})^2 = 3.85, \text{ from Figure 3.6, } G_2 = 0.20$$

$$R_{ga} = (0.96 - 0.58) \div 1.4 \text{ Btu/h·ft·°F} = 0.271 \text{ h·ft·°F/Btu}$$

$$R_{gm} = (0.58 - 0.20) \div 1.4 \text{ Btu/h·ft·°F} = 0.264 \text{ h·ft·°F/Btu}$$

$$R_{gst} = 0.20 \div 1.4 \text{ Btu/h·ft·°F} = 0.143 \text{ h·ft·°F/Btu}$$

Determine the thermal resistances of the bore. Using the equation in Table 3.3 to find the estimated flow through each U-tube during cooling (loop transfers q_{cond} = –372,600 Btu/h),

$$\text{Flow/U-tube (gpm)} = -372{,}600 \text{ Btu/h} \div [500 \times (85°\text{F} - 95°\text{F}) \times 25 \text{ U-tubes}] = 2.98 \text{ gpm}$$

At 68°F, the Reynolds number (Re) for water flowing at 3 gpm in a 1.0 in. DR 11 tube is 8500. Re will be higher at the 90°F average water temperature. So the bore resistance will be found based on the turbulent flow value of 10,000 used in Table 3.3. If the flow rate is adjusted during the final design phase, the results should be reconfirmed.

For k_{grout} = 0.8 Btu/h·ft·°F, turbulent flow, 5 in. bore, location B: R_b = 0.23 h·ft·°F/Btu

For k_{grout} = 1.2 Btu/h·ft·°F, turbulent flow, 5 in. bore, location B: R_b = 0.18 h·ft·°F/Btu

Via interpolation for k_{grout} = 1.0 Btu/h·ft·°F, R_b = 0.205 h·ft·°F/Btu

For k_{grout} = 0.8 Btu/h·ft·°F, turbulent flow, 5 in. bore, location C: R_b = 0.17 h·ft·°F/Btu

For k_{grout} = 1.2 Btu/h·ft·°F, turbulent flow, 5 in. bore, location C: R_b = 0.14 h·ft·°F/Btu

Via interpolation for k_{grout} = 1.0 Btu/h·ft·°F, R_b = 0.155 h·ft·°F/Btu

The average bore resistance value for locations B and C is applied:

R_b = 0.18 h·ft·°F/Btu for location BC, k_{grout} = 1.0 Btu/h·ft·°F, turbulent flow, 5-in. bore

The ground-loop differential temperature is 10°F (ELT = 85°F, LLT = 95°F), thus the short-circuiting heat loss factor (F_{sc}) is 1.04 as indicated in Figure 3.7.

The required total bore length for cooling is computed using Equation 3.5. The procedure for determining long-term ground temperature change (t_p = 0) is presented later in this chapter. To complete this example, a value of 2.0°F is assumed.

$$L_c = \frac{(-42{,}700 \times 0.271) - 372{,}600 \times (0.18 + 0.28 \times 0.264 + 1.04 \times 0.143)}{65°\text{F} - \dfrac{85°\text{F} + 95°\text{F}}{2} + 2°\text{F}} = 7025 \text{ ft}$$

$$= 7025 \text{ ft} \div 25 \text{ bores} = 281 \text{ ft/bore}$$

The process is repeated using Equation 3.6 to find the bore length for heating (L_h); the design bore length is the larger value of L_c and L_h.

EXAMPLE 3.3— VERTICAL GROUND HEAT EXCHANGER DESIGN—SI

Find the required vertical ground heat exchanger for the building described.

- Office in Ottawa, Ontario, Canada, with eight zones
- Cooling block load (q_{lc}) = 75 kW
- Heating block load (q_{lh}) = 90 kW
- Design month (January) part-load factor (PLF_m) = 0.31
- Vertical U-tube = 32 mm, DR 11, HDPE, 125 mm (0.125 m) borehole diameter
- 5 × 4 square grid (20 vertical bores) with 6 m borehole separation
- Heat pump ELT = 0°C
- Heat pump LLT = –3.0°C
- Heat pump cooling efficiency (COP_c) = 4.8
- Heat pump heating efficiency (COP_h) = 3.5
- Twenty year (7300 day), one month (30 day), and four hour (0.167 day) heat pulse analysis

- $EFLH_c = 450$ h
- $EFLH_h = 900$ h
- A thermal property test provided the following information:
    - Ground temperature (t_g) = 8°C
    - Ground conductivity (k_g) = 2.0 W/m·K
    - Ground diffusivity (α_g) = 0.08 m^2/day
    - Borehole fill conductivity (k_b) = 1.4 W/m·K

Solution

Determine the ground heat transfer rates in cooling and heating and net annual heat to and from the ground (Equations 3.2, 3.3, and 3.4):

$$q_{cond} = q_{lc} \times \frac{COP_c + 1.0}{COP_c} = -75 \text{ kW} \times \frac{4.8 + 1}{4.8} = -90.6 \text{ kW } (-90,600 \text{ W})$$

$$q_{evap} = q_{lh} \times \frac{COP_h - 1.0}{COP_h} = 90 \text{ kW} \times \frac{3.5 - 1}{3.5} = 64.3 \text{ kW } (64,300 \text{ W})$$

$$q_a = \frac{q_{cond} \times EFLH_c + q_{evap} \times EFLH_h}{8760 \text{ h}}$$

$$= \frac{-90.6 \text{ kW} \times 450 \text{ h} + 64.3 \text{ kW} \times 900 \text{ h}}{8760 \text{ h}} = 1.95 \text{ kW } (1950 \text{ W})$$

Determine the thermal resistances of the ground for the three prescribed heat pulses (Equations 3.11, 3.12, and 3.13):

$$Fo_f = 4 \times 0.08 \text{ m}^2/\text{day} \times 7330.167 \text{ days} \div (0.125 \text{ m})^2 = 150{,}100, \text{ from Figure 3.6, } G_f = 0.96$$

$$Fo_1 = 4 \times 0.08 \text{ m}^2/\text{day} \times (7330.167 - 7300) \div (0.125 \text{ m})^2 = 618, \text{ from Figure 3.6, } G_1 = 0.58$$

$$Fo_2 = 4 \times 0.08 \text{ m}^2/\text{day} \times (7330.167 - 7330) \div (0.125 \text{ m})^2 = 3.42, \text{ from Figure 3.6, } G_2 = 0.20$$

$$R_{ga} = (1.02 - 0.56) \div 2.0 \text{ W/m·K} = 0.23 \text{ m·K/W}$$

$$R_{gm} = (0.56 - 0.19) \div 2.0 \text{ W/m·K} = 0.185 \text{ m·K/W}$$

$$R_{gst} = 0.19 \div 2.0 \text{ W/m·K} = 0.095 \text{ m·K/W}$$

Determine the thermal resistances of the bore. Using the equation in Table 3.3b to find the estimated flow through each U-tube during heating (loop transfers $q_{evap} = 64.3$ kW),

$$\text{Flow/U-tube (L/min)} = 64.3 \text{ kW} \div [0.0692 \times (0°\text{C} - -3.0°\text{C}) \times 25 \text{ U-tubes}) = 12.4 \text{ L/min}$$

At 0°C, the Reynolds number (Re) for a 20% propylene glycol solution flowing at 10 L/min in a 1.0 in. DR 11 tube is 2011 and at 20 L/min is 4080. Re will be just over 2500 at 12.4 L/min, which is transition flow. So the bore resistance will be found based on the transition flow but the value for bore resistance will be interpolated between laminar and transition values. If the flow rate is adjusted during the final design phase, the results should be reconfirmed. Also note the 0.0692 multiplier for the equation above is based on water and the value for antifreeze solutions will be slightly lower, thus making the flow rate higher.

At 0°C, the Reynolds number (Re) for a 20% propylene glycol solution flowing at 10 L/min in a 1.0 in. DR 11 tube is 2011 and at 20 L/min is 4080. Re will be just over 2500 at 12.4 L/min, which is transition flow. So the bore resistance will be found based on the transition flow but the value for bore resistance will be interpolated between laminar and transition values. If the flow rate is adjusted during the final design phase, the results should be reconfirmed. Also note the 0.0692 multiplier for the equation above is based on water and the value for antifreeze solutions will be slightly lower, thus making the flow rate higher.

For k_{grout} = 1.4 W/m·K, laminar flow, 125 mm, location B: R_b = 0.17 m·K/W

For k_{grout} = 1.4 W/m·K, transition flow, 125 mm, location B: R_b = 0.14 m·K/W

For k_{grout} = 1.4 W/m·K, laminar flow, 125 mm, location C: R_b = 0.13 m·K/W

For k_{grout} = 1.4 W/m·K, transition flow, 125 mm, location C: R_b = 0.10 m·K/W

Via double interpolation, the average bore resistance is

R_b = 0.135 m·K/W for location BC, k_{grout} = 1.4 W/m·K, laminar/transition flow, 125 mm bore

The ground-loop differential temperature is 3°C,thus the short-circuiting heat loss factor (F_{sc}) is 1.01, as indicated in Figure 3.7.

The required total bore length for heating is computed using Equation 3.6. The procedure for determining long-term ground temperature change (t_p) is presented later in this chapter. To complete this example, a value of –0.5°C is assumed.

$$L_h = \frac{(1950 \times 0.23) + 64{,}300 \times (0.135 + 0.31 \times 0.185 + 1.01 \times 0.095)}{8°\text{C} - \dfrac{0°\text{C} + -3°\text{C}}{2} + -0.5°\text{C}} = 7025 \text{ ft}$$

$$= 2110 \text{ m} \div 25 \text{ bores} = 84 \text{ m/bore}$$

The process is repeated using Equation 3.6 to find the bore length for cooling (L_c); the design bore length is the larger value of L_c and L_h.

3.5 GCHP SITE ASSESSMENT: GROUND THERMAL PROPERTIES

The resistance to heat flow imposed by the ground is complex because of the variations in soils and operational patterns of the building being heated and cooled. Estimation of ground temperature (t_g), thermal conductivity (k_g), and diffusivity (α_g) is a necessary but unfamiliar task to HVAC engineers. Converting geological information to meaningful thermal properties is challenging. Test methods are now available that provide improved accuracy compared to estimating properties from tables, maps, and well logs. However, these traditional methods remain an alternative for residential and small commercial projects where the cost of the thermal property tests are likely to exceed the cost of using conservative estimates to size the ground heat exchanger. Thus, tables of thermal properties and groundwater temperature maps are provided in this section in addition to a presentation of recommended field tests that can more accurately determine local ground thermal properties.

The variations in soil composition and thermal properties are extreme, as can be noted by examination of Tables 3.4 and 3.5. How moisture content in sands and clay affects thermal conductivity is extremely important. Sands (grain size greater than 0.075 mm) and clays (grain size less than 0.075 mm) affect both thermal conductivity and diffusivity. However, soils do not have to be saturated with moisture to provide good thermal conductivity, as shown in Table 3.4. Note that sandy soils, which have courser grain sizes compared to clays, have higher thermal conductivities. Soils are typically a combination of fine-grain clays and coarse-grain sands. A sieve analysis can be conducted to determine the percentage of the components that are coarse grain and fine grain. A weighted average can be calculated and the value of thermal conductivity can be interpolated between the 100% coarse-grain and 100% fine-grain soils in the tables.

To obtain accurate values for thermal properties, a detailed geological site survey is required. Although some uncertainty can be eliminated by conducting sieve analysis and by weighing the excavated material and applying the equations summarized by Farouki (1982), accuracy is still limited.

Table 3.5 lists thermal properties of rocks common in the earth's crust. The variation in thermal conductivity is even greater than in soils. The references for the table (Toulokian et al. 1981; Robertson 1988; Carmichael 1989) contain a vast number of samples from the United States.

The local undisturbed deep ground temperature can be obtained from local water well logs and geological surveys. A second, but less accurate, source is temperature contour maps prepared by state geological surveys, similar to that shown in Figure 3.8. A third source that can yield ground temperatures within ±6°F (±3.3°C) is a U.S. map with contours, such as that shown in Figure 3.9. Comparison of Figures 3.8 and 3.9 indicates the complex variations that would not be accounted for if detailed contour maps are not used.

For residential and small commercial applications, it may be acceptable to estimate soil and rock thermal properties using information from sources similar to Tables 3.4 and 3.5 in combination with local water well logs that contain groundwater temperatures. Conservative estimates of thermal properties may result in larger-than-optimum heat

Table 3.4 Thermal Conductivity (k) and Diffusivity (α) of Sand and Clay Soils—Values Indicate Ranges Predicted by Five Independent Methods (Farouki 1982)

Dry Density		Moisture	Sands: 0.075 to 5 mm (> #200 Standard Sieve)				Clays: < 0.075 mm (< #200 Standard Sieve)			
			Thermal Conductivity (±20%)		Thermal Diffusivity (±20%)		Thermal Conductivity (±20%)		Thermal Diffusivity (±20%)	
lb/ft^3	kg/m^3	%	Btu/h·ft·°F	W/m·°C	ft^2/day	m^2/day	Btu/h·ft·°F	W/m·°C	ft^2/day	m^2/day
80	1280	5	0.80	1.38	0.95	0.088	0.40	0.69	0.48	0.045
80	1280	10	0.85	1.47	0.85	0.079	0.42	0.73	0.42	0.039
80	1280	15	0.90	1.56	0.75	0.070	0.47	0.81	0.40	0.037
80	1280	20	0.95	1.64	0.71	0.066	0.50	0.87	0.37	0.034
100	1600	5	1.10	1.90	1.04	0.097	0.55	0.95	0.53	0.049
100	1600	10	1.45	2.51	1.03	0.096	0.55	0.95	0.44	0.041
100	1600	15	1.40	2.42	1.00	0.093	0.65	1.13	0.42	0.039
100	1600	20	1.55	2.68	0.92	0.086	0.70	1.21	0.48	0.045
120	1920	5	1.55	2.68	1.23	0.114	0.70	1.21	0.56	0.052
120	1920	10	1.70	2.94	1.12	0.104	0.70	1.21	0.46	0.043
120	1920	15	1.90	3.29	1.06	0.099	0.95	1.64	0.55	0.051

exchanger lengths. The added costs for smaller systems are likely to be lower than the price of performing a thermal property test. Typically, the cost for a test is equivalent to the installation cost for three to four vertical heat exchangers. However, the heat exchanger used for the test can be used in the ground-loop system, thus reducing the net cost to two to three vertical heat exchangers.

Another method is to estimate the ground temperature at various depths using seasonal air temperature variations, which are available from weather data. Equation 5.25 is provided to determine the ground temperature for any depth and day of the year if the ground thermal diffusivity is available. (It appears in the discussion in Chapter 5 on direct cooling with surface water as ground temperature impacts shallow header heat gain between the reservoir and building.) The accuracy of the equation has limitations for shallow-earth applications because near-surface thermal properties vary with moisture content (rainfall). It also has limitations for vertical deep-bore applications because of variations in the thermal gradient from the earth core to the surface. This can be observed

Table 3.5 Ranges of Thermal Properties of Rocks at 77°F (25°C) (Toulokian et al. 1981; Robertson 1988; Carmichael 1989)

	Thermal Conducivity (k), Btu/h·ft·°F (W/m·K)		Specific Heat, Btu/lb·°F (kJ/kg·K)		Density, lb/ft^3 (kg/m^3)		Thermal Diffusivity (α), ft^2/day	m^2/day
Rock Type	Low	High	Low	High	Low	High	Midrange	
Igneous Rocks								
Granite (10% quartz)	1.1(1.9)	3.0 (5.2)	0.21 (0.88)		165 (2640)		1.10	0.10
Granite (25% quartz)	1.5 (2.6)	2.1 (3.6)	0.21 (0.88)		165 (2640)		1.20	0.11
Amphibolite	1.5 (2.6)	2.2 (3.8)	—		175 (2800)	195 (3120)	—	—
Andesite	0.9 (1.6)	1.4 (2.4)	0.12 (0.50)		160 (2560)		1.40	0.13
Basalt	1.2 (2.1)	1.4 (2.4)	0.17–0.21 (0.71–0.88)		180 (2880)		0.80	0.07
Gabbro (Cen. Plains)	0.9 (1.6)	1.6 (2.8)	0.18 (0.75)		185 (2960)		0.90	0.08
Gabbro (Rocky Mtns.)	1.2 (2.1)	2.1 (3.6)	0.18 (0.75)		185 (2960)		1.20	0.11
Diorites	1.2 (2.1)	1.7 (2.9)	0.22 (0.92)		180 (2880)		0.85	0.08
Grandiorites	1.2 (2.1)	2 (3.5)	0.21 (0.88)		170 (2720)		1.10	0.10
Sedimentary Rocks								
Claystone	1.1 (1.9)	1.7 (2.9)	—		—	—	—	—
Dolomite	1.6 (2.8)	3.6 (6.2)	0.21 (0.88)		170 (2720)	175 (2800)	1.70	0.16
Limestone	1.0 (1.7)	3.0 (5.2)	0.22 (0.92)		150 (2400)	175 (2800)	1.20	0.11
Rock Salt	—	3.7 (6.4)	0.20 (0.84)		130 (2080)	135 (2160)	—	—
Sandstone	1.2 (2.1)	2.0 (3.5)	0.24 (1.0)		160 (2560)	170 (2720)	0.95	0.09
Siltstone	0.8 (1.4)	1.4 (2.4)	—		—	—	—	—
Wet shale (25% quartz)	1.0 (1.7)	1.8 (3.1)	0.21 (0.88)		130 (2080)	165 (2640)	—	—
Wet shale (no quartz)	0.6 (1.0)	2.3 (4.0)	0.21 (0.88)		130 (2080)	165 (2640)	0.55	0.05
Dry shale (25% quartz)	0.8 (1.4)	1.4 (2.4)	0.21 (0.88)		130 (2080)	165 (2640)	0.85	0.08
Dry shale (no quartz)	0.5 (0.9)	0.8 (1.4)	0.21 (0.88)		130 (2080)	165 (2640)	0.50	0.05
Metamorphic Rocks								
Gneiss	1.3 (2.2)	2.0 (3.5)	0.22 (0.92)		160 (2560)	175 (2800)	1.05	0.10
Marble	1.2 (2.1)	3.2 (5.5)	0.22 (0.92)		170 (2720)		1.00	0.09
Quarzite	3.0 (5.2)	4.0 (6.9)	0.20 (0.84)		160 (2560)		2.60	0.24
Schist	1.2 (2.1)	2.6 (4.5)	—		170 (2720)	200 (3200)	—	—
Slate	0.9 (1.6)	1.5 (2.6)	0.22 (0.92)		170 (2720)	175 (2800)	0.75	0.07

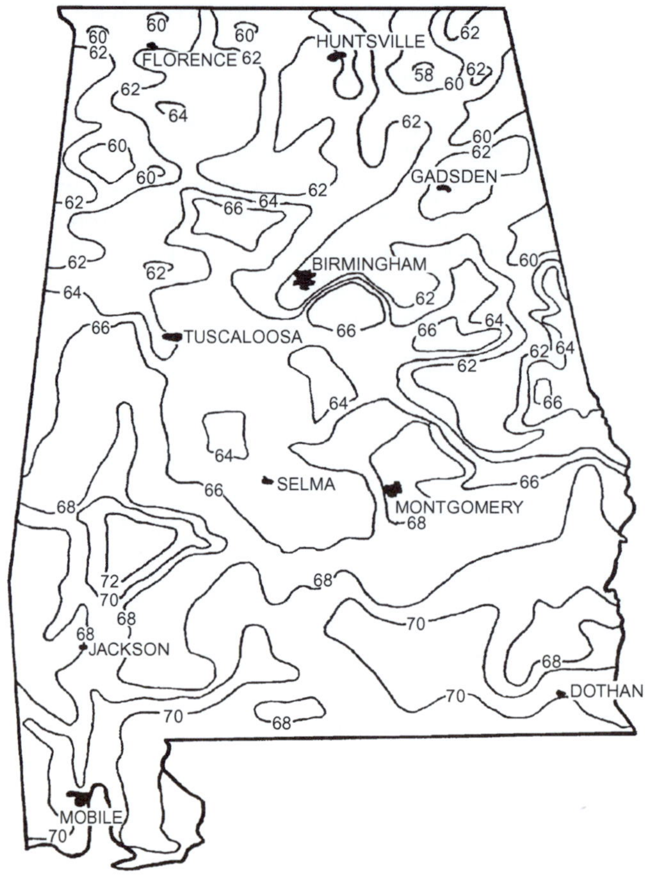

Figure 3.8 Groundwater Temperature (°F) Profiles for One State (Chandler 1987)

by noting the groundwater temperature variations in Figure 3.8. Note in particular the much warmer values a short distance southwest of Selma, Alabama, compared to the much cooler temperature just a few miles northwest. The spreadsheet tool *Ground Temp&Resist.xlsm*, available with this book at www.ashrae.org/GSHP, can be used to estimate the temperature change in horizontal headers located in shallow ground.

One additional alternative method of obtaining thermal property information is to search for databases that contain results of previous tests. An example is a utility that provided thermal property test funding as an incentive and also made all test results available to the public (TVA 2002).

3.6 GCHP SITE EVALUATION: THERMAL PROPERTY TESTS

For larger GCHP systems, an accurate knowledge of soil and rock thermal properties is critical to optimum ground heat exchanger design. Properties can be more accurately determined at each site by following ASHRAE (2011) recommendations for using a test apparatus similar to the one shown in Figure 3.10. A ground heat exchanger must be installed to the approximated bore depth for the site. This depth can be estimated by performing a preliminary calculation based on the required building cooling and heating loads, available ground area, and estimated thermal properties of expected formations at

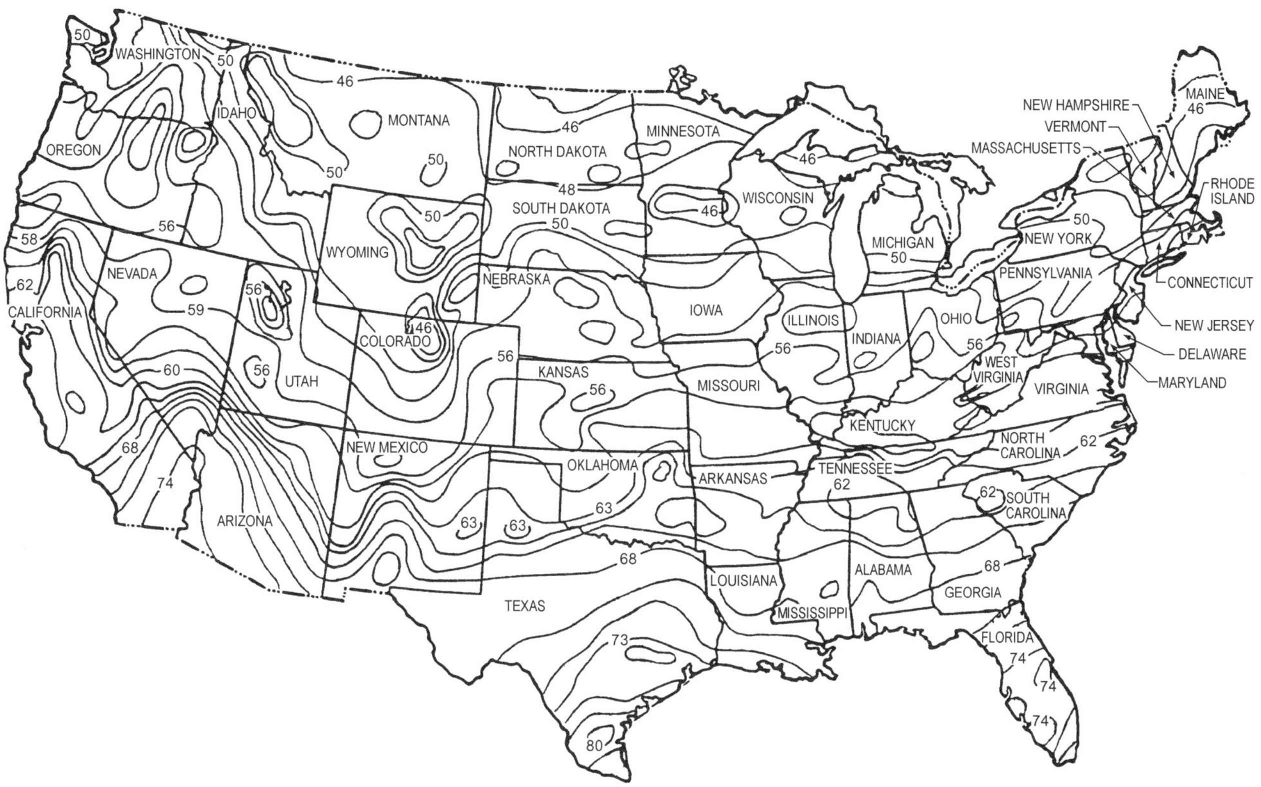

Figure 3.9 Approximate Groundwater Temperatures (°F) in the USA (Collins 1925)

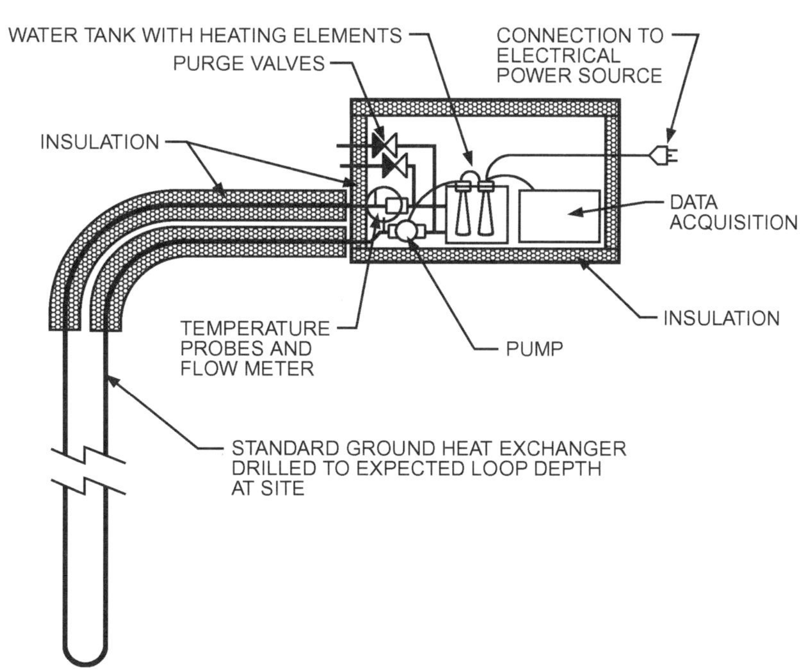

Figure 3.10 Formation Thermal Properties Test Apparatus (ASHRAE 2011)

the site. Soil and rock types can often be found in state and county water well logs. It is also prudent to consult local ground-loop contractors concerning the range of optimal drilling depths.

It is highly recommended that thermal property tests be conducted by an independent third-party individual rather than a drilling contractor or engineer of record. This maintains a degree of separation that ensures the contractor does not bias the results while also protecting both the drilling contractor and engineer of record should disputes arise in the future. A drilling log, as discussed in Chapter 7, should also be requested to reduce the uncertainty of drilling conditions for contractors bidding for ground-loop installations.

The following specifications for conducting thermal property tests adhere to the recommendations of ASHRAE RP-1118 (2001):

1. Thermal property test should be performed for 36 to 48 h.
2. The heat rate is to be 15 to 25 W/ft (50 to 80 W/m) of bore. These heat rates are the expected peak loads on the U-tubes for an actual heat pump system.
3. The standard deviation of input power is to be less than ±1.5% of the average value and peaks less than ±10% or resulting temperature variation less than ±0.5°F (0.3°C).
4. The accuracy of the temperature measurement and recording devices is to be ±0.5°F (0.3°C).
5. The accuracy of the power transducer and recording device is to be ±2% of the reading.
6. Flow rates are to be in the range to provide a differential loop temperature of 6°F to 12°F (3.5°C to 7°C). This is the temperature differential for an actual heat pump system.
7. A waiting period of five days is recommended for low-conductivity soils ($k <$ 1.0 Btu/h·ft·°F [1.7 W/m·K]) after the ground loop has been installed and grouted (or filled) before the thermal conductivity test is initiated. A delay of three days is recommended for higher-conductivity formations ($k >$ 1.0 Btu/h·ft·°F [1.7 W/m·K]).
8. The initial ground temperature measurement is to be made at the end of the waiting period by direct insertion of a probe inside a liquid-filled ground heat exchanger at three locations representing the average or by the measurement of temperature as the liquid exits the loop during the period immediately following start-up.
9. Data collection should be at least once every 10 minutes.
10. All aboveground piping is to be insulated with a minimum of 0.5 in. (1.25 cm) closed-cell insulation or equivalent. Test rigs are to be enclosed in a sealed cabinet that is insulated with a minimum of 1.0 in. (25 mm) fiberglass insulation or equivalent.
11. If retesting a bore is necessary, the loop temperature should be allowed to return to within 0.5°F (0.3°C) of the pretest initial ground temperature. This typically corresponds to a 10- to 12-day delay in mid- to high-conductivity formations and a 14-day delay in low-conductivity formations if a complete 48 hour test has been conducted. Waiting periods will be proportionally reduced if test terminations occurred after shorter periods.
12. Any of the public-domain software programs tested in conjunction with ASHRAE RP-1118, with the exception of the line-source method that only ignores the first 0.08 h of data, can be used to evaluate thermal conductivity. It is suggested that multiple programs be used to further enhance reported accuracy.

The line-source method of analysis is the simplest approach to determine the thermal conductivity of formations. Carslaw and Jaeger (1947) recast the equation for the temperature change for a constant line heat source in an infinite medium. Because the information gathered during the test are the bore length, the heat rate, and the average temperature of the loop $t_{avg} = (t_{in} + t_{out})/2$ over time, the unknown is the thermal conductivity. The inverse method takes the form of $\Delta t = \text{slope} \times \ln(\tau) + B$, where $k = q/(4\pi L_{bore} \times \text{slope})$; thus,

$$k = \frac{W}{4\pi L_{bore} \times \text{slope}} \tag{3.14}$$

where

Δt	=	difference in average loop temperatures at end of test and beginning of test
k	=	thermal conductivity of formation
L_{bore}	=	length of test bore
q	=	heat rate into test apparatus
τ	=	time from start of test
slope	=	slope of linear plot of average loop temperature versus natural log of time (τ)
W	=	power input into heating elements and pump

The limitations of using the line-source method (Ingersoll et al. 1954) are that the heat rate must be constant (specification 3 in the list above) and that the test length must be extended to minimize the error of assuming a line heat source rather than a pipe/cylinder of grout (specification 1 in the list above).

Figure 3.11 shows a plot of the average loop temperature from a 44 h thermal property test performed on a 300 ft (91 m) deep, 5.5 in. (140 mm) diameter bore with a nominal 1 in. (32 mm) HDPE U-tube. The loop temperature at the start of the test was 60.5°F (16°C) and the average power input to the bore from the heating elements and circulation pump was 6114 W.

Figure 3.12 shows the same information plotted versus the natural log of time with the first eight hours of test data removed. The result is a straight line with a slope of 3.5723 (°F). Also note the absence of any significant variation from the trend line and measured data, which provides an indication of quality results.

Equation 3.14 is applied to determine the formation thermal conductivity:

$$k = \frac{W}{4\pi L_{bore} \times \text{slope}} = \frac{3.412 \text{ Btu/Wh} \times 6114 W}{4\pi \times 300 \text{ ft} \times 3.5273°\text{F}} = 1.57 \text{ Btu/h·ft·°F} \qquad \text{(I-P)}$$

$$k = \frac{W}{4\pi L_{bore} \times \text{slope}} = \frac{6114 W}{4\pi \times 91.4 \text{ m} \times 1.960°\text{C}} = 2.72 \text{ W/m·K} \qquad \text{(SI)}$$

The remaining unknown thermal property of diffusivity ($\alpha = k/\rho c_p$) is estimated using values of density (ρ) and specific heat (c_p) from tables such as Tables 3.4 and 3.5 in conjunction with the value of thermal conductivity. Because specific heat and density are not typically measured in the field there will be a range of uncertainty. However, thermal conductivity has a much greater impact on heat exchanger design calculations. It can be demonstrated that uncertainty in diffusivity has a marginal impact on results. Computations can be conducted using the range of possible values for specific heat and density to demonstrate the impact upon heat exchanger lengths.

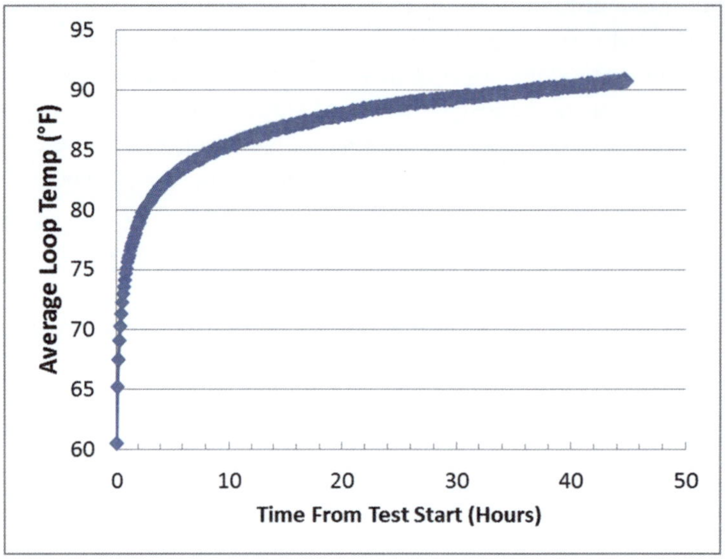

Figure 3.11 Average Loop Temperature Data for 300 ft (91 m) Test Bore

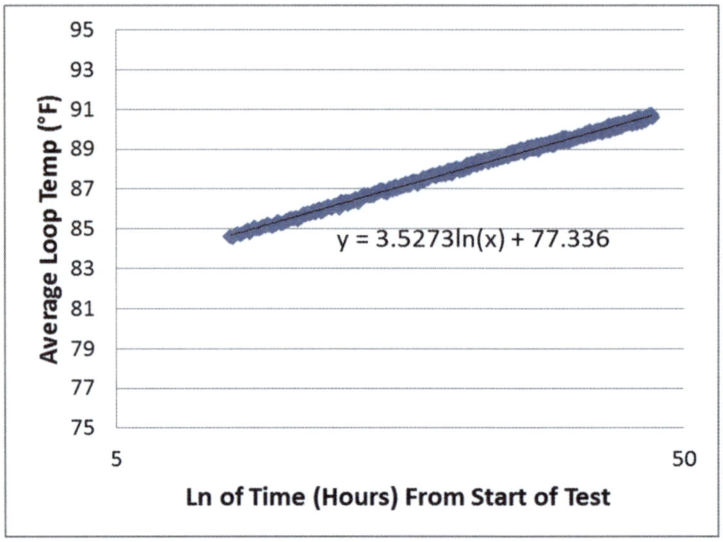

Figure 3.12 Average Loop Temperature Data vs Natural Log of Time—Hours 8 to 44

The values used should also treat the formation as a combination of soil or rock and moisture. Thus,

$$c_{p\text{-}Formation} = (1 - \%\text{Moisture}) \times c_{p\text{-}soil.rock} + \%\text{Moisture} \times c_{p\text{-}water} \quad (3.15)$$

$$\rho_{Formation} = (1 - \%\text{Moisture}) \times \rho_{soil.rock} + \%\text{Moisture} \times \rho_{water} \quad (3.16)$$

Specific heat values for dry soils and rocks vary little from 0.2 Btu/lb·°F (0.84 kJ/kg·°C). When Equations 3.15 and 3.16 are applied to the resulting product ($\rho \times c_p$), the impact of the higher specific heat of water is offset by the lower density of water compared to soils and rocks.

EXAMPLE 3.4—
ESTIMATION OF THERMAL DIFFUSIVITY

Estimate the range of thermal diffusivities for a limestone formation whose thermal conductivity is determined from information given in Figures 3.11 and 3.12 assuming a moisture content of 10%.

Solution

Table 3.5 indicates the midrange specific heat of dry limestone is 0.22 Btu/lb·°F (0.92 kJ/kg·K) and the dry density range is from 150 to 175 lb/ft^3 (2480 to 2800 kg/m^3). For 10% moisture, the specific heat and densities for the lower and upper ranges are

$$c_p = (1 - 0.1) \times 0.22\ \text{Btu/lb·°F} + 0.1 \times 1.0\ \text{Btu/lb·°F} = 0.298\ \text{Btu/lb·°F} \quad \text{(I-P)}$$

$$\rho_{low} = (1 - 0.1) \times 150\ \text{lb/ft}^3 + 0.1 \times 62.3\ \text{lb/ft}^3 = 141\ \text{lb/ft}^3 \quad \text{(I-P)}$$

$$\rho_{high} = (1 - 0.1) \times 175\ \text{lb/ft}^3 + 0.1 \times 62.3\ \text{lb/ft}^3 = 164\ \text{lb/ft}^3 \quad \text{(I-P)}$$

$$c_p = (1 - 0.1) \times 0.92\ \text{kJ/kg·K} + 0.1 \times 4.2\ \text{kJ/kg·K} = 1.25\ \text{kJ/kg·K} \quad \text{(SI)}$$

$$\rho_{low} = (1 - 0.1) \times 2400\ \text{kg/m}^3 + 0.1 \times 998\ \text{kg/m}^3 = 2260\ \text{kg/m}^3 \quad \text{(SI)}$$

$$\rho_{high} = (1 - 0.1) \times 2800\ \text{kg/m}^3 + 0.1 \times 998\ \text{kg/m}^3 = 2620\ \text{kg/m}^3 \quad \text{(SI)}$$

The resulting thermal diffusivities are as follows (recall that in SI units, W = J/s):

$$\alpha_{high} = \frac{k}{\rho c_p} = \frac{1.57\ \text{Btu/h·ft·°F} \times 24\ \text{h/day}}{0.298\ \text{Btu/lb·°F} \times 141\ \text{lb/ft}^3} = 0.90\ \text{ft}^2/\text{day} \quad \text{(I-P)}$$

$$\alpha_{low} = \frac{k}{\rho c_p} = \frac{1.57\ \text{Btu/h·ft·°F} \times 24\ \text{h/day}}{0.298\ \text{Btu/lb·°F} \times 164\ \text{lb/ft}^3} = 0.77\ \text{ft}^2/\text{day} \quad \text{(I-P)}$$

$$\alpha_{high} = \frac{k}{\rho c_p} = \frac{2.72\ \text{J/s·m·K} \times 3600\ \text{s/h} \times 24\ \text{h/day}}{1.25\ \text{kJ/kg·K} \times 1000\ \text{J/kJ} \times 2260\ \text{kg/m}^3} = 0.083\ \text{m}^2/\text{day} \quad \text{(SI)}$$

$$\alpha_{low} = \frac{k}{\rho c_p} = \frac{2.72\ \text{J/s·m·K} \times 3600\ \text{s/h} \times 24\ \text{h/day}}{1.25\ \text{kJ/kg·K} \times 1000\ \text{J/kJ} \times 2620\ \text{kg/m}^3} = 0.072\ \text{m}^2/\text{day} \quad \text{(SI)}$$

3.7 LONG-TERM GROUND TEMPERATURE CHANGE

A final temperature to consider is defined as the temperature penalty (t_p) resulting from imbalances between the amount of heat added to the ground in cooling and removed from the ground in heating. The fundamental equations used to develop Equations 3.5 and 3.6 assume a single cylinder heat source in an infinite medium. Thus, adjustments must be made to account for thermal interference from adjacent bores. The designer is faced with selecting a separation distance that is reasonable in order to minimize required land area without causing large increases in the required bore length. The suggested approach is to assume some reasonable temperature penalty value (±1°F to 5°F over a 10- or 20-year period), apply Equations 3.5 and 3.6, calculate the actual penalty based on the bore lengths as discussed below, modify the separation distance and/or adjust the bore lengths if desired, and recalculate the bore lengths based on the calculated temperature penalty.

The line-source heat solution used is acceptable for determining temperature penalty since the error between a line and a cylindrical heat source is small when the length of time is extended (Ingersoll et al. 1954). Only the annual net heat transfer to the ground (q_a) is necessary to calculate the temperature change over an extended period of time. A vertical bore surrounded by other bores is not able to diffuse the heat beyond one-half the bore separation distance. Therefore, the cylinder of earth surrounding the vertical bore will rise in temperature if the annual heat rejected is greater than the heat absorbed. This temperature will decline if the heat absorbed is greater.

Groundwater movement can have a large impact in mitigating the long-term temperature rise in that it can replenish moisture that has been evaporated as ground temperature rises. The evaporative cooling effect is significant compared to the thermal capacity of the ground, although the amount of impact has not been thoroughly studied. So the design engineer is left with establishing a range of design lengths, one based on minimal groundwater movement as in very tight clay soils with poor percolation rates and a second based on higher rates characteristic of porous formations.

The worst-case scenario assumes the earth is a solid and conduction is the only mode of heat transfer. The line heat source solution (discussed later) is used to develop a temperature profile at points of increasing radii from a single constant heat source in an infinite medium. If the line source is surrounded by other heat sources (as is the case in a vertical-loop field), heat cannot be diffused beyond one-half the separation distance (S_{bore}) to adjacent heat sources of equal magnitude. The heat must be stored in the earth surrounding the line heat source (or borehole).

The amount of heat that is stored in the surrounding soil can be estimated by using the temperature profile of the single heat source. The volume of incremental round cylinders [$= \rho L_{bore}(r_o^2 - r_i^2)$] of earth at increasing radii beyond $S_{bore}/2$ is multiplied by the thermal capacity of the earth (ρc_p) and the single source temperature increase above the undisturbed earth temperature (t_g) at the midpoint of the cylinder [$(r_o + r_i)/2$].

$$Q_{stored} = \sum_{r = S_{bore}/2}^{\infty} \left[\rho c_p \pi L_{bore}(r_o^2 - r_i^2) \times \left(t@\frac{r_o + r_i}{2} - t_g\right)\right] \quad (3.17)$$

The number of cylinders required to provide a reasonable substitute for ($r = \infty$) is dependent on the moisture content and porosity characteristics of the soil surrounding the line source of heat. Porous soils with high moisture content may require the cylinders of influence to a radii equal to a single bore separation (S_{bore}), while low-porosity soil may require computation for a radius more than 5 times S_{bore}.

For square-grid borehole arrangements, temperature change (t_{p1}) is computed using a square cylinder of earth surrounding the bore:

$$t_{p1} = \frac{Q_{stored}}{\rho c_p S_{bore}^2 L_{bore}} \quad \text{(Square grid)} \quad (3.18a)$$

For a staggered-grid arrangement, the volume surrounding the bore is reduced to nearly the volume of the round cylinder of earth:

$$t_{p1} = \frac{4Q_{stored}}{\pi\rho c_p S_{bore}^2 L_{bore}} \quad \text{(Staggered grid)} \tag{3.18b}$$

Consider a grid in which vertical bores are separated by 20 ft (6 m). A square cylinder with 20 ft (6 m) sides must store all the heat normally diffused beyond a distance of 10 ft (3 m) from the bore. The impact of a monthly heat pulse would be small at this distance. However, an annual imbalance could result in a change of several degrees. To compute this amount, the line heat source solution is used to find the temperature change 12.5 ft (3.8 m) from a single bore after 10 years of net heat rejection. The amount of heat stored in a hollow cylinder with an outside radius of 15 ft (4.6 m) and an inside radius of 10 ft (3.0 m) is found by multiplying the temperature change at 12.5 ft (3.8 m) by the heat storage capacity (ρc_p) and the cylinder volume. This process is repeated for hollow cylinders of increasing radii until the temperature rise at distance from the ground-loop perimeter is negligible (< 0.5°F [0.3°C]). At this distance any heat storage effect is normally offset with the evaporative cooling and moisture recharge mechanisms shown in Figure 3.3. The heat-stored term for Equation 3.18a is found by summing the totals in all the cylinders.

Application of the line-source solution is similar to that of the cylindrical heat source solution (Figure 3.6). A dimensionless term is used to relate soil thermal diffusivity ($\alpha = k/\rho c_p$), time of operation (τ), and distance from the heat source (r). Ingersoll et al. (1954) use the term

$$X = \frac{r}{2\sqrt{\alpha\tau}} \tag{3.19}$$

The difference between the undisturbed ground temperature and the temperature at a distance r from the line heat source is

$$\Delta t_r = \frac{q_a \times I(X)}{2\pi k_g L_{bore}} \tag{3.20}$$

The values for $I(X)$ are determined from Figure 3.13 or with the equation shown in Figure 3.13.

The field temperature penalty is prorated based upon the number of bores adjacent to only one, two, or three other bores. For example, in Example 3.2 the five bore wide (N_{Wide}) by five bore long (N_{Long}) vertical grid with 200 ft (61 m) bores would have 9 internal bores (N_{Int}) adjacent to 4 other bores, 12 bores on the perimeter surrounded by 3 adjacent bores (N_{Side}), and 4 corner bores (N_{Corner}) with 2 adjacent bores for a total number of 25 bores (N_{Bores}). A single-row 25-bore field will have two end-row bores (N_{End}) with 1 adjacent bore and the remainder of the bores in the row (N_{Midrow}) with 2 adjacent bores.

The temperature penalty must also be corrected for the heat flow from the bottom of the bore field. The bore field with 20 ft (6 m) bore separation (S_{bore}) would have four vertical planes each 80 ft (24 m) in width by 200 ft (61 m) in depth (L_{Bore}) for a total vertical area of 64,000 ft^2 (5950 m^2). The area comprised by the bottom of the loop field is 80 × 80 ft (24 × 24 m) for a horizontal area of 6400 ft^2 (595 m^2).

Equation 3.21 is the corrected temperature penalty value.

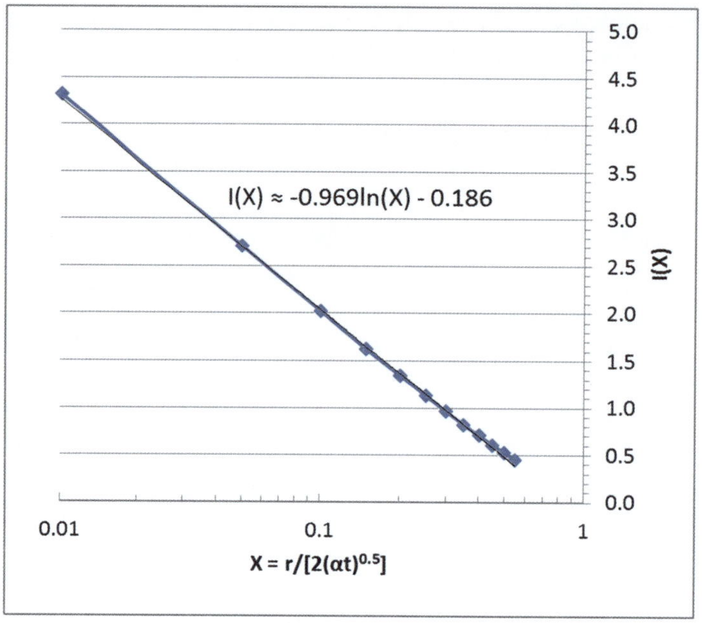

Figure 3.13 Chart and Equation for Determining $I(X)$ (Ingersoll et al. 1954)

$$t_p = \frac{N_{Int} + 0.75N_{Side} + 0.5N_{Corner} + 0.5N_{Midrow} + 0.25N_{End}}{\text{Total number of bores} \times C_{fHoriz}} \times t_{p1} \tag{3.21}$$

where t_{p1} is the penalty for a bore surrounded on all four sides by other bores and

$$C_{fHoriz} = \frac{[L_{bore} \times 2 \times (W_{Field} + L_{Field})] + [W_{Field} \times L_{Field}]}{L_{Bore} \times 2 \times (W_{Field} + L_{Field})}$$

$$W_{Field} = (N_{Wide} - 1) \times S_{bore} \quad \text{and} \quad L_{Field} = (N_{Long} - 1) \times S_{bore}$$

Caution is advised because excessive moisture migration will drive down the thermal conductivity of granular soils and porous formations (Kusuda and Achenbach 1965; Salomone and Marlowe 1989). Placing vertical bores in close proximity increases the possibility of reducing moisture content below a critical point within a single season before the regenerative effects of heating-mode operation can occur. Until more field data suggests otherwise, the minimum recommended vertical bore separation distance is 20 ft (6 m).

EXAMPLE 3.5— TEMPERATURE PENALTY CALCULATION

Compute the 10-year temperature penalty for the system described in Example 3.2. Assume the ground temperature change at a distance of 30 ft (9 m) from the bore field perimeter is negligible. Recalculate the required cooling length if the temperature penalty is different from the value assumed in Example 3.2.

Solution

The assumption requires that t_{p1} be computed to a distance of 30 ft (9 m) from the center of a single U-tube bore. This can be accomplished using three radii of earth beginning at 10 ft (3 m), each with a thickness of 5 ft (1.5 m) as shown in Figure 3.14. The first step is to calculate the amount of heat diffused beyond 10 ft (3 m). This is the heat that would be stored in the inner 10 ft (3 m) radius cylinder, thereby causing a change in temperature. The inner radius represents the cylinder in which heat must be stored, while the outer circles are hollow cylinders in which heat would normally be stored if adjacent U-bends did not block the diffusion of heat.

The amount of heat stored in a hollow cylinder with an outside radius of 15 ft (4.5 m) and an inside radius of 10 ft (3 m) can be computed by multiplying the heat storage capacity (c_p × volume) by the average change in temperature (which can be approximated by the temperature change at 12.5 ft (3.8 m). This can be repeated for hollow cylinders until a distance of 30 ft (9 m) is reached. The total amount of heat in all cylinders is summed. Equation 3.11 is applied to find the temperature rise for a single U-tube that is surrounded on all four sides by U-tubes 20 ft (6 m) away. Equation 3.17 is then applied to prorate the average penalty for the entire grid.

Equations 3.19 and 3.20 and Figure 3.13 are used to find the change in temperature in the ground around a single U-tube with no adjacent bores. The annual average heat rate to the ground (q_a) and the 20 years plus one month (7330 day) time frame is used. The dimensionless factor needed to find the temperature change at 12.5 ft (3.8 m) is

$$X = \frac{r}{2\sqrt{\alpha\tau}} = \frac{12.5 \text{ ft}}{2\sqrt{1.0 \text{ ft/day} \times 7330 \text{ days}}} = 0.073$$

From the equation in Figure 3.13, $I(X) = -0.969 \times \ln(0.073) - 0.186 = 2.35$, and

$$\Delta t_{12.5} = \frac{42{,}700 \times 2.35}{2\pi \times 1.4 \text{ Btu/h·ft·°F} \times 7025 \text{ ft}} = 1.62\text{°F}$$

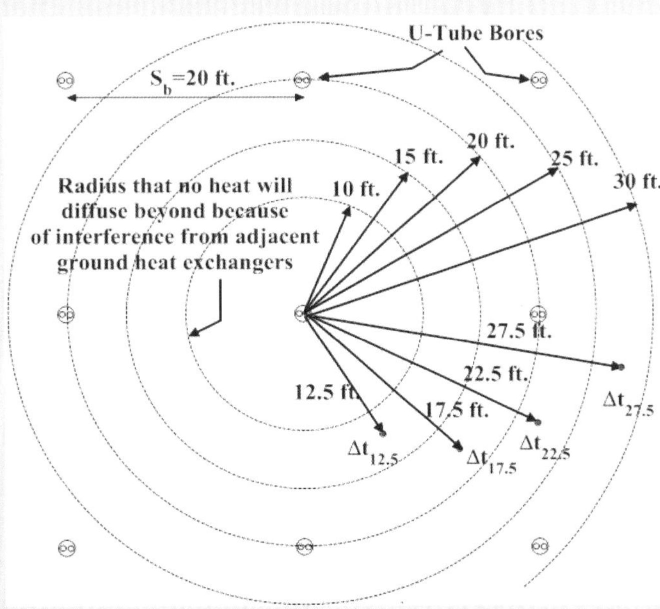

Figure 3.14 Representative Earth Cylinders for Heat Storage

Repeating for $r = 17.5$ ft: $X_{17.5} = 0.102$, $I(X)_{17.5} = 2.02$, $\Delta t_{17.5} = 1.40°F$
Repeating for $r = 22.5$ ft: $X_{22.5} = 0.131$, $I(X)_{22.5} = 1.78$, $\Delta t_{22.5} = 1.23°F$
Repeating for $r = 27.5$ ft: $X_{27.5} = 0.161$, $I(X)_{27.5} = 1.59$, $\Delta t_{27.5} = 1.10°F$

Equation 3.17 is applied to determine total heat stored in the three hollow cylinders.

$$Q_{stored} = Q_{15-10} + Q_{20-15} + Q_{25-20} + Q_{30-25}$$

Recall that $\rho c_p = k/\alpha$. Therefore, $\rho c_p = 1.4$ Btu/h·ft·°F ÷ 1.0 ft²/day × 24 h/day = 33.6 Btu/ ft³·°F.

$$Q_{15-10} = (33.6 \text{ Btu/ft}^3\cdot°\text{F})\pi \times 7025 \text{ ft } (15 \text{ ft}^2 - 10 \text{ ft}^2) \times 1.62°\text{F} = 150.5 \times 10^6 \text{ Btu}$$

$$Q_{20-15} = (33.6 \text{ Btu/ft}^3\cdot°\text{F})\pi \times 7025 \text{ ft } (20 \text{ ft}^2 - 15 \text{ ft}^2) \times 1.40°\text{F} = 181.5 \times 10^6 \text{ Btu}$$

$$Q_{25-20} = (33.6 \text{ Btu/ft}^3\cdot°\text{F})\pi \times 7025 \text{ ft } (25 \text{ ft}^2 - 20 \text{ ft}^2) \times 1.23°\text{F} = 205.3 \times 10^6 \text{ Btu}$$

$$Q_{30-25} = (33.6 \text{ Btu/ft}^3\cdot°\text{F})\pi \times 7025 \text{ ft } (30 \text{ ft}^2 - 25 \text{ ft}^2) \times 1.10°\text{F} = 223.2 \times 10^6 \text{ Btu}$$

$$Q_{stored} = 150.5 \times 10^6 + 181.5 \times 10^6 + 205.3 \times 10^6 + 223.2 \times 10^6 = 760.5 \times 10^6 \text{ Btu}$$

Equation 3.18a is now applied to find the increase in temperature in a 20 ft square cylinder of ground if 760.5×10^6 Btu were rejected over a period of 10 years. This represents the temperature change if the U-tube was surrounded on all four sides by adjacent U-tubes separated by 20 ft.

$$t_{p1} = \frac{-760.5 \times 10^6 \text{ Btu}}{33.6 \text{ Btu/ft}^3\cdot°\text{F} \times 20 \text{ ft}^2 \times 7025 \text{ ft}} = 8.04°\text{F}$$

Equation 3.21 and the correction for heat transfer from the bottom of the loop field are applied to the 5 × 5 vertical grid to find the corrected temperature penalty.

$$W_{Field} = (N_{Wide} - 1) \times S_{bore} = (5 - 1) \times 20 = 80 \text{ ft}$$

and

$$L_{Field} = (N_{Long} - 1) \times S_{bore} = (5 - 1) \times 20 = 80 \text{ ft}$$

$$C_{fHoriz} = \frac{[L_{Bore} \times 2 \times (W_{Field} + L_{Field})] + [W_{Field} \times L_{Field}]}{L_{Bore} \times 2 \times (W_{Field} + L_{Field})}$$

$$= \frac{2 \times 281 \times (80 + 80) + 80 \times 80}{2 \times 281 \times (80 + 80)} = 1.07$$

$$t_p = \frac{9 + 0.75 \times 12 + 0.5 \times 4 + 0.5 \times 0 + 0.25 \times 0}{25 \text{ bores} \times 1.07} \times t_{p1} = 0.75 \times 8.04°\text{F} = 6.0°\text{F}$$

The value of –6.0°F replaces the originally assumed value of –2.0°F in Equation 3.2.

$$L_c = \frac{(-42{,}700 \times 0.271) - 372{,}600 \times (0.18 + 0.28 \times 0.264 + 1.04 \times 0.143)}{65°\text{F} - \frac{85°\text{F} + 95°\text{F}}{2} + 6.0°\text{F}} = 8460 \text{ ft}$$

$$= 8460 \text{ ft} \div 25 \text{ bores} = 338 \text{ ft/bore}$$

However, now that the bore length has increased, a second iteration can be performed recognizing the bore field now has 20% greater thermal storage capacity because of the 20% increase in length from 281 ft/bore to 338 ft/bore. Thus, the temperature penalty would be reduced by 20% to 4.7°F. In this case the bore length in cooling is

$$L_c = \frac{(-42{,}700 \times 0.271) - 372{,}600 \times (0.18 + 0.28 \times 0.264 + 1.04 \times 0.143)}{65°\text{F} - \dfrac{85°\text{F} + 95°\text{F}}{2} + 4.7°\text{F}} = 7960 \text{ ft}$$

$$= 7960 \text{ ft} \div 25 \text{ bores} = 318 \text{ ft/bore}$$

Additional iterations would result in a bore length of 320 ft and a temperature rise of 5.0°F.

This length is the required value assuming minimal groundwater movement and vertical percolation of water through the ground coil field. If high rates of moisture recharge occur, the temperature penalty would be substantially reduced due to the mechanisms shown in Figure 3.3. Although no concerted efforts have been published, residential systems in many cases provide ground-loop temperatures near the undisturbed ground temperature when initially starting up in the heating mode after being off for several weeks. This, along with the information summarized in Figure 3.2, indicates the magnitude of temperature penalty will be overstated if calculations do not consider the impact of ground moisture phase change (evaporation, freezing, condensation) and moisture migration. It should also be noted that high-velocity groundwater movement across the vertical ground heat exchangers has minimal impact on performance. The benefit of groundwater movement is the enhancement of the thermal properties of the soil itself.

Even when groundwater movement is prevalent, it is not prudent to assume the temperature penalty is zero. Extended periods of drought mitigate the impacts of moisture for one or possibly two years of operation. In areas where formations have multiple layers that can produce groundwater flow in wells, the temperature penalty will likely be moderate. In these cases it is suggested that an appropriate temperature penalty would result if a value of one year (365 days) were substituted for the 20-year assumption used in Example 3.5. The resulting temperature penalty and required bore length for cooling are as follows:

$$t_p = 1.3°\text{F } (0.7°\text{C})$$

$$L_c = 6820 \text{ ft } (273 \text{ ft/bore}) \ (2080 \text{ m } [83 \text{ m/bore}])$$

for high rates of ground moisture recharge

The process for calculating the required length for the nondominant mode, which in Example 3.5 is for heating, is somewhat simplified. Because the annual heat balance favors cooling mode heat rejection and ground temperature tends to increase with system life, the long-term required heating length will be less than the heating length in year one. **Thus, the design conditions for the nondominant mode should be determined with the temperature penalty (t_p) and the net annual heat transfer to the ground (q_a) set to zero.**

Table 3.6 is provided as an alternatives to calculating long-term temperature change using Equations 3.17 through 3.21. The calculation shown in Example 3.5 assumed Δt_g = 0°F (0°C) at 30 ft (9 m) for a 20 ft (6 m) bore separation. Table 3.6 indicates a 4.4°F (2.4°C) rise in temperature for a system with 300 ft (90 m) bores, $EFLH_c$ = 1000, $EFLH_h$ = 500, and a cooling load 1.33 times as large as the heating load. A correction factor of 1.05 is applied for 300 ft (90 m) bores arranged in a 5 × 5 grid. Thus, the estimated temperature rise is 4.6°F (2.6°C). In Example 3.5, the values are near those listed above and there is reasonable agreement with the results using the extended calculation.

Table 3.6 Twenty-Year Temperature Change for 10 x 10* Vertical Bore Ground Heat Exchanger for Moisture Recharge Estimates, EFLH Ratio, and Building Loads

20 Years			Low Water Recharge		Mild Water Recharge		High Water Recharge	
$EFLH_c$,	$EFLH_h$,	Bore Sep.,	Δt_g = 0°F at 40 ft		Δt_g = 0°F at 30 ft		Δt_g = 0°F at 20 ft	
h/yr	h/yr	ft	200 ft/ton	300 ft/ton	200 ft/ton	300 ft/ton	200 ft/ton	300 ft/ton
250	1250	20	–8.4	–5.9	–5.1	–3.7	–2.3	–1.6
$q_{lc} = 0.5 \times q_{lh}$		25	–4.8	–3.5	–4.1	–2.1	–1.1	–0.8
500	1000	20	–3.1	–2.2	–1.9	–1.3	–0.8	–0.6
$q_{lc} = 0.75 \times q_{lh}$		25	–1.8	–1.3	–1.1	–0.8	–0.4	–0.3
750	750	20	3.8	2.7	2.4	1.7	1.0	0.7
$q_{lc} = q_{lh}$		25	2.2	1.6	1.3	0.9	0.5	0.4
1000	500	20	10.1	7.2	6.2	4.4	2.7	1.9
$q_{lc} = 1.33 \times q_{lh}$		25	5.8	4.2	3.5	2.5	1.3	0.9
		30	3.5	2.6	2	1.5	0.6	0.4
1250	250	20	16.9	1.1	10.0	6.7	4.4	2.9
$q_{lc} = 2 \times q_{lh}$		25	9.5	6.4	5.7	3.8	2.1	1.4
		30	6	4	3.3	2.2	1	0.7

**Correction factors for other grids:*
200 ft bores: C_f (5x5) = 0.95, C_f (2x10) = 0.85, C_f (1x10) = 0.6
300 ft bores: C_f (5x5) = 1.05, C_f (2x10) = 1.0, C_f (1x10) = 1.0
C_f for grids >10x10 will be less than 1.0 due to relative increase in downward heat dissipation.

20 Years			Low Water Recharge		Mild Water Recharge		High Water Recharge	
$EFLH_c$,	$EFLH_h$,	Bore Sep.,	Δt_g = 0°C at 12 m		Δt_g = 0°C at 9 m		Δt_g = 0°C at 6 m	
h/yr	h/yr	m	15 m/kW	25 m/kW	15 m/kW	25 m/kW	15 m/kW	25 m/kW
250	1250	6	–5.0	–3.4	–3.0	–2.1	–1.4	–0.9
$q_{lc} = 0.5 \times q_{lh}$		7.5	–2.9	–2.0	–2.6	–1.3	–0.7	–0.5
500	1000	6	–1.9	–1.3	–1.1	–0.8	–0.5	–0.3
$q_{lc} = 0.75 \times q_{lh}$		7.5	–1.1	–0.8	–0.7	–0.5	–0.2	–0.2
750	750	6	2.3	1.6	1.4	1.0	0.6	0.4
$q_{lc} = q_{lh}$		7.5	1.3	0.9	0.8	0.5	0.3	0.2
1000	500	6	6.0	4.2	3.7	2.6	1.6	1.1
$q_{lc} = 1.33 \times q_{lh}$		7.5	3.5	2.4	2.1	1.5	0.8	0.5
		9	2.1	1.5	1.2	0.9	0.4	0.2
1250	250	6	11.8	1.6	6.0	3.9	2.7	1.7
$q_{lc} = 2 \times q_{lh}$		7.5	5.7	3.8	3.5	2.2	1.3	0.8
		9	3.6	2.4	2.0	1.3	0.6	0.4

**Correction factors for other grids:*
60 m bores: C_f (5x5) = 0.95, C_f (2x10) = 0.85, C_f (1x10) = 0.6
60 m bores: C_f (5x5) = 1.05, C_f (2x10) = 1.0, C_f (1x10) = 1.0
C_f for grids >10x10 will be less than 1.0 due to relative increase in downward heat dissipation.

3.8 COMMENTS ON THE DESIGN OF VERTICAL GROUND HEAT EXCHANGERS

Several cautions should be applied while using this method of ground coil design. It is important to *maintain an adequate separation distance*. If the computations performed in Sections 3.4 and 3.7 are repeated for a smaller separation distance, such as 10 or 15 ft (3 or 4.5 m), greater bore length requirements will result even if the thermal properties of the surrounding soils are not affected. In some cases, the reduced amount of thermal capacity available in bore fields with small separation distances will more than likely be insufficient to prevent unwanted reductions in thermal conductivities. The probability of drilling through an adjacent bore (cross drilling) will increase with smaller bore separation distance and greater depths. (This has occurred.) A small difference in the angle at which the drill rig is set up or a small deflection in the drilling angle caused by a hard obstruction could easily cause the drill bit to be several feet away from the desired point at the bottom of a deep bore. In this situation two bores will be lost and the HDPE pipe is unlikely to release the drill stem around which it is wrapped.

Oversizing of heating and cooling systems by engineers is a common practice to offset uncertainties in building construction and equipment installation quality. The incremental cost of oversizing a conventional system is small (a 4 ton unit is not double the cost of a 2 ton unit). However, ground coil costs are almost nearly directly proportional to equipment size for a larger building. Thus, *oversizing escalates GCHP costs much more than those of conventional systems*.

Some designers have used rules of thumb for coil sizing that produce loop lengths substantially shorter than those recommended using the procedures described in the previous sections. It is also a false conventional wisdom that higher-rated-efficiency equipment will require shorter ground lengths. Multicapacity and variable-speed heat pumps typically have lower efficiencies at peak conditions compared to equivalent constant-speed units (see Tables 2.3a and 2.3b). This impact is typically small, and note that Equations 3.2, 3.3, 3.4, 3.5, and 3.6 used to determine heat exchanger size include the system efficiency. No matter how high the rated efficiency of a heat pump, smaller ground heat exchangers will result in higher loop temperatures and a corresponding decrease in efficiency in cooling. In heating, the result will be lower loop temperatures and a corresponding decrease in heating capacity, which may result in auxiliary heat activation.

3.9 REFERENCES

Allan, M.L. 1996. Improvement of cementitious grout thermal conductivity for GHP applications. Preliminary Report, Brookhaven National Laboratory, U.S. Department of Energy Contract DE-AC02-76CH00016, June.

ASHRAE. 2001. *Investigation of methods for determining soil formation thermal characteristics from short term field tests*. RP-1118 Final Report, ASHRAE, Atlanta.

ASHRAE. 2011. *ASHRAE Handbook—HVAC Applications*, Chapter 34, Geothermal Energy. Atlanta: ASHRAE.

Carlson, S. 2001. Development of equivalent full load heating and cooling hours for GCHPs applied to various building types and locations. ASHRAE RP-1120, Final Report. Atlanta: ASHRAE.

Carmichael, R.S. 1989. *Physical Properties of Rocks and Minerals*. Boca Raton, FL: CRC Press.

Carslaw, H.S., and J.C. Jaeger. 1947. *Conduction of Heat in Solids.* Oxford: Claremore Press.

Chandler, R.V. 1987. *Alabama streams, lakes, springs, and ground waters for use in heating and cooling.* Geological Survey of Alabama, Bulletin 129, Tuscaloosa.

Claesson, J., and P. Eskilson. 1987. *Thermal Analysis of Heat Extraction Boreholes.* Lund, Sweden: Lund Institute of Technology.

Collins, W.D. 1925. *Temperature of water available for industrial use in the United States*. U.S. Geological Survey Paper 520-F, Washington, DC.

EIS. 2009. Ground source heat pump system designer. Northport, AL: Energy Information Services. www.geokiss.com/software/Ver50Inst5-12.pdf

Farouki, O.T. 1982. Evaluation of methods for calculating soil thermal conductivity. U.S. Army Cold Regions Research and Engineering Laboratory Report 82-8, Hanover, NH.

GPI. 2014. GeoPro Grouts. Elkton, SD: GeoPro, Inc. www.geoproinc.com/products.html

Hellström, G. 1991. Ground heat storage—Thermal analyses of duct storage systems. PhD thesis, University of Lund, Lund, Sweden.

Ingersoll, L.R., O.J. Zobel, and A.C. Ingersoll. 1954. *Heat Conduction: With Engineering and Geological Applications,* 2nd ed. New York: McGraw Hill.

Kavanaugh, S.P. 1984. Simulation and experimental verification of vertical ground-coupled heat pump systems. PhD Dissertation, Oklahoma State University, Stillwater.

Kavanaugh, S.P. 1992. Simulation of ground-coupled heat pumps with an analytical solution. *Proceedings of the ASME International Solar Energy Conference.*

Kavanaugh, S.P. 2010. Determining thermal resistance: Ground heat exchangers. *ASHRAE Journal* 52(8).

Kavanaugh, S.P., and J.S. Kavanaugh. 2012. Long-term commercial GSHP performance, part 3: Loop temperatures. *ASHRAE Journal* 54(9).

Kusuda, T., and P.R. Achenbach. 1965. Earth temperatures and thermal diffusivity at selected stations in the U.S. *ASHRAE Transactions* 71(1).

Philippe, M., M.A. Bernier, and D. Marchio. 2010. Vertical geothermal borefields. *ASHRAE Journal* 52(7).

Remund, C. 1999. Borehole thermal resistance: Laboratory and field studies. *ASHRAE Transactions* 105(1).

Robertson, E.C. 1988. *Thermal properties of rocks*. U.S. Geological Survey Open File Report 88-411, Washington DC.

Salomone, L.A., and J.I. Marlowe. 1989. *Soil and rock classification for the design of ground-coupled heat pumps*. EPRI CU-6600, Electric Power Research Institute, Palo Alto, CA.

Toulokian, Y.S., W.R. Judd, and R.F. Roy. 1981. *Physical Properties of Rocks and Minerals.* New York: McGraw-Hill/Cintas.

TVA. 2002. Mapping the results of thermal conductivity testing performed in the Tennessee Valley. Project Closure Report, Tennessee Valley Authority, Knoxville, TN. www.tva.com/commercial/TCStudy/index.htm

4 Applied Ground-Coupled Heat Pump System Design

4.1 SYSTEM DESIGN OVERVIEW

Quality GSHPs are designed as a system and tend to be simple. Merely attaching a ground heat exchanger to a conventional HVAC system will typically result in poor economic value because unnecessary components add costs, energy consumption, and demand. This can be demonstrated by viewing the unitary water-to-air heat pump system in Figure 2.16 that has a full-load system energy efficiency ratio (EER) of 14.6 Btu/Wh (COP_c = 4.3) as shown in Table 2.8. This compares with the chilled-water variable-air-volume (VAV) GSHP system shown in Figure 2.17 that has a full-load system EER of 7.8 Btu/Wh (COP_c = 2.3) as shown in Table 2.9. It is also recommended that comparisons at part load be conducted using the program used to generate Table 2.8 and 2.9, *HVAC SystemEff.xlsx*, because heat pump systems and chilled-water systems are rated by two entirely different methods. (*HVAC SystemEff.xlsx* is available with this book at www.ashrae.org/GSHP.)

Quality design engineers assume responsibility for the entire system and have a vested interest in optimum performance of each component and system interaction. The building envelope, lighting, and ancillary loads impact the size of the ground heat exchanger, and optimization of economic values requires interaction with the building owner and architect. Simple equipment options that can minimize the cost and complexity of controls, piping loops, and air distribution systems will likely enhance long-term performance and minimize maintenance requirements.

If the building structure, internal loads, and interior HVAC components of the GSHP system have been optimized, the engineer is in a much better position to conduct the task of designing a high-quality, economically viable ground heat exchanger. Quality design engineers familiarize themselves with ground-loop installation practices and procedures and do not relegate the ground heat exchanger design to others. Section 9.5 includes suggestions for providing evidence of quality engineering practices and lists the characteristics of successful GSHP design firms.

Design of the ground heat exchanger is the responsibility of the mechanical engineer of record. While the consultation of experienced GSHP specialists is encouraged, design should not be performed by nonprofessional engineers (PEs), including

- nonengineer certified geothermal designers (CGDs),
- ground-loop contractors,

- equipment vendors,
- ground-loop pipe vendors, and
- other nonengineer "certified" professionals.

The value of the engineer of record taking responsibility for the design of the ground heat exchanger is supported by the ASHRAE Code of Ethics (2013a), which states "Our products and services shall be offered only in areas where our competence and expertise can satisfy the public need."

The recommended design steps for ground-coupled heat pump (GCHP) systems provided here are an update of previous versions provided in an *ASHRAE Transactions* paper (Kavanaugh 2008) and the Geothermal Energy chapter of *ASHRAE Handbook—HVAC Applications* (2011).

1. Calculate peak zone cooling and heating requirements and provide a summary that can be reviewed by building owners and architects.
2. Provide suggestions to reduce building envelope, lighting, and ancillary loads with estimates of reduction in HVAC and ground-loop costs.
3. Estimate off-peak, monthly, and annual cooling and heating requirements so that the annual heat addition to and removal from the loop field can be determined (Equation 3.4) to account for potential ground temperature change.
4. Select the preliminary loop operating temperatures and flow rate to begin optimization of first cost and efficiency (selecting temperatures near the normal ground temperature will result in high efficiencies but larger and more costly ground loops).
5. Correct heat pump performance at rated conditions to actual design conditions (Section 2.3).
6. Select heat pumps to meet cooling and heating loads and locate units to minimize duct cost, fan power, and noise.
7. Arrange heat pumps into ground-loop circuits to minimize system cost, pump energy, and demand (see Figures 1.6, 1.7, 1.8, and 1.9).
8. Conduct a detailed site survey to determine ground thermal properties and drilling conditions (Section 3.6).
9. Determine and evaluate possible loop field arrangements that are likely to be optimum for the building and site (bore depth, separation distance, completion methods, annulus grout/fill, and header arrangements). Include subheader circuits (typically 5 to 15 U-tubes on each) with isolation valves to permit air and debris flushing of sections of the loop field through a set of full-port purge valves.
10. Determine ground heat exchanger dimensions (Sections 3.4 and 3.7). Recognize one or more alternatives (depth, number of bores, grout/fill material, hybrid designs, etc.) that provide equivalent performance and that may yield more competitive bids.
11. Evaluate alternative designs: loop field arrangements, operating temperatures, flow rates, heat exchanger depths/number of bores/materials, grout/fill materials, etc.
12. Lay out interior piping and exterior piping network, compute head loss through the critical path, and select pump(s) to provide recommended flow rates.
13. Verify system efficiency of the final design as outlined in Section 2.4 of this book. If the system cooling EER is less than 12 Btu/Wh ($COP_c < 3.5$) or system heating coefficient of performance (COP) is less than 3.5 at design conditions, consider the following options:
    - Modify the water distribution system if pump demand exceeds 10% of the total system demand.

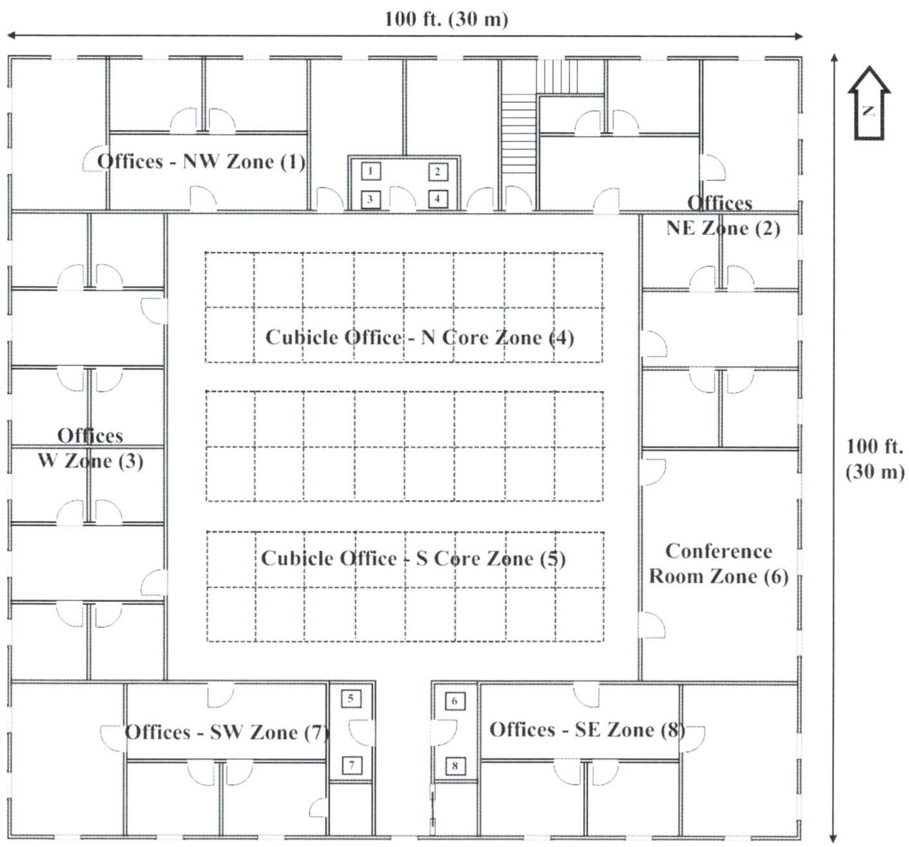

Figure 4.1 Eight-Zone Office Building in St. Louis, Missouri

- Revise the air distribution system if fan demand exceeds 15% of the total system demand.
- Replace the heat pumps if they do not meet the recommendations listed in Table 2.10.
- Redesign the ground heat exchanger to improve entering liquid temperatures (ELTs).

These recommended steps are demonstrated in the following sections for the example 10,000 ft^2 (930 m^2) office building shown in Figure 4.1. Step 12 is not discussed in detail in this chapter; the details of this step are presented in Chapter 6. Step 13 is performed in this chapter with the assumption that the pump power is less than 10% of the total power.

4.2 APPLIED DESIGN PROCEDURE FOR VERTICAL GCHPs (STEPS 1–10)

4.2.1 Step 1—Calculate Building Cooling and Heating Requirements

The conditions used to compute the cooling and heating requirements are as follows:

Outdoor conditions:

- 95°F/76°F (35°C/24°C) dry-/wet-bulb temperatures (max dry bulb)
- 85°F/78°F (29°C/26°C) dry-/wet-bulb temperatures (max humidity ratio)
- 2°F (–17°C)

Indoor conditions:

- 75°F/63°F (24°C/17°C) dry-/wet-bulb temperatures (cooling)
- 70°F (21°C) (heating)

Envelope:

- R_{wall} = 15 h·ft^2·°F/Btu (2.6 m^2·K/W)
- R_{roof} = 25 h·ft^2·°F/Btu (4.4 m^2·K/W),
- R_{window} = 2.0 h·ft^2·°F/Btu (0.35 m^2·K/W)
- SHGF = 0.63

Occupancy:

- 84 people, 5 days per week, 8:00 a.m. to 5:00 p.m., 10% occupancy, 5:00 to 9:00 p.m.

Lighting, plug load:

- 1.0 W/ft^2 (10.8 W/m^2), 7770 W (0.78 W/ft^2 [8.4 W/m^2])

Ventilation air:

- 1300 cfm (610 L/s) (15.5 cfm/person [7.3 L/s·person])
- Requirements based on dedicated outdoor air system (DOAS)

Table 4.1 presents the total cooling loads and heat losses for each building zone at four periods (10:00 a.m., 3:00 p.m., 6:00 p.m., and 2:00 a.m.) of the design day for the ASHRAE-recommended outdoor conditions (ASHRAE 2013b). The maximum total building load and loss for each time period are also provided. The maximum cooling load is 266 kBtu/h (78 kW) or 22 tons. The maximum total heat loss is 191 kBtu/h (56 kW). These calculations were performed with *TideLoad10.xlsm*, a program based off of cooling load temperature difference/cooling load factor (CLTD/CLF) and detailed in *HVAC Simplified* (Kavanaugh 2006). The program is not intended to replace more sophisticated and automated methods but it does conduct zone-by-zone psychrometric analysis and prepares the off-peak loads, total heat losses, and net heat losses (total loss – internal heat gain) necessary to estimate ground heat transfer.

4.2.2 Step 2—Provide Alternatives to Reduce Loads, Losses, and Ground-Loop Costs

In newer buildings, lighting efficacy and office equipment power consumption improvements have significantly reduced sensible and total building cooling loads. However, latent loads generated by occupants and ventilation air remain largely unchanged. In conducting psychrometric load analysis at the maximum dry bulb and maximum humidity ratio (HR), the sensible heat ratio (SHR) in the morning for the example building was well below what most cooling coils can provide at 0.51. A 1300 cfm (610 L/s) energy recovery unit (ERU) is proposed with a 70% sensible effectiveness and 60% latent effectiveness with a fan able to provide a total static pressure of 3.0 in. H_2O (750 Pa). The unit includes an auxiliary cooling coil (either air-cooled direct expansion or hydronic with a water-to-water heat pump) because the building morning SHR at maximum HR conditions will be low (0.64) even with the ERU in operation. The coil is also able to provide adequate moisture removal should the ERU become inoperative and require service.

The maximum cooling load is reduced to 227 kBtu/h or 19 tons (67 kW) and the heat loss is lowered to 121 kBtu/h (36 kW), as shown in Table 4.2. The improvements are made by the addition of the ERU, which requires a small supplemental coil for adequate dehumidification during very humid periods. A savings is available to offset the cost of

Table 4.1 Results of Initial Cooling Load and Heat Loss Calculation for Example Building

	Zone	Cooling Loads, kBtu/h				Total Heat Loss, kBtu/h				Net Heat Requirement, kBtu/h			
		8 a.m.– Noon	Noon– 4 p.m.	4 p.m.– 8 p.m.	8 p.m.– 8 a.m.	8 a.m.– Noon	Noon– 4 p.m.	4 p.m.– 8 p.m.	8 p.m.– 8 a.m.	8 a.m.– Noon	Noon– 4 p.m.	4 p.m.– 8 p.m.	8 p.m.– 8 a.m.
N. West	1	19.1	28.6	14.9	3.9	20.3	15.8	8.8	10.4	13.2	10.3	8.1	9.6
N. East	2	28.6	29.4	15.4	4.8	22.6	17.6	10.4	12.3	15.2	11.8	9.6	11.4
West	3	21.3	37.0	22.1	5.1	21.8	17.0	8.3	9.9	14.3	11.2	7.6	9.0
N. Core	4	34.2	44.7	11.8	4.7	32.0	24.9	5.3	6.3	19.6	15.2	4.0	4.8
S. Core	5	34.2	44.7	11.8	4.7	32.0	24.9	5.3	6.3	19.6	15.2	4.0	4.8
Conf	6	35.1	38.3	8.2	4.2	33.3	26.0	5.5	6.5	26.5	20.6	4.8	5.7
S.West	7	13.8	21.7	14.2	3.9	13.9	10.8	7.0	8.3	9.2	7.2	6.6	7.8
S.East	8	18.3	21.6	13.1	3.9	15.4	12.0	8.2	9.7	10.3	8.1	7.7	9.1
Total Building		205	266	112	35	191	149	59	70	128	100	52	62
	Zone	Cooling Loads, kW				Total Heat Loss, kW				Net Heat Requirement, kW			
		8 a.m.– Noon	Noon– 4 p.m.	4 p.m.– 8 p.m.	8 p.m.– 8 a.m.	8 a.m.– Noon	Noon– 4 p.m.	4 p.m.– 8 p.m.	8 p.m.– 8 a.m.	8 a.m.– Noon	Noon– 4 p.m.	4 p.m.– 8 p.m.	8 p.m.– 8 a.m.
N. West	1	5.6	8.4	4.4	1.1	5.9	4.6	2.6	3.0	3.9	3.0	2.4	2.8
N. East	2	8.4	8.6	4.5	1.4	6.6	5.2	3.0	3.6	4.4	3.5	2.8	3.3
West	3	6.3	10.9	6.5	1.5	6.4	5.0	2.4	2.9	4.2	3.3	2.2	2.6
N. Core	4	10.0	13.1	3.5	1.4	9.4	7.3	1.5	1.8	5.7	4.5	1.2	1.4
S. Core	5	10.0	13.1	3.5	1.4	9.4	7.3	1.5	1.8	5.7	4.5	1.2	1.4
Conf	6	10.3	11.2	2.4	1.2	9.8	7.6	1.6	1.9	7.8	6.0	1.4	1.7
S.West	7	4.0	6.4	4.2	1.1	4.1	3.2	2.1	2.4	2.7	2.1	1.9	2.3
S.East	8	5.4	6.3	3.8	1.2	4.5	3.5	2.4	2.8	3.0	2.4	2.2	2.7
Total Building		60	78	33	10	56	44	17	20	37	29	15	18

the ERU as a result of the reduction of the cooling-mode ground-loop requirement from 22 to 19 tons (78 to 67 kW).

4.2.3 Step 3—Estimate Off-Peak, Monthly, and Annual Cooling and Heating Requirements

The values for design-day off-peak cooling and heating requirements are provided with many load calculation programs, such as *TideLoad10.xlsm*, which was used to generate Tables 4.1 and 4.2. The values that require the highest level of accuracy are the peak cooling and heating requirements of each zone. Errors in these values have almost a one-to-one impact on required ground heat exchanger length. Off-peak, monthly, and annual requirements affect loop length, but errors in these values have a smaller impact than errors in peak requirements. These effects can be verified by adjusting values when applying Equations 3.5 and 3.6 (see Example 3.2 or 3.3).

Therefore, estimates for off-peak, monthly, and annual cooling and heating requirements are acceptable and provide more than adequate accuracy. The recommended procedure for the cooling mode is as follows:

1. Find the maximum load for each zone (e.g., for zone 1 = 25.7 kBtu/h) and multiply by 24 hours per day and 7 days per week:

$$25.7 \times 24 \times 7 = 4318 \text{ kBtu/week}$$

Table 4.2 Results of Revised Cooling Load and Heat Loss Calculation for Example Building

	Zone	Cooling Loads, kBtu/h				Total Heat Loss, kBtu/h			
		8 a.m.–Noon	Noon–4 p.m.	4 p.m.–8 p.m.	8 p.m.–8 a.m.	8 a.m.–Noon	Noon–4 p.m.	4 p.m.–8 p.m.	8 p.m.–8 a.m.
N. West	1	16.8	25.7	14.9	3.8	15.0	11.7	8.4	10.0
N. East	2	26.3	26.5	15.5	4.7	17.4	13.5	10.0	11.8
West	3	18.4	33.4	22.2	5.1	15.3	11.9	7.9	9.3
N. Core	4	27.2	36.1	11.9	4.6	16.3	12.7	4.2	4.9
S. Core	5	27.2	36.1	11.9	4.6	16.3	12.7	4.2	4.9
Conf	6	27.8	29.3	8.3	4.0	17.0	13.2	4.3	5.1
S.West	7	12.6	20.3	14.3	3.8	11.3	8.8	6.8	8.1
S.East	8	17.1	20.1	13.1	3.9	12.8	10.0	8.0	9.4
Total Building		173	227	112	35	121	95	54	64

	Zone	Cooling Loads, kW				Total Heat Loss, kW			
		8 a.m.–Noon	Noon–4 p.m.	4 p.m.–8 p.m.	8 p.m.–8 a.m.	8 a.m.–Noon	Noon–4 p.m.	4 p.m.–8 p.m.	8 p.m.–8 a.m.
N. West	1	4.9	7.5	4.4	1.1	4.4	3.4	2.5	2.9
N. East	2	7.7	7.8	4.5	1.4	5.1	4.0	2.9	3.5
West	3	5.4	9.8	6.5	1.5	4.5	3.5	2.3	2.7
N. Core	4	8.0	10.6	3.5	1.3	4.8	3.7	1.2	1.4
S. Core	5	8.0	10.6	3.5	1.3	4.8	3.7	1.2	1.4
Conf	6	8.2	8.6	2.4	1.2	5.0	3.9	1.3	1.5
S.West	7	3.7	5.9	4.2	1.1	3.3	2.6	2.0	2.4
S.East	8	5.0	5.9	3.9	1.1	3.8	2.9	2.3	2.8
Total Building		51	67	33	10	36	28	16	19

2. Find the total kBtu for each zone by multiplying the values in the 8:00 a.m.–noon, noon–4:00 p.m., and 4:00–8:00 p.m. columns by 4 hours, multiplying the values in the 8:00 p.m.–8:00 a.m. column by 12 hours, and summing these products for zone 1:

$$Q_{Zone\ 1\text{-}clg} = 16.8 \times 4 + 25.7 \times 4 + 14.9 \times 4 + 3.8 \times 12 = 275.2 \text{ kBtu/day}$$
$$= 275.2 \text{ kBtu/day} \times 5 \text{ occupied days} = 1376 \text{ kBtu}$$

3. Find the part-load factor (PLF) for each zone by dividing the values in step 2 by the values in step 1:

$$\text{PLF zone 1} = 1376/4318 = 0.32$$

4. Obtain a weighted average for the entire building by multiplying all zone PLFs by the zone maximum load and summing them. Then obtain the building PLF by dividing this total by the sum of the maximum loads for each zone. In reality this is a weekly PLF, but it will essentially be the same if the computation was performed for four weeks or using monthly values. Results are shown in the left four columns of Table 4.3.

The procedure for the heating mode is modified to include the contribution of the building internal load. Cooling loads are computed to include these loads, but in heating these loads are not included because peak heating requirements typically occur at morn-

Table 4.3 Results of Monthly Part-Load Factor (PLF) Calculation for Example Building (Occupied 5 Days/Week)

Zone	Cooling: Max q_{lc}	Cooling: PLF_m	Cooling: q × PLF	Heating: Max q_{lh}	Heating: PLF_m	Heating: q × PLF
1	25.7	0.32	8.2	15.0	0.39	5.9
2	26.5	0.37	9.8	17.4	0.41	7.1
3	33.4	0.32	10.6	15.3	0.36	5.5
4	36.1	0.29	10.6	16.3	0.15	2.4
5	36.1	0.29	10.6	16.3	0.15	2.4
6	29.3	0.31	9.2	17.0	0.24	4.1
7	20.3	0.34	7.0	11.3	0.43	4.9
8	20.1	0.37	7.4	12.8	0.44	5.7
Total	227.4		73.4	121.4		37.9
	Cool PLF = 0.32			Heat PLF = 0.31		

Table 4.4 Comparison of Total Heat Losses to Net Heat Losses for Example Building

Total Heat Loss, kBtu/h				Net Heat Requirement, kBtu/h			
8 a.m.–Noon	Noon–4 p.m.	4 p.m.–8 p.m.	8 p.m.–8 a.m.	8 a.m.–Noon	Noon–4 p.m.	4 p.m.–8 p.m.	8 p.m.–8 a.m.
15.0	11.7	8.4	10.0	7.9	6.2	7.7	9.1
17.4	13.5	10.0	11.8	9.9	7.7	9.2	10.9
15.3	11.9	7.9	9.3	7.8	6.1	7.1	8.4
16.3	12.7	4.2	4.9	3.9	3.0	2.9	3.4
16.3	12.7	4.2	4.9	3.9	3.0	2.9	3.4
17.0	13.2	4.3	5.1	10.1	7.9	3.7	4.3
11.3	8.8	6.8	8.1	6.6	5.1	6.4	7.6
12.8	10.0	8.0	9.4	7.7	6.0	7.5	8.8
121	95	54	64	58	45	47	56
Total Heat Loss, kW				**Net Heat Requirement, kW**			
8 a.m.–Noon	**Noon–4 p.m.**	**4 p.m.–8 p.m.**	**8 p.m.–8 a.m.**	**8 a.m.–Noon**	**Noon–4 p.m.**	**4 p.m.–8 p.m.**	**8 p.m.–8 a.m.**
4.4	3.4	2.5	2.9	2.3	1.8	2.3	2.7
5.1	4.0	2.9	3.5	2.9	2.3	2.7	3.2
4.5	3.5	2.3	2.7	2.3	1.8	2.1	2.5
4.8	3.7	1.2	1.4	1.1	0.9	0.9	1.0
4.8	3.7	1.2	1.4	1.1	0.9	0.9	1.0
5.0	3.9	1.3	1.5	3.0	2.3	1.1	1.3
3.3	2.6	2.0	2.4	1.9	1.5	1.9	2.2
3.8	2.9	2.3	2.8	2.3	1.8	2.2	2.6
36	28	16	19	17	13	14	16

ing start-up. Because of building thermal mass effects, internal loads only partially contribute to warming the space during morning start-up, so their contribution to reducing the heating requirement is not typically considered at this critical period. However, these loads are available after morning warm-up has been satisfied. They provide useful input to satisfy the building heating requirement and reduce the amount of heat required from the ground loop. Values in Table 4.3 are adjusted in Table 4.4 to consider the contributions of these internal loads.

The heating mode procedure is similar to that for cooling in that the maximum heating requirement for each zone is selected from the zone total heat loss. However, the net heating losses (total loss – internal loads) are used in step 2 for heating to determine monthly PLF rather than total losses. These total and net heating requirements for the example building are compared in Table 4.4. As an example, the total kBtu and PLF for zone 1 are

$$Q_{Zone\ 1\text{-}Htg} = 5 \text{ days} \times (7.9 \times 4 + 6.2 \times 4 + 7.7 \times 4 + 9.1 \times 12) = 982 \text{ kBtu/week}$$

$$\text{PLF}_{Zone\ 1\text{-}Htg} = 982 \text{ kBtu/week} \div (15.0 \text{ kBtu/h} \times 24 \text{ h/day} \times 7 \text{ days/week}) = 0.39$$

Equivalent full-load hours (EFLH) are used to account for the annual heat into and out of the ground as an alternative to a detailed hour-by-hour building energy simulation. Table 4.5 provides the results of an ASHRAE-sponsored research project to develop annual cooling and heating EFLH values for a variety of locations and occupancies (Carlson 2001).

For the office building located in St. Louis, the range of equivalent full-load hours for cooling ($EFLH_c$) is 680 to 1100 h and for heating ($EFLH_h$) the range is 710 to 800 h. It is suggested that average values be used—$EFLH_c = 890$ and $EFLH_h = 755$. Conservative design would use the upper end of the range for cooling ($EFLH_c = 1100$) and the lower end of the range for heating ($EFLH_h = 710$) because the building cooling load is greater than the heat loss.

4.2.4 Step 4—Conduct a Site Survey to Determine Ground Thermal Properties and Drilling Conditions

If the designer is not familiar with the drilling conditions in the area it is prudent to survey potential drilling contractors to determine the optimum drilling depths and borehole sizes for their equipment, personnel, and local geology. This example assumes the results indicate drilling depths 200 to 300 ft (60 to 90 m) are optimum and that the drill bits they prefer are 4 5/8 in. (120 mm) diameter, which typically produce a 5 in. (130 mm) diameter bore. Drilling deeper requires a larger bit, which reduces drilling speed because larger U-tubes are typically required to overcome pumping head losses in the longer tubes (see Chapter 6).

The example design is based on a 300 ft (90 m) borehole being completed with a nominal 1.0 in. (32 mm) U-tube. After a three-day waiting period a thermal property test was conducted by an independent testing firm; results indicated the initial formation temperature was 59°F (15°C), thermal conductivity of 1.3 Btu/h·ft·°F (2.25 W/m·K), and thermal diffusivity of 0.85 ft^2/day (0.079 m^2/day). The drilling log indicated the bore was drilled with a mud rotary drilling rig and the formation was primarily clay and sandy clay with occasional layers of sand and sandstone to a depth of 260 ft (79 m). At this depth, hard rock was encountered and progress with the mud rotary rig was much slower. The standing water column level was 55 ft (17 m) below grade.

As previously mentioned, it is highly recommended that thermal property tests be conducted by independent third-party individuals rather than a drilling contractor or engineer. This maintains a degree of separation that ensures the contractor does not bias the results and also protects both the drilling contractor and the engineer of record should disputes arise in the future.

Table 4.5 Equivalent Full-Load Cooling and Heating Hours (Carlson 2001)

Building Type:	Nine-to-Ten-Month School		Office, 8:00 a.m. to 5:00 p.m., Five Days per Week		Retail, 8:00 a.m. to 10:00 p.m., Seven Days per Week	
Occupied Hours:	**1300–1500**		**2200–2400**		**2800–3600**	
Location	**Cooling**	**Heating**	**Cooling**	**Heating**	**Cooling**	**Heating**
Atlanta	590–830	200–290	950–1360	480–690	1300–1860	380–600
Baltimore	410–610	320–460	690–1080	720–890	880–1480	570–770
Bismarck	150–250	460–500	250–540	950–990	340–780	810–900
Boston	300–510	450–520	450–970	960–1000	610–1380	760–870
Charleston, WV	430–570	310–440	620–1140	770–840	820–1600	620–730
Charlotte	510–730	200–320	940–1340	530–780	1280–1830	420–670
Chicago	280–410	390–470	420–780	820–920	550–1090	670–810
Dallas	620–890	120–200	1100–1580	340–520	1460–2090	280–440
Detroit	230–360	400–480	390–820	970–1020	530–1170	790–900
Fairbanks, AK	25–50	560–630	60–200	1050–1170	110–320	930–1090
Great Falls, MT	130–220	360–430	210–490	820–890	290–710	680–800
Hilo, HI	970–1390	0	1800–2580	15–25	2260–3370	0–20
Houston	670–1000	90–130	1240–1770	250–350	1600–2290	190–300
Indianapolis	380–560	400–480	560–1000	840–920	730–1410	690–820
Los Angeles	610–910	80–160	1140–1670	370–580	1650–2350	250–440
Louisville	470–670	290–430	770–1250	710–830	1000–1720	570–720
Madison	210–310	390–470	320–640	840–900	420–900	700–800
Memphis	580–830	170–240	950–1350	420–600	1250–1780	330–510
Miami	950–1300	10	1500–2150	35–45	1920–2740	25–40
Minneapolis	200–300	420–500	320–610	860–950	430–870	720–860
Montgomery	630–910	120–180	1060–1510	330–470	1390–1990	250–400
Nashville	520–740	250–320	830–1280	590–680	1030–1710	470–590
New Orleans	690–990	70–110	1200–1720	230–320	1570–2240	160–260
New York	360–550	350–440	540–1040	790–870	720–1480	630–760
Omaha	310–440	330–400	480–820	720–800	610–1130	600–720
Phoenix	710–1020	70–110	1130–1610	210–290	1430–2090	170–250
Pittsburgh	300–530	470–500	440–920	910–950	600–1310	750–840
Portland, ME	190–300	400–480	310–630	880–980	410–900	710–870
Richmond, VA	510–730	270–410	880–1310	660–820	1110–1770	520–710
Sacramento	600–850	220–360	1000–1430	640–990	1390–2020	480–830
Salt Lake City	410–710	520–540	510–1090	1040–1060	660–1520	830–930
Seattle	260–460	460–650	440–1200	1270–1370	710–1860	960–1170
St. Louis	390–550	280–400	680–1100	710–800	850–1500	570–700
Tampa	780–1110	40–60	1440–2000	140–190	1780–2560	100–160
Tulsa	540–770	240–300	830–1300	560–620	1030–1730	450–540

4.2.5 Step 5—Select Loop Operating Temperatures and Flow Rates to Optimize First Cost and Performance Trade-Off

As stated in Chapter 3, the optimal trade-off between system efficiency and ground-loop length typically occurs when the maximum value for the heat pump ELT in the cooling mode is 20°F to 30°F (11°C to 17°C) greater than the undisturbed ground temperature (t_g). The optimum tends to be on the lower end of this range for warmer climates (t_g > 60°F [15°C]) and toward the upper end of the range for cooler climates. For heating, the optimum value for the ELT is typically 8°F to 15°F (5°C to 8°C) less than the undisturbed ground temperature (t_g). Buildings in warmer climates or those with high internal cooling loads tend to have optimal values on the lower end of this range, whereas buildings in cold climates with high heat losses compared to heat gains tend to have optimum values on the higher end of this range.

Note that a standard cooling-mode heat pump ELT is 86°F (30°C), which is 27°F (15°C) above the 59°F (15°C) ground temperature for the St. Louis office building. This is within the typical optimal range suggested above. Note also that a standard heating-mode heat pump ELT is 50°F (10°C), which is 9°F (5°C) below the local ground temperature. This also is within the typical optimal range suggested above. These values for ELT of 86°F (30°C) in cooling and 50°F (10°C) for heating are used for the initial example calculation.

As mentioned in Chapter 3, optimum liquid flow rates for closed-loop systems are typically in the 2.5 to 3.0 gpm/ton (2.7 to 3.2 L/min·kW) range. The following estimates can be used with good accuracy for the heat pump leaving liquid temperatures (LLTs). These values assume water is the fluid; values will be 3% to 5% higher for typical antifreeze solutions used with GSHPs (see Appendix F for properties of antifreeze solutions).

- For a flow rate of 3.0 gpm/ton (3.2 L/min·kW) the LLT will be approximately 10°F (5.6°C) higher than the ELT in cooling and 6°F (3.3°C) less than the ELT in heating.
- For a flow rate of 2.5 gpm/ton (2.7 L/min·kW), the LLT will be approximately 12°F (6.7°C) higher than the ELT in cooling and 7.2°F (4°C) less than the ELT in heating.
- For a flow rate of 2.0 gpm/ton (2.15 L/min·kW), the LLT will be approximately 15°F (6.7°C) higher than the ELT in cooling and 9°F (5°C) less than the ELT in heating.

The example calculation uses the heat pumps listed in Table 2.3, which all appear to be rated with a flow rate of approximately 3.0 gpm/ton (3.2 L/min·kW). A flow rate of 3.0 gpm/ton (3.2 L/min·kW) based on maximum block load (not installed capacity) is used, so the LLT for the building is 10°F (5.6°C) higher than the ELT in cooling and 6°F (3.3°C) less than the ELT in heating. The building total peak block load is 227 kBtu/h (19 tons, 70 kW), resulting in a design flow rate of 57 gpm (220 L/min). The peak block load of the north cluster of zones is 122 kBtu/h (10.2 tons, 36 kW), resulting in a flow rate of 31 gpm (117 L/min). The peak block load of the south cluster of zones is 106 kBtu/h (8.8 tons, 31 kW), resulting in a flow rate of 27 gpm (102 L/min).

4.2.6 Step 6—Correct Heat Pump Performance at Rated Conditions to Design Conditions

Note: The correction of heat pump rated capacity and efficiency to actual values is a time-consuming ordeal. The spreadsheet tool discussed in Chapter 2, *WAHPCorrector.xlsm*, can assist designers with the process of correcting heat pump performance. A short-cut alternative is to apply the multipliers to full-load total cooling (TC), EER, heating capacity (HC), and COP values to correct performance to conditions and constraints likely to occur in actual applications. These conditions are as follows:

- Cooling indoor air temperatures of 75°F db/63°F wb (24°C db/17°C wb) (from 80.6°F/66.2°F [27°C/19C°])
- Heating indoor air temperatures of 70°F db (21°C db) (from 68°F [20°C])
- Includes fan power/heat required to distribute air through average duct/filter systems

The correction factors from AHRI/ASHRAE ISO Standard 13256-1 (ASHRAE 2012a) rating conditions are as follows:

- **Multiply rated TC by 0.93**
- **Multiply rated EER by 0.80**
- **Multiply rated HC by 1.03**
- **Multiply rated COP by 0.89**

This applies to rated TC and EER for ELTs at 86°F, 77°F, and 59°F (30°C, 25°C, and 15°C) but not to part-load values at 68°F (20°C) and to rated HC and COP for ELTs at 68°F, 50°F, and 32°F (20°C, 10°C, and 0°C) but not for part-load values at 41°F (5°C). These corrections do not account for added pump power, which must be included for total system efficiency.

The following paragraphs describe the more detailed process of correcting performance. The selection of an ELT of 86°F (30°C) for cooling, an ELT of 50°F (10°C) for heating, and a flow rate of 3.0 gpm/ton (3.2 L/min·kW) results in Table 2.3 values only needing to be corrected for return air temperatures, fan heat, and fan power. Had the ELTs been different, the cooling capacity, EER, HC, and COP values would be found by interpolation using values at the other ELTs in Table 2.3.

The building loads shown in Table 4.2 indicate the cooling load will dictate heat pump size. Requirements range from 20 to 36 kBtu/h (6 to 11 kW). Table 2.3 indicates the zones are likely to require models 22, 30, 36, or 42 if the single-speed heat pumps are specified. The capacities and efficiencies for each unit can be verified by correcting for return air temperatures, fan heat, and fan power.

Consider model 36, which has a rated TC of 34.5 kBtu/h (10.1 kW) and an EER of 19.6 Btu/Wh (COP_c = 5.7). The first step is to correct for an entering air wet-bulb temperature (EATWB) from 66.2°F to 63°F (19°C to 17.2°C). Table 2.5 indicates the TC correction factor (Cf_{TC}) is 0.962 and the cooling power correction factor (Cf_{CP}) is 0.997. Thus,

$$TC_{63} = Cf_{TC\text{-}66.2\rightarrow 63} \times TC_{66.2} = 0.962 \times 34.5 \text{ kBtu/h} = 33.2 \text{ kBtu/h} \qquad \text{(I-P)}$$

$$TC_{17.2} = Cf_{TC\text{-}19\rightarrow 17.2} \times TC_{19} = 0.962 \times 10.1 \text{ kW} = 97 \text{ kW} \qquad \text{(SI)}$$

And noting that the units for EER can be either Btu/Wh or kBtu/kWh,

$$kW_{66.2} = TC_{66.2} \div EER_{66.2} = 34.5 \text{ kBtu/h} \div 19.6 \text{ kBtu/kWh} = 1.76 \text{ kW}$$

$$kW_{63} = Cf_{CP\text{-}66.2\rightarrow 63} \times kW_{66.2} = 0.997 \times 1.76 \text{ kW} = 1.75 \text{ kW}$$

The second correction is to deduct the heat generated by the fan from the cooling capacity. Since the fan and motor are located in the airstream, all of the input power is converted to heat through motor losses, fan losses, and air distribution system fiction losses. Figure 4.2 provides a method of determining the heat from duct and filter losses that are not included in the rated performance. The assumption is made that the air distribution system will be designed to limit the duct and filter losses of 0.8 in. H_2O (174 Pa). The heat pump fan wheels are forward-curved (squirrel cage) impellers driven by electronically commutated motors (ECMs). This combination typically results in wire-to-air efficiencies of 30% (Kavanaugh 2012). For this type of fan at the assumed duct and filter losses, Figure 4.2 indicates the reduction in TC is 3.6%. Thus,

$$TC_{63,0.8} = TC_{63} \times (1 - Cf_{FanHeat}) = 33.2 \text{ kBtu/h} \times (1 - 0.036) = 32.0 \text{ kBtu/h}$$

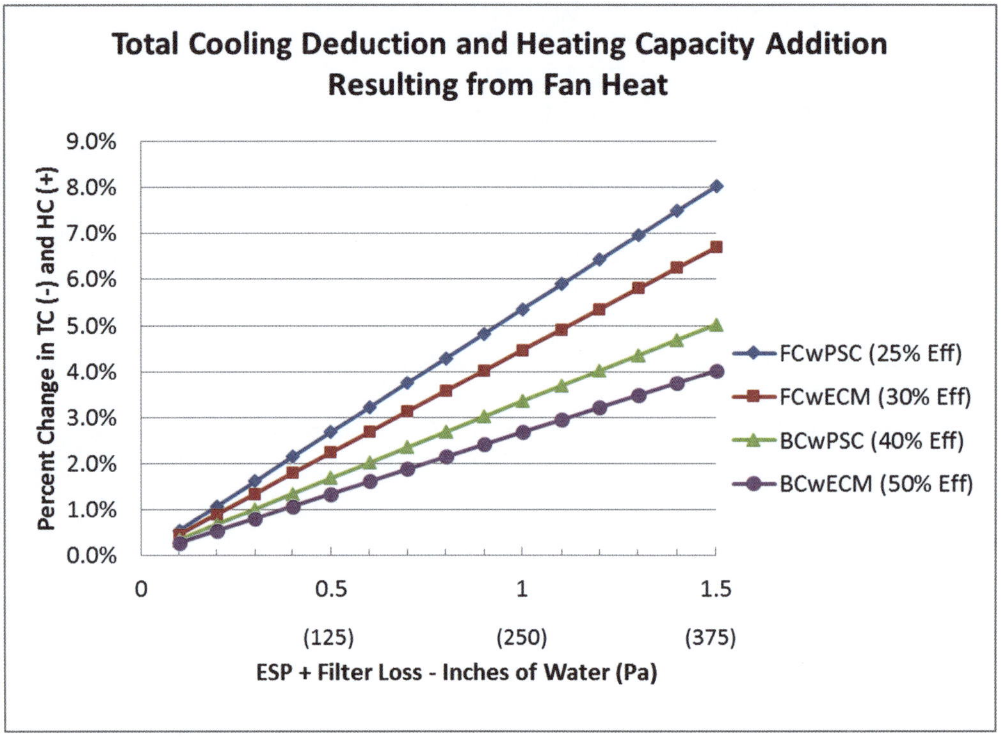

Figure 4.2 Capacity Correction for Fan Heat Based on 400 cfm/ton (54 L/s·kW) for Unitary Heat Pumps with Permanent Split Capacitor and Electrically Commutated Motors and Forward- and Backward-Curved Blades

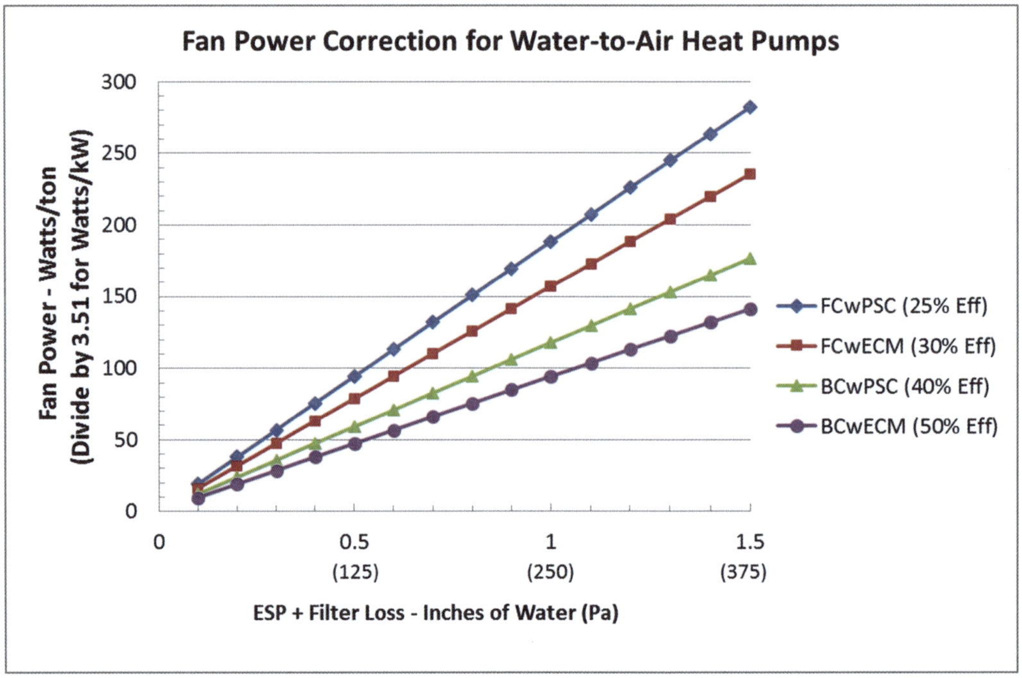

Figure 4.3 Fan Power Addition Based on 400 cfm/ton (54 L/s·kW) for Unitary Heat Pumps with Permanent Split Capacitor and Electronically Commutated Motors and Forward- and Backward-Curved Blades

The third correction is to include the fan power that is used to compute the overall heat pump unit EER (not including the pump). Figure 4.3 indicates the resulting fan power correction is 125 W (0.125 kW) per ton for the forward-curved fan with an ECM at 0.8 in. H_2O (174 Pa). The TC of model 36 is 2.67 tons (= 32.0 kBtu/h ÷ 12 kBtu/ton·h).

Therefore, the input power for external static pressure (ESP), filter loss, and EATWB is

$$kW_{63,0.8} = kW_{63} + Cf_{FanPower} \times TC\ (tons) = 1.75\ kW + 0.125\ kW/ton \times 2.67\ tons$$
$$= 2.09\ kW$$

The heat pump EER (EER_{HP}) is found using the corrected cooling capacity and power input:

$$EER_{HP} = TC_{63,0.8} \div kW_{63,0.8} = 32.0\ kBtu/h \div 2.09\ kW = 15.3\ kBtu/kWh$$
$$= 15.3\ Btu/Wh$$

The process for heating is similar, but unit corrections are necessary because unlike EER the rated COP is dimensionless, the return air temperature correction is based on dry-bulb temperature, and the fan heat is added to the heating capacity. Table 2.6 indicates the HC correction factor (Cf_{HC}) is 0.995 and the heating power correction factor (Cf_{HP}) is 1.025 when correcting from the rated entering air dry-bulb temperature (EATDB) of 68°F (20°C) to the design entering air temperature (EAT) of 70°F (21°C).

The rated values for the model 36 unit at ELT = 86°F (30°C) and EAT = 68°F (20°C) are HC_{68} = 30.3 kBtu/h (8.9 kW) and COP_{68} = 5.2. Thus,

$$HC_{70} = Cf_{HC\text{-}68\rightarrow 70} \times HC_{68} = 0.995 \times 30.3\ kBtu/h = 30.1\ kBtu/h$$

$$kW_{68} = HC_{68} \div (3.412 \times COP) = 30.3\ kBtu/h \div (3.412\ kBtu/kWh \times 5.2) =$$
$$1.71\ kW$$

$$kW_{70} = Cf_{HP\text{-}68\rightarrow 70} \times kW_{68} = 1.025 \times 1.71\ kW = 1.75\ kW$$

In heating, the amount of heat generated by the fan is of the same magnitude as in cooling, but it is added to HC. The fan power is also the same that is added to the rated power input corrected for EAT. The actual heat pump COP (COP_{HtPmp}) is found using the corrected capacity and power.

$$HC_{70,0.8} = HC_{70} \times (1 + Cf_{FanHeat}) = 30.1\ kBtu/h \times (1 + 0.036) = 31.2\ kBtu/h$$
$$= 31.2 \div 12 = 2.6\ tons$$

$$kW_{70,0.8} = kW_{70} + Cf_{FanPower} \times HC\ (tons) = 1.75\ kW + 0.125\ kW/ton \times (31.2/12)\ tons$$
$$= 2.08\ kW$$

$$COP_{HtPmp} = HC_{70,0.8} \div (3.412 \times kW_{70,0.8}) = 31.2\ kBtu/h \div (3.412\ kBtu/kWh \times 2.08\ kW)$$
$$= 4.4$$

The correction process is laborious but necessary given that ELT, EAT, and fan power have a significant impact on cooling capacity, cooling efficiency, and heating efficiency. In the example above the corrected TC is 7% lower, the corrected EER is 22% lower, and the corrected COP is 15% lower than the rated values. The process is even more critical with central air distribution systems that typically have much higher fan pressure requirements and often include return air fans and fan-powered variable-air-volume (FPVAV) terminals. In these systems cooling capacity reductions in excess of 20% can be experienced, with even greater reductions in system EER and COP. Although fan heat results in additional heating capacity, it is added at very low efficiency (COP = 1) and can exacerbate imbalances in ground heat exchange. When cooling is the critical mode, additional fan heat will result in warmer loops unless the ground heat exchanger is increased in size.

The final correction in the process is to include the pump power. The ground-loop pump has only a small indirect effect on the cooling and heating capacities. It does reduce the system EER and COP. This example assumes each heat pump has a single 200 W (0.20 kW) circulator pump. An alternative would be to set a limit for pump power, as suggested in Chapter 6, of 5% (excellent design) to 15% (poor design) of heat pump power. The EER and the cooling mode COP_c with the pump power included are

$$\begin{aligned} EER_{wPump} &= TC_{63,0.8} \div (kW_{63,0.8} + kW_{Pump}) = 32.0 \text{ kBtu/h} \div (2.09 \text{ kW} + 0.20 \text{ kW}) \\ &= 14.0 \text{ kBtu/kWh} \end{aligned}$$

$$COP_{c\text{-}wPump} = 14.0 \text{ kBtu/kWh} \div 3.412 \text{ kBtu/kWh} = 4.1$$

The heating-mode COP with the pump power included is

$$\begin{aligned} COP_{h\text{-}wPump} &= HC_{70,0.8} \div (3.412 \times kW_{70,0.8} + kW_{Pump}) \\ &= 31.2 \text{ kBtu/h} \div (3.412 \text{ kBtu/kWh} \times 2.08 + 0.20 \text{ kW}) = 4.3 \end{aligned}$$

Table 4.6 provides the corrected performance for all four heat pumps considered for the example building. Values can be generated using a spreadsheet that repeats the preceding calculations for model 36. To substantiate the importance of the performance correction process, note that the uncorrected TC of the model 22 heat pump would be sufficient to meet the cooling requirements of zones 7 and 8 (Table 4.2) in the example building but that the corrected capacity would be insufficient.

Table 4.6 Heat Pump Performance Corrected for Air Temperatures, Fan Power, and Pump Power

Cooling:	**Rated Values at ELT = 86°F (30°C)**			**Wet-Bulb Correction**		**Fan Heat Correction**			**Pump Included**
Model	**TC, kBtu/h (kW)**	**EER**	**kW**	**TC, kBtu/h (kW)**	**kW**	**TC, kBtu/h (kW)**	**kW**	**EER**	**EER**
22	20.7 (6.1)	17.5	1.18	19.9 (5.8)	1.18	19.2 (5.6)	1.38	13.9	12.2
30	28.3 (8.3)	19.2	1.47	27.2 (8.0)	1.47	26.2 (7.7)	1.74	15.1	13.5
36	34.5 (10.1)	19.6	1.76	33.2 (9.7)	1.75	32.0 (9.4)	2.09	15.3	14.0
42	40.6 (11.9)	19.2	2.11	39.1 (11.5)	2.11	37.7 (11.0)	2.50	15.1	13.9
Heating:	**Rated Values at ELT = 50°F (10°C)**			**Dry-Bulb Correction**		**Fan Heat Correction**			**Pump Included**
Model	**HC, kBtu/h (kW)**	**COP**	**kW**	**HC, kBtu/h (kW)**	**kW**	**HC, kBtu/h (kW)**	**kW**	**COP**	**COP**
22	19.8 (5.8)	5.3	1.09	19.7 (5.8)	1.12	20.4 (6.0)	1.33	4.5	3.9
30	25.8 (7.6)	5	1.51	25.7 (7.5)	1.55	26.6 (7.8)	1.83	4.3	3.8
36	30.3 (8.9)	5.2	1.71	30.1 (8.8)	1.75	31.2 (9.1)	2.08	4.4	4.0
42	34.9 (10.2)	5.2	1.97	34.7 (10.2)	2.02	36.0 (10.6)	2.39	4.4	4.1

4.2.7 Step 7—Select Heat Pumps to Meet Cooling and Heating Loads and Locate Units to Minimize Duct Cost, Fan Power, and Noise

The second, third, fourth, and fifth columns of Table 4.7 list the maximum cooling and heating requirements of each zone taken from Table 4.2. The sixth column lists the model number of the smallest heat pump that can satisfy both the cooling and heating requirements of each zone. These numbers correspond to the "nominal" cooling capacity of the units in kBtu/h at AHRI/ASHRAE ISO Standard 13256-1 (ASHRAE 2012a) ground-loop heat pump (GLHP) rating conditions (77°F [25°C] ELT). The four middle columns show the corrected cooling and heating capacities and efficiencies of the selected units, and the right four columns provide the specified airflow and water flow rates.

Figure 4.1 includes the recommended location for each unit. The units are located in closets either in or near the zones they serve. The duct runs will be relatively short, which reduces fan power and installation costs. Closet locations also minimize the level of noise to occupants. Units are accessible for service without ladders and with minimum disruption to occupants.

Note that the psychrometric analysis is omitted in this example. The procedure to ensure the heat pumps are able to satisfy both the total and latent heat requirements is discussed in Chapter 2 and in more detail in HVAC texts such as *HVAC Simplified* (ASHRAE 2006). In office buildings, satisfying both the total and latent heat requirements is often possible to accomplish with heat pumps alone because the ventilation air requirements are modest in many cases. However, in densely populated buildings such as schools, supplemental treatment of the outdoor ventilation air is necessary to reduce latent loads.

The rating standards do not require the publication of sensible heat capacity for heat pumps. This complicates psychrometric analysis, as published data may or may not contain performance corrected for fan power. It is suggested that designers solicit this information in writing directly from engineers at the factory.

In this example it would be prudent to solicit this information because Figure 4.7 indicates the cooling capacities of the heat pumps are rated at airflow rates above 400 cfm/ton (54 L/s·kW), which may result in unacceptable latent performance. This can be countered by reducing airflow rates, which will also slightly reduce total cooling capacity. Table 2.7 provides both total and sensible cooling correction factors that can be applied to ensure adequate latent capacity is available.

Table 4.7 Zone Cooling and Heating Requirements with Heat Pumps and Specifications

Zone	Cooling Required		Heating Required		Model No.	TC		HC		Airflow		Water Flow	
	kBtu/h	kW	kBtu/h	kW		kBtu/h	kW	kBtu/h	kW	cfm	L/s	gpm	L/s
1	25.7	7.5	11.7	3.4	30	26.2	7.7	26.6	7.8	900	425	8	30
2	26.5	7.8	13.5	4.0	36	32.0	9.4	31.2	9.1	1200	580	9	34
3	33.4	9.8	11.9	3.5	42	37.7	11.0	36.0	10.6	1300	610	11	42
4	36.1	10.6	12.7	3.7	42	37.7	11.0	36.0	10.6	1300	610	11	42
5	36.1	10.6	12.7	3.7	42	37.7	11.0	36.0	10.6	1300	610	11	42
6	29.3	8.6	17.0	5.0	36	32.0	9.4	31.2	9.1	1200	580	9	34
7	20.3	5.9	11.3	3.3	30	26.2	7.7	26.6	7.8	900	425	8	30
8	20.1	5.9	12.8	3.8	30	26.2	7.7	26.6	7.8	900	425	8	30
Total	**228**	**67**	**104**	**30**		**256**	**75**	**250**	**73**	**9000**	**4265**	**75**	**284**

It is also suggested that calculations for cooling and heating requirements be repeated using the maximum humidity ratio and dehumidification conditions (ASHRAE 2013) in humid and mildly humid areas (design outdoor air wet-bulb temperatures > 70°F [21°C]) that have ventilation requirements greater than 10% of supply airflow. In some cases the total cooling requirement using the maximum dehumidification conditions will exceed the requirement using the maximum dry-bulb conditions. The maximum humidity conditions will result in higher latent loads and will alter the situation in which supplemental latent cooling is required for the outdoor ventilation air.

4.2.8 Step 8—Arrange Heat Pumps into Ground-Loop Circuits to Minimize System Cost, Pump Energy, and Demand

The location of the heat pumps in two clusters in the building provides the opportunity to minimize indoor piping. Two common-loop GCHP circuits (see Figure 1.8) each connected to four heat pumps is a prudent option. The interior piping would be limited to a small area of the building and the purge values shown in Figure 1.8 could be located in the closets rather than outdoors. However, the use of ground-loop close headers, shown in Figure 1.8, is one of several options, including the standard reverse-return (Figure 1.7) or modified reverse-return (Figure 1.9) options.

The liquid flow rates of 31 gpm (148 L/s) to the north cluster of heat pumps and 27 gpm (136 L/s) to the south cluster are within the recommended flow rates for 2 in. (60 mm) nominal DR 11 high-density polyethylene (HDPE) pipe. The final piping design in Step 12 may dictate that slightly oversized pipe is required in order to have sufficient flow rate with only one circulator pump on each heat pump. Additional pumps will reduce system efficiency 8% to 12%. In some cases, the cost savings of fewer pumps would offset the higher cost of the larger pipe. Additionally, the number of bores required for each cluster will likely be 8 or 10, so only one ground-loop circuit will be required.

The use of a central loop in this example does not reduce the required ground-loop size because there is no cooling load diversity in the building (Table 4.2). It adds to the interior piping cost and requires multiple ground-loop circuits and additional isolation valves, as shown in Figure 1.9, because 15 to 25 bores will be required. There will also be additional head loss because of the added piping lengths.

Another option is individual ground loops (see Figure 1.6), which would require the minimum amount of equipment and the most reliable control method (on-off pumps and no check or flow control valves). Multiple small-diameter headers would be required, and bore depths might have to be varied in order to optimize each individual heat pump.

A final option to consider is the use of a one-pipe system (see Figure 1.7), which could consist of two loops connected to each four-heat-pump cluster or a single one-pipe loop for all eight heat pumps. Like the individual loop system, this method has reliable control (on-off pumps) but does require a central pump to operate continuously. The central pump control can be optimized with a variable-speed drive or multiple central pumps to minimize energy use when few heat pumps are operating. If the option of two one-pipe loops is selected, the compact location of the four heat pumps in each cluster permits relatively simple control to turn the central pump off if none of the four units are operating.

4.2.9 Step 9—Determine and Evaluate Possible Loop Field Arrangements

The peak cooling load of the example building is 227 kBtu/h (67 kW) or 19 tons. A typical starting point for the vertical bore field layout is one bore per ton of load (~ four bores per kW). It is prudent to have a balanced number of bores, as a prime number of bores is not able to be subdivided into equal numbers of bores per parallel path. In the example case, two clusters of four heat pumps will each be connected to a ground loop.

Thus, eight or ten bores for each of the two clusters would be approximately one bore per ton of load (~ four bores per kW). The initial design uses eight bores (16 total for the building) arranged in a reverse-return ground circuit as shown in Figure 4.4. This layout maybe be altered pending the results of Steps 10, 11, and 12. The results of the example thermal property test indicate drilling was more difficult below 260 ft (79 m). If the design result in Step 10 indicates a depth greater than this is required, it would be prudent to increase the number of bores to ten if space permits. The increase in the number of bores will reduce the length of each bore to 80% of the original length and decrease the flow rate through each bore to 80%. The combined effect will result in a ground-loop head loss approximately 52% ($[8/10]^3$) of the original because the reduction due to the shorter length is linear and the reduction due to the flow is a function of the rate squared. Recall the loop length will be 80% shorter and the flow rate will be 80% less, and head loss is approximately a function of flow rate squared. However, the eight-bore option requires less ground area.

The initial design assumes flow can be provided by a single nominal 1/6 hp (200 W input) circulator pump on each pump (800 W total). The EER of the system will be adjusted accordingly so the ground loop is able to handle the additional heat of the pump.

4.2.10 Step 10—Determine Ground Heat Exchanger Dimensions

The ground heat exchanger can be designed following the procedure used in Example 3.2 in Chapter 3. This example designs the ground loop serving the north cluster of heat pumps shown in Figure 4.4. The building cooling and heating loads are taken from

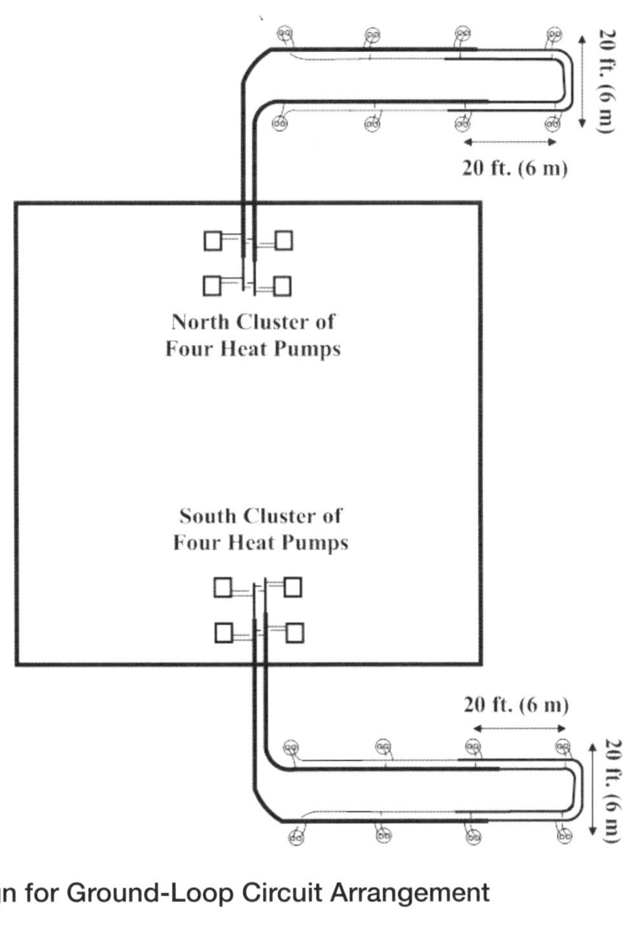

Figure 4.4 Initial Design for Ground-Loop Circuit Arrangement

Table 4.2, and the EER and COP values are found in Table 4.6. The weighted average EER of the eight heat pump is 13.8 and the weighted average of the COP is 4.0. The average EFLH for an office in St. Louis are 890 hours in cooling and 755 hours in heating. Thus,

$$q_{cond} = q_{lc} \times \frac{\text{EER} + 3.412}{\text{EER}} = -122{,}000 \text{ Btu/h} \times \frac{13.8 + 3.412}{13.8} = -152,200 \text{ Btu/h}$$

$$q_{evap} = q_{lh} \times \frac{\text{COP} - 1}{\text{COP}} = 64{,}000 \text{ Btu/h} \times \frac{4.0 - 1}{4.0} = 48{,}000 \text{ Btu/h}$$

$$q_a = \frac{q_{cond} \times \text{EFLH}_c + q_{evap} \times \text{EFLH}_h}{8760 \text{ h}}$$

$$= \frac{-152{,}200 \text{ Btu/h} \times 890 \text{ h} + 48{,}000 \text{ Btu/h} \times 755 \text{ h}}{8760 \text{ h}}$$

$$= -11{,}300 \text{ Btu/h}$$

Determine the thermal resistances of the ground for the three prescribed heat pulses (Equations 3.11, 3.12, and 3.13 or *GfactorCalc.xlsm*, a spreadsheet tool that is available with this book at www.ashrae.org/GSHP) using the ground properties shown in Step 4 and a 5 in. (13 cm) bore diameter.

$$\text{Fo}_f = 4 \times 0.85 \text{ ft}^2\text{/day} \times 7330.167 \text{ days} \div (5 \text{ in.} \div 12 \text{ in./ft})^2 = 143{,}600,$$
from Figure 3.6, $G_f = 1.00$

$$\text{Fo}_1 = 4 \times 0.85 \text{ ft}^2\text{/day} \times (7330.167 - 7300) \div (5 \text{ in.} \div 12 \text{ in./ft})^2 = 591,$$
from Figure 3.6, $G_1 = 0.58$

$$\text{Fo}_2 = 4 \times 0.85 \text{ ft}^2\text{/day} \times (7330.167 - 7330) \div (5 \text{ in.} \div 12 \text{ in./ft})^2 = 3.27,$$
from Figure 3.6, $G_2 = 0.19$

$$R_{ga} = (1.00 - 0.58) \div 1.3 \text{ Btu/h·ft·°F} = 0.323 \text{ h·ft·°F/Btu}$$

$$R_{gm} = (0.58 - 0.19) \div 1.3 \text{ Btu/h·ft·°F} = 0.30 \text{ h·ft·°F/Btu}$$

$$R_{gst} = 0.19 \div 1.3 \text{ Btu/h·ft·°F} = 0.147 \text{ h·ft·°F/Btu}$$

Determine the thermal resistances of the bore using the 31 gpm for the four heat pumps (see Step 8). The initial design specifies a 1.0 in. DR 11 HDPE tube, water without antifreeze, and a thermally enhanced grout with a thermal conductivity of 0.90 Btu/h·ft·°F (four parts silica sand, one part bentonite grout; Table 3.2).

Flow/U-tube (gpm) = 31 gpm ÷ 8 U-tubes = 3.9 gpm

Table 3.3 indicates that for water flowing at 3 gpm in a 1.0 in. DR 11 tube at 68°F the Reynolds number (Re) is 8500 and at 5 gpm it is 14,200. Re will be higher at the 91°F average water temperature. So the bore resistance is found based on the turbulent flow value of 10,000 used in the table. If the flow rate is adjusted during the final design phase, the results should be reconfirmed.

For k_{grout} = 0.8 Btu/h·ft·°F, turbulent flow, 5 in. bore, location B: R_b = 0.23 h·ft·°F/Btu

For k_{grout} = 1.2 Btu/h·ft·°F, turbulent flow, 5 in. bore, location B: R_b = 0.18 h·ft·°F/Btu

Via interpolation for k_{grout} = 0.9 Btu/h·ft·°F, R_b = 0.218 h·ft·°F/Btu

For k_{grout} = 0.8 Btu/h·ft·°F, turbulent flow, 5 in. bore, location C: R_b = 0.17 h·ft·°F/Btu

For k_{grout} = 1.2 Btu/h·ft·°F, turbulent flow, 5 in. bore, location C: R_b = 0.14 h·ft·°F/Btu

Via interpolation for k_{grout} = 0.9 Btu/h·ft·°F, R_b = 0.163 h·ft·°F/Btu

The average bore resistance value for locations B and C is applied,

R_b = 0.191 h·ft·°F/Btu for location BC, k_{grout} = 0.9 Btu/h·ft·°F, turbulent flow, 5 in. bore.

The ground loop differential temperature is 10°F (6°C) [ELT = 85°F, LLT = 95°F]; thus, the short-circuit heat loss factor (F_{sc}) is 1.04 as indicated in Figure 3.7.

The monthly part-load factor for cooling of 0.32 provided in Table 4.3 for the entire building is approximately the same for the four zones served by the north ground loop.

In lieu of the extended procedure for computing long-term temperature penalty, this example demonstrates a procedure for extending Table 3.6 to conditions slightly different than those listed. The calculations are conducted assuming mild water recharge, which assumes the ground temperature at a distance of 30 ft (18 m) from the vertical U-tubes on the perimeter of the ground loop is equal to the undisturbed ground temperature. In this example, the values for EFLH$_c$ (890) and EFLH$_h$ (755) are nearly the same. However, the building cooling load (228 kBtu/h) is nearly twice the heating requirement (121 kBtu/h). The ratio of the product of cooling load (Q_c) and EFLH$_c$ to the product of heating requirement (Q_h) and EFLH$_h$ is

$$Q_c/Q_h = \frac{228 \text{ kBtu/h} \times 890 \text{ h}}{121 \text{ kBtu/h} \times 755 \text{ h}} = 2.2$$

Table 3.6 includes values for temperature penalty when the EFLH are the same (750), but note the results are based on the cooling load and heating requirement being the same (Q_c/Q_h = 1.0). However, the table includes values for EFLH$_c$ = 1000 and EFLH$_h$ = 500 with a cooling load 33% greater than the heating requirement. The total operating hours (1500) are also near the operating hours of the example (890 + 755 = 1645). In this case,

$$Q_c/Q_h = \frac{1.33 \times q_{lh} \times 1000 \text{ h}}{q_{lh} \times 500 \text{ h}} = 2.67$$

If mild water recharge and a 300 ft (90 m) bore is assumed, the temperature penalty of 3.9°F can be *estimated* by interpolating between the value of 1.7°F for Q_c/Q_h = 1.0 and 4.4°F for Q_c/Q_h = 2.67.

The required total bore length for cooling is computed using Equation 3.5 with the temperature penalty of –3.9°F (–2.2°C) assumed:

$$L_c = \frac{(11{,}300 \times 0.323) - 152{,}200 \times (0.191 + 0.32 \times 0.30 + 1.04 \times 0.147)}{59°\text{F} - \frac{86°\text{F} + 96°\text{F}}{2} - (-3.9°\text{F})} = 2512 \text{ ft}$$

$$= 8 \text{ bores @ } 314 \text{ ft/bore}$$

This length exceeds the depth at which drilling becomes more difficult. As mentioned previously, the head loss through a vertical bore that is 80% of the original length with 80% of the flow rate of the original design results in a head loss of approximately 52% (0.8^3) of the eight-loop head loss. Therefore, it is prudent to adjust the number of bores to 10 and redesign the length. There will be some adjustment to the temperature penalty given the change in number of bores. If the process above is repeated for 10 bores, the bore length is 253 ft (77 m). If the total length of the eight-bore calculation (2512 ft [766 m]) is divided by the number of bores, the length is 251 ft (77 m). The results are rounded up to 10 bores at 255 ft (78 m) each.

The process is repeated using Equation 3.6 to find the bore length for heating (L_h), and the design bore length is the larger value of L_c and L_h. But recall, the critical condition for the nondominant mode, in this case heating, will be in year one because the long-term temperature rise tends to improve with the warmer ground. **The design conditions for the nondominant mode should be determined with the temperature penalty (t_p) and the net annual heat transfer to the ground (q_a) set to zero.**

$$L_h = \frac{q_{evap} \times (R_b + \text{PLF}_{mh} \times R_{gm} + F_{sc} \times R_{gst})}{t_g - \frac{\text{ELT} + \text{LLT}}{2}}$$

$$= \frac{48{,}900 \times (0.191 + 0.31 \times 0.30 + 1.04 \times 0.147)}{59°\text{F} - \frac{50°\text{F} + 44°\text{F}}{2}} = 1780 \text{ ft}$$

$$= 10 \text{ bores @ } 178 \text{ ft/bore}$$

Repeating the design process for the south cluster of zones yields three options based on the cooling load:

- 8 bores at 280 ft (85 m)
- 9 bores at 250 ft (76 m)
- 10 bores at 226 ft (69 m)

The final arrangement is shown in Figure 4.5. The result is a total bore length requirement of 4800 ft (1463 m) when the north and south bore fields are added.

4.3 DESIGN ALTERNATIVES (STEP 11)

Step 11 in the design procedure is to evaluate other alternatives to the common-loop option shown in Figure 4.5. Several alternatives are presented in this section with detailed calculations. A summary table of results is provided at the end of this section.

The initial two design alternatives to consider are the unitary loop and the one-pipe loop. The options not only are the simplest alternatives, but field tests indicate they outperform other alternatives (Kavanaugh and Kavanaugh 2012). They should be considered as the primary alternatives for ground-coupled systems. (GWHPs and SWHPs are typically not as well suited to unitary loops.) The simplicity of these two alternatives and the

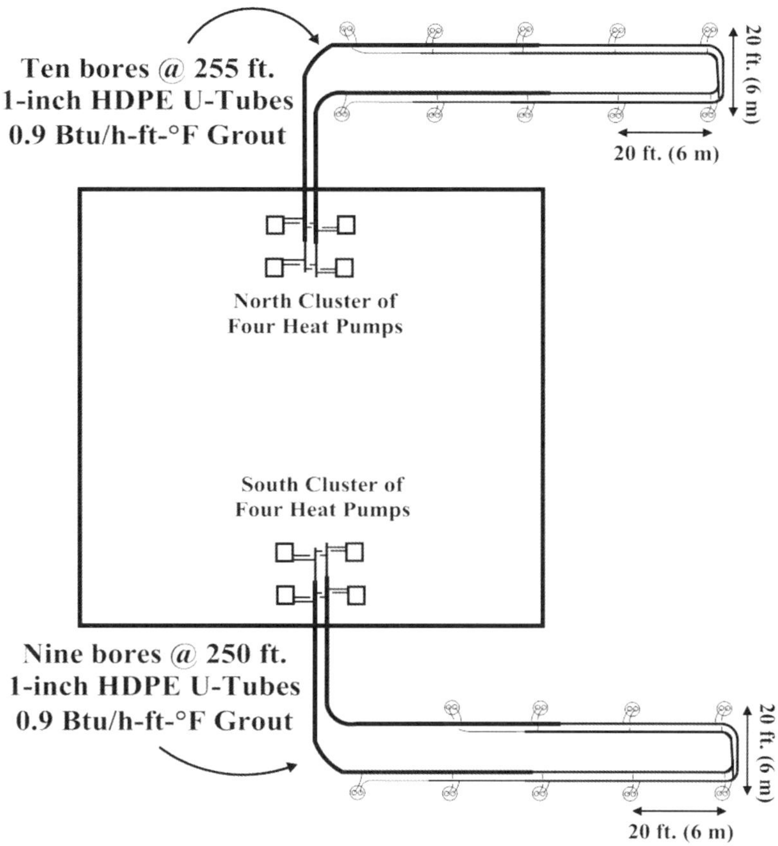

Figure 4.5 Final Design for Common Ground-Loop Circuit Arrangement

common (subcentral) loop described in the previous section are well suited to schools and buildings that have limited personnel and resources for maintenance.

Note that the following design options are evaluated using the same ground-loop water recharge assumption used in the preceding section.

4.3.1 Unitary Loop System

Figure 4.6 depicts a unitary loop system with a single circulator pump for each unit that has the highest average ENERGY STAR rating of systems surveyed (see Section 9.1 for an explanation of ENERGY STAR ratings). This option is often best for one- and two-story buildings with large footprints. Installation costs are minimized by the absence of long interior runs of large-diameter headers. The required pump head is less than the original design because of the short header runs, as shown in Figure 4.6. The four smaller units can be served by 150 W pumps. The total loop length for this option is nearly the same as the initial design because there is no load diversity (see Table 4.2). In applications in which diversity is present, an option is to serve zones with a common loop and the areas without diversity with unitary loops.

Although it is prudent to always consider this option in the initial design phase, it is not universally the best option. Multistory buildings with more compact footprints would not have long interior runs (main headers are short vertical runs with each floor having short-run headers). Thus, savings would not be noteworthy compared to common (subcentral) or central loops.

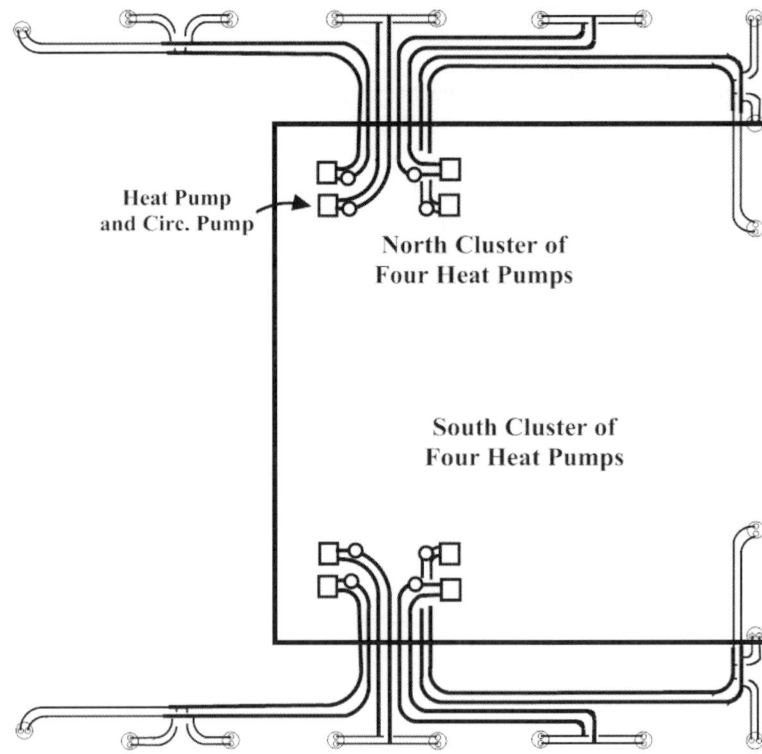

Figure 4.6 Unitary-Loop System

4.3.2 One-Pipe-Loop System

Figure 4.7 depicts the one-pipe-loop system, which performed almost as well as unitary loops in terms of ENERGY STAR ratings of systems surveyed (Kavanaugh and Kavanaugh 2012). The one-pipe loop is also very simple but can take advantage of load diversity when it is present. Staged or variable-speed main pumps provide continuous flow through the building. Flow rate is controlled to maintain favorable ground-loop return temperature. Low-head circulator pumps are activated with each individual heat pump's operation and draw water from the single pipe loop and discharge it downstream. These pumps only need to provide sufficient head to circulate water through the heat pump and connections. Therefore, they are smaller than the circulator pumps used with common-loop and unitary-loop systems. When combined with the main pump, the demand is increased by about 600 W compared to common-loop and unitary-loop pumps.

4.3.3 Central Ground Loop, Building Loop, and Pumps

Although the central ground loop is perhaps the most common option for commercial and institutional buildings, field surveys indicate it is far from the most energy efficient option (see Section 1.6). There are also indications that the potential cost savings in ground-loop reductions due to load diversity are offset by higher cost for interior piping, ground-loop header piping, and controls. There are applications in which the central system is a viable option, such as tall buildings with small footprints, applications with a central heat/cool source (groundwater, lake loops, waste heat stream, etc.), hybrid GCHPs, and applications with large load diversities. The point of this discussion is that there are multiple building types and the central loop option in many cases is not the best choice.

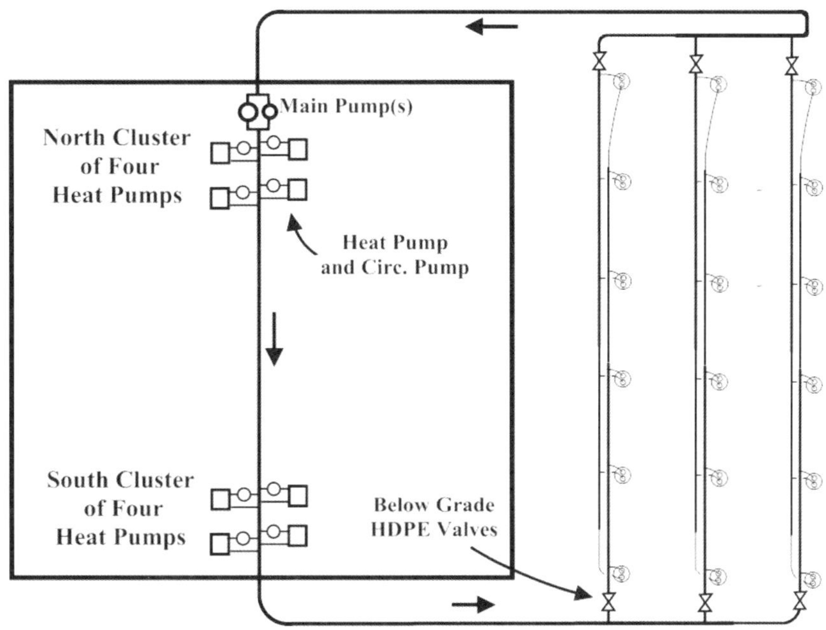

Figure 4.7 One-Pipe Loop System

Figure 4.8 demonstrates a typical central ground-loop option for the example building. An 18-bore loop results in a depth of 270 ft (82 m) for each U-tube for a total length of 4860 ft (1480 m). Flow to the loop field is split into two reverse-return circuits with nine U-tubes on each circuit. Total loop flow is 57 gpm (216 L/min), which results in 3.2 gpm (12 L/min) per U-tube. Isolation valves on each circuit allow loop purging/flushing to be performed one circuit at a time with a smaller, less expensive purge pump as discussed in Chapter 6. The circuits are split in an equipment room with adequately sized purge valves nearby for convenient access.

To save energy, a variable-speed pump is likely a viable option. This requires a two-way valve on each heat pump and a signal to control pump speed when building load is mild or nonexistent. Details are discussed in Chapter 6.

If the pump is properly sized, power for this relatively small central loop is likely to be about the same as for the system with the small circulators. There is some increase in the pump head required by the central loop, but this is offset by the improved efficiency typical of larger pumps and motors. Energy use will be higher if the pump drive is allowed to operate continuously when no heat pumps are operating.

The example building has little load diversity, so the reduction in ground-loop length does not occur. In fact, the total bore length increases by 1.3% to 4860 ft (1481 m) because the three-row grid pattern results in a slightly higher long-term temperature penalty compared to the two-row designs of the common-loop option.

4.3.4 Advanced Piping Materials and Enhanced Grout/Fill

Products are available that have improved thermal conductivities compared to HDPE. Piping arrangements, such as two U-tubes in each bore, offer improved thermal performance. Likewise, graphite-based bentonite mixtures are available with advertised values of up to 1.6 Btu/h·ft·°F (2.8 W/m·K). In jurisdictions where the use of porous fills (i.e., sands, gravel, etc.) in combination with surface seals are permitted with high water tables, enhanced performance is possible. In these cases, water movement permits in-bore natu-

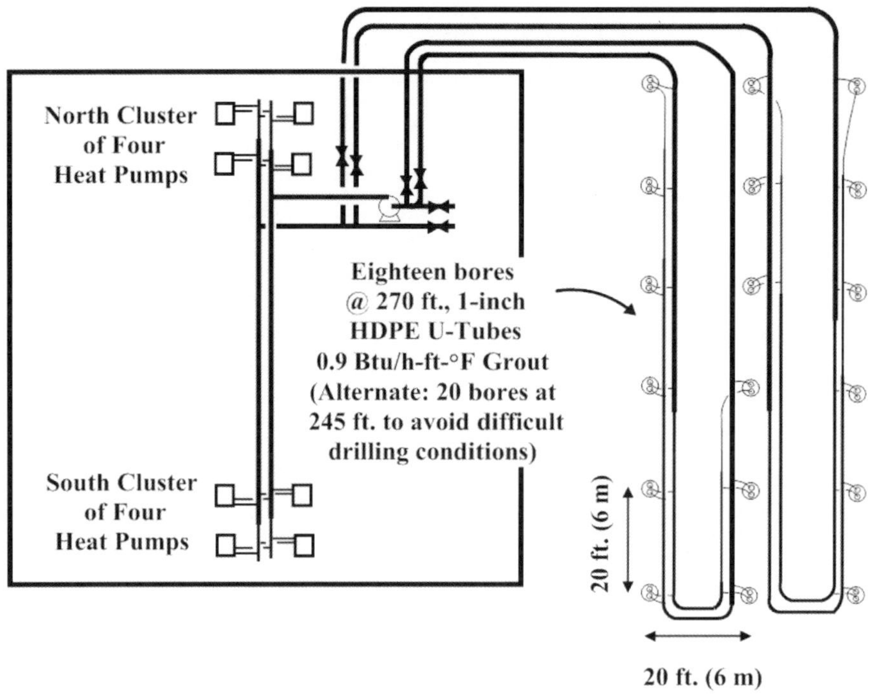

Figure 4.8 Central Ground Loop, Building Loop, and Pumps

ral convection heat transfer, which results in high "equivalent" thermal conductivity. These options result in cost premiums that may or may not offset the reduced drilling cost for shorter bore lengths. There will also be a higher long-term temperature penalty due to the shorter bores and reduced thermal storage in the loop field. Any grouting or fill material that increases installation time or difficulty must be evaluated to include labor cost, product cost, and the cost of any specialized equipment required for installation.

The computations that follow compare the reductions in bore lengths to the single bore field option shown in Figure 4.8. The ground loop of this option consists of 18 bores, 270 ft (82 m) in depth, with a nominal 1 in. (32 mm) DR 11 HDPE and a grout thermal conductivity of 0.9 Btu/h·ft·°F (1.6 W/m·K). This results in a bore resistance of 0.19 h·ft·°F/Btu (0.11 m·K/W).

- Substituting a grout with a thermal conductivity of 1.5 Btu/h·ft·°F (2.6 W/m·K) for the 0.9 Btu/h·ft·°F (1.6 W/m·K) product reduces the required bore length from 270 to 245 ft (82 to 75 m), which is a **9.3% reduction**. The bore resistance is reduced to 0.14 h·ft·°F/Btu (0.08 m·K/W).
- Inserting a double nominal 1 in. (32 mm) DR 11 HDPE U-tube reduces the required bore length from 270 to 236 ft (82 to 72 m), which is a **12.6% reduction**. The bore resistance is reduced to 0.12 h·ft·°F/Btu (0.07 m·K/W).
- Inserting a double nominal 1 in. (32 mm) DR 11 HDPE U-tube and substituting a grout with a thermal conductivity of 1.5 Btu/h·ft·°F (2.6 W/m·K) for the 0.9 Btu/h·ft·°F (1.6 W/m·K) product reduces the required bore length from 270 to 222 ft (82 to 68 m), which is a **17.8% reduction**. The bore resistance is reduced to 0.09 h·ft·°F/Btu (0.05 m·K/W).
- Substituting a piping material with a thermal conductivity of 0.44 Btu/h·ft·°F (0.76 W/m·K) for the 0.22 Btu/h·ft·°F (1.6 W/m·K) product reduces the required

bore length from 270 to 250 ft (82 to 75 m), which is a **7.4% reduction**. The bore resistance is reduced to 0.15 h·ft·°F/Btu (0.09 m·K/W).

- Substituting a piping material with a thermal conductivity of 0.44 Btu/h·ft·°F (0.76 W/m·K) for the 0.22 Btu/h·ft·°F (1.6 W/m·K) product in conjunction with using a grout having a thermal conductivity of 1.5 Btu/h·ft·°F (2.6 W/m·K) reduces the required bore length from 270 to 231 ft (82 to 68 m), which is a **14.4% reduction**. The bore resistance is reduced to 0.11 h·ft·°F/Btu (0.06 m·K/W).

There is a diminishing return on enhancements in piping and grouting materials because the thermal resistance of the ground dominates the total resistance. If U-tubes were constructed of copper and the grout was enhanced to three times what is currently available, the reduction in bore length would be 21%. This is currently the absolute limit of possible bore length reductions. Designers should be wary of technologies that claim greater savings.

4.3.5 Decrease Grout Thermal Conductivity

The material handling workload is substantially increased with bentonite grouts that are thermally enhanced with silica sand. Table 3.2 indicates that 50 lb (23 kg) of sodium bentonite when mixed with water yield 27 gal (0.10 m^3) of grout with a thermal conductivity of approximately (±10%) 0.42 Btu/h·ft·°F (0.73 W/m·K). However, mixing 50 lb (23 kg) of sodium bentonite with 200 lb (91 kg) of silica sand and water yields only 31 gal (0.12 m^3) but provides a thermal conductivity of approximately (±10%) 0.90 Btu/h·ft·°F (1.56 W/m·K).

The amount of material handled with the enhanced grout is five times the weight compared to conventional grout with only a 15% increase in yield. In some cases loop contractors may request alternates to bid without thermally enhanced grout.

Substituting a grout with a thermal conductivity of 0.42 Btu/h·ft·°F (73 W/m·K) for the 0.9 Btu/h·ft·°F (1.6 W/m·K) product increases the required bore length for the example building from 270 to 332 ft (82 to 101 m), which is a **23% increase in required bore length**. The bore resistance is increased to 0.32 h·ft·°F/Btu (0.08 m·K/W).

4.3.6 Increase or Decrease Bore Separation Distance

The impact of bore separation distance is highly dependent on the moisture recharge over several years of operation. The preceding design calculations assume a mild moisture recharge rate. This assumption is repeated when demonstrating the impact of increasing and decreasing the bore separation distance. A second set of comparisons is presented for a low moisture recharge formation. All calculations are based on results after 20 years of operation and a three by six (18-bore) grid.

- Increasing the vertical bore separation of the example system from 20 to 25 ft (6 to 7.6 m) reduces the required bore length from 270 to 249 ft (82 to 76 m) assuming a mild moisture recharge formation. This is a **7.8% reduction in required bore length**.
- If a low moisture recharge formation is assumed, the required bore length with a 20 ft (6 m) bore separation is 305 ft (93 m). This increase, compared to the mild recharge assumption with a 20 ft (6 m) separation, is to be expected. If the bore separation distance is increased to 25 ft (7.6 m), the required bore length is decreased to 268 ft (82 m). This is a **12% reduction in required bore length**.
- Decreasing the vertical bore separation of the example system from 20 to 15 ft (6 to 4.6 m) increases the required bore length from 270 to 328 ft (82 to 100 m)

assuming a mild moisture recharge formation. This is a **21% increase in required bore length**.

- If a low moisture recharge formation is assumed, the required bore length with a 20 ft (6 m) bore separation is 305 ft (93 m). If the bore separation distance is decreased to 15 ft (7.6 m), the required bore length is increased to 440 ft (134 m). This is a **44% increase in required bore length**.

4.3.7 Lower or Raise Heat Pump Cooling-Mode Entering Liquid Temperature

The preceding example designs were performed using a cooling-mode ELT = 86°F (30°C). Increasing this value will reduce the required bore length but will result in lower heat pump and system efficiencies. Inversely, lowering the design cooling-mode ELT increases the required ground heat exchanger length but improves efficiency. The program *WAHPCorrector.xlsm*, which is available with this book at www.ashrae.org/GSHP, was used to adjust the efficiencies of the heat pumps for other ELT values.

Table 4.6 demonstrates the average efficiency of selected heat pumps is EER = 15.2 Btu/Wh (COP_c = 4.5) when the pump power is not included and EER = 14.0 Btu/Wh (COP_c = 4.1) if 200 W pumps are used on each unit. If the cooling-mode ELT is raised to 95°F (35°C), the system efficiencies (pump power included) would decrease from EER = 14.0 to 12.1 Btu/Wh (COP_c = 4.1 to 3.5). However, the required bore length will be reduced from 270 to 216 ft (82 to 66 m). **This represents a 20% reduction in the required bore length, which results in an estimated 13% decline in system cooling efficiency.** Note that performance characteristics of heat pumps at ELTs above 86°F (30°C) are estimates because this is the highest rating condition.

If the cooling-mode ELT is lowered to 77°F (25°C), the system efficiencies (pump power included) would increase from EER = 14.0 to 15.3 Btu/Wh (COP_c = 4.1 to 4.5). However, the required bore length is increased from 270 to 372 ft (82 to 114 m). **This represents a 38% increase in the required bore length, which results in a 9% improvement in system cooling efficiency.**

4.3.8 Hybrid System

Many commercial and institutional buildings have larger cooling requirements than heating requirements. This, coupled with the fact that in cooling the heat transfer rates per ton (kW) are 40% to 60% greater than the heating mode heat transfer rates per ton (kW), results in most ground loops being designed to meet the cooling requirements. In some applications an option is a hybrid system in which the ground loop is sized to meet the heating requirement and the cooling load is satisfied by using the ground loop in parallel with a fluid cooler (as shown in Figure 4.9) or a cooling tower with an isolation heat exchanger. Additional details are available in the final report of ASHRAE RP-1384 (Hackel et al. 2009).

The positive benefits of a hybrid system include the following:

- It is a viable option in applications where sufficient land surface area is not available, drilling costs are high, and/or the required heating length is substantially less than the required cooling length.
- The fluid cooler could be operated not only to reduce the ground loop load during peak cooling periods but also to balance the annual heat load on the ground loop. In this mode, the cooler can be operated during periods of low outdoor air wet-bulb temperatures, when capacity is much greater, while the parasitic fan and pump power can be minimized.

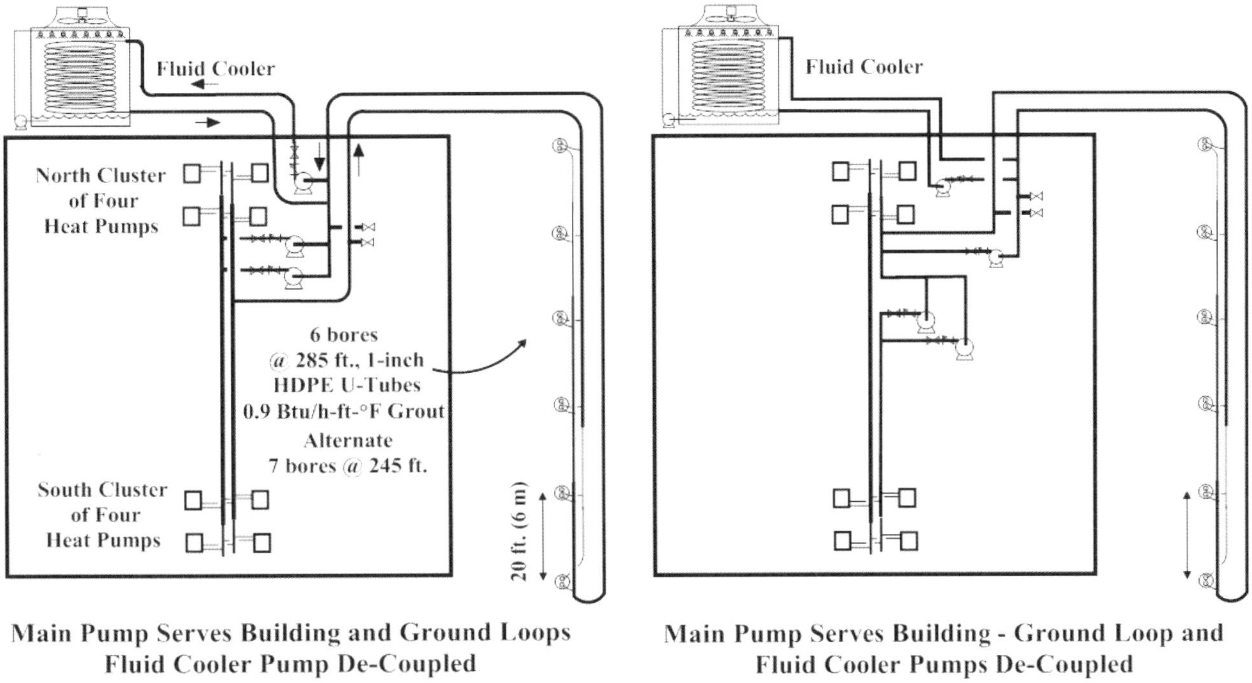

Figure 4.9 Hybrid Fluid Cooler—GSHP System

- In buildings with high internal loads (core office zones, computer rooms, etc.), the cooler can be operated when the outdoor air wet-bulb temperature is low to reduce the loop temperature so that free cooling is possible via hydronic coils.
- It has frequently been used to supplement poorly performing (overheated) ground loops.

The potential downsides of hybrid systems include the following:

- There are added maintenance costs for the fluid cooler or cooling tower that can be significant given raw outdoor air is drawn into a device where moisture is being introduced and circulated.
- There are potential health risks to occupants of poorly maintained systems (ASHRAE 2000).
- System efficiency is typically lower (unless the heat pump ELT is substantially reduced) due to the added parasitic power of the cooler fan(s), circulation pump, and spray pump or open cooling tower sump pump.
- The significant reliability/serviceability advantage that results from having no outdoor equipment is eliminated.
- Water is consumed (see Equation 4.1), which may be a significant cost or limited by legal restrictions in some locations.

The minimum amount of water (m_w) required for cooling is a direction function of the amount of condenser capacity (q_{cond}) displaced by the cooler and the hours of operation (τ_{Oper}):

$$m_w = \frac{q_{cond} \times \tau_{Oper}}{h_{fg}} = \frac{Q_{cond}}{h_{fg}} \tag{4.1}$$

where

h_{fg} = heat of vaporization (for water)

Q_{cond} = total amount of heat rejected by the condenser during period of operation

The makeup water for open cooling towers will contain minerals. As water is evaporated, concentrations will increase in the basin water. Thus, periodic blowdown will be required to reduce mineral concentrations to acceptable levels.

The example building is not an ideal candidate for selecting a hybrid system because of its relatively small size; also, the percentage savings would not be as great as it would be for a larger building. However, the example building will be used to highlight the design process and provide insight into the possible percent reduction of ground-loop size.

Table 4.6 indicates the heating requirements for all of the zones are lower than the heating capacities of the heat pumps selected to meet the cooling requirement. This presents the potential of lowering the heating design ELT so that the required loop length is less but the equipment is able to maintain comfort. An ELT of 45°F (7°C) is suggested to avoid the use of an antifreeze mixture. The program *WAHPCorrector.xlsm*, which is available with this book at www.ashrae.org/GSHP, is used to predict the heating capacities of the heat pumps for ELT = 45°F (7°C). HC for the model 30 units is 24.9 kBtu/h (7.3 kW), the model 35 is 29.6 kBtu/h (8.7 kW), and the model 42 is 33.9 kBtu/h (9.9 kW). All of these models are able to meet the zone heating requirements. However, the values of COP_h for the units are reduced to 4.1, 4.2, and 4.2, respectively. This reduces the system COP_h to 3.8, which is **a 7% reduction** from the initial design.

The program *GchpCalc2014.xlsm*, which is available with this book at www.ashrae.org/GSHP, is applied using the lower ELT and COPs. The bore length results based on no long-term temperature change are used because the critical condition for the nondominant mode (heating) occurs in year one. The vertical grid arrangement must also be altered to provide bore depths that are typical, in the range of 200 to 300 ft (60 to 90 m). Results indicate six bores at 285 ft (87 m) or seven bores at 245 ft (75 m) will meet load at a heating-mode total length of 1715 ft (520 m).

The required fluid cooler size (q_{fc}) to replace the ground-loop capacity that is no longer available can be determined from the differences in the heating length (L_h), cooling length (L_c) for the central-loop nonhybrid design, and the condenser capacity (q_{cond}). The condenser capacity can be determined with Equation 4.2, which is arrived at by rearranging Equation 3.3 using the cooling efficiency and the cooling load (q_{lc}).

$$q_{cond} = \frac{\text{EER} + 3.412}{\text{EER}} \times q_{lc} = \frac{13.9 + 3.412}{13.9} \times 227\ \text{kBtu/h} = 283\ \text{kBtu/h} \qquad \text{(I-P)}$$

$$q_{cond} = \frac{\text{COP} + 1}{\text{COP}} \times q_{lc} = \frac{4.1 + 1}{4.1} \times 66.5\ \text{kW} = 82.7\ \text{kW} \qquad \text{(SI)}$$

Thus,

$$q_{fc} = \frac{L_c - L_h}{L_c/q_{cond}} = \frac{4860\ \text{ft} - 1715\ \text{ft}}{4860\ \text{ft}/283\ \text{kBtu/h}} = 183\ \text{kBtu/h} \qquad \text{(I-P)} \qquad (4.2a)$$

$$q_{fc} = \frac{L_c - L_h}{L_c/q_{cond}} = \frac{1481\ \text{m} - 523\ \text{m}}{1481\ \text{m}/82.9\ \text{kW}} = 54\ \text{kW} \qquad \text{(SI)} \qquad (4.2b)$$

The required flow rate can be estimated using Equation 4.3, which is a conversion of a fundamental relationship for heat transfer rate as a function of mass flow rate of water (m_w) and differential temperature loop temperatures, which are ELT = 86°F (30°C) and LLT = 96°F (36°C) for the example design.

$$q = m_w c_p (t_{wo} - t_{wi}) = \rho c_p Q_w (t_{wo} - t_{wi}) \tag{4.3}$$

When values of density and specific heat are applied with the common I-P volumetric flow rate unit of gpm, the equation becomes

$$q\ (\text{Btu/h}) = \rho\ (\text{lb/ft}^3) \times c_p\ (\text{Btu/lb·°F}) \times Q_w \times (t_{wo} - t_{wi})$$

$$= \frac{62.3\ \text{lb/ft}^3 \times 1.0\ \text{Btu/lb·°F} \times 60\ \text{min/h}}{7.48\ \text{gal/ft}^3} \times Q_w\ (\text{gpm}) \times (t_{wo} - t_{wi}\ (\text{°F}))$$

$$= 500\ \text{Btu·min/h·gal·°F} \times Q_w\ (\text{gpm}) \times (t_{wo} - t_{wi}\ (\text{°F}))$$

(Note that 488 should be substituted for 500 for 20% propylene glycol-water solutions and 481 for 20% methanol-water solutions.)

$$q\ (\text{kBtu/h}) = 0.500\ \text{kBtu·min/h·gal·°F} \times Q_w\ (\text{gpm}) \times (t_{wo} - t_{wi})\ \text{°F}$$

When SI values of density and specific heat are applied with the volumetric flow rate of L/s, a coefficient of 4.15 results, as shown in the following equation:

$$q\ (\text{kW}) = 4.15\ \text{kW·s/L·°C} \times Q_w\ (\text{L/s}) \times (t_{wo} - t_{wi})\ \text{°C}$$

(Note that 4.05 should be substituted for 4.15 for 20% propylene glycol-water solutions and 4.0 for 20% methanol-water solutions.)

The required water flow rate for the fluid cooler is

$$Q_w\ (\text{gpm}) = \frac{q_{cond}\ (\text{kBtu/h})}{0.500\ \text{kBtu·min·h·gal·°F} \times (\text{LLT} - \text{ELT})\ \text{°F}}$$

$$= \frac{183\ (\text{kBtu/h})}{0.500\ \text{kBtu/min·h·gal·°F} \times (96\text{°F} - 86\text{°F})} = 37\ \text{gpm} \qquad \text{(I-P)}$$

$$Q_w\ (\text{L/min}) = \frac{q_{cond}\ (\text{kW})}{4.15\ \text{kW·s·L·°C} \times (\text{LLT} - \text{ELT})\ \text{°C}}$$

$$= \frac{54\ (\text{kW})}{4.15\ \text{kW·s·L·°C} \times (35.6\text{°C} - 30\text{°C})} = 2.32\ \text{L/s} = 139\ \text{L/min} \qquad \text{(SI)}$$

The pump power to the smaller ground loop and the fluid cooler pump should be approximately the same as the pump power required for the nonhybrid ground loop. The added parasitic powers are the cooler fan motor and spray pump. Many fluid cooler manufacturers provide sizing programs that typically require the following information:

Water flow rate: 37 gpm (2.32 L/s or 139 L/min)

Water inlet and outlet temperatures: 96°F and 86°F (36°C and 30°C)

Design wet-bulb temperature: 79°F (26°C)

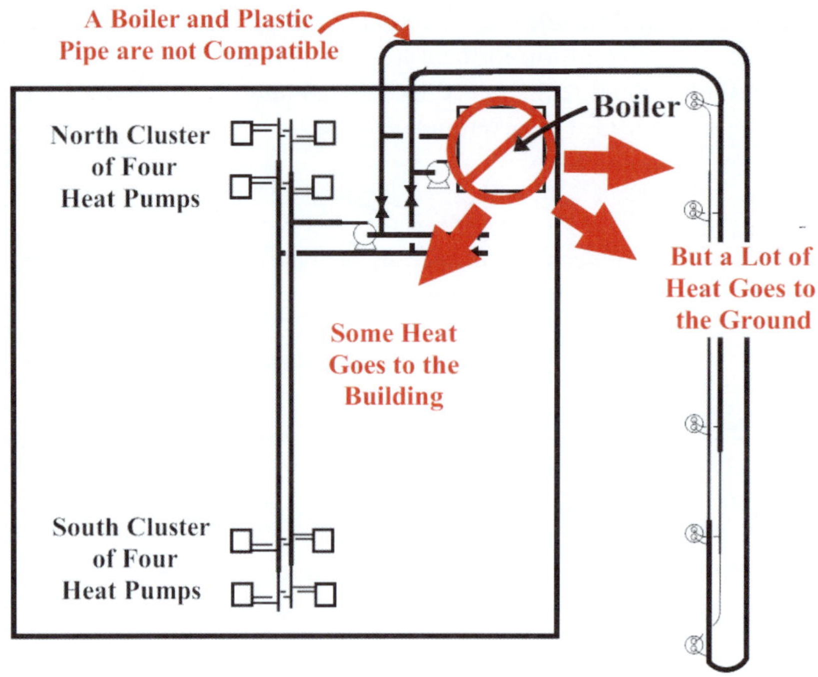

Figure 4.10 Hybrid System with Boiler Connected to Ground Loop

The results indicate a nominal 15 ton (53 kW) fluid cooler is required with a with a 1.5 hp (1.2 kW) fan motor (1.3 kW input) and a 0.25 kW spray pump. This added power reduces the system EER from 14.0 to 12.8 Btu/Wh (COP_c = 4.1 to 3.8), which is **a reduction of 9%** compared to the original design. It is possible to optimize the size and operation mode of the fluid cooler to enhance system efficiency. A larger cooler could be used to reduce loop temperature so that heat pump efficiency is improved. Overall efficiency can be improved if the added power of larger fan and spray pump motors is minimized.

There are many other options for hybrid systems with other types of GSHPs in addition to the vertical GCHP example hybrid discussed here. It is important to know the characteristics of the building loads and the cost of not only the GSHP loop but also the additional equipment and controls.

In some instances the hybrid concept has been extended to supplementing the heating capacity, as shown in Figure 4.10. This practice is discouraged, as a significant percentage of the heat is dissipated to the ground and is not recovered in the heating mode. It could, however, negatively impact the ground-loop cooling capacity in the following season. It is much more effective to apply a conventional approach of supplying auxiliary heat directly to the building. This can be done with a boiler connected to a conventional hot-water distribution system. It can also be applied at a much lower installed cost using conventional electric resistance coils if the amount of supplemental heat is small. This is often the case in commercial buildings when the cooling and heating loads are carefully calculated.

Furthermore, the connection of a boiler to HDPE piping is a risk. The ground loop is an expensive investment that will likely outlast two or more heat pump systems. One control malfunction or an override in a well-intended attempt to increase thermal comfort could damage the plastic-pipe heat exchanger.

Table 4.8 provides a summary of the relative ground-loop sizes, efficiency differences, and important elements of the alternative designs presented in this section.

Table 4.8 Impact of Design Alternatives

Original Design: System EER = 13.9, COP = 4.0, ELT(clg) = 86°F (30°C), ELT(htg) = 50°F (10°C), 19 vertical bores at 4800 ft (1460 m) total, 1 in. (32 mm) nominal HDPE U-tubes, 20 ft (6 m) bore separation, two ground-loop circuits (10 bore and 9 bore), 0.90 Btu/h·ft·°F (1.56 W/m·K) grout conductivity, eight 200 W pumps, on-off controls with check valves

Design Alternative	Ground Loop Size	Efficiency	Other
Eight unitary loop	No change	1% Increase	Check valves no longer required
One-pipe loop 1.5 hp (1.1 kW) and 12 circulator pumps	1% Increase	1% Decrease	Central pump(s) added
Central loop with single 2 hp (1.5 kW) pump	1% Increase	No change	Two-way heat pump valves, variable-speed pump
Increase grout conductivity to 1.5 Btu/h·ft·°F (2.6 W/m·K)	9.3% Decrease	No change	Higher material cost
Use double U-tubes in vertical bores	12.6% Decrease	No change	Additional header fittings required
Double U-tubes and 1.5 Btu/h·ft·°F (2.6 W/m·K) grout	17.8% Decrease	No change	Additional header fittings required
Double U-tube conductivity (to 0.44 Btu/h·ft·°F) (0.76 W/m·K)	12.6% Decrease	No change	Higher material cost
Double U-tube conductivity and 1.5 Btu/h·ft·°F (2.6 W/m·K) grout	14.4% Decrease	No change	Much higher material cost
Reduce grout conductivity to 0.42 Btu/h·ft·°F (0.73 W/m·K)	23% Increase	No change	Grout material weight reduced 400%
Increase bore separation distance to 25 ft (7.6 m)	8 to 12% Decrease	No change	Increase in required ground area by 56%
Decrease bore separation distance to 15 ft (4.6 m)	21 to 44% Increase	No change	Increased possibility of cross-drilling
Increase ELT(clg) to 95°F (35°C)	20% Decrease	13% Decrease	Heat pumps only rated to ELT = 86°F (30°C)
Decrease ELT(clg) to 77°F (25°C)	38% Increase	9% Increase	Much higher ground-loop cost
Hybrid system (fluid cooler)	60% Decrease	9% Decrease	Much higher maintenance cost
Copper U-tubes and 5.0 Btu/h·ft·°F (8.7 W/m·K) grout	21% Decrease	No change	Much higher cost, grout not available

4.4 PERFORMANCE VERIFICATION AND NECESSARY DOCUMENTS

The final step in the design process is to verify system efficiency and check for excessive fan and pump power requirements. The values for fan power are included with unitary heat pump selection, as shown in Section 4.2. For systems with fan-coils and air-handling units, the fan power calculation can be conducted independent of the ground heat exchanger design. However, Step 12 (piping design/pump selection) is dependent on the ground heat exchanger, is quite involved, and requires an entire chapter (Chapter 6) to discuss. The process suggested here, therefore, is to assume the pump power falls within the recommended limit (10% of total system power) and proceed to Step 13. Once completed, Step 12 is conducted to find the actual pump power and then Step 13 can be completed with a more accurate result. The specific verifications are listed here:

- System EER > 12.0 Btu/Wh (COP_c >3.5)
    - EER_{sys} = 13.9 Btu/Wh (COP_c = 4.1) for initial design
    - EER_{sys} = 12.8 Btu/Wh (COP_c = 3.8) for hybrid design with ELT = 85°F (30°C)
    - EER_{sys} = 12.1 Btu/Wh (COP_c = 3.5) for design with ELT = 95°F (35°C)

- System COP_h > 3.5 Btu/Wh (COP_c > 3.5)
  - COP_{sys} = 4.1 for initial design
  - COP_{sys} = 3.8 for hybrid design with ELT = 45°F (7°C)
- Pump power < 10% of total power (for initial design, $kW_{sys} = q_{lc} \div EER = 227 \div 13.9 = 16.3$ kW)
  - W_e = 8 pumps × 0.2 kW = 1.6 kW (10% of total) for common-loop design
  - W_p = 57 gpm (216 L/min or 3.6 L/s) pump at 60 ft (180 kPa) head = 1.3 kW (8% of total) for central loop design
- Fan power < 15% of total power
  - In a unitary heat pump system, fan power is listed as part of heat pump efficiency and cannot be checked. Because the system efficiencies are above minimum recommendations, it is assumed that the fan power is within the suggested limit.

Documents necessary to adequately describe a GCHP installation include at a minimum the following (ASHRAE 2011):

- Heat pump specifications at rated conditions.
- Pump specifications, expansion tank size, and air separator specification.
- Fluid specifications (system volume, inhibitors, antifreeze concentration if required, water quality, etc.).
- Design operating conditions (entering and leaving ground-loop temperatures, return air temperatures [including wet bulb in cooling], airflow rates, and liquid flow rates.
- Pipe header details with ground-loop layout, including pipe diameters, spacing, and clearance from building and utilities.
- Bore depth, approximate bore diameter, approximate bore separation, and grout/fill specifications (thermal conductivity, acceptable placement methods to eliminate any voids).
- Piping material specifications and visual inspection and pressure testing requirements.
- Purge provisions and flow requirements to ensure removal of air and debris without reinjection of air when switching to adjacent subheader circuits.
- Instructions on connections to building loop(s) and coordination of building and ground-loop flushing.
- Sequence of operation for controls.

4.5 REFERENCES

ASHRAE. 2000. ASHRAE Guideline 11, *Minimizing the Risk of Legionellosis Associated with Building Water Systems*. Atlanta: ASHRAE.

ASHRAE. 2011. *ASHRAE Handbook—HVAC Applications*, Geothermal Energy, p. 34.13. Atlanta: ASHRAE.

ASHRAE. 2012a. ANSI/AHRI/ASHRAE ISO Standard 13256-1: 1998 (RA 2012), *Water-Source Heat Pumps-Testing and Rating for Performance—Part 1: Water-to-Air and Brine-to-Air Heat Pumps*. Arlington, VA: Air-Conditioning, Heating, and Refrigeration Institute.

ASHRAE. 2013a. ASHRAE Code of Ethics. Atlanta: ASHRAE. www.ashrae.org/about-ashrae/ashrae-code-of-ethics

ASHRAE. 2013b. *ASHRAE Handbook—Fundamentals*, Climatic Design Information. Atlanta: ASHRAE.

Carlson, S. 2001. *Development of equivalent full load heating and cooling hours for GCHPs applied to various building types and locations*. ASHRAE RP-1120 Final Report. Atlanta: ASHRAE.

Hackel, S., G. Nellis, and S. Klein. 2009. *Optimization of cooling dominated ground-coupled heat pump systems (RP-1384)*. RP-1384 Final Report. Atlanta: ASHRAE.

Kavanaugh, S.P. 2006. *HVAC Simplified*. Atlanta: ASHRAE.

Kavanaugh, S.P. 2008. A 12-step method for closed-loop ground-source heat pump design. *ASHRAE Transactions* 114(2).

Kavanaugh, S.P. 2012. Backward-curved fans. *ASHRAE Journal* 54(5).

Kavanaugh, S.P., and J.S. Kavanaugh. 2012. Long-term commercial GSHP performance, part 1: Project overview and loop circuit types. *ASHRAE Journal* 54(6).

5 Surface-Water Heat Pumps

5.1 INTRODUCTION

Surface-water heat pump (SWHP) systems are a viable and potentially modest-cost GSHP option. Lakes, streams, bays, and even oceans can be very good heat sources and sinks for GSHP systems if properly utilized. Many successful systems are currently in operation, and some design recommendations have been developed. Additional information needs to be assembled and published based on the measurement of the performance of installed systems to supplement the design tools that have been developed from fundamental heat transfer concepts and laboratory experiments. This is especially true regarding environmental impact, degree of fouling, minimum lake size, and depth required to avoid poor performance and prevent unwanted changes (excessive evaporation, heat buildup, disruption of natural temperature gradients, etc.), especially to smaller bodies of water. Several of these issues may be addressed by ASHRAE RP-1385 (2009), which is currently in progress, and readers are encouraged to consult the final report when it becomes available.

This chapter presents information regarding the thermal behavior of surface-water systems, provides examples of successful systems in operation, discusses some existing design methods and tools, and briefly describes installation practices.

Figures 5.1 and 5.2 illustrate the primary systems possible. A closed-loop system is shown in Figure 5.1. Water-to-air heat pumps are linked to a surface-water heat exchanger (SWHE). Heat is exchanged to (cooling mode) or from (heating mode) the reservoir with the fluid (usually a water/antifreeze mixture) circulating inside the SWHE. Heat pumps are then used to transfer heat to or from the air in the building. Figure 5.1 also shows a central loop in the building connected to a network of loose-bundle high-density polyethylene (HDPE) coils. Another popular option is stainless steel or titanium plate exchangers.

In an open-loop system, shown in Figure 5.2, water is pumped from near the bottom of the surface-water reservoir through an intermediate heat exchanger. A closed piping loop connects the building heat pumps to the other side of the intermediate heat exchanger. Heat exchangers are similar to those recommended for use with groundwater heat pumps (see Chapter 8). Water is returned to the lake some distance from the point at which it was removed. Pumps can be either located slightly above or submerged below the lake water level.

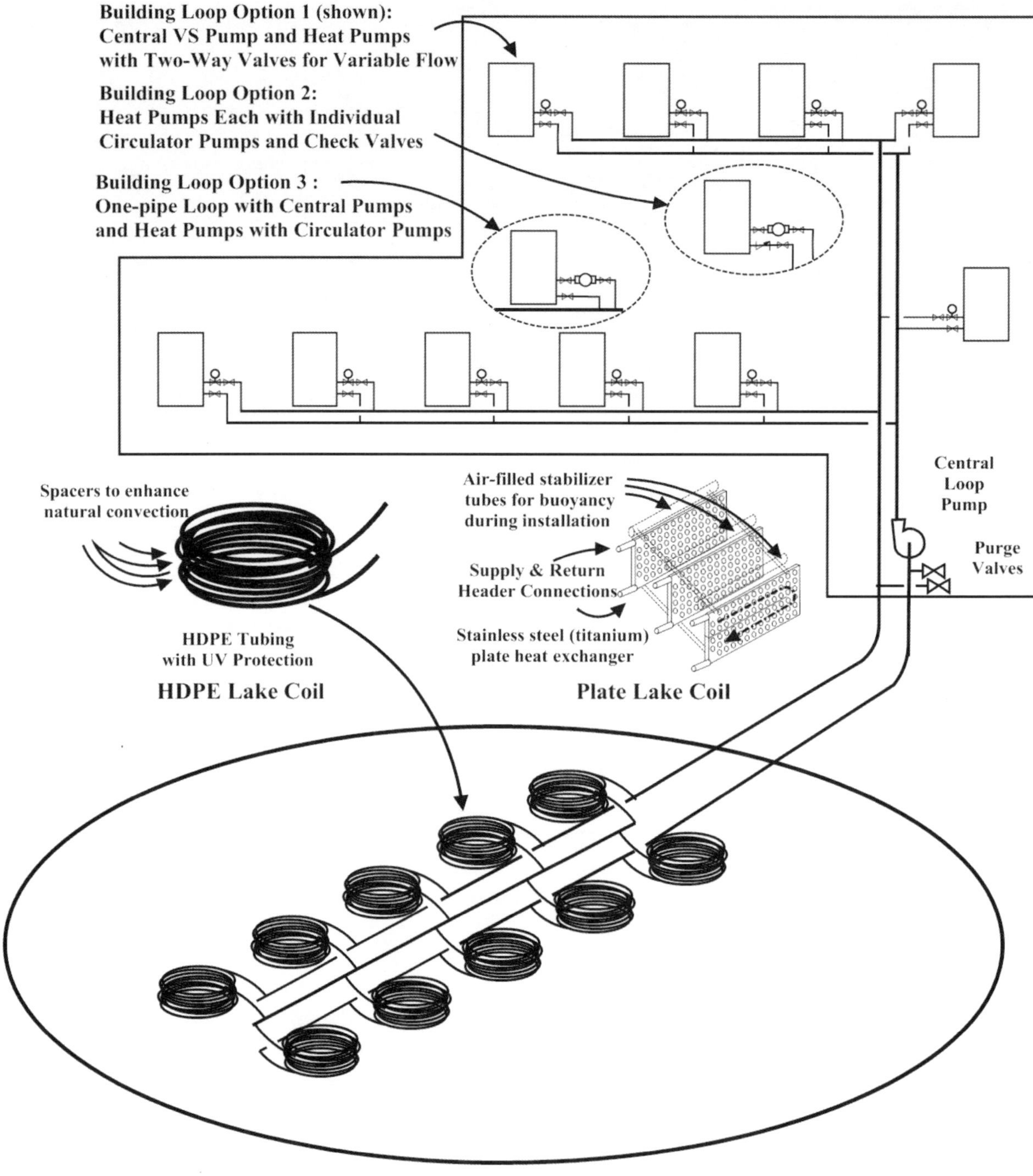

Figure 5.1 Closed-Loop Surface-Water Heat Pump System with HDPE and Plate SWHEs

Open systems are restricted for use in warmer climates or for buildings in colder climates with cooling-only or modest heating requirements. In colder climates, lake temperatures may be less than 40°F (4.4°C). Typical liquid flow rates of 3 gpm/ton (3.2 L/min·kW) result in a heat pump heating-mode leaving liquid temperature (LLT) 6°F (3.3°C) below the entering liquid temperature (ELT). Because the outside surface temperature of the heat exchanger must be lower than the water temperature to remove heat, freezing will occur on the outside surfaces of the SWHEs as the LLT approaches 32°F

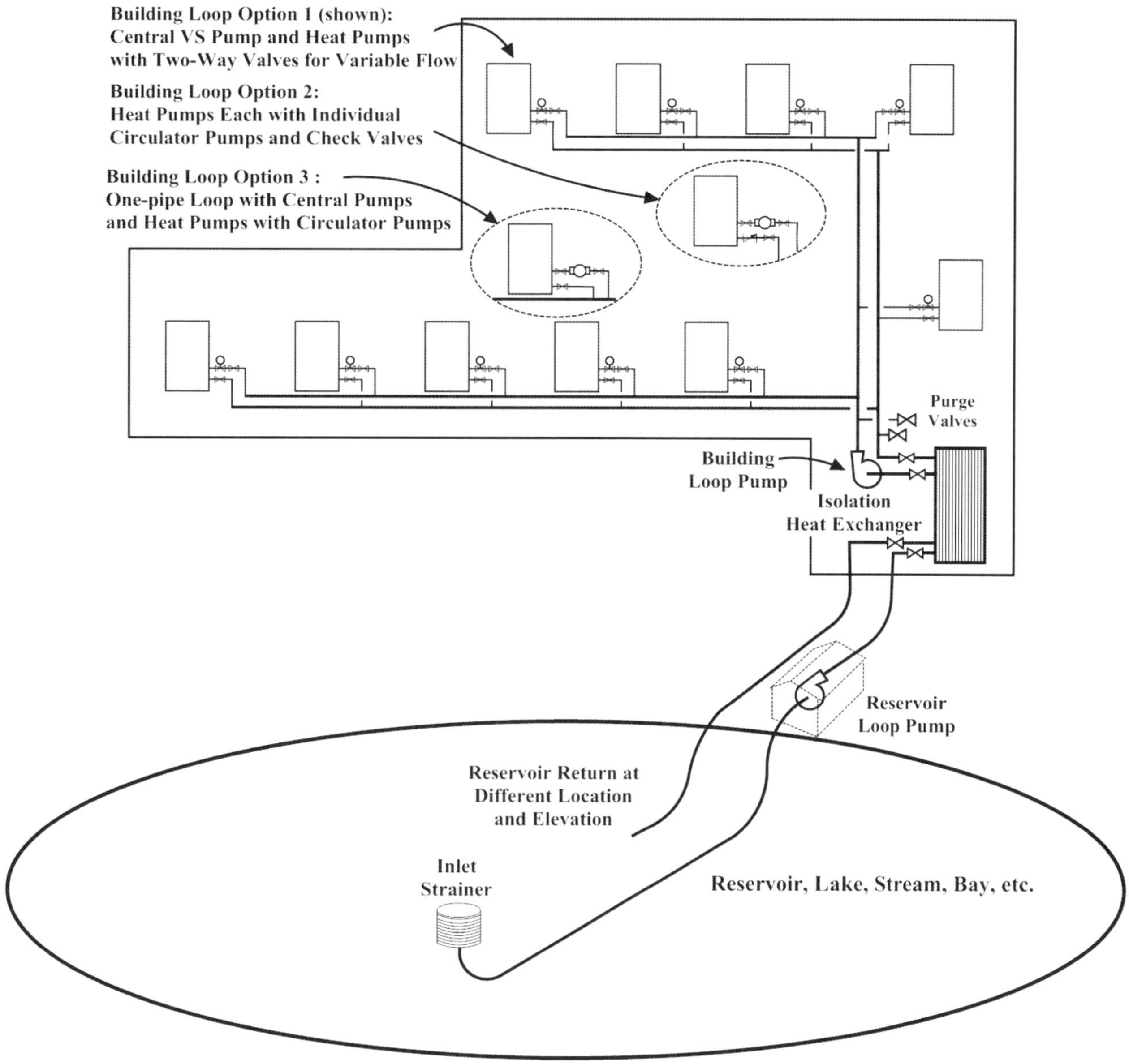

Figure 5.2 Open-Loop System for Cooling-Only or Modest Heating Applications

(0°C) when freshwater reservoirs are used. Ice buildup impedes heat transfer and eventually causes equipment to shut down because of low heat pump ELT. In extreme cases, the ice buildup can become sufficient to cause the SWHE to float to the surface. Even in warmer climates, caution is necessary to verify adequate open-loop flow rates and that the reservoir size and depth are sufficient to ensure the heat exchanger ELT is above 42°F (6°C) at all times during heating conditions.

Thermal stratification of water often results in large quantities of cold water remaining undisturbed near the bottom of deep lakes. In these cases, the building loop may be cold enough to precool (or cool) building return air or ventilation air by being circulated through finned-tube air coils. After leaving these coils, the water can be routed to the heat pumps that are operating before returning to the SWHE (closed-loop systems) or reservoir (open-loop systems).

Reservoir water temperature variations are somewhat more complex and difficult to predict than ground or groundwater temperatures. Therefore, a discussion of the heating and cooling mechanisms in lakes, as well as typical thermal patterns, is necessary before proceeding to system performance and design.

5.2 HEAT TRANSFER IN RESERVOIRS

Figure 5.3 demonstrates the variety of reservoir heat (and mass) transfer mechanisms. Currents and thermal gradients transport heat within reservoirs. As expected, the relative amount of each component varies considerably. A heat rate balance on the reservoir takes the form of

$$q_{solar} + q_{evap} + q_{conv} + q_{grn} + q_{ice} + q_{inflow} + q_{outflow} + q_{leak} + q_{swhp} + q_{rain} = \rho c_p V(\Delta t/\Delta\tau) \tag{5.1}$$

where

q_{xxx} = heat transfer rates for items shown in Figure 5.3, Btu/h (kW)
ρ = density, lb/ft^3 (kg/m^3)
c_p = specific heat, Btu/lb·°F (kJ/kg·K)
V = volume of reservoir, ft^3 (m^3)
Δt = temperature change, °F (°C)
$\Delta\tau$ = time period over which temperature change occurs, h (s)

Because several of these terms are dependent on the temperature of the reservoir, the equation must be solved simultaneously. The equation must also be solved repetitively, as almost all the terms are transient. Ice formation and evaporation, which also result in a loss of mass, complicate the prediction of reservoir temperatures. The difficulty of solving Equation 5.1 is further compounded by the uncertainty of weather patterns. Thus, the incorporation of a weather-data-driven computer model of the reservoir is necessary to predict the bulk temperature change in the reservoir.

Typically, the primary heat input modes are radiant energy from the sun (q_{solar}), inflow (q_{inflow}), convective heat transfer from the surrounding air (q_{conv}), and ground conduction ($q_{grn.\ cond}$). Solar radiation is a dominant heating mechanism, but it occurs primarily in the upper portion of the reservoir. At midday the heat rate can exceed 300 Btu/h·ft^2 (0.95 kW/m^2). Average daily values on a horizontal surface range from near 3000 Btu/day·ft^2 (34 MJ/day·m^2) in June in clear climates to less than 500 Btu/day·ft^2 (6 MJ/day·m^2) on average winter days in higher-latitude cloudy climates. A portion of the incident radiation is reflected at the surface. As shown in Equation 5.2, the remainder is

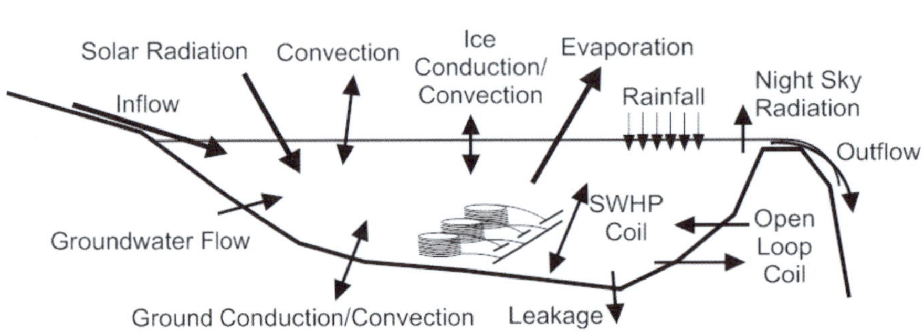

Figure 5.3 Reservoir Heat Transfer Modes

transferred into the reservoir (q_{solar}) and about 40% of this total is absorbed at the surface (Holman 1986). Approximately 93% of the remaining energy is absorbed at depths visible to the human eye. Therefore, almost all the solar radiation is absorbed in the upper portion of all lakes (except very clear ones), so the amount of heat transfer to a reservoir of surface area (A_s) is

$$q_{solar} = (1.0 - \text{Surface Reflectance}) \times A_s \times (q/A)_{horz.\ insolation} \qquad (5.2)$$

Back radiation or night-sky radiation can also contribute to reservoir cooling. Back radiation typically occurs at night when the sky is clear. The relatively warm water surface radiates heat to the cooler sky. For example, a cooling rate of up to 50 Btu/h·ft^2 (160 W/m^2) can be experienced from a lake during a clear night (Duffie and Beckman 1980).

Duffie and Beckman (1980) also provide information on both the calculation of horizontal insolation and data for various cities. Holman (1986) provides an introduction to solar radiation and information on reflectance of water surfaces as a function of the angle of incidence. Siegel and Howell (1981) provide a much more detailed discussion. Weather data are available from a variety of sources, including Dengelman (1986) and InterEnergy (1999).

Convection heat transfer (q_{conv}) to the reservoir occurs when the water surface temperature is lower than the air temperature. Wind speed increases the rate of heat transfer to the lake, but maximum heat gain by convection is usually only 10% to 20% of maximum solar heat gain. Convective cooling or heating in warmer months contributes only a small percentage of the total because of the relatively small temperature difference between the air and lake surfaces.

Inflow heat transfer (q_{inflow}) and accompanying mass transfer include contributions from surface-water flow, groundwater flow, and rainfall. These values are difficult to quantify in terms of both temperatures and flow rates. Heat transfer with the ground (q_g) is likewise difficult to predict given the uncertainty of the makeup (and conductivity) of the materials on the bottom of the reservoir. However, ground conduction appears to be an important mode of heating in a lake that is frozen at the surface.

Evaporative heat transfer (q_{evap}) at the surface is a primary contributor to cooling of reservoirs. Evaporative cooling is dependent on the lake water surface temperature, the wind speed, and the amount of moisture in the surrounding air. A warm lake in a dry climate can be cooled at a rate approximately equal to the heat gained by maximum solar radiation. The rate of cooling increases rapidly as the surface temperature of the water rises because of the increasing vapor pressure difference between the water surface and the air. For example, heat transfer from a reservoir with 85°F (29.4°C) surface-water temperature is approximately 50% greater on a warm day compared to that of an 80°F (26.7°C) surface. Wind speed also has a great influence on cooling rate.

A good deal of empirical data can be found regarding the rate of evaporation (E) from the surfaces of lakes, which is typically expressed in the units of level change per day. When the other effects of lake level (inflow/outflow, leakage, rainfall) are minimized, the change in lake level has been expressed as (USGS 1952)

$$E = 0.122 \times (e_o - e_a) \times (0.417 + 0.096\,u_a) \qquad \text{(I-P)} \qquad (5.3a)$$

$$E = 0.0054 \times (e_o - e_a) \times (0.259 + 0.060\,u_a) \qquad \text{(SI)} \qquad (5.3b)$$

where

E = reduction in water level, ft/day (m/day)

e_o = saturated vapor pressure of water at surface temperature, lb/in.^{2} (kPa)
e_a = vapor pressure of surrounding air, lb/in.^{2} (kPa)
u_a = wind speed, mph (km/h)

The evaporative heat rate can be calculated from

$$q_{evap} = E \times A_s \times h_{fg} @ t_s \times \rho_w @ t_s \tag{5.4}$$

where
h_{fg} = latent heat of vaporization , Btu/lb (kJ/kg)
ρ_w = density of water, lb/ft^3 (kg/m^3)
t_s = reservoir surface temperature, °F (°C)

Additional evaporation will occur as a result of heat pump condenser rejection. This may be problematic in smaller lakes with large condenser heat transfer loads (q_{cond}) and operating hours (τ_{Oper}). The maximum amount of makeup water (m_{water}) required is computed using Equation 4.1 and assuming evaporation is the only mode of cooling for the additional heat pump load:

$$m_{water} = \frac{q_{cond} \times \tau_{Oper}}{h_{fg}} = \frac{Q_{cond}}{h_{fg}}$$

EXAMPLE 5.1— DETERMINING SURFACE-WATER EVAPORATION AND HEAT TRANSFER RATES

Problem and Solution in I-P Units

Calculate the level change and evaporative heat transfer rate from a 1 acre lake when the surface water temperature is 80°F, wind speed is 5 mph, and air temperature is 95°F db/75°F wb. Repeat for an 85°F water surface temperature. Compare this with the level change induced by the addition of a heat pump with a 10 ton (35 kW) cooling capacity with an EER = 13.6 Btu/Wh that operates 8 h/day. Assume evaporation is the only mode of heat transfer.

The vapor pressure of water at 80°F is 0.50744 psia and at 85°F is 0.59656 psia. Recall the water vapor pressure of air is equal to the vapor pressure of saturated air at the dew-point temperature. The dew point of 95°F/75°F air is 67°F. The saturated vapor pressure of water at 67°F is 0.32777 psia. The enthalpy of vaporization (h_{fg}) at 80°F is 1048 Btu/lb, and the density of liquid water is 62.2 lb/ft^3. These values are 1045 Btu/lb and 62.2 lb/ft^3 at 85°F (ASHRAE 2013a).

For 80°F surface water temperature:

$$E = 0.122 \times (0.50744 \text{ psia} - 0.32777 \text{ psia}) \times [0.417 + (0.096 \times 5 \text{ mph})] = 0.0197 \text{ ft/day}$$

$$\begin{aligned} q_{evap} &= 0.0197 \text{ ft/day} \times 1048 \text{ Btu/lb} \times 62.2 \text{ lb/ft}^3 \div 24 \text{ h/day} = 53.5 \text{ Btu/h·ft}^2 \\ &= 53.5 \text{ Btu/h·ft}^2 \times 43{,}560 \text{ ft}^2\text{/acre} = 2.33 \times 10^6 \text{ Btu/h per acre} \end{aligned}$$

For 85°F surface water temperature:

$$E = 0.122 \times (0.59656 \text{ psia} - 0.32777 \text{ psia}) \times [0.417 + (0.096 \times 5 \text{ mph})] = 0.0294 \text{ ft/day}$$

$$\begin{aligned} q_{evap} &= 0.0294 \text{ ft/day} \times 1045 \text{ Btu/lb} \times 62.2 \text{ lb/ft}^3 \div 24 \text{ h/day} = 79.6 \text{ Btu/h·ft}^2 \\ &= 79.6 \text{ Btu/h·ft}^2 \times 43{,}560 \text{ ft}^2\text{/acre} = 3.47 \times 10^6 \text{ Btu/h per acre} \end{aligned}$$

Note: For comparison, a typical average solar flux on a water surface in June is 80 × 10^6 Btu/day per acre, an average of 77 Btu/h·ft^2 over an entire 24 h day.

Equation 3.2 is used to find the heat rate and total amount of heat delivered to the reservoir by the heat pump.

$$q_{cond} = q_{lc} \times \frac{\text{EER} + 3.412}{\text{EER}} = 10 \text{ tons} \times 12{,}000 \text{ Btu/ton·h} \times \frac{13.6 + 3.412}{13.6} = 150{,}000 \text{ Btu/h}$$

$$Q_{cond} = 150{,}000 \text{ Btu/h} \times 8 \text{ h} = 1.2 \times 10^6 \text{ Btu}$$

The amount of water (m_{water}) that will be evaporated per day (assuming 8 h/day operation), assuming that evaporation is the only cooling mechanism, is

$$m_{water} = \frac{Q_{cond}}{h_{fg}} = \frac{1.2 \times 10^6 \text{ Btu}}{1045 \text{ Btu/lb}} = 1150 \text{ lb}$$

$$\text{Volume} = \frac{m_{water}}{\rho_{water}} = \frac{1150 \text{ lb}}{62.2 \text{ lb/ft}^3} = 18.5 \text{ ft}^3$$

The decline in the reservoir level due to the heat pump (E_{HP}) is found by dividing the volume of the evaporated liquid by the surface area of the reservoir:

$$E_{HP} = \frac{\text{Volume}}{A_{surface}} = \frac{18.5 \text{ ft}^3}{43{,}560 \text{ ft}^2} = 0.00042 \text{ ft}$$

Note: This level decline is 2% of the decline calculated for the naturally occurring decline with an 80°F lake surface temperature.

Problem and Solution in SI Units

Calculate the level change and evaporation rate for a 5000 m^2 reservoir at 27°C and air with a 20°C dew point and 10 km/h wind speed. Values for water vapor pressure, density, and enthalpy of vaporization are found in the SI edition of *ASHRAE Handbook–Fundamentals* (2013c).

$$e_o = 3.5679 \text{ kPa}$$

$$e_a = 2.3392 \text{ kPa}$$

$$\rho_w = 996 \text{ kg/m}^3$$

$$h_{fg} = 2437 \text{ kJ/kg}$$

$$E = 0.\ 0054 \times (3.5679 - 2.3392) \times [0.259 + (0.060 \times 10 \text{ km/h})] = 0.0057 \text{ m/day}$$

$$q_{evap} = 0.0057 \text{ m/day} \times 5000 \text{ m}^2 \times 2437 \text{ Btu/lb} \times 996 \text{ kg/m}^3$$

$$q_{evap} = 69.2 \times 10^6 \text{ kJ/day} = 2.88 \times 10^6 \text{ kJ/h}$$
$$= 2.88 \times 10^6 \text{ kJ/h} \times 1000 \text{ W/kW} \div 5000 \text{ m}^2 \div 3600 \text{ s/h} = 160 \text{ W/m}^2$$

The relative impact of the heat transfer from surface-water heat pumps (q_{swhp}) should be viewed in perspective of the naturally occurring heat transfer rates in reservoirs. Consider the example 10,000 ft^2 (930 m^2) office building discussed in Chapter 4, which was conditioned by a 20 ton (70 kW) heat pump system and had a daily part-load factor of 32% to 40%. If this system is connected to a 1 acre (43,560 ft^2) (4050 m^2) lake, the amount of heat rejected to the lake for an EER = 15 (COP$_c$ = 4.4) system at peak load would be

$$\frac{q_{swhp}}{A} = \frac{q_{lc}}{A} \times \frac{q_{cond}}{q_{lc}} = \frac{q_{lc}}{A} \times \frac{\text{EER} + 3.412}{\text{EER}}$$

$$\frac{q_{swhp}}{A} = \frac{20 \text{ tons} \times 12{,}000 \text{ Btu/h·ton}}{43{,}560 \text{ ft}^2} \times \frac{15 + 3.412}{15} = 6.8 \text{ Btu/h·ft}^2 \qquad \text{(I-P)} \qquad (5.5a)$$

$$\frac{q_{swhp}}{A} = \frac{q_{lc}}{A} \times \frac{q_{cond}}{q_{lc}} = \frac{q_{lc}}{A} \times \frac{\text{COP}_c + 1.0}{\text{COP}_c}$$

$$\frac{q_{swhp}}{A} = \frac{70 \text{ kW}}{4050 \text{ m}^2} \times \frac{4.4 + 1.0}{4.4} = 0.021 \text{ kw/m}^2 = 20 \text{ W/m}^2 = 21 \text{ J/s·m}^2 \qquad \text{(SI)} \qquad (5.5b)$$

This would be approximately 2% of the peak solar radiation incident on the lake surface in the cooling season.

On a daily basis this would be

$$\frac{q_{swhp}}{A}/\text{day} = \frac{q_{lc}}{A} \times \text{PLF} \times 24 \text{ h/day} = 6.76 \text{ Btu/h·ft}^2 \times 0.32 \times 24 \text{ h/day} = 52 \text{ Btu/ft}^2\text{·day} \qquad \text{(I-P)}$$

$$\frac{q_{swhp}}{A}/\text{day} = \frac{q_{lc}}{A} \times \text{PLF} \times 24 \text{ h/day} = \frac{21 \text{ J/s·m}^2 \times 0.32 \times 86{,}400 \text{ s/day}}{1000 \text{ J/kJ}} = 580 \text{ kJ/m}^2\text{·day} \qquad \text{(SI)}$$

This would also be approximately 2% of the clear-day insolation in the cooling season.

It is clear that accurate modeling of reservoirs, even with very sophisticated simulation tools, is nearly impossible given the uncertainty, variation, and unavailability of required input information. The following section offers an alternative that suggests the measured historical reservoir temperature data is a more appropriate resource in lieu of a futile quest to accurately model the behavior of Mother Nature.

5.3 THERMAL PATTERNS IN RESERVOIRS AND STREAMS

The impact of SWHPs on reservoirs is important to evaluate in terms of the relative amount of heat added (Equation 5.5) or extracted and the potential change in sometimes critical water levels (Equation 5.3). It is also important to evaluate the potential change in thermal patterns that can occur when a significant amount of heat is rejected through coils in very cold water near the bottom of a stratified reservoir.

Primary input data for any SWHP design procedure are reservoir temperature versus depth profiles at various times of the year. Typically, the critical periods are late winter and summer, when reservoirs reach their minimum and maximum temperatures. The large thermal mass of a water body results in the more extreme temperatures occurring late in the seasons. Water has several unusual characteristics. Common knowledge is that water expands upon freezing (solidifying), unlike most other materials. Equally odd is that the maximum density of water occurs at 39.2°F (4°C), not at the freezing point of water. This behavior, when coupled with the normal modes of heat transfer to and from reservoirs, results in temperature profiles advantageous to efficient heat pump operation. Figure 5.4 shows seasonal temperature versus depth plots for a stagnant lake in a location that has both high summer temperatures and sufficient winter temperatures to form ice on the lake surface (Peirce 1964).

In the winter the coldest water is at the surface. Because water at 32°F (0°C) is less dense than water in the 35°F to 45°F (2°C to 7°C) range, it tends to remain at the surface and freeze. The bottom of a deep lake will remain a few degrees warmer than the surface. This condition is referred to as *winter stagnation*. The warmer water serves as a better heat source than the colder water at the surface. In colder climates, a shallow lake tends to be a better heat source after it has frozen because the ice tends to insulate the water from the disturbances of cold, windy weather.

As spring approaches, surface water is warmed until the temperature approaches the maximum density point of 39.2°F (4°C). The winter lake stratification becomes unstable and circulation loops begin to develop from top to bottom. This condition is referred to as the *spring overturn*. The lake temperature is fairly constant at all levels.

Later in the spring, as the water temperature rises, the circulation loops tend to stay in the upper portion of the lake. This is a result of the warmer water near the surface (heated by solar radiation) having a lower density than the cooler water that begins to settle at the

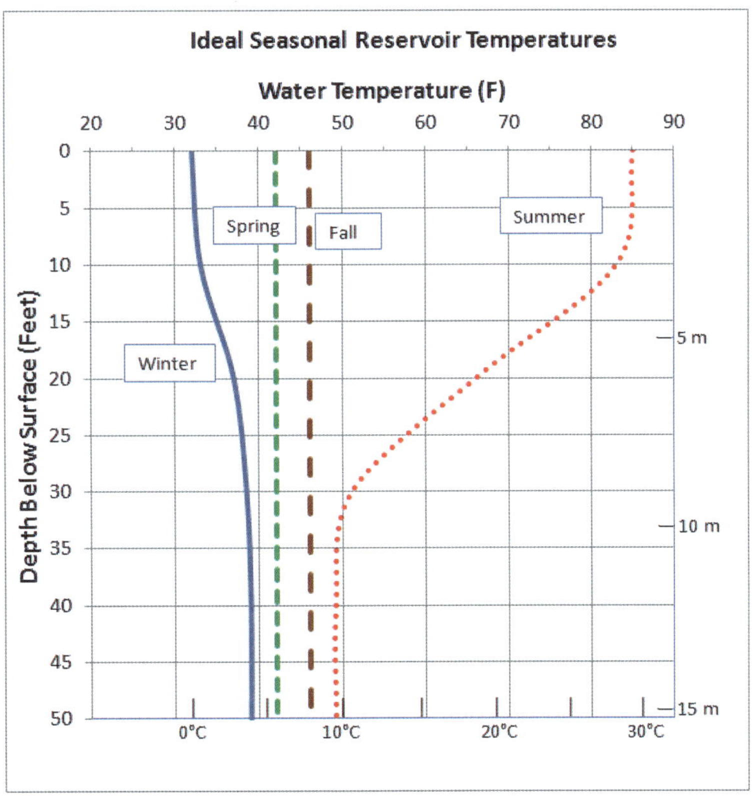

Figure 5.4 Reservoir Depth vs Temperature for Four Seasons

bottom of the lake. This pattern continues throughout the summer. The upper portion of the lake remains relatively warm, with evaporation cooling the lake and solar radiation warming it. The lower portion of the lake remains cold because most radiation is absorbed in the upper zone, circulation loops do not penetrate to the lower zone, and conduction to the ground is relatively small. The result is that in deeper lakes with small to medium inflows, the upper zone in summer is 70°F to 90°F (21°C to 32°C), the lower zone is 40°F to 55°F (4°C to 13°C), and the intermediate zone (thermocline) has a sharp change in temperature within a small change in depth. This condition is referred to as *summer stagnation*.

As the fall season begins, the water surface begins to cool by back radiation and evaporation. The convection loops begin to extend deeper and deeper into the lower zone since the surface water is now denser. Eventually the convection loops extend to the bottom of the lake and the stratification is destroyed. The entire lake is approximately the same temperature. This condition is referred to as the *fall overturn*. As winter approaches, the upper portion begins to cool and approach the freezing point and the lower levels approach the maximum density temperature of 39.2°F (4°C).

The ideal temperature patterns shown in Figure 5.4 hold the promise of high-efficiency heat pump performance. Summer cooling with water at 40°F to 55°F (4°C to 13°C) offers heat pump operation, precooling, or total cooling with efficiency far exceeding the most efficient conventional refrigeration equipment. In northern climates, a 39.2°F (4°C) heat source would be a big improvement over much colder air.

Many water bodies do exhibit near ideal temperature profiles. However, a variety of circumstances disrupt these profiles. These include high rates of inflows/outflows, insufficient depth for stratification, level fluctuations, wind, and lack of enough cold weather to establish sufficient amounts of cold water necessary for summer stratification. Therefore, it is suggested that thermal surveys of reservoirs be conducted or that previous surveys in similar geographic locations be consulted.

Figures 5.5 to 5.7 show results (temperature-depth plots) of thermal surveys conducted in Alabama (Peirce 1964). Alabama has a relatively mild winter climate and hot and humid summers. However, even with these conditions a vast amount of cold water at 45°F (7°C) is available in August, as demonstrated in Figure 5.5. The upper 20 ft (6 m) of the lake in this figure is between 80°F and 86°F (27°C and 30°C) at this time. The fall overturn appears to occur between 45°F and 50°F (7°C and 10°C), as indicated by the Dec 8 temperature-depth profile. The lake varies from the ideal profile because the lack of severe cold weather prevents the establishment of ice or water below 45°F (7°C). Thus there is no winter stagnation. The lake is used as a water reservoir for Birmingham, Alabama. When the thermal survey was conducted in 1961–62, the average outflow was relatively small, at 72 ft^3/s (2 m^3/s), compared to its size (1540 acres [620 ha]) and depth.

Figure 5.6 is included to demonstrate temperature profiles in rivers or lakes with high inflows/outflows. The data were taken at a reservoir south of Birmingham that is used for hydroelectric power generation. Although the lake has moderate depth (60 ft [18 m]), it is relatively narrow (1800 ft [550 m]) at the survey point and 15 mi (24 km) in length. The lake flow rate is between 11,600 and 13,500 ft^3/s (330 and 380 m^3/s). This high flow is the primary reason that no summer stratification occurs. The temperature of the entire body of water for each season is near the monthly average air temperature. The winter temperature profile is very similar to the temperatures of the more stagnant lake presented in Figure 5.5.

Figure 5.7 shows the temperature profile of a shallow lake near Tuscaloosa, Alabama, that is relatively stagnant. Again there is no significant summer thermal stratification. In this case, the lack of cold water in the summer is a result of shallow depth and limited

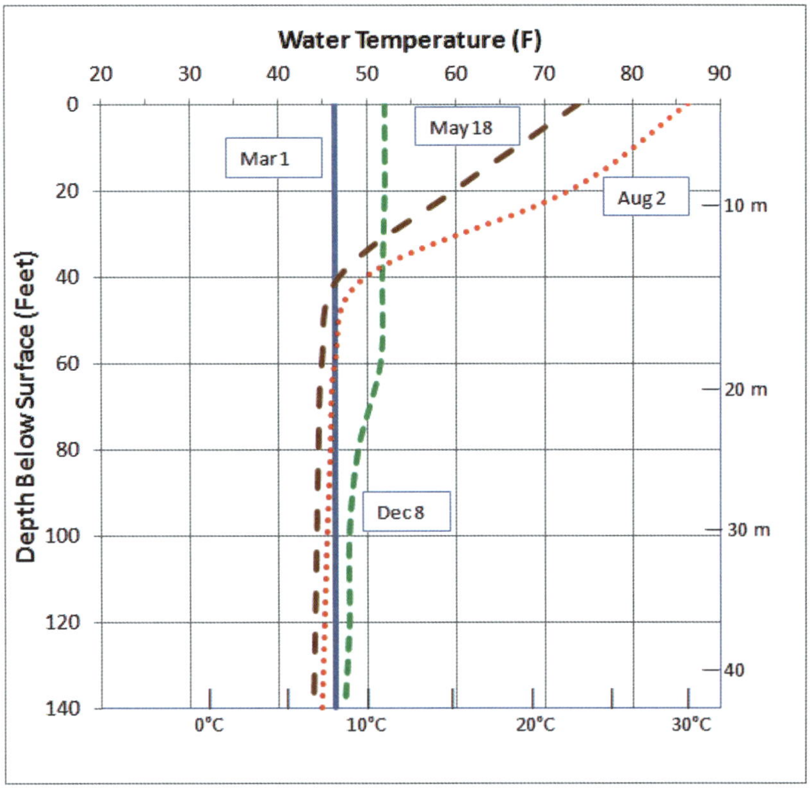

Figure 5.5 Temperature Profiles for a Deep Lake in North Alabama (Peirce 1964)

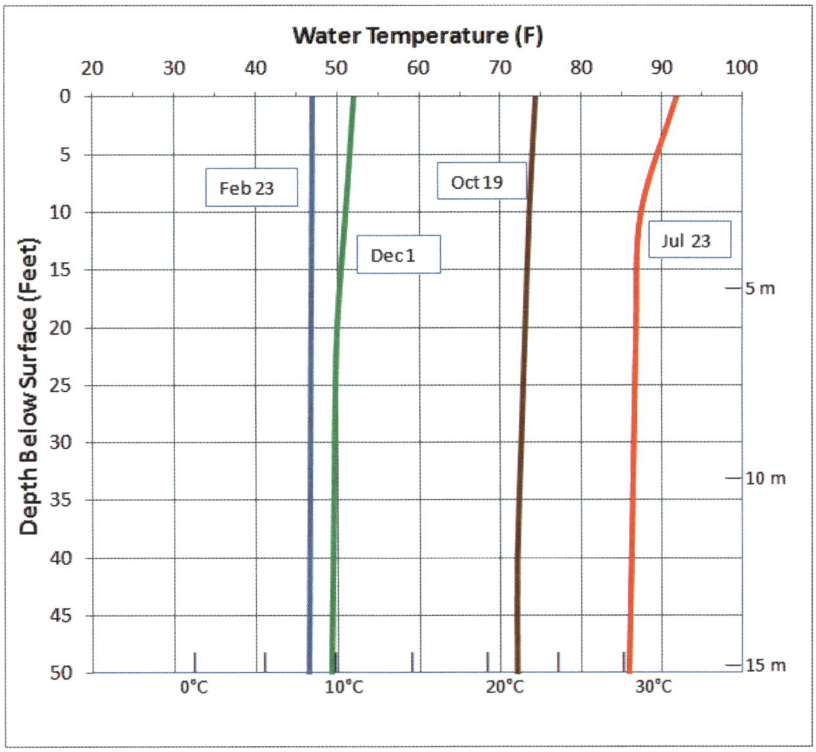

Figure 5.6 River Temperature Profiles in Central Alabama (Peirce 1964)

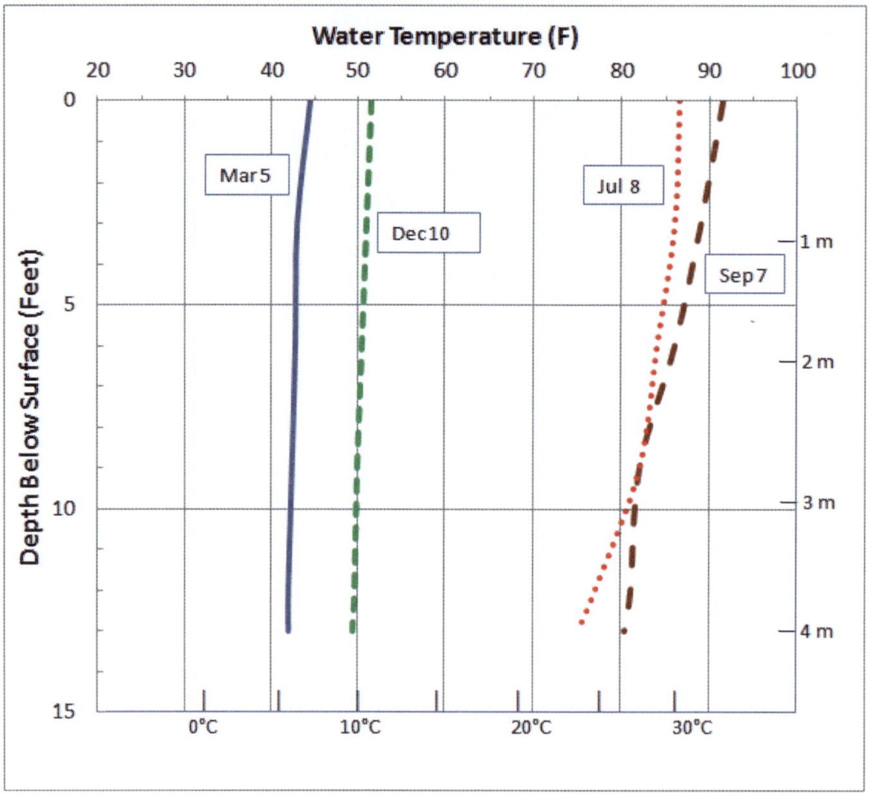

Figure 5.7 Shallow Lake Temperature Profiles in Central Alabama (Peirce 1964)

thermal mass. Radiation penetrates to the lower depths and warms the entire lake to above 80°F (27°C) by August. Mixing, resulting from wave action, also contributes to the warming of the lower levels. In the winter, the temperature is very similar to the deeper lakes except that it is slightly lower (42°F [6°C]) in the latter months because of the smaller thermal mass of the lake.

Figure 5.8 illustrates temperature profiles for a deep lake bordering Seattle, Washington. Although it is one of the northernmost cities in the continental United States (latitude = 48°N), it has a relatively mild climate in both summer and winter. The water in the lower half of this deep lake remains between 45°F and 48°F (7°C and 9°C), which could provide excellent heat pump performance in both cooling and heating. In spite of the perception that the local climate is cloudy and wet, the relative humidity in the cooling season is surprisingly low. This improves the potential for direct cooling and/or precooling with lake water in many buildings. Even the upper portions of the lake have excellent potential for high heat pump cooling efficiency since the temperatures at this location remain below 70°F (21°C).

Figure 5.9 displays the temperature profiles of a deep, high-flow reservoir that is one of the largest flood control/power generation impoundments on the Tennessee River. The reservoir is located north of Knoxville, Tennessee, at latitude of 36°N. The lake demonstrates a summer thermocline with the lower portion remaining below 55°F (13°C) throughout the summer months. It is interesting to note the significant effects of thermal mass evidenced by the November temperatures being warmer than the summer temperatures below a depth of 40 ft (12 m). The late winter temperatures of the lake remain nearly 45°F (7°C) at all depths.

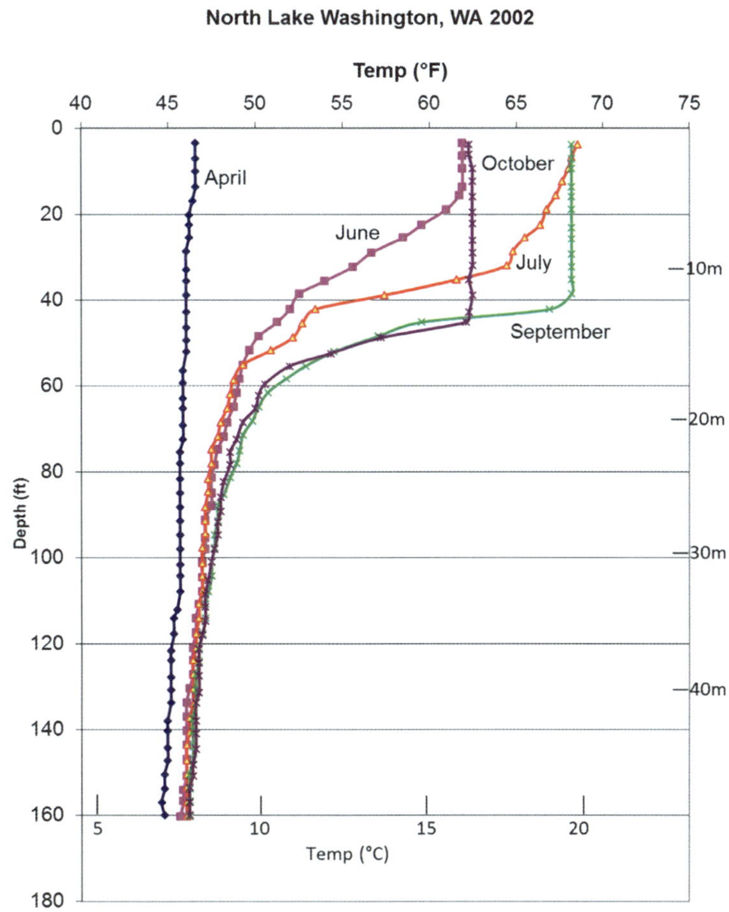

Figure 5.8 Deep Lake Temperatures in Temperate Climate (Hattemer and Kavanaugh 2005)

Temperatures in deeper northern lakes, such as those shown in Figure 5.10 for Lake Grindstone in Minnesota, more closely match the ideal profiles in Figure 5.4. The near-surface winter temperature of 32°F (0°C) indicates ice formation, as expected in this climate. In February, at a depth of 10 ft (3 m) the temperature ranges from 37°F to 39°F (3°C to 4°C) at the 90 ft (27 m) depth. The lower half of the lake remains near 42°F (6°C) during the fall, spring, and summer. Winter profiles in shallow, cold-climate lakes have similar profiles but in many cases are slightly colder, as shown in Figure 5.11. As expected, April through November temperature variations in the lower portions of shallow lakes are greater than those in lakes with greater volumes of water. This creates concern in cold-climate applications with regard to excessive temperature variations when large amounts of heat are withdrawn from small lakes to support heat pump operation. It is possible that ASHRAE RP-1385 (2009) will address this issue when the final report is completed.

A final point to consider is the depth to the summer thermocline (the portion of the lake with a pronounced change in temperature). The reservoir temperature profiles shown in Figures 5.5 through 5.11 indicate the depths of the summer thermoclines are between 25 and 50 ft (7.5 and 15 m). The conclusion that lakes deeper than 25 to 40 ft (8 to 12 m) will have summer temperatures below 50°F (10°C) cannot automatically be drawn. In murky lakes, solar radiation is blocked at shallow depths and the thermoclines are shallow. In clear bodies of water, radiation warms deeper waters and the thermocline may be

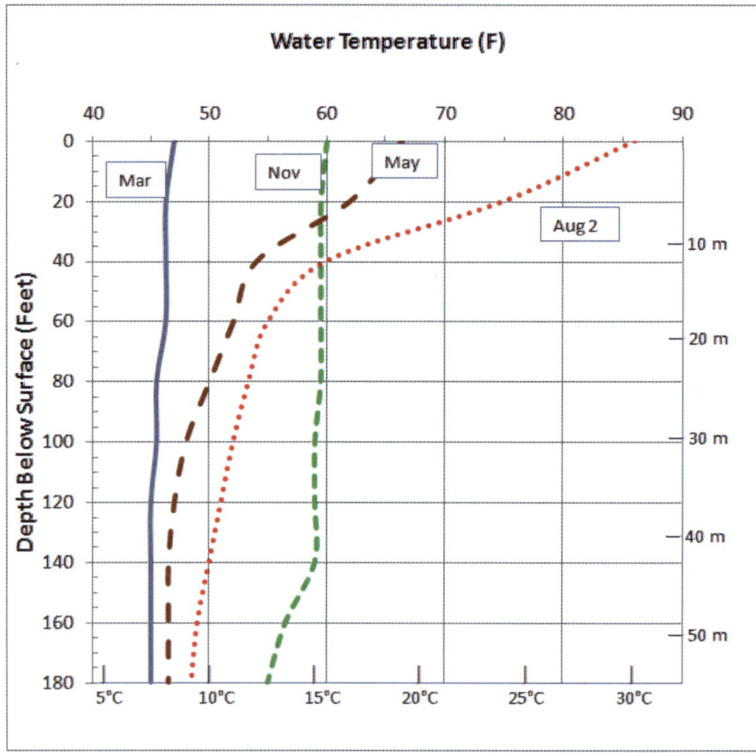

Figure 5.9 High-Flow Reservoir Temperatures in Tennessee (Hattemer and Kavanaugh 2005)

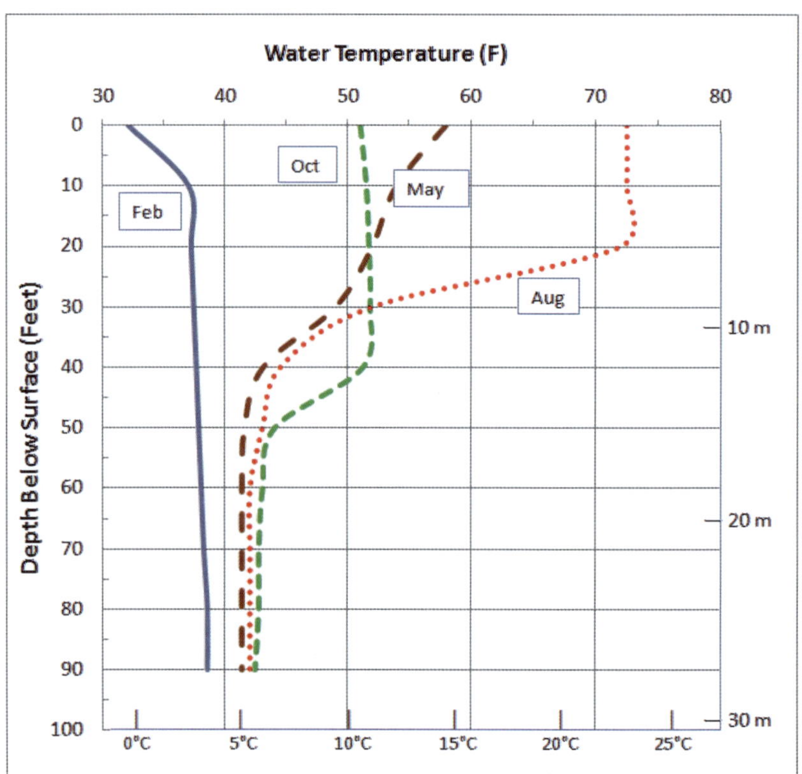

Figure 5.10 Deep Lake Temperatures in Minnesota (Hattemer and Kavanaugh 2005)

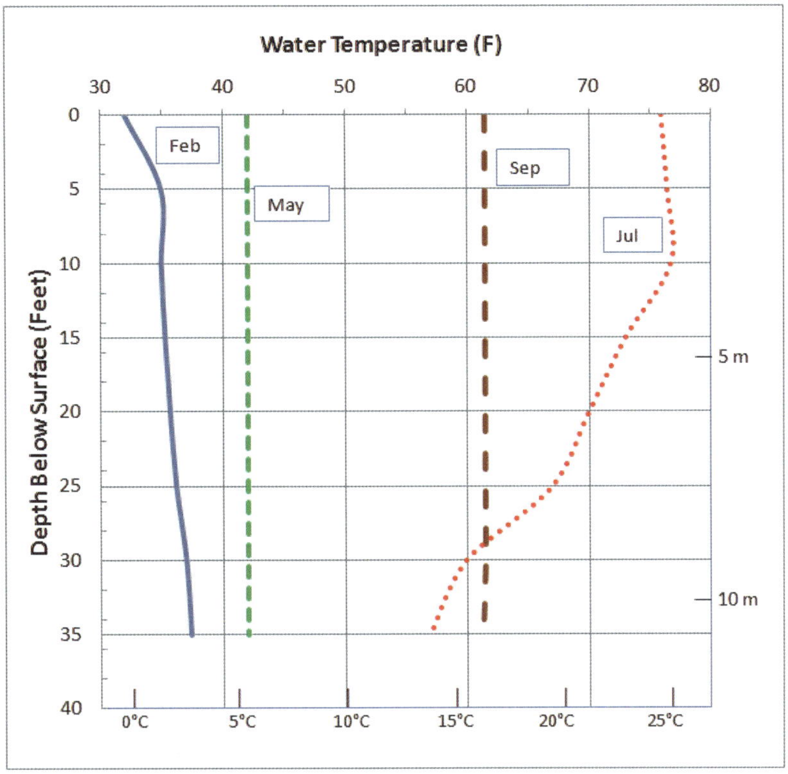

Figure 5.11 Shallow Lake Temperatures in Minnesota (Hattemer and Kavanaugh 2005)

much deeper. Pezent and Kavanaugh (1990) provide information on the use of a high-contrast Secchi disk for predicting the depth of solar radiation penetration.

Additional temperature profiles and information sources are referenced by Hattemer and Kavanaugh (2005). This includes a reference (EIS 2014) with more than 40 temperature profiles in a format like Figure 5.8. Additional information is also provided by Hattemer (2005).

5.4 FUNDAMENTALS OF CLOSED-LOOP SURFACE-WATER HEAT EXCHANGERS

The closed-loop SWHP system as shown in Figure 5.1 has three primary advantages. The most obvious is the reduced fouling resulting from the circulation of clean water (or water/antifreeze solution) through the heat pump. A less evident advantage is the reduced pumping power requirement. Closed-loop pumping systems can be designed to operate with less than 60 W/ton (16 W_e/kW_t). This results from the negligible elevation head from the lake surface to the heat pump. The third advantage of closed-loop systems is that open-loop systems are not recommended for heating when winter lake temperatures below 42°F (6°C) are possible. The leaving water temperature (LWT) will be about 6°F (3.3°C) below the entering water temperature (EWT) for a 3 gpm/ton (3.2 L/min·kW) flow. Furthermore, the surface of the heat pump water coil must be several degrees below the LWT in order remove the necessary heat from the water. Thus, the heat pump LWT must be 36°F (2°C) or higher to avoid frost on the water coil, which suggests the heat pump EWT from the reservoir should be 42°F (6°C) or higher for open-loop systems to operate with some margin of safety. Closed-loop systems with environmentally acceptable antifreeze solutions can operate with a heat pump leaving liquid temperature (LLT) below the freeze point of water as long as the heat exchanger in the reservoir is large enough to prevent the outside coil surface (reservoir water side) from falling below 32°F (0°C).

In addition to the potential for ice buildup on the outside of an undersized SWHE, there are several disadvantages of closed-loop systems, most of which can be avoided or minimized with quality design:

- An obvious disadvantage of the closed-loop system is the possibility of damage to coils located in public reservoirs.
- There is a possibility of fouling on the outside of the SWHE, which would more likely be an issue in murky lakes or in installations in which coils are located on or near the reservoir bottom.
- The performance of the heat pumps is slightly reduced because ELTs are several degrees higher (cooling mode) or lower (heating mode) when compared to the reservoir temperature.
- There may be regulations that either prohibit SWHPs or raise the cost of compliance to unreasonable values so that SWHPs are not economically viable.
- The reservoir is of insufficient size or depth to support the heat pump system, which could result in system shutdown, inadequate performance, or unacceptable temperature changes.

There are currently acceptable options for SWHE materials. HDPE (PE 3406, PE 3408, or PE 4710) is a recommended choice in terms of performance, durability, and economics. These plastic pipes typically have protection from ultraviolet radiation, but protection above standard practice is suggested if headers are exposed in shallow water near the shore. All connections must be thermally fused. Stainless steel plate heat exchangers are also acceptable. Polyvinyl chloride (PVC) pipe and plastic pipe with mechanical fastener joints are not acceptable or recommended for SWHEs. Copper tubing has been used in some applications, but the relative impact of fouling is much greater because the surface area is likely to be much less than that of SWHEs with HDPE tubing. Ice formation would be more problematic in colder climates.

The design approach begins with calculations for a single-pipe SWHE placed horizontally in the reservoir. The required coil total length is found by rearranging the more familiar equation for heat transfer rate based on surface area to one based on length of tubing. As shown in Figure 5.12, the coefficients are transformed into thermal resistances for the fluid film inside the pipe (R_i), the pipe resistance (R_p), the fluid film outside the pipe (R_o), and the fouling resistance on the outside surface (R_f). These terms are summed to find the overall thermal resistance (R_{ov}).

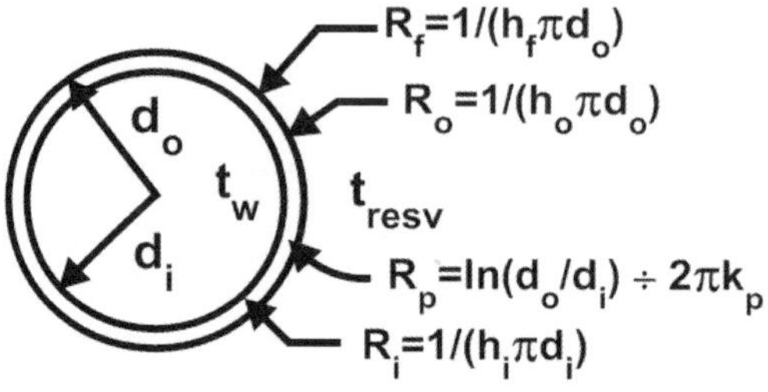

Figure 5.12 Thermal Resistance per Unit Length for Single SWHE Coil

SWHP coils must be arranged in parallel flow paths to minimize head loss and pump size while maintaining adequate fluid velocity for satisfactory heat transfer. However, Equation 5.6 is suggested to determine the required overall length (L_{swhe}) of a single-pipe SWHE and is typically used for design purposes rather than sizing each individual coil.

$$L_{swhe} = \frac{q_{hp} \times R_{ov}}{\text{LMTD}} = \frac{q_{hp} \times (R_i + R_p + R_o + R_f)}{\dfrac{(\text{LLT} - t_{resv}) - (\text{ELT} - t_{resv})}{\ln[(\text{LLT} - t_{resv}) \div (\text{ELT} - t_{resv})]}} \tag{5.6}$$

where

q_{hp} = heat pump heat transfer rate (q_{cond} in cooling, q_{evap} in heating), Btu/h (W)
R_{ov} = overall thermal resistance of per-unit-length SWHE coil, h·ft·°F/Btu (m·K/W)
R_i = $1/h_i \pi d_i$ = thermal resistance of inside fluid film, h·ft·°F/Btu (m·K/W)
R_p = $\ln(d_o/d_i)/2\pi k_p$ = thermal resistance of pipe wall, h·ft·°F/Btu (m·K/W)
R_o = $1/h_o \pi d_o$= thermal resistance of outside fluid film, h·ft·°F/Btu (m·K/W)
R_f = $1/h_f \pi d_o$ = thermal resistance of fouling factor, h·ft·°F/Btu (m·K/W)
h_i = inside heat transfer coefficient, Btu/h·ft^2·°F (W/m^2·K)
k_p = thermal conductivity of pipe, Btu/h·ft·°F (W/m·K)
h_o = outside heat transfer coefficient, Btu/h·ft^2·°F (W/m^2·K)
h_f = inside heat transfer coefficient, Btu/h·ft^2·°F (W/m^2·K)
t_{resv} = reservoir temperature at depth of coil, °F (°C)
LLT = leaving liquid temperature of SWHE, °F (°C)
ELT = entering liquid temperature of SWHE, °F (°C)
LMTD = log mean temperature difference, °F (°C)

The inside heat transfer coefficients for forced convection are determined by a variety of equations that were developed for the three flow regimes of laminar, transition, and turbulent. The appropriate flow regime is identified by the Reynolds number (Re):

$$\text{Re} = \frac{\rho d_i V}{\mu} = \frac{d_i V}{\nu} \tag{5.7}$$

where

ρ = fluid density, lb/ft^3 (kg/m^3)
d_i = inside diameter, ft (m)
V = fluid velocity, ft/s (m/s)
μ = dynamic viscosity, lb/ft·s (kg/m·s, centipoise ≡ 0.001 kg/m·s)
ν = kinematic viscosity, ft^2/s (m^2/s)

Laminar flow (Re < 2300 ±200) is characterized by layers of fluid sliding along in the direction of flow without mixing. Layers near the pipe wall move at a low velocity, and maximum velocity occurs at the center of the pipe. Laminar flow thermal resistance is high because the fluid is not mixing and heat transfer through the fluid boundary layer at the pipe wall is via conduction. Laminar flow occurs when the fluid velocity (flow rate) is low and/or the fluid has a high viscosity. In SWHE applications, laminar flow impacts heat transfer in the heating mode because fluid viscosities are elevated at lower temperatures and with the addition of antifreeze solutions. This can be offset by increasing SWHE length or liquid flow rate. Note that laminar flow at lower heat pump part-load factors is not problematic since the greater-than-required length of the SWHE more than offsets the higher inside-film thermal resistance.

Laminar flow heat transfer coefficients are commonly determined from theoretical equations for fully developed flow (long pipes), which are very simple. Equation 5.8 assumes a constant heat rate (Holman 1986):

$$h_i = 4.36\frac{k}{d_i} \tag{5.8}$$

where

k = thermal conductivity, Btu/h·ft·°F (W/m·K)

d_i = inside diameter, ft (m)

However, almost all heat transfer correlations used in the industry were developed from empirical data from carefully controlled experiments, and theoretical correlations rarely match measured results. Many of the classical equations were developed from experiments conducted by Sieder and Tate (1936). The results are published in graphical format of j-factor versus Re, and heat transfer coefficients are determined using Equation 5.9:

$$h_i = \frac{j\rho c_p V}{(\mathrm{Pr}^{2/3})(\mu_w/\mu_b)^{0.14}} \tag{5.9}$$

where

ρ = fluid density, lb/ft^3 (kg/m^3)

c_p = fluid specific heat, Btu/lb·°F (kJ/kg·K)

V = fluid velocity, ft/s (m/s)

Pr = Prandtl Number $\equiv c_p\mu/k$

μ_b = dynamic viscosity at bulk fluid temperature, lb/ft·s (kg/m·s, centipoise ≡ 0.001 kg/m·s)

μ_w = fluid dynamic viscosity at pipe wall, lb/ft·s (kg/m·s, centipoise ≡ 0.001 kg/m-s)

In the laminar regime (Re < 2300 ±200) for long tubes ($L/d_i > 400$), the j-factor can be expressed as

$$j = 0.268\,\mathrm{Re}^{-0.675} \tag{5.10}$$

Heat transfer coefficients for the transition flow regime are highly uncertain. In this regime, eddy currents that improve heat transfer begin to develop and may disappear, but the boundary layer near the pipe wall where heat transfer is via conduction becomes thinner. Thus, flow is a combination of laminar and turbulent, where mixing occurs and boundary layer thickness declines. However, the exact values of Re where transition begins and fully turbulent flow begins is dependent on a variety of complex fluid property and flow phenomena. Thus, the equations for transition are almost nonexistent in the literature. It is suggested that in the transition regime (2300 < Re < 4000 to as high as 10,000), the j-factor of Sieder and Tate (Equation 5.9) be coupled with a curve-fit of their reported measurements to *estimate* the heat transfer coefficient:

$$j = 8.536 \times 10^{-15}\mathrm{Re}^3 - 2.386 \times 10^{-10}\mathrm{Re}^2 + 2.163 \times 10^{-6}\mathrm{Re} - 0.002 \tag{5.11}$$

In the fully turbulent flow regime (Re > 10,000), Sieder and Tate (1936) suggest using an alternative to the j-factor approach with the following equation:

$$h_i = 0.027\frac{k}{d_i} \times \text{Re}^{0.8}\text{Pr}^{1/3}(\mu_b/\mu_w)^{0.14} \quad (5.12)$$

Use of Equations 5.11 and 5.12 results in a discontinuity at Re = 10,000, which can be smoothed out by using a weighted average value of h_i calculated with each equation between Re = 4000 and Re = 10,000.

$$h_i = [((10{,}000 - \text{Re})/6000) \times h_{i(transition)} \text{ (Use Equation 5.11 for Re = 4000)}] + [((\text{Re} - 4000)/6000) \times h_{i(turbulent)} \text{ (Use Equation 5.12 for Re = 10,000)}] \quad (5.13)$$

Once the appropriate equation for the inside heat transfer coefficient is applied, the thermal resistance of the inside fluid film is calculated using Equation 5.14:

$$R_i = 1/h_i \pi d_i \quad (5.14)$$

Calculating the thermal resistance of the pipe wall using Equation 5.15 is more exact when the thermal conductivity of the pipe (k_p) is well established. Table 5.1 provides the thermal properties for the two general classifications of HDPE that are recommended for GSHP applications. Because the thermal conductivity of HDPE is low, the pipe resistance is the largest factor and the uncertainties of the inside film, outside film, and fouling factor are less influential to the uncertainty of the total resistance.

$$R_p = \ln(d_o/d_i)/2\pi k_p \quad (5.15)$$

Calculation of the outside thermal resistance also has a degree of uncertainty; therefore, the calculation must be supplemented by measured data to improve accuracy. This is especially true for reservoirs that are stagnant and in which heat is transferred via natural convection. Equations for natural convection are a function of the temperature difference between the surrounding water and the SWHE outside surface ($t_{resv} - t_o$). Since this value is also a function of the heat transfer rate, an iterative calculation is required in which a value of $t_{resv} - t_o$ is assumed and the outside resistance, overall resistance, and SWHE length (or area) for the first iteration are found. Equation 5.16 is applied to find the resulting value of $t_{resv} - t_o$. The iteration is repeated until the assumed value matches the resulting value of Equation 5.16:

$$t_{resv} - t_o = \frac{q_{hp} \times R_o}{L_{swhe}} \quad (5.16)$$

The accuracy of the calculation is also complicated by the fact that SWHEs are arranged in bundles (or flat plates in close proximity), and data for natural convection coefficients in these arrangements are sparse.

Table 5.1 Thermal Properties of HDPE Pipe (PPI 2014)

Thermal Property	PE3xxx	PE4xxx
Thermal Conductivity	0.25 Btu/h·ft·°F (0.43 W/m·K)	0.26 Btu/h·ft·°F (0.45 W/m·K)
Specific Heat	0.46 Btu/lb·°F (1.93 kJ/kg·K)	0.46 Btu/lb·°F (1.93 kJ/kg·K)
Coefficient of Expansion	9x10^{-15} in./in.·°F (16 × 10^{-15} m/m·K)	8x10^{-15} in./in.·°F (14 × 10^{-15} m/m·K)

The equations for outside heat transfer coefficients for flowing water can provide a higher degree of accuracy, even for tube bundles, but unfortunately the velocity of the water is often unknown and/or highly variable. Thus, the following discussion is based on the more conservative worst-case scenario of natural convection (stagnant reservoirs). If the water velocity can be well established, introductory heat transfer texts such as that by Holman (1986) provide equations for forced convection coefficients through tube bundles.

The Rayleigh number (Ra) is the dimensionless number for the natural convection coefficient that serves a similar function as the Reynolds number for forced convection. It represents a ratio of the buoyancy forces to the viscous forces. In a similar approach, the equations used to determine the heat transfer coefficients differ according to Rayleigh number and physical geometry. Many equations take the form of Equation 5.17 for horizontal tubes, with different values of the coefficients C and m being dependent on Ra and fluid type (gas, liquid) (Holman 1986):

$$h_o = \frac{k_w}{d_o} C(\mathrm{Ra})^m = \frac{k_w}{d_o} C\left(\frac{g\beta\rho^2 c_p}{\mu k}(t_o - t_{resv})d_o^3\right)^m \tag{5.17}$$

where

k_w = thermal conductivity of water, Btu/h·ft·°F (W/m·K)
g = acceleration of gravity, 32.2 ft/s^2 (9.81 m/s^2)
β = volumetric coefficient of expansion, °R^{-1} (K^{-1})
ρ = fluid density, lb/ft^3 (kg/m^3)
c_p = fluid specific heat, Btu/lb·°F (kJ/kg·K)
μ = fluid dynamic viscosity, lb/ft·s (kg/m·s, centipoise ≡ 0.001 kg/m·s)

Note: Fluid properties are evaluated at the average film temperature, $(t_o + t_{resv})/2$

For typical diameters and temperatures in SWHE applications, two sets of coefficients for Equation 5.17 apply:

$$C = 0.85 \text{ and } m = 0.188 \text{ for } 10^2 < \mathrm{Ra} < 10^4$$

$$C = 0.53 \text{ and } m = 0.25 \text{ for } 10^4 < \mathrm{Ra} < 10^9$$

Calculation of the Rayleigh number is simplified by combining all of the fluid properties of freshwater into a single number shown in the right two columns of Table 5.2.

Little information has been discovered concerning fouling factors for SWHEs in reservoirs. The fouling factors for low-velocity tube-and-shell heat exchangers are suggested as a substitute until field data is available (TEMA 1978). These values and equivalent fouling factors for layers of mud and biological growth are provided in Table 5.3.

5.5 CLOSED-LOOP SURFACE-WATER HEAT EXCHANGERS

Direct application of fundamental equations for straight horizontal tubes or flat-plate heat exchangers must be adjusted for conditions and arrangements characteristic of SWHEs. As noted in Section 5.4, there is a high degree of uncertainty for the inside tube, outside tube, and outside fouling coefficients. The coils are circular (not straight) and often arranged in tube bundles with random spacing. There is likely little variation between the inside coefficients between coiled and straight tubes and the outside coeffi-

Table 5.2 Properties of Water (Holman 1986)

Temperature		Conductivity (k)		Density (ρ)		Viscosity (μ)		Specific Heat (c_p)		$g\beta\rho^2c_p/\mu k$	
°F	°C	Btu/h·ft·°F	W/m·K	lb/ft³	kg/m³	lb/ft·s	kg/m·s	Btu/lb·°F	kJ/m·K	1/ft³·°F	1/m³·°C
40	4.4	0.332	0.575	62.4	1000	1.04×10^{-3}	1.55×10^{-3}	1.00	4.21	0.30×10^{8}	0.19×10^{10}
50	10.0	0.338	0.585	62.4	999	0.88×10^{-3}	1.31×10^{-3}	1.00	4.21	1.00×10^{8}	0.63×10^{10}
60	15.6	0.344	0.595	62.3	999	0.75×10^{-3}	1.12×10^{-3}	1.00	4.19	1.70×10^{8}	1.08×10^{10}
70	21.1	0.349	0.604	62.3	998	0.66×10^{-3}	0.98×10^{-3}	1.00	4.18	2.3×10^{8}	1.46×10^{10}
80	26.7	0.355	0.614	62.2	996	0.58×10^{-3}	0.86×10^{-3}	1.00	4.18	3.0×10^{8}	1.91×10^{10}
90	32.2	0.360	0.623	62.1	995	0.51×10^{-3}	0.77×10^{-3}	1.00	4.17	3.9×10^{8}	2.48×10^{10}
100	37.8	0.364	0.630	62.0	993	0.46×10^{-3}	0.68×10^{-3}	1.00	4.17	5.2×10^{8}	3.30×10^{10}

Table 5.3 Approximate Fouling Factors* for SWHE Coils

Water Type/Fouling Condition	Btu/h·ft²·°F	W/m²·K
Clean river/reservoir	500	2800
Muddy river/reservoir	300	1700
Sanitary sewer water	125	700
Spray pond—Untreated	300	1700
Mud layer—1/16 in. (1.5 mm)	200	1150
Mud layer—1/8 in. (3 mm)	100	570
Biological growth—1/8 in. (3 mm)	40	230

**Tables in some cases report for fouling resistances, the inverse of fouling factors.*

cients between slinky-style coils to single straight tubes. While there are well-developed correlations for outside coefficients for tube bundles, they are typically for higher-velocity forced-convection applications rather than natural-convection situations for SWHEs. There is also little information on fouling factors in these applications. Fortunately, for HDPE SWHEs the pipe wall thermal resistance is not only predictable but is almost always the largest resistance. Therefore, uncertainties in the other resistances tend to have a lower impact on the overall uncertainty.

Two additional concerns in the heating mode are the need for antifreeze solutions and the potential for coil freezing, especially in smaller reservoirs whose temperatures may be affected by heat pumps. Propylene glycol is the most acceptable solution in terms of environment, health, safety, and corrosion, but it has a higher pumping cost than most other alternatives (ASHRAE 2011). Therefore, care must be taken to use concentrations that ensure adequate freeze protection but also minimize pump energy while maintaining non-laminar flow at near full heating load conditions.

While ice formation on all types of SWHEs is possible, metallic SWHEs (copper tubes, stainless steel flat plates, etc.) typically operate with lower surface temperatures in the heating mode. Since the pipe/tube resistance is low in metal SWHEs, most of the thermal resistance is in the outside film. This means the temperature difference across this surface relative to the total temperature difference will be much greater compared to plastic SWHEs. Thus, outside surface temperature (t_o) tends to be lower and more likely to be below the freeze point.

Experimental data and field measurements on SWHEs are more limited than those for ground heat exchangers. A small number of projects that addressed the design of SWHEs have been completed or are currently in progress. The final report for ASHRAE RP-1385,

Development of Design Tools for Surface Water Heat Pump Systems (2009), may provide information when it is available.

A master's thesis funded by ASHRAE RP-1385 provides a summary of correlations for both the inside and outside heat transfer coefficients for straight, curved, and helical pipe (Hansen 2011). The internal coefficients are limited to fully turbulent flow. Tests were conducted on nominal 3/4 in. (19 mm), 1 in. (25 mm), and 1 1/4 in. (32 mm) DR 11 HDPE tubing in a variety of bundle, helical, and slinky coil arrangements. Tests were also conducted on a bundled stainless steel vertical flat-plate SWHE. All tests were performed with clean SWHEs, so the impact of fouling resistance was not considered. Table 5.4 summarizes the relative results of the three other thermal resistances. While tests in the heating mode may be available at a later date, these measurements were taken in the cooling mode.

Note that heating-mode overall thermal resistances tend to be higher than cooling-mode values because of both reduced natural convection effects in colder water and lower inside heat transfer coefficients with higher-viscosity antifreeze fluids at the lower heating-mode temperatures. In many cases it is a challenge to prevent the inside flow from becoming laminar at full load, especially when excessive concentration of antifreeze solutions are employed.

Hansen (2011) noted the influence of reservoir temperature on the overall SWHE thermal resistance. The measured trends closely match the trends predicted by Equation 5.17, which primarily results from increasing reservoir water viscosity at lower temperatures. The higher viscosity of the fluids inside the SWHEs plays a minor role in the trend. The higher viscosity results in longer SWHEs in colder reservoirs for a fixed approach temperature ($t_{app} = \mathrm{LLT}_{swhe} - t_{resv}$) in cooling.

A limited amount of testing was performed on flat-plate SWHEs (Hanson 2011). The outside heat transfer coefficients agreed well with coefficients predicted with equations of the form of Equation 5.17 when the characteristic length of plate height and vertical plate values for C and m are substituted. Tests were conducted with clean plates, no antifreeze solution, and in the cooling mode only at a variety of reservoir temperatures, so results cannot be directly applied to SHWE design. Thus, designers must rely on flat-plate SWHE manufacturers for recommendations.

Hattemer (2005) performed tests on a nominal 3/4 in. (19 mm) DR 11 HDPE slinky coil in which the tubes were separated to minimize tube-to-tube interference, as shown in Figure 5.13. Tests were repeated with bundled coils with three closely controlled separation distances as shown in Figure 5.14. Tests were conducted in both cooling and heating modes. All tests were conducted with clean tubing, so the impact of fouling resistance was not considered. Test results were provided in the format of correction factors for the bundled coils relative to the slinky coil. Comparisons were also made to the sizing charts provided in *Ground-Source Heat Pumps: Design of Geothermal Systems for Commercial and Institutional Buildings* (Kavanaugh and Rafferty 1997).

Table 5.4 Cooling-Mode Resistances of Clean SWHEs with Turbulent Flow (Hansen 2011)

SWHE Type	Inside Resistance (Turbulent Flow)	Tube or Plate Resistance	Outside Resistance
3/4 in. (19 mm) HDPE	4%	58%	38%
1 in. (25 mm) HDPE	3%	68%	29%
1 1/4 in. (32 mm) HDPE	3%	72%	25%
Stainless Steel Flat Plate	11%	1%	88%

Results indicate that in cooling, bundle coils with spacing between tubes (S_{tube}) of at least one-fourth the outside coil diameter ($S_{tube} > 0.25d_o$) performed nearly the same as slinky coils arranged in the expanded arrangement shown in Figure 5.13. In heating, the bundle coils required approximately 20% more length to match the performance of the expanded slinky SWHE. It was also noted that coefficients and convection currents declined as the water near the coil approached the temperature of 39°F (4°C), where density variations (the driving force for natural convection) are small. This manifests itself when SWHEs are placed in small ponds or in confined areas of larger reservoirs, where downward convection currents are constrained (local reservoir temperatures near or below 39°F [4°C]). When possible, SWHEs in colder reservoirs should be placed near but not at the bottom to allow downward natural convection flows, as shown in Figure 5.15.

Based on the suggestions of Hattemer (2005) and Hansen (2011) and the improved correlations outlined in Equations 5.6 through 5.17, the design length recommendations

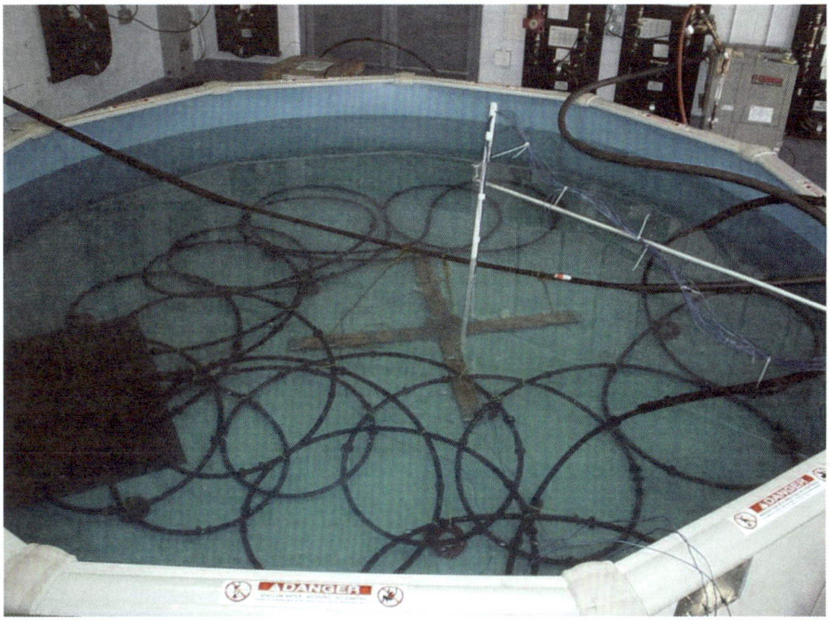

Figure 5.13 Slinky Coil Test Arrangement (Hattemer, 2005)

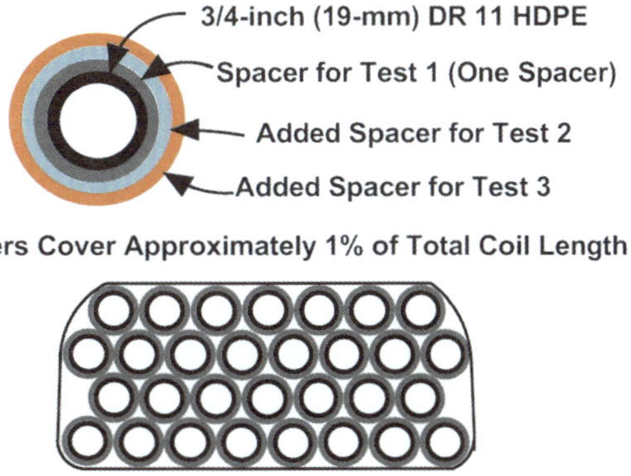

Figure 5.14 Test Arrangement of Bundled Coils with Spacers

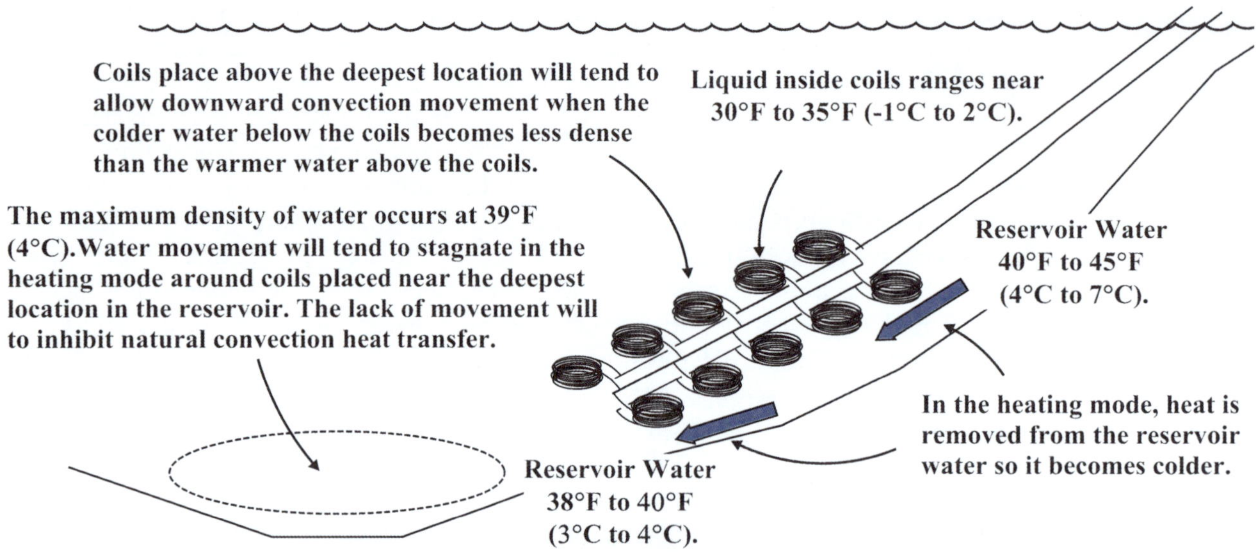

Figure 5.15 Suggested Cold-Reservoir SWHE Location

for HDPE SWHE have been updated from those provided in *Ground-Source Heat Pumps: Design of Geothermal Systems for Commercial and Institutional Buildings* (Kavanaugh and Rafferty 1997). Correction factors for reservoir temperature, heat pump system efficiency (EER and COP), antifreeze concentrations, and coil arrangement and location are included. Figures 5.16 and 5.17 show the results of these revisions for the cooling mode in I-P and SI units, respectively; Figures 5.18 and 5.19 show the updated charts for the heating mode in I-P and SI units, respectively. It is important to note that these charts are based on a fixed range of coils per ton (kW) arranged in parallel flow paths. A range of 0.75 to 1.25 parallel coils per ton (3 to 4 coils per kW) results when creating a balance between minimizing pump energy (head loss) and providing adequate velocity for acceptable inside heat transfer coefficients. SWHEs with small approach temperatures (t_{app}) will be longer and the number of parallel coils should be in the upper range, while coils with larger approach temperatures should be in the lower range of parallel coils per ton (coils per kW).

Laminar flow did occur in the heating mode in some cases (Hattemer 2005), but the curves in Figures 5.18 and 5.19 do account for this situation. More detailed optimization is possible by using software based on Equations 5.6 through 5.17 that could include situations for which no correction factors are available, such as larger fouling factor values. The optimization is particularly challenging when the heating requirements dictate SWHE length. The higher viscosity of antifreeze concentrations at low temperatures increases head loss while lowering heat transfer performance. Though propylene glycol is the recommended fluid in terms of environmental risk, health risk, fire risk, and safety (ASHRAE 2011), glycol is more viscous than methanol (but only slightly more viscous than ethanol). In most cases in commercial buildings, adequate protection can be obtained with antifreeze concentrations below values recommended by vendors.

The cooling-mode design flow rate and the viscosity (which is much less than the viscosity in the heating mode) should be used to determine the size of the SWHE required for cooling and the system head loss. The heating-mode design flow rate and the viscosity should be used to determine the size of the SWHE required for heating and the system head loss. Using the heating-mode viscosity with the cooling-mode flow rate will result in

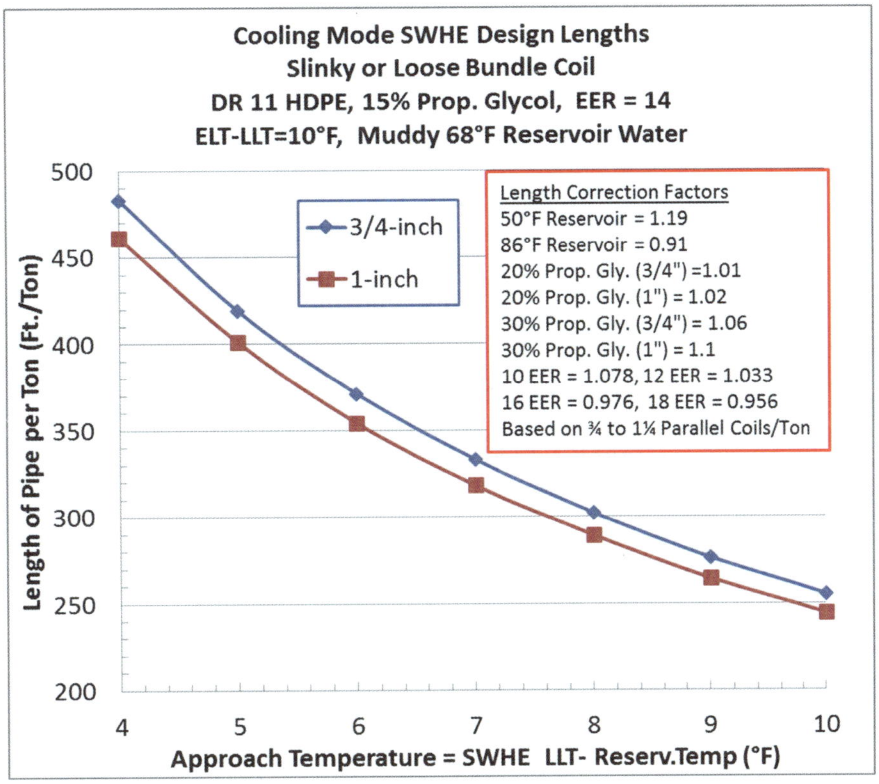

Figure 5.16 Cooling-Mode Design Lengths for HDPE SWHEs (I-P Units)

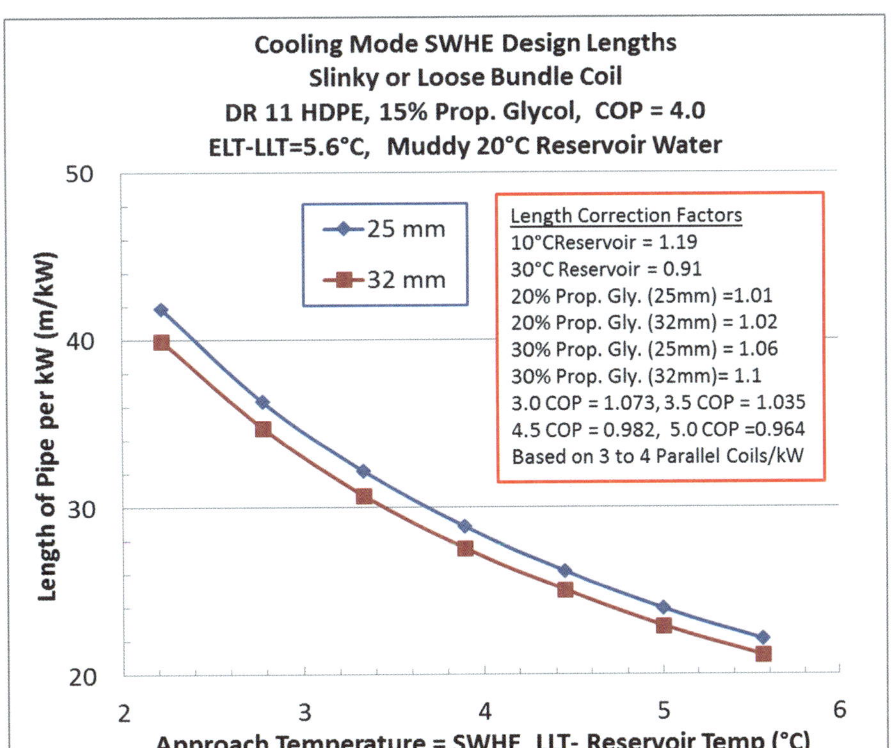

Figure 5.17 Cooling-Mode Design Lengths for HDPE SWHEs (SI Units)

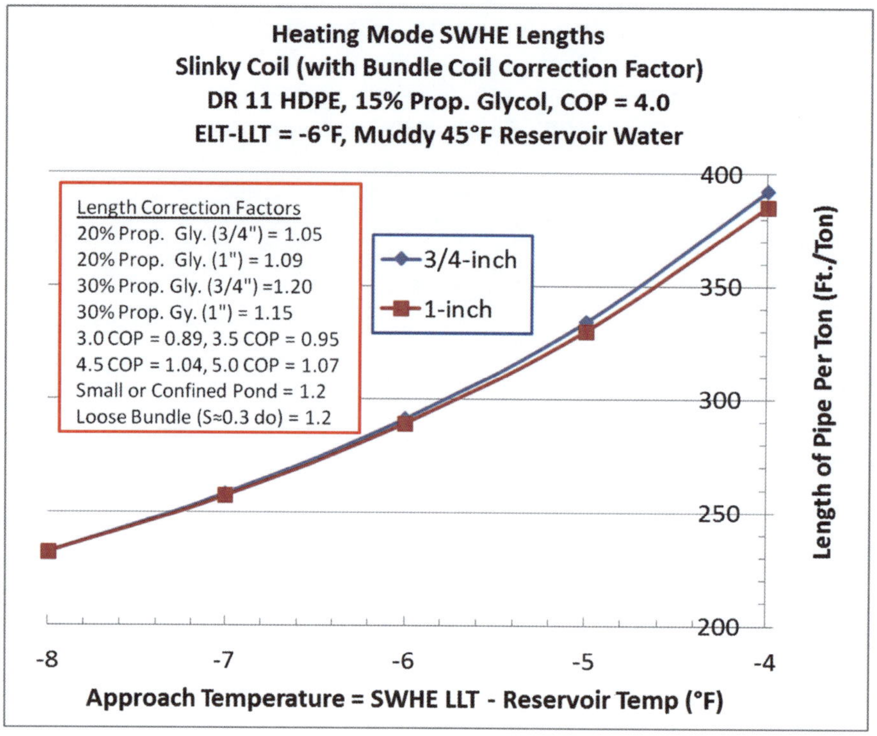

Figure 5.18 Heating-Mode Design Lengths for HDPE SWHEs (I-P Units)

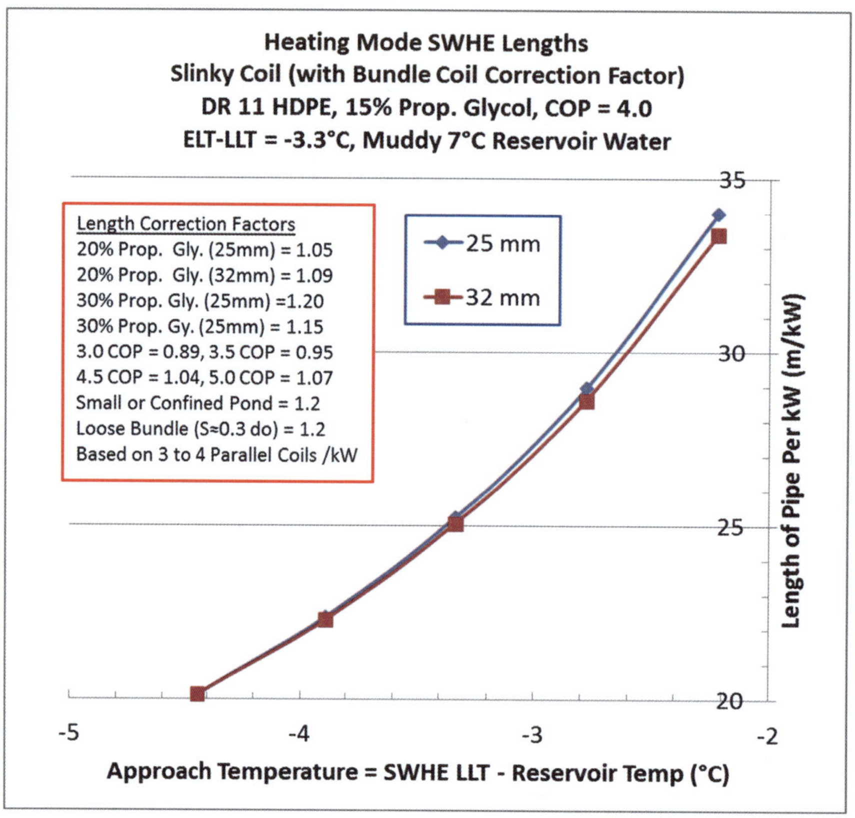

Figure 5.19 Heating-Mode Design Lengths for HDPE SWHEs (SI Units)

an significantly oversized pump when the SWHE size for cooling is larger than the SWHE size for heating.

The required SWHE cooling length ($L_{c\text{-}swhe}$) and heating length ($L_{h\text{-}swhe}$) are found by multiplying the size of the heat pump by the length per unit capacity estimated from Figures 5.16 through 5.19, with the appropriate correction factors applied as shown in Equation 5.18. For cooling, the correction factors include ones for reservoir temperature [CF(t_{resv})], antifreeze solution [CF(AF)], and system efficiency [CF(EER) or CF(COP)]. For heating, the factors in Equation 5.19 are for antifreeze solution [CF(AF)], system efficiency [CF(COP)], coil type ([CF(CoilType)], and location [CF(Loc)].

$$L_{c\text{-}swhe} = L_c/\text{ton (kW)} \times \text{TC} \times \text{CF}(t_{resv}) \times \text{CF(AF)} \times \text{CF(EER or COP)} \quad (5.18)$$

$$L_{h\text{-}swhe} = L_h/\text{ton (kW)} \times \text{TC} \times \text{CF(AF)} \times \text{CF(COP)} \times \text{CF(CoilType)} \times \text{CF(Loc)} \quad (5.19)$$

The heating-mode correction factors for coil type and location [CF(CoilType) and CF(Loc)] may be somewhat conservative. Tests conducted by Hattemer (2005) showed little difference between slinky coils and loose bundle coils in cooling mode performance. However, results in the heating mode suggest that differences are dramatic because of the small variation in viscosity of water near 39°F (4°C), a frequent SWHE condition. It is suggested that the correction factors for coil type and location [CF(CoilType) and CF(Loc)] in Figures 5.18 and 5.19 be applied until more extensive cold-temperature field tests can confirm laboratory results.

EXAMPLE 5.2— COOLING-MODE SWHE DESIGN

Conduct a comparative cooling-mode design for a 20 ton (70 kW) bundle coil SWHP system to be placed in a lake at a 50 ft (15 m) depth where the maximum late-summer water temperature is 60°F (16°C). With a liquid flow rate of 60 gpm (3.8 L/s), the EER of the system is 16 Btu/Wh (COP = 4.7) with an ELT of 65°F (18°C); it is 15 Btu/Wh (COP = 4.4) with an ELT of 70°F (21°C). Compute the added cost of the higher-efficiency system based on a 1/4 in. (25 mm) DR 11 HDPE cost of \$0.60/ft (\$2.00/m) and a propylene glycol cost of \$15/gal (\$4.00/L) using a 20% by volume solution. Assume the headers between the lake and building are insulated so the SWHE LLT is equal to the heat pump ELT, and the fouling factor is for a muddy lake.

Solution

15 EER (COP = 4.4) System

$$t_{app} = \text{LLT}_{swhe} - t_{resv} = 70°\text{F} - 60°\text{F} = 10°\text{F}\ (21.1°\text{C} - 15.6°\text{C} = 5.5°\text{C})$$

From Figure 5.16 for t_{app} = 10°F (6°C): L_c/ton = 255 ft/ton (L_c/kW = 22.1 m/kW)

CF(t_{resv}) = 1.08 (interpolated between 1.19 for 50°F [10°C] lake and 1.0 for 68°F [20°C] lake)

CF(AF) = 1.01 (20% propylene glycol)

CF(EER [COP]) = 0.99 (interpolated between 0.976 for EER = 16 [COP = 4.7] and 1.0 for EER = 14 [COP = 4.1])

Applying Equation 5.18,

$$L_{c\text{-}swhe} = L/\text{ton} \times \text{TC} \times \text{CF}(t_{resv}) \times \text{CF(AF)} \times \text{CF(EER)}$$
$$= 255 \text{ ft/ton} \times 20 \text{ tons} \times 1.08 \times 1.01 \times 0.99 = 5510 \text{ ft} \quad \text{(I-P)}$$

$$L_{c\text{-}swhe} = L/\text{kW} \times \text{TC} \times \text{CF}(t_{resv}) \times \text{CF(AF)} \times \text{CF(COP)}$$
$$= 22.1 \text{ m/kW} \times 70 \text{ kW} \times 1.08 \times 1.01 \times 0.99 = 1670 \text{ m} \quad \text{(SI)}$$

For a 20 ton (70 kW) system and high approach temperature, it is suggested the number of parallel coils per ton be on the lower end of the range (0.75 coils/ton) at 15 or 16. Because 400 ft (≈125 m) is a standard length for coils, the recommended length total length would be 6000 ft (≈1800 m) (15 coils at 400 ft [≈125 m] each).

It would also be possible to meet the total length requirement with 5600 ft (≈1700 m) (14 coils at 400 ft [≈125 m] each), but a head loss calculation is suggested to ensure pump size is within limits, as discussed in Section 5.6 and in Chapter 6.

16 EER (COP = 4.7) System

$$t_{app} = LLT_{swhe} - t_{resv} = 65°\text{F} - 60°\text{F} = 5°\text{F} \; (18.3°\text{C} - 15.6°\text{C} = 2.7°\text{C})$$

From Figure 5.16 for t_{app} = 5°F (2.7°C): L_c/ton = 420 ft/ton L_c/kW = (36.4 m/kW)

$\text{CF}(t_{resv})$ = 1.08 (interpolated between 1.19 for 50°F [10°C] lake and 1.0 for 68°F [20°C] lake)

CF(AF) = 1.01 (20% propylene glycol)

CF(EER [COP]) = 0.976

Applying Equation 5.18,

$$L_{c\text{-}swhe} = L/\text{ton} \times \text{TC} \times \text{CF}(t_{resv}) \times \text{CF(AF)} \times \text{CF(EER)}$$
$$= 420 \text{ ft/ton} \times 20 \text{ tons} \times 1.08 \times 1.01 \times 0.976 = 8940 \text{ ft} \quad \text{(I-P)}$$

$$L_{c\text{-}swhe} = L/\text{kW} \times \text{TC} \times \text{CF}(t_{resv}) \times \text{CF(AF)} \times \text{CF(COP)}$$
$$= 36.4 \text{ m/kW} \times 70 \text{ kW} \times 1.08 \times 1.01 \times 0.976 = 2710 \text{ m} \quad \text{(SI)}$$

For a 20 ton (70 kW) system and low approach temperature, it is suggested the number of parallel coils per ton be on the mid to upper end of the range (1.0 to 1.25 coils/ton) at 20 to 25. The standard coil length of 400 ft (≈125 m) yields 23 parallel coils for a total of 9200 ft (≈2800 m). It would also be possible to meet the total length requirement with 9000 ft (≈2750 m) (18 coils at 500 ft [≈150 m] each), but a head loss calculation is suggested to ensure pump size is within limits, as discussed in Section 5.6 and Chapter 6.

The added cost of the HDPE pipe based on the difference in pipe length is

$$\text{Added pipe cost} = (9200 \text{ ft} - 6000 \text{ ft}) \times \$0.60/\text{ft} = \$1920 \quad \text{(I-P)}$$

$$\text{Added pipe cost} = (2800 \text{ m} - 1800 \text{ m}) \times \$2.00/\text{m} = \$2000 \quad \text{(SI)}$$

Appendix G indicates 3/4 in. (25 mm) DR 11 pipe contains 3.0 gal/100 ft (33 L/100 m). Using a volume percentage of 20%, the added cost of the propylene glycol is

$$\text{Added glycol cost} = (9200 \text{ ft} - 6000 \text{ ft}) \times 3.0 \text{ gal/100 ft} \times 20\% \times \$15/\text{gal} = \$288 \quad \text{(I-P)}$$

$$\text{Added glycol cost} = (2800 \text{ m} - 1800 \text{ m}) \times 33 \text{ L/100 m} \times 20\% \times \$4/\text{L} = \$264 \quad \text{(SI)}$$

Thus, the added cost for the pipe and propylene glycol of the larger SWHE is

$$\text{Added cost of SWHE for 16 EER system} = \$1920 + \$288 = \$2208 \quad \text{(I-P)}$$

$$\text{Added cost of SWHE for 4.7 COP system} = \$2000 + \$264 = \$2264 \quad \text{(SI)}$$

Cooling Mode Condition				
Installed Heat Pump Specifications				
Water=W	ELT	LLT	Heat Rej.	Liq. Flow
Brine = B	°F	°F	kBtu/h	gpm
B	80.0	90.1	684.3	135.0
Lake/Pond Temp.	70.0	°F	Warmest	
Entering Δt	20.1	°F		
Leaving Δt	10.0	°F		
LMTD	14.5	°F		
Option for Plate Combinations				
	Wth x Hgt	Area/PHE	Flow/PHE	Plates
Model #	(ft x ft)	(sq. ft.)	(gpm)	Required
TiGR-02T	2 x 5	10	4.7	29
TiGR-03T	4 x 5	20	7.9	17
TiGR-04T	4 x 6	24	9.0	15
TiGR-05T	4 x 8	32	12.3	11
TiGR-05Tc	2 x 15	30	11.3	12
TiGR-08T	4 x 12	48	16.9	8
TiGR-10T	4 x 15	60	22.5	6

Figure 5.20 Manufacturer's Cooling-Mode Design Results for Flat-Plate SWHEs (AWEB 2014)

Plate SWHE manufacturers typically offer custom design software for products. Figures 5.20 and 5.21 are examples of output results provided to designers. The example shown is for a water-to-water heat pump application in which the heating mode is critical; it dictates the required SWHE dimensions. This particular manufacturer used the total installed capacity of the heat pump equipment rather than the building load to size the SWHEs. It is suggested that the designer request the cooling-mode conditions be adjusted until the plate dimensions match the heating-mode design. This allows the operating conditions in the noncritical mode to be known.

A variety of plate sizes are available, so multiple combinations are possible. The plates are assembled in frames with flotation devices for installation. The plates are separated at much wider spacing than conventional plate heat exchangers, as shown in Figures 5.1, 5.25, 5.26, and 5.27.

Heating Mode Condition				
Installed Heat Pump Specifications				
Water=W	ELT	LLT	Heat Rej.	Liq. Flow
Brine = B	°F	°F	kBtu/h	gpm
B	32.0	26.6	362.7	135.0
Freeze Inhibiting Solution Required				
Lake/Pond Temp.		41.0	°F	Coldest
Entering Δ t		14.4	°F	
Leaving Δ t		9.0	°F	
LMTD		11.5	°F	
Option for Plate Combinations (Heating Mode Dominant)				
	Wth x Hgt	Area/PHE	Flow/PHE	Plates
Model #	(ft x ft)	(sq. ft.)	(gpm)	Required
TiGR-02T	2 x 5	10	3.6	38
TiGR-03T	4 x 5	20	5.9	23
TiGR-04T	4 x 6	24	7.1	19
TiGR-05T	4 x 8	32	9.0	15
TiGR-05Tc	2 x 15	30	8.4	16
TiGR-08T	4 x 12	48	13.5	10
TiGR-10T	4 x 15	60	16.9	8

Figure 5.21 Manufacturer's Heating-Mode Design Results for Flat-Plate SWHEs (AWEB 2014)

Engineers should use these programs with caution because some assumptions may not be apparent. The results shown in Figures 5.20 and 5.21 appear to assume a particular antifreeze type and concentration that is required by the heat pump manufacturer used by this plate heat exchanger manufacturer. If the packaged software does not include the surface temperature on the reservoir side of the plate heat exchanger, the value should be provided by the manufacturer to ensure adequate protection from ice formation.

5.6 CIRCUITS AND LAYOUT OF SURFACE-WATER HEAT EXCHANGERS

The piping networks of closed-loop SWHP systems resemble systems used in ground-coupled heat pumps. Most frequently, a single set of supply and return headers connects the building heat pump loop and SWHEs. Like vertical ground loops, the individual SWHEs must be piped in multiple parallel loops. When the required number of individual SWHEs (bundle coil, slinky coil, plate heat exchanger) exceeds 10 to 20, flow is split into multiple parallel circuits with up to 20 individual SWHEs on each circuit. These circuits must have isolation valves so each can be purged of debris and air at start-up.

Because the water body is typically at a remote distance from the building, the option of multiple unitary loops does not usually have a cost or energy consumption advantage over a central loop. However, equipment is available that can straighten large coils of HDPE from a diameter of 2 in. up to 6 in. (60 mm up to 170 mm) (see Figure 6.27). This eliminates the need to thermally fuse multiple pieces of straight pipe, which is typically 20 or 40 ft (6 or 12 m) in length. When this equipment is available, multiple header and common (subcentral) loops may be more cost-effective than a single central loop.

Three bundle SWHE coils with spacers are shown in Figure 5.22. The coil is split into multiple parallel slinky loops in the reservoir. These loops are separated by 10 to 20 ft

Figure 5.22 HDPE Bundle Coil SWHEs with Spacers

Figure 5.23 Slinky Coil SWHEs Delivered to Site in Shipping Bundles

(3 to 6 m) to limit thermal interference, hot spots, or cold pockets. Many contractors simply unbind plastic pipe coils and rebind them in a looser and randomly spaced coil. It is not recommended that the pipe coils be submerged in unseparated shipping bundles because performance is reduced by up to 60% in cooling (Hansen 2011) and possibly by a greater percentage in heating.

Figure 5.23 shows several slinky coils rolled into shipping bundles that are to be placed in a municipal wastewater pond. Figure 5.24 demonstrates the parallel arrangement with the unbundled coils with reverse-return headers. There are eight parallel slinky coils and two circuits. Note the insert in Figure 5.24, which is a side view of the two sets of supply and return headers. Recommended practice is to bury the supply and return headers below grade, where they enter the reservoir for thermal and physical protection. In this application, the pond bank could not be penetrated due to potential environmental issues with the wastewater stream. Therefore, the headers were placed above the surface, insulated, and weighted with concrete inserts.

Figure 5.25 shows a reservoir plate SWHE being placed into the water. The plates are vertical and should remain in a nearly vertical position to attain rated performance. There are six floats made of capped PVC pipe that allow the SWHEs to be maneuvered into the proper location and sunk. Figure 5.26 shows a plate SWHE being installed in a cold climate. This design is intended for applications in rivers or high-flow locations with a deflector to protect the heat exchanger from debris and ice. Figure 5.27 shows a plate SWHE that was installed before the human-made lake was filled.

In applications where the heating mode dictates the SWHE size and liquid flow rate, it is more of a challenge to optimize the trade-off between the heat transfer and pump power requirements. The high viscosity of low-temperature antifreeze results in an

Figure 5.24 Slinky Coil SWHEs Being Floated In Place

Figure 5.25 Nominal 50 ton (175 kW) Flat-Plate SWHE (AWEB 2014)

increased need for high liquid flow for good heat transfer, but it is also accompanied by increased pump power at design conditions.

Tables 5.5a and 5.5b provide the head/pressure losses and Reynolds numbers for six different antifreeze solutions at various flow rates for average liquid temperatures of 32°F and 0°C, respectively. The design process is to provide an optimum number of coils to meet the following constraints:

- Meet or slightly exceed the total SWHE length requirement.
- Minimize the head loss across the coils.
- Avoid laminar flow at full-load design conditions (2300 > Re > 3000 is tolerable, Re > 3000 is good).
- Select standard coil lengths to avoid waste and/or higher-cost nonstandard lengths.
- Use an antifreeze solution to provide freeze protection (5°F [3°C] < design SWHE ELT is marginal, 10°F [6°C] < design SWHE ELT is good]).

Figure 5.26 Flat-Plate SWHE with Deflector for River Application (AWEB 2014)

Figure 5.27 Nominal 24 ton (84 kW) SWHE Installed Before Lake is Filled (AWEB 2014)

Table 5.5a Head Losses and Reynolds Numbers for SWHE Coils with Antifreeze Solutions at 32°F (CRC 1970; Dow 1990)

Solution	Percent by Volume	Freeze Point, °F	HDPE Pipe		Liquid Flow Rate, gpm 2	3	4	5
Propylene glycol	20	19	3/4 in. DR 11	Δh/100 ft	0.9	2.5	4.5	NR
				Re	1870	2800	3730	NR
			1 in. DR 11	Δh/100 ft	NR	0.6	1.5	2.3
				Re	NR	2230	2980	3720
	25	14	3/4 in. DR 11	Δh/100 ft	1.3	1.9	4.3	NR
				Re	1360	2040	2730	NR
			1 in. DR 11	Δh/100 ft	NR	0.76	1.1	2.2
				Re	NR	1360	2180	2720
Methanol	15	17	3/4 in. DR 11	Δh/100 ft	1.0	2.5	4.1	NR
				Re	2670	4010	5340	NR
			1 in. DR 11	Δh/100 ft	0.26	0.85	1.4	2.1
				Re	2130	3200	4260	5330
	20	11	3/4 in. DR 11	Δh/100 ft	0.88	2.6	4.2	NR
				Re	2480	3610	4820	NR
			1 in. DR 11	Δh/100 ft	0.29	0.83	1.5	2.1
				Re	1920	2890	3850	4810
Ethanol	15	22	3/4 in. DR 11	Δh/100 ft	0.92	2.5	4.6	NR
				Re	1850	2770	3700	NR
			1 in. DR 11	Δh/100 ft	0.37	0.58	1.5	2.3
				Re	1480	2210	2950	3690
	20	17	3/4 in. DR 11	Δh/100 ft	1.1	1.8	4.7	NR
				Re	1510	2270	3020	NR
			1 in. DR 11	Δh/100 ft	0.46	0.69	1.14	2.4
				Re	1210	1810	2410	3020

For head loss interpolation at other flow rates, use $\Delta h_{Actual} = \Delta h_{Table} \times (Q_{Actual}/Q_{table})^2$.
For pressure loss interpolation at other flow rates, use $\Delta p_{Actual} = \Delta p_{Table} \times (Q_{Actual}/Q_{table})^2$.

EXAMPLE 5.3— SWHE CIRCUIT DESIGN WITH HEATING MODE DOMINANT

Select a circuit arrangement for the SWHE coil described in Example 5.2 (16 EER [4.7 COP_c] system) for heating-mode temperatures of ELT = 30°F (–1°C) and LLT = 36°F (2°C). Propylene glycol is the required antifreeze solution. Assume the required liquid flow rate in heating is also 60 gpm (3.8 L/s).

Solution

The required total length of the Example 5.2 16 EER (4.7 COP_c) system is 8940 ft (2720 m) of 3/4 in. (19 mm) DR 11 HDPE pipe, and the liquid flow rate is 60 gpm (3.8 L/s). The design ELT for the SWHE is 30°F (–1°C); therefore, a 20% propylene glycol-80% water solution with a freeze point of 19°F (–7°C) is acceptable. Standard lengths of tubing are 300, 400, and 500 ft. The options for the arrangements are as follows:

a. Thirty 300 ft coils (9000 ft total) at 2 gpm per coil (60 gpm/30 coils)
b. Twenty-three 400 ft coils (9200 ft total at 2.6 gpm per coil (60 gpm/23 coils)
c. Eighteen 500 ft coils (9000 ft total) at 3.3 gpm per coil (60 gpm/18 coils)

Table 5.5b Head Losses and Reynolds Numbers for SWHE Coils with Antifreeze Solutions at 0°C (CRC 1970; Dow 1990)

Solution	Percent by Volume	Freeze Point, °C	HDPE Pipe		Liquid Flow Rate, L/s 0.125	0.1875	0.250	0.3125
Propylene glycol	20	−7	25 mm DR 11	Δh/100 m	0.09	0.25	0.45	NR
				Re	1995	2988	3980	NR
			32 mm DR 11	Δh/100 m	NR	0.06	0.15	0.23
				Re	NR	2328	3111	3884
	25	−10	25 mm DR 11	Δh/100 m	0.13	0.19	0.43	NR
				Re	1451	2177	2913	NR
			32 mm DR 11	Δh/100 m	NR	0.08	0.11	0.22
				Re	NR	1420	2276	2840
Methanol	15	−8	25 mm DR 11	Δh/100 m	0.1	0.25	0.41	NR
				Re	2849	4279	5698	NR
			32 mm DR 11	Δh/100 m	0.026	0.09	0.14	0.21
				Re	2224	3341	4447	5565
	20	−12	25 mm DR 11	Δh/100 m	0.09	0.26	0.42	NR
				Re	2646	3852	5143	NR
			32 mm DR 11	Δh/100 m	0.03	0.08	0.15	0.21
				Re	2004	3017	4019	5022
Ethanol	15	−6	25 mm DR 11	Δh/100 m	0.09	0.25	0.45	NR
				Re	1974	2956	3948	NR
			32 mm DR 11	Δh/100 m	0.04	0.06	0.15	0.23
				Re	1545	2307	3080	3852
	20	−8	25 mm DR 11	Δh/100 m	0.11	0.18	0.46	NR
				Re	1611	2422	3222	NR
			32 mm DR 11	Δh/100 m	0.05	0.07	0.11	0.24
				Re	1263	1890	2516	3153

For head loss interpolation at other flow rates, use $\Delta h_{Actual} = \Delta h_{Table} \times (Q_{Actual}/Q_{table})^2$.
For pressure loss interpolation at other flow rates, use $\Delta p_{Actual} = \Delta p_{Table} \times (Q_{Actual}/Q_{table})^2$.

Results for each option are as follows:

a. From Table 5.5, the head loss for 300 ft of 3/4 in. DR 11 HDPE at 2 gpm is

$$\Delta h_{2\text{gpm}} = 0.9 \text{ ft water/100 ft} \times 300 \text{ ft} = 2.7 \text{ ft water and Re} = 1870$$

b. The head loss for 400 ft of 3/4 in. DR 11 HDPE at 2.61 gpm is (using the interpolation equation at the bottom of Table 5.5)

$$\Delta h_{2.61\text{gpm}} = \Delta h_{3\text{gpm}} \times (2.61 \text{ gpm}/3 \text{ gpm})^2 \times 400 \text{ ft}$$
$$= 2.5 \text{ ft}/100 \text{ ft} \times (2.61/3.0)^2 \times 400 \text{ ft}$$
$$= 7.6 \text{ ft of water and Re} = 2440 \text{ (by interpolation)}$$

c. The head loss for 500 ft of 3/4 in. DR 11 HDPE at 3.33 gpm is (using the interpolation equation at the bottom of Table 5.5)

$$\Delta h_{3.33\text{gpm}} = \Delta h_{3\text{gpm}} \times (3.33 \text{ gpm}/3 \text{ gpm})^2 \times 500 \text{ ft}$$
$$= 2.5 \text{ ft}/100 \text{ ft} \times (3.33/3.0)^2 \times 500 \text{ ft}$$
$$= 15.4 \text{ ft water and Re} = 3370 \text{ (by interpolation)}$$

The Reynolds numbers for options a and b are low. The Reynolds number for option c indicates transition flow and the head loss is palatable.

In cooling mode, higher liquid temperature and corresponding low viscosity (even with antifreeze solutions) provide greater flexibility to minimize SWHE head loss while maintaining good inside heat transfer coefficients. In many commercial buildings the cooling requirement is much larger than the heating requirement, even in cold climates. This is especially true for modern buildings in which improved building envelopes and the increased use of energy recovery units (ERUs) for ventilation air tend to cause a greater reduction in heating requirements compared to cooling. The practice of using the design cooling-load flow rate for the heating-mode fluid conditions results in system overdesign. Example 5.4 demonstrates the recommended procedure of sizing the system using the larger of the two loads, which in this case is cooling. The cooling-mode flow rate is used with the cooling-mode fluid conditions. The procedure is repeated using the heating-mode (lesser of the two loads) flow rate with the heating-mode fluid conditions. The example building used in Chapter 4 serves as the model since the cooling load is larger than the heating requirement.

EXAMPLE 5.4—
SWHE CIRCUIT DESIGN WITH COOLING MODE CRITICAL

Calculate the required pump head for the SWHP shown in Figure 5.28, which has a 20 ton (70 kW) cooling requirement and a 10 ton (35 kW) heating requirement.

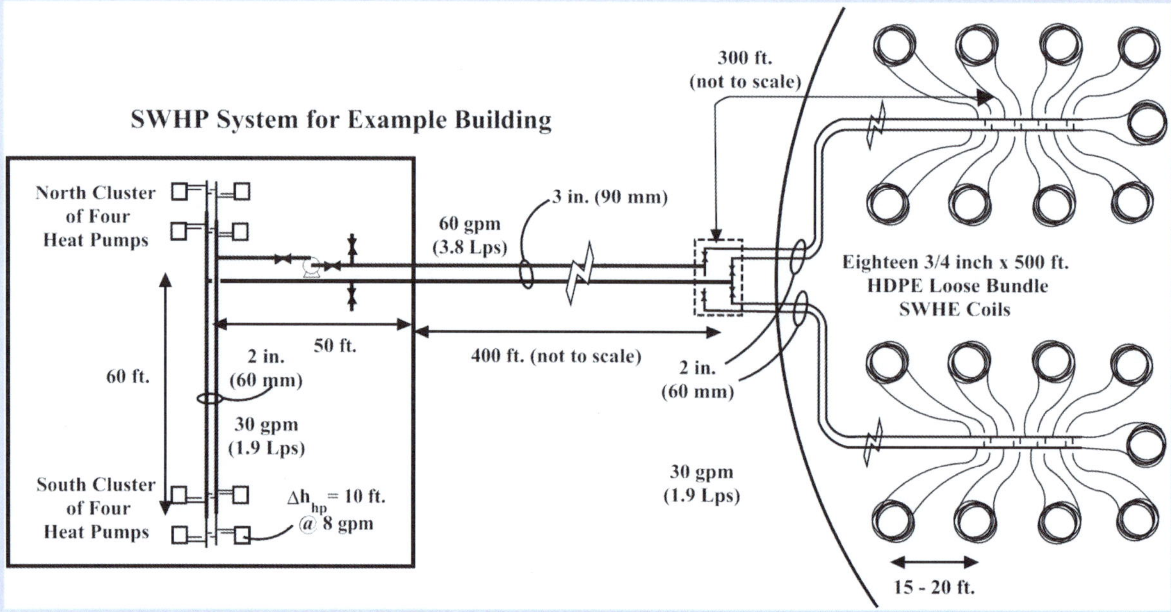

Figure 5.28 SWHP System: 20 Ton (70 kW) Cooling Load and 10 Ton (35 kW) Heat Loss

Solution

Figure 5.29 shows a screenshot from the water distribution system design software discussed in more detail in Chapter 6. The procedure begins with design for the larger cooling-mode load. The SWHE consists of two circuits, each with nine 500 ft (150 m) 3/4 in. (25 mm) HDPE loose bundle coils. Note the fluid properties for a 20% propylene glycol-water mixture at the cooling-mode temperatures are used for the calculation. The design is based on limiting head loss to less than 3 ft of liquid /100 ft (0.3 kPa/m) of pipe. The main headers require a 3 in. (90 mm) pipe, which results in a total loss of 10.4 ft of liquid (31 kPa), while the circuit headers are designed at 2 in. (60 mm) pipe, which results in a loss of 17.4 ft of liquid (52 kPa). The SWHE coil loss is 13.4 ft of liquid (40 kPa). Note the Reynolds number (seventh column from left in Figure 5.29) for the 3/4 in. (25 mm) tubing is 6403, which indicates turbulent flow.

The total head loss for the 60 gpm (3.8 L/s) system flow is 62 ft of water (186 kPa). Procedures are discussed in Chapter 6 that demonstrate this requires a 1.5 hp (1.1 kW) pump assuming a pump efficiency of 70%.

The design procedure is repeated for the heating mode, which is the smaller of the two requirements, with the results shown in Figure 5.30. A flow rate of 30 gpm (1.9 L/s) is used since the load is equivalent to 10 tons (35 kW). Applying the more viscous heating-mode fluid conditions, the total head loss is only 30.4 ft of liquid (1.9 L/s). Therefore, the cooling mode is critical and dictates design parameters. Although flow in the 3/4 in. (25 mm) HDPE loose bundle coils is laminar (Re = 1621), the differential temperature across the inside will be small since the heat transfer rate is much smaller than the cooling-mode heat transfer rate. The pump power requirement is less than 1/2 hp (0.37 kW), and increasing the rate to eliminate laminar flow in the heating mode is counterproductive to system efficiency. If the system design had used the cooling-mode flow rate of 60 gpm (3.8 L/s) with the heating-mode fluid conditions, the head loss would have been 73 ft of liquid (220 kPa) and the required pump size would be 2 hp (1.5 kW).

Liquid	20% Prop Glyc		➔ Percent by Volume	
Temp	70	°F		
Den	63.6	lbm/ft³		
Vis	1.30E-03	lbm/ft-s	1.93	cps

Coils	Flow gpm	Rated Flow gpm	Rated Δh @ 60°F ft. water	Inlet Size inches	Inlet Vel fps	Re(in)	Rated Vel fps	Re(rated)	Δh Ft. Liquid
Heat Pump	8	8	10	1	3.3	13335	3.3	21932	10.5
					0.0	0	0.0	0	0.0
	0	0	0	0	0.0	0	0.0	0	0.0
								Coil sub-total	10.5

HDPE Piping

Flow gpm	Nom. Dia. Inches	DR OD ÷ t	I.D. in.	Roughness ft.	Vel fps	Re	Δh(ft) 100 ft.	Length ft.	Fitting Type	Leqv ft	Qty.	Fitting Type	Leqv ft	Qty.	Fitting Type	Leqv ft	Qty.	Δh Ft. Liquid
60	3	13.5	2.98	7.0E-05	2.8	33545	1.13	800	ButtTeeBr	26	4	ButtRed	10	2	ButtTeeRun	5	0	10.4
30	2	11	1.94	7.0E-05	3.2	25734	2.57	600	ButtRed	7	2	5-LoopHdrLas	30	2	ButtRed	7	0	17.4
3.3	0.75	11	0.86	7.0E-05	1.8	6403	2.65	500	Butt90	5	0	ButtRed	4	0	5-LoopHdrLas	30	0	13.2
60	3	11	2.86	7.0E-05	3.0	34925	1.37	100	ButtTeeBr	26	2	ButtRed	10	0	5-LoopHdrLas	30	0	2.1
30	2	11	1.94	7.0E-05	3.2	25734	2.57	120	ButtRed	7	2	Butt90	17	4	5-LoopHdrLas	30	0	5.2
0	1	11	1.08	7.0E-05	0.0	0	0.00	0	Butt90	8	0	ButtRed	4	0	5-LoopHdrLas	30	0	0.0
0	0.75	11	0.86	7.0E-05	0.0	0	0.00	0	Butt90	5	0	ButtRed	4	0	5-LoopHdrLas	30	0	0.0
0	1	11	1.08	7.0E-05	0.0	0	0.00	0	Butt90	8	0	ButtRed	4	0	5-LoopHdrLas	30	0	0.0
															HDPE sub-total			48.3

Non-HDPE Pipe & Fittings

Flow gpm	Nom. Dia. Inches	Sch	I.D. in.	Drawn Tube oughness in	Vel fps	Re	Δh(ft) 100 ft.	Length ft.	·itting Selecto	Leqv ft	Qty.	·itting Selecto	Leqv ft	Qty.	·itting Selecto	Leqv ft	Qty.	Ft. Liquid
60	3	40	3.07	Brass-Bronz	2.6	32599	2.58	0	Gate Valve	8.3	2	T-Straight	10.0	0	Gate Valve	8.3	0	0.4
30	2	40	2.07	Steel-New	2.9	24193	2.04	0	90 L	5.6	2	T-Straight	6.7	0	Gate Valve	5.6	0	0.2
0	3	40	3.07	Steel-New	0.0	0	0.00	0	90 L	8.3	2	T-Straight	10.0	0	Gate Valve	8.3	0	0.0
															Non-HDPE subtotal			0.7

Other Fittings & Valves	Flow gpm	Cv @ 60°F gpm	Quanity	Inlet Size inches	Inlet Vel fps	Re(in)	Rated Vel fps	Re(rated)		Δh Ft. Liquid
ball valve	8	35	2	1	3.3	13335	14.3	95954		0.3
zone valve	8	13.9	1	1	3.3	13335	5.7	38107		0.8
3' X1"Hose	8	16.4	2	1	3.3	13335	6.7	44961		1.2
	0	0	0	2	0.0	0	0.0	0		0.0
	0	0	0	2	0.0	0	0.0	0		0.0
								Fitting sub-total		2.3
						Open Systerr Only	►►	Elevation	(Feet)	0
								Total Loss	(Ft. Liquid)	61.8

Figure 5.29 *E-PipeAlator14.xlsm* Head Loss Results for SWHP System with 20 Ton (70 kW) Cooling Requirement

								Rated	Rated Δh	Inlet	Inlet		Rated					
Liquid	20% Prop Glyc	➔	Percent by Volume			Coils	Flow	Flow	@ 60ºF	Size	Vel	Re(in)	Vel		Re(rated)			Δh
Temp	34	ºF					gpm	gpm	ft. water	inches	fps		fps					Ft. Liquid
Den	64.1	lbm/ft^3				Heat Pump	8	8	10	1	3.3	6673	3.3		21932			11.7
Vis	2.61E-03	lbm/ft-s	3.89	cps							0.0	0	0.0		0			0.0
							0	0	0	0	0.0	0	0.0		0			0.0
	HDPE Piping														**Coil sub-total**			**11.7**
Flow	Nom. Dia.	DR	I.D.	Roughness	Vel	Re	Δh(ft)	Length	Fitting Type	Leqv	Qty.	Fitting Type	Leqv	Qty.	Fitting Type	Leqv	Qty.	Δh
gpm	Inches	OD ÷ t	in.	ft.	fps		100 ft.	ft.		ft			ft			ft		Ft. Liquid
30	3	13.5	2.98	7.0E-05	1.4	8393	0.39	800	ButtTeeBr	26	4	ButtRed	10	2	ButtTeeRun	5	0	3.6
15	2	11	1.94	7.0E-05	1.6	6439	0.90	600	ButtRed	7	2	5-LoopHdrLas	30	2	ButtRed	7	0	6.1
1.67	0.75	11	0.86	7.0E-05	0.9	1621	0.73	500	Butt90	5	0	ButtRed	4	0	5-LoopHdrLas	30	0	3.7
30	3	11	2.86	7.0E-05	1.5	8738	0.48	100	ButtTeeBr	26	2	ButtRed	10	0	5-LoopHdrLas	30	0	0.7
15	2	11	1.94	7.0E-05	1.6	6439	0.90	120	ButtRed	7	2	Butt90	17	4	5-LoopHdrLas	30	0	1.8
0	1	11	1.08	7.0E-05	0.0	0	0.00	0	Butt90	8	0	ButtRed	4	0	5-LoopHdrLas	30	0	0.0
0	0.75	11	0.86	7.0E-05	0.0	0	0.00	0	Butt90	5	0	ButtRed	4	0	5-LoopHdrLas	30	0	0.0
0	1	11	1.08	7.0E-05	0.0	0	0.00	0	Butt90	8	0	ButtRed	4	0	5-LoopHdrLas	30	0	0.0
Non-HDPE Pipe & Fittings															**HDPE sub-total**			**15.9**
Flow	Nom. Dia.	Sch	I.D.	Roughness	Vel	Re	Δh(ft)	Length	Fitting Selecto	Leqv	Qty.	Fitting Selecto	Leqv	Qty.	Fitting Selecto	Leqv	Qty.	
gpm	Inches		in.	ft.	fps		100 ft.	ft.		ft			ft			ft		Ft. Liquid
30	3	40	3.07	Brass-Bronz	1.3	8156	0.68	0	Gate Valve	8.3	2	T-Straight	10.0	0	Gate Valve	8.3	0	0.1
15	2	40	2.07	Steel-New	1.4	6053	0.69	0	90 L	5.6	2	T-Straight	6.7	0	Gate Valve	5.6	0	0.1
0	3	40	3.07	Steel-New	0.0	0	0.00	0	90 L	8.3	2	T-Straight	10.0	0	Gate Valve	8.3	0	0.0
						Other		Cv		Inlet	Inlet		Rated		**Non-HDPE subtotal**			**0.2**
						Fittings	Flow	@ 60ºF	Quanity	Size	Vel	Re(in)	Vel		Re(rated)			Δh
						& Valves	gpm	gpm		inches	fps		fps					Ft. Liquid
						ball valve	8	35	2	1	3.3	6673	14.3		95954			0.3
						zone valve	8	13.9	1	1	3.3	6673	5.7		38107			0.9
						3' X1"Hose	8	16.4	2	1	3.3	6673	6.7		44961			1.4
							0	0	0	2	0.0	0	0.0		0			0.0
							0	0	0	2	0.0	0	0.0		0			0.0
															Fitting sub-total			**2.6**
												Open Systen	Only	►►	**Elevation**	(Feet)		0
															Total Loss	(Ft. Liquid)		30.4

Figure 5.30 *E-PipeAlator14.xlsm* Head Loss Results for SWHP System with 10 Ton (35 kW) Heat Loss

5.7 OPEN-LOOP SURFACE-WATER HEAT PUMP SYSTEMS

Information on open-loop SWHP systems for buildings is more limited than for closed-loop systems. Fouling and protection of the piping systems and heat exchanger equipment presents a challenge for small-building owners. Additionally, caution is necessary when heating with open-loop systems because the water temperature leaving the heat exchangers must be several degrees above the water freeze point to prevent freezing. Thus, systems are often surface water cooling-only (SWC).

However, the cold temperatures of large, deep reservoirs provide the potential for direct cooling (without refrigeration equipment) or very high cooling heat pump efficiency (especially with return air precooling). Total (sensible and latent) cooling of outdoor ventilation air is also possible with cool water temperatures that would normally be too warm to dehumidify room air.

Many of the components discussed in Chapters 7 and 8 for groundwater heat pumps (GWHPs) can be applied to open-loop SWHP and SWC systems. Larger commercial buildings typically employ indirect methods that have a heat exchanger between the surface-water loop and the building loop to which the cooling coils or heat pumps are connected. Direct systems, in which the water is pumped from the reservoir through the heat pumps, are also possible, but the level of required maintenance is highly dependent on the quality of the water and filtration system.

A major difference between open-loop reservoir and groundwater systems is the type and location of the pump. Possible pump options are a vertical shaft pump (motor above

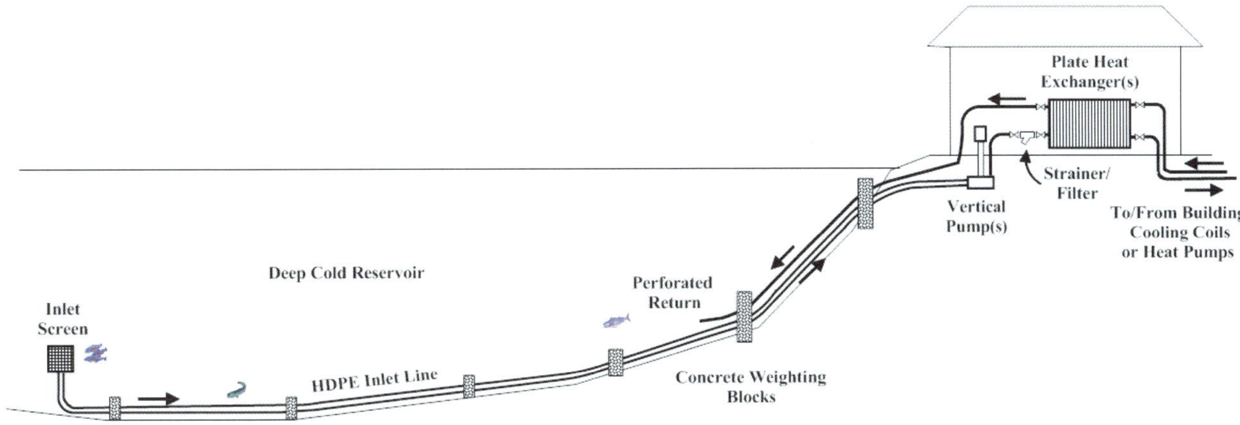

Figure 5.31 Open-Loop Surface-Water Cooling System (with Heating for 42°F+ [6°C+] Lakes)

the water level with the impeller below) or an above-surface horizontal shaft pump with some means of maintaining suction. Figure 5.31 shows the primary components for a surface-water cooling system with a vertical shaft pump. For large systems, water enters the inlet pipe through a screen or grate that is elevated off the reservoir bottom (CUFS 2014). Filtration may require multiple stages to remove large items (logs, fish, etc.) and smaller particles that clog or build up in heat exchangers. Provisions should be provided to periodically backwash/clean the screen or grate.

HDPE has proven to be the piping material of choice due to its cost and corrosion resistance (Heffernan 2001). HDPE density requires that weights, typically concrete collars, be installed to keep the pipe from floating. Protection from damage is required when the pipe is located near the surface.

A wide variety of vertical pumps are available since the application is similar to those that use drainage pumps or pumps that provide cooling water to process coolers and condensers from rivers and cooling ponds. The constraint on the standard design is the long run of inlet pipe that can create pump suction pressures below the required net positive suction head ($NPSH_R$). For both vertical and horizontal shaft designs, the net positive suction head available ($NPSH_A$) of the pump must be greater than the $NPSH_R$ required by the system as given in Equation 5.20:

$$NPSH_R \text{ (ft water)} = 34 \text{ ft} - \text{Elevation (ft)} - \Delta h_{suction} \text{ (ft)} \qquad \text{(I-P)} \qquad (5.20a)$$

$$NPSH_R \text{ (m water)} = 10.4 \text{ m} - \text{Elevation (m)} - \Delta h_{suction} \text{ (m)} \qquad \text{(SI)} \qquad (5.20b)$$

where elevation is the vertical distance between the pump impeller and minimum lake level and $\Delta h_{suction}$ is the head loss in feet of water (metre) across the suction filter, pipe, and foot valve. The $NPSH_A$ of the pump is found from pump curves and is a function of flow rate. Should additional filtration be necessary, care should be taken when suction strainers are incorporated not to add additional loss ($\Delta h_{suction}$ in Equation 5.20). Cavitation is possible, especially when filters are dirty.

For smaller applications, submersible pumps with well screens and casings located off the reservoir body can be a viable alternative. This requires electrical service to the pump, which is likely to be problematic if pumps are installed in the reservoir near the screen. An option is installing the pump beneath a dock or a protected, limited-access location. Some designs require that the pump be placed vertically to avoid bearing failure.

However, the $NPSH_A$ may result in large suction line sizes to avoid excess inlet losses and cavitation. A means of backwashing the screens requires an additional line if the standard option check valve remains in the pump.

5.8 DIRECT COOLING AND PRECOOLING WITH SURFACE-WATER SYSTEMS

It is also possible to provide cooling or precooling without mechanical refrigeration, which is sometimes referred to as *free* cooling, although pumps and fans are necessary. The sensible and latent cooling loads can be satisfied with entering water temperatures (EWTs) slightly above temperatures of conventional chilled-water systems (~44°F [7°C]) with a combination of low air velocity in primary air coils and dedicated air coils for ventilation air conditioning. This is especially true in mild and dry climates. In more humid climates, precooling or supplemental cooling can enhance the capacity and efficiency of heat pump systems. Warm, humid outdoor ventilation air can even be cooled and dehumidified with EWTs above 55°F (13°C). This reduces the latent load on the primary return air coil, which under many conditions may only need to provide sensible cooling, which can be accomplished with higher EWTs.

Two large direct surface-water cooling systems have been successfully operating for more than a decade in New York and Toronto. Cornell University has been using the concept for more than 50 years (CUFS 2014). In 2000, the concept was also applied to a district cooling system that provides 20,000 tons (70,000 kW) of cooling capacity to more than 4.5 million ft^2 (420,000 m^2) of campus buildings. The inlet screen is similar in design to the schematic in Figure 5.31 and is made from 2 mm wedge wire screen. A maximum flow rate of 33,000 gpm (2080 L/s) is drawn from a depth of 250 ft (75 m) through a 2 mi (3.2 km) long HDPE pipe. Heat is transferred to the campus district cooling system via a bank of plate exchangers. The lake water is returned to Cayuga Lake 500 ft (150 m) offshore at a depth of 10 ft (3 m) through a perforated HDPE pipe with an end cap.

In 2001 the concept was applied to a district cooling system that provides downtown Toronto with 39,000 tons (137,000 kW) to 34 million ft^2 (2.2 million m^2) of buildings. Water from Lake Ontario is drawn from a depth of 272 ft (73 m) through three 3.5 mi (5.6 km) long, 63 in. (1.6 m) diameter HDPE intake pipes (SUNY 2011). Heat is transferred to the district cooling system via a bank of plate exchangers. The water is used to provide domestic water for the city and is not returned to the lake.

Several advantages accompany open-loop systems:

- In deeper lakes with temperatures in the 40°F to 50°F (4°C to 10°C) range, direct cooling is possible, thus the major energy-use component of conventional cooling systems is unnecessary.
- Precooling of return air is also a possibility with water in the 50°F to 59°F (10°C to 15°C) range, which substantially increases the efficiency and capacity of the heat pump system.
- Total cooling of outdoor ventilation air can be accomplished with 50°F to 59°F (10°C to 15°C) water.
- Greater heat pump capacity is possible when compared to a closed-loop system since the water to the heat pump is 5°F to 10°F (3°C to 6°C) warmer in the winter and 8°F to 15°F (4°C to 8°C) cooler in the summer.
- Open systems can be designed to limit disturbance (compared to closed systems) to the natural thermocline of deeper lakes. The warmer water that exits the

building in the cooling mode can be reinjected closer to the surface. This reduces the adverse circulation loops that would result if warmer water were injected in the colder regions of the lake.

Figure 5.32 shows a schematic arrangement of an outdoor ventilation air coil in parallel with the primary return air coil. In applications with high ventilation air requirements, such as schools, the greatest latent load is often due to this component. In low-activity classrooms or offices, the latent load from occupants is much lower than the outdoor air load. High levels of humidity can be removed from the outdoor airstream with cool reservoir water, groundwater, or even liquid from a closed-loop SWHE.

Figure 5.33 shows a schematic arrangement of a chilled-water coil in series with a heat pump evaporator coil. The water coil can serve as either a precooler or a direct cooling coil. In some applications, the EWT is low enough to manage the total cooling load during most hours of operation and the heat pump can serve as a second-stage cooling device during the more extreme conditions. It can be activated by a humidistat when room humidity levels rise above the desired setpoint and/or when the room temperature cannot be maintained by the chilled-water coil. Note that water flow rate can be minimized with a three-way valve by routing the water stream leaving coil to the heat pump when necessary or returned to the reservoir when the heat pump is not operating. In heating mode, another three-way valve is used to route the flow directly to the heat pump.

The feasibility of these approaches is enhanced by using lower-than-conventional air coil face velocities. This increases dehumidification and also reduces fan friction losses, which are critical when a precooling coil is placed in series with the primary (heat pump) coil.

Figure 5.34 shows a set of total and sensible cooling coil performance curves for two EWTs that span the upper range of acceptable values for direct or precooling applica-

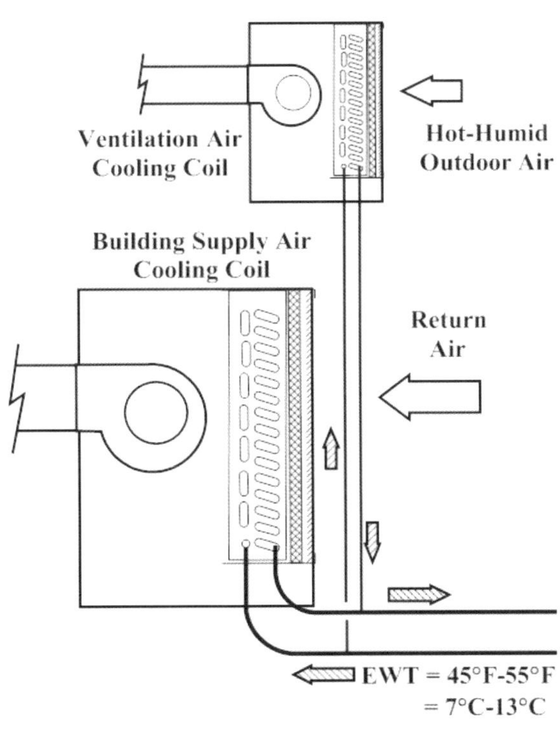

Figure 5.32 Air Coil Arrangement for Surface-Water or Groundwater Direct Cooling Systems

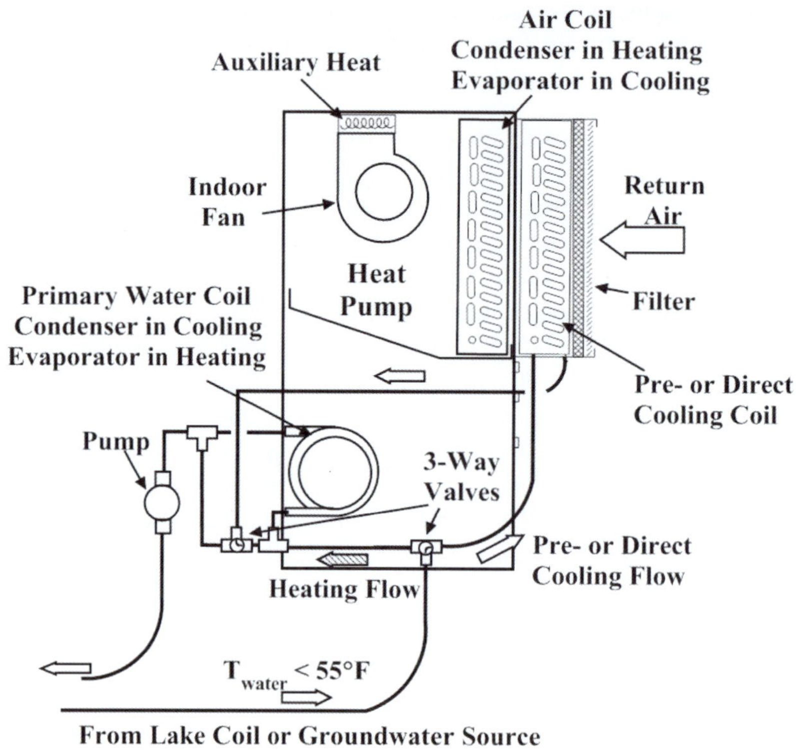

Figure 5.33 Schematic Arrangement of Direct/Precooling Water Coil and Heat Pump

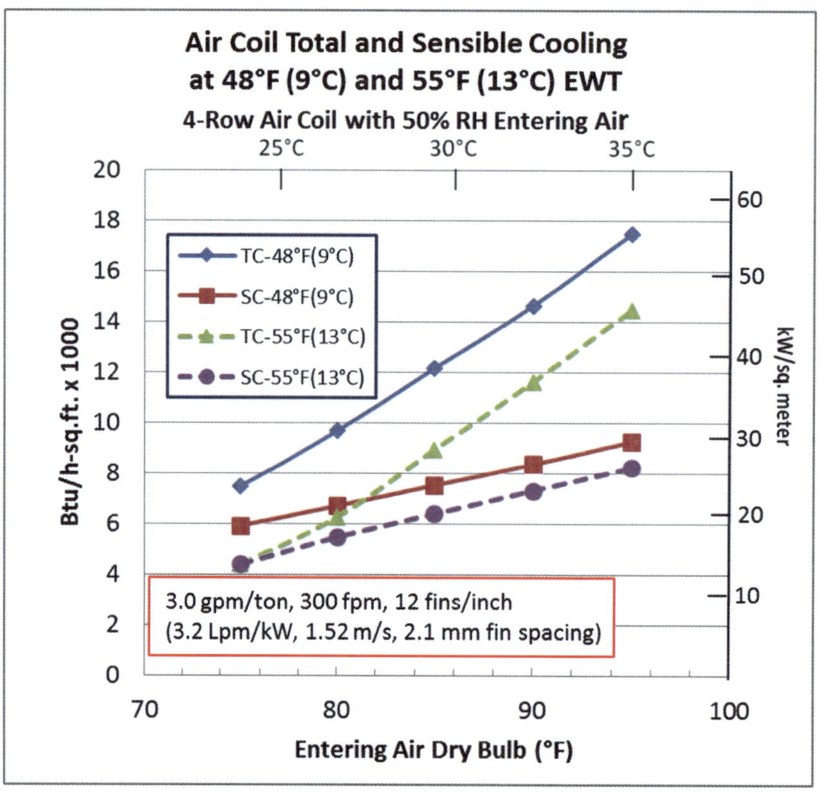

Figure 5.34 Total and Sensible Capacities of Four-Row Chilled-Water Coil

tions. The figure can be used when manufacturers' coils are not available for a broad range of EWTs and entering air temperatures (EATs). The curves are based on unit face area (kBtu/h·ft^2 [kW/m^2]) and are restricted to four-row coils, 12 fins/in. (fin spacing = 2.1 mm), 3 gpm/ton (3.2 L/min·kW), and 50% entering air relative humidity. The information is sufficient, however, to demonstrate the potential benefits of direct cooling and precooling of buildings with low-temperature reservoir water and groundwater.

EXAMPLE 5.5— AIR COIL DESIGN FOR RESERVOIR FREE COOLING

A building has total and sensible cooling loads of 50,000 Btu/h (14.7 kW) and 42,000 Btu/h (12.3 kW). The outdoor ventilation air cooling load adds 12,000 Btu/h (3.5 kW) total with 6000 Btu/h (1.8 kW) sensible. Outdoor conditions are 95°F (35°C) with 50% rh and indoor conditions are 77°F (25°C) and 50% rh. Water at 52°F (11°C) is available from a closed-loop SWHE. Select a building supply air coil and outdoor air ventilation coil to meet load requirements and specify necessary airflow and water flow rates.

Solution

The combined building and outdoor air loads are as follows:

$$TC = TC_{bldg} + TC_{oa} = 50{,}000 + 12{,}000 = 62{,}000 \text{ Btu/h} = 62 \text{ kBtu/h} \quad \text{(I-P)}$$

$$SC = SC_{bldg} + SC_{oa} = 42{,}000 + 6{,}000 = 48{,}000 \text{ Btu/h} = 48 \text{ kBtu/h} \quad \text{(I-P)}$$

$$TC = TC_{bldg} + TC_{oa} = 14.7 + 3.5 = 18.2 \text{ kW} \quad \text{(SI)}$$

$$SC = SC_{bldg} + SC_{oa} = 12.3 + 1.8 = 14.1 \text{ kW} \quad \text{(SI)}$$

Thus the combined load sensible heat ratio (SHR) is

$$SHR_{load} = SC/TC = 48{,}000/62{,}000 = 0.77 \quad \text{(I-P)}$$

$$SHR_{load} = SC/TC = 14.1/18.2 = 0.77 \quad \text{(SI)}$$

Figure 5.34 indicates via interpolation for 52°F (11°C) EWT that the four-row coil will provide 7.1 kBtu/h·ft^2 (22.4 kW/m^2) total cooling and 5.3 kBtu/h·ft^2 (16.7 kW/m^2) sensible cooling with 77°F (25°C) and 50% rh entering air. To meet the building total cooling load, the face area of the building supply air coil would be

$$A_{sac} = 50 \text{ kBtu/h} \div 7.1 \text{ kBtu/h·ft}^2 = 7.0 \text{ ft}^2 \quad \text{(I-P)}$$

$$A_{sac} = 14.7 \text{ kW} \div 22.4 \text{ kW/m}^2 = 0.66 \text{ m}^2 \quad \text{(SI)}$$

The sensible cooling capacity of this coil will be

$$SC_{sac} = 7.0 \text{ ft}^2 \times 5.3 \text{ kBtu/h·ft}^2 = 37.1 \text{ kBtu/h} \quad \text{(I-P)}$$

$$SC_{sac} = 0.66 \text{ m}^2 \times 16.7 \text{ kW/m}^2 = 11.0 \text{ kW} \quad \text{(SI)}$$

Figure 5.34 also indicates via interpolation for 52°F (11°C) EWT that the four-row coil will provide 15.8 kBtu/h·ft^2 (50 kW/m^2) total cooling and 11.5 kBtu/h·ft^2 (36 kW/m^2) sensible cooling with 95°F (25°C) and 50% rh entering air. To meet the ventilation total cooling load, the face area of the outdoor air coil would be

$$A_{oac} = 12 \text{ kBtu/h} \div 15.8 \text{ kBtu/h·ft}^2 = 0.76 \text{ ft}^2 \quad \text{(I-P)}$$

$$A_{oac} = 3.5 \text{ kW} \div 50 \text{ kW/m}^2 = 0.07 \text{ m}^2 \quad \text{(SI)}$$

The sensible cooling capacity of the outdoor air coil will be

$$SC_{oac} = 0.76 \text{ ft}^2 \times 11.5 \text{ kBtu/h·ft}^2 = 8.7 \text{ kBtu/h} \quad \text{(I-P)}$$

$$SC_{oac} = 0.07 \text{ m}^2 \times 36 \text{ kW/m}^2 = 2.5 \text{ kW} \quad \text{(SI)}$$

To maintain comfort (humidity level/latent capacity), the combined SHR_{coil} of the supply air and outdoor air coils must be less than or equal to the combined SHR_{load} of the loads.

$$SHR_{coil} = (SC_{sac} + SC_{oac}) \div (TC_{sac} + TC_{oac}) = (37.1 + 8.7) \div (50 + 12) = 0.74$$

The condition of $SHR_{coil} \leq SHR_{load}$ is satisfied at full load.

The required airflow rates for the coils are as follows:

$$Q_{a\text{-}sac} = A_{sac} \times V_{face} = 7.0 \text{ ft}^2 \times 300 \text{ ft/min} = 2100 \text{ cfm} \quad \text{(I-P)}$$

$$Q_{a\text{-}sac} = A_{sac} \times V_{face} = 0.66 \text{ m}^2 \times 1.52 \text{ m/s} = 1.0 \text{ m}^3\text{/s or } 3600 \text{ m}^3\text{/h} \quad \text{(SI)}$$

$$Q_{a\text{-}oac} = A_{oac} \times V_{face} = 0.76 \text{ ft}^2 \times 300 \text{ ft/min} = 230 \text{ cfm} \quad \text{(I-P)}$$

$$Q_{a\text{-}oac} = A_{oac} \times V_{face} = 0.07 \text{ m}^2 \times 1.52 \text{ m/s} = 0.11 \text{ m}^3\text{/s or } 396 \text{ m}^3\text{/h} \quad \text{(SI)}$$

The required water flow rates for the coils are as follows:

$$Q_{w\text{-}sac} = TC_{sac} \times Q_w\text{/ton} = 50 \text{ kBtu/h} \div 12 \text{ kBtu/t·h} \times 3 \text{ gpm/ton} = 12.5 \text{ gpm} \quad \text{(I-P)}$$

$$Q_{w\text{-}sac} = 14.7 \text{ kW} \times 3.2 \text{ L/min·kW} = 47 \text{ L/min or } 0.78 \text{ L/s} \quad \text{(SI)}$$

$$Q_{w\text{-}oac} = TC_{oac} \times Q_w\text{/ton} = 12 \text{ kBtu/h} \div 12 \text{ kBtu/t·h} \times 3 \text{ gpm/ton} = 3.0 \text{ gpm} \quad \text{(I-P)}$$

$$Q_{w\text{-}oac} = 3.5 \text{ kW} \times 3.2 \text{ L/min·kW} = 11.2 \text{ L/min or } 0.19 \text{ L/s} \quad \text{(SI)}$$

Several cautions are advised before universally applying the procedures in Example 5.5:

- Sensible heat ratios at part load are typically less than sensible heat ratios at full load; thus, the condition of $SHR_{coil} \leq SHR_{load}$ should be verified at part-load, humid-day conditions. In many cases, the latent capacities of the chilled-water coils can be improved by reducing the face velocity (airflow) below the 300 fpm (1.52 m/s) assumed in Figure 5.34.

- Adequate system dehumidification was achieved because the very warm, humid outdoor air was delivered to the outdoor air coil before mixing with the building return air. Had the ventilation air mixed with the return air before being delivered to the main coil, adequate dehumidification for the combined load could not have been accomplished with 52°F (11°C) EWT.
- The procedure assumed the EWT to the coils is equal to the LLT of the SWHE. The temperature rise in the supply-line liquid from the ground and the portion of the line in the warm, upper regions of the reservoir can be minimized by adding pipe insulation. The spreadsheet tool *GroundTemp&Resist.xlsm*, available with this book at www.ashrae.org/GSHP, can be used to estimate the temperature change in horizontal headers located in shallow ground.

5.9 HEAT TRANSFER IN GSHP HEADERS

This section addresses heat transfer from horizontal headers connecting ground loops or SWHEs and building heat pumps. It includes heat transfer to the ground and heat transfer between the supply and return headers. Example 5.5 assumes minimal heat loss or gain in the header between the heat pump and the coil in the reservoir. This assumption is good for large systems located near the reservoir because the heat loss relative to the flow rate results in minimal temperature change. However, the heat gain in headers located in shallow ground and in the upper portions of stratified lakes could be significant in the cooling mode for small flow rates (<50 gpm [3 L/s]) and/or long headers (>200 ft [60 m]). In the heating mode, the heat transfer from the soil could be beneficial when burial depths are 3 ft (1 m) and deeper, because the soil is likely warmer than the reservoir in the winter. For open-loop SWHPs, Equation 5.21 is used to estimate the heat pump ELT:

$$\text{ELT} = t_{resv} + t_{app} + \Delta t_{resv\ header} + \Delta t_{grn\ header} \tag{5.21}$$

For closed-loop SWHPs, Equation 5.22 is applied:

$$\text{ELT} = \text{LLT}_{swhe} + t_{app} + \Delta t_{resv\ header} + \Delta t_{grn\ header} \tag{5.22}$$

The temperature change in the header ($\Delta t_{resv\ header}$) should be minimal except for the case of a cooling-mode operation with a cold lake. Equation 5.23 should be applied only to the return from the reservoir (supply to the heat pump) portion of the header located above the thermocline. Equation 5.24 applies to the header between the reservoir and the heat pump.

$$\Delta t_{resv\ header} = C_{resv} \times [t_{resv} - t_{coil}] \times L_{resv\ header}\ (\text{ft [m]}) \div Q \tag{5.23}$$

$$\Delta t_{\text{grn header}} = C_{grn} \times [t_{grn} - t_{coil}] \times L_{grn\ header}\ (\text{ft [m]}) \div Q \tag{5.24}$$

The coefficients for Equations 5.23 and 5.24 (C_{resv}, C_{grn}) can be found in Table 5.6. They were developed for DR 11 HDPE headers but can be used with acceptable accuracy for other types of plastic pipe. The temperatures in the equations refer to the temperature in the reservoir above the thermocline where the header passes (t_{resv}), the liquid temperature inside the coil (t_{coil}), and the temperature of the ground (t_{grn}) surrounding the return header. Local ground temperatures for various depths below grade and days of the year can be found from Equation 5.25 or by adding the temperature variations in Figure 5.35 to the local deep ground temperature.

Table 5.6 Coefficients for Reservoir and Ground Header Heat Transfer

Nominal Diameter, in.	C_{resv}, gpm/ft Pipe Insulation Thickness, in.			C_{grn}, gpm/ft Pipe Insulation Thickness, in.		
	0	0.5	1	0	0.5	1
1.5	0.0087	0.00055	0.00034	0.0017	0.00045	0.00029
2	0.0093	0.00066	0.00039	0.0019	0.00052	0.00034
3	0.0103	0.00091	0.00053	0.0020	0.00067	0.00044
4	0.0109	0.00120	0.00065	0.0021	0.00079	0.00052

Values based on insulation k = 0.02 Btu/h·ft·°F (0.24 Btu·in./h·ft^2·°F)

Nominal Diameter, mm	C_{resv}, L/s·m Pipe Insulation Thickness, mm			C_{grn}, L/s·m Pipe Insulation Thickness, mm		
	0	12.5	25	0	12.5	25
50	0.0018	0.00011	0.00007	0.00035	0.00009	0.00006
63	0.0019	0.00014	0.00008	0.00039	0.00011	0.00007
90	0.0021	0.00019	0.00011	0.00041	0.00014	0.00009
125	0.0023	0.00025	0.00013	0.00043	0.00016	0.00011

Values based on insulation k = 0.035 W/m·K

The temperature of the ground at shallow (>30 ft [10 m]) depths can be determined for any day of the year using Equation 5.25 (Remund 2009). Figure 5.35 is a graphic plot for four depths in a soil that has average values of thermal conductivity, density, and specific heat.

$$t_{grn}(d, \tau_d) = t_{mean} - A_s \times e^{\{[-d(\pi/365\alpha)^{0.5}]\cos[(2\pi/365)(\tau_d - \tau_{min} - 0.5d(365/\pi\alpha)^{0.5})]\}} \tag{5.25}$$

where

t_{mean} = mean earth temperature at surface or average annual air temperature (available as the Annual [column d] Monthly Climatic Design Conditions [ASHRAE 2013b])

A_s = annual daily average temperature variation at surface above and below t_{mean} (if not available, use the maximum and minimum values for Monthly Climatic Design Conditions [ASHRAE 2013b])

d = depth below surface

α = thermal diffusivity

τ_d = days after January 1 (Julian day)

τ_{min} = number of days after January 1 when minimum earth (or air) temperature occurs (if not available, use the 15th day of the month with the lowest Monthly Climatic Design Conditions [ASHRAE 2013b])

In rare cases, designers have specified that supply and return headers be placed in separate trenches to minimize short-circuit heat transfer (q_{ss}). Note that U-tube vertical heat exchangers continue to be very effective in spite of the fact that the supply and return tubes are in very close proximity. However, simple steady-state calculations in the form of shape factors (S_f) can be used to estimate heat transfer between buried headers, as given in Equation 5.26 (Holman 1986):

$$q_{ss} = k_g \times S_f \times (t_{supply} - t_{return}) \tag{5.26}$$

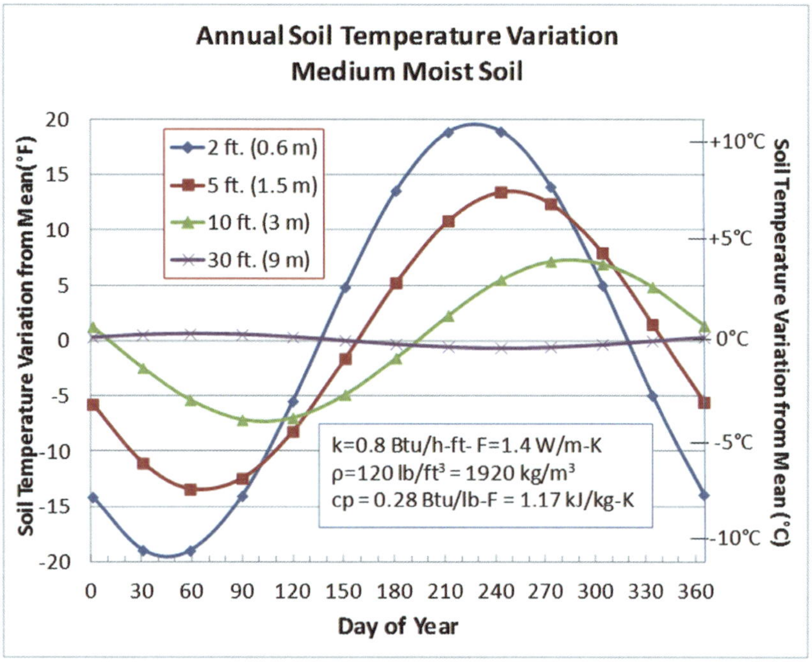

Figure 5.35 Ground Temperature Variation from Local Mean for Damp, Medium-Density Soil

EXAMPLE 5.6— CALCULATION OF RESERVOIR AND GROUND HEADER TEMPERATURE RISE

Find the temperature rise and heat pump ELT in August in an uninsulated 3 in. (90 mm) header that flows at a rate of 50 gpm (3.15 L/s) from a 50°F (10°C) lake to a set of building heat pumps. The header passes through 200 ft (61 m) of shallow water at 80°F (26.7°C) and 600 ft (183 m) of ground 5 ft (1.5 m) beneath the surface. The earth temperature at the surface varies from 35°F to 85°F (2°C to 29°C) over the annual cycle with a mean annual temperature of 60°F (16°C). Assume soil conditions similar to those shown in Figure 5.35.

Solution

$$\Delta t_{resv\ header} = C_{resv} \times [t_{grn} - t_{coil}] \times L_{header} \div Q$$
$$= 0.0103\ (\text{gpm/ft}) \times [80°\text{F} - 50°\text{F}] \times 200\ \text{ft} \div 50\ \text{gpm} = 1.24°\text{F} \qquad \text{(I-P)}$$

$$\Delta t_{resv\ header} = C_{resv} \times [t_{grn} - t_{coil}] \times L_{header} \div Q$$
$$= 0.0021\ (\text{L/s·m}) \times [26.7°\text{C} - 10°\text{C}] \times 61\ \text{m} \div 3.15\ \text{L/s} = 0.68°\text{C} \qquad \text{(SI)}$$

The temperature (t_{lrh}) of liquid leaving the portion of the header in the reservoir is

$$t_{lrh} = t_{resv} + \Delta t_{resv\ header} = 50 + 1.24 = 51.24°\text{F} \qquad \text{(I-P)}$$

$$t_{lrh} = t_{resv} + \Delta t_{resv\ header} = 10 + 0.68 = 10.68°\text{C} \qquad \text{(SI)}$$

Figure 5.35 indicates the ground temperature at a 5 ft (1.5 m) depth is 14°F (8°C) above the average earth temperature of 60°F (16°C), which is 74°F (23.3°C). Thus,

$$\Delta t_{grn\ header} = C_{resv} \times [t_{grn} - t_{lrh}] \times L_{header} \div Q$$
$$= 0.0020\ (\text{gpm/ft}) \times [74°\text{F} - 51.24°\text{F}] \times 600\ \text{ft} \div 50\ \text{gpm} = 0.55°\text{F} \qquad \text{(I-P)}$$

$$\Delta t_{grn\ header} = C_{resv} \times [t_{grn} - t_{lrh}] \times L_{header} \div Q$$
$$= 0.00041\ (\text{L/s·m}) \times [23.3°\text{C} - 10.68°\text{C}] \times 183\ \text{m} \div 3.15\ \text{L/s} = 0.30°\text{C} \quad \text{(SI)}$$

The temperature of the liquid leaving the ground header (t_{lgh}) and entering the heat pumps (ELT) is

$$\text{ELT} = t_{resv} + \Delta t_{resv\ header} + \Delta t_{resv\ header} = 50 + 1.24 + 0.55 = 51.8°\text{F} \quad \text{(I-P)}$$

$$\text{ELT} = t_{resv} + \Delta t_{resv\ header} + \Delta t_{resv\ header} = 10 + 0.68 + 0.3 = 11.0°\text{C} \quad \text{(SI)}$$

EXAMPLE 5.7—SHORT-CIRCUIT HEAT TRANSFER IN HORIZONTAL HEADERS

Find the temperature rise in a nominal 6 in. (170 mm) buried steel supply header that is 12 in. (0.3028 m) center-to-center from the return header. The headers are 100 ft (30 m) in length, are placed in soil with a thermal conductivity of 0.7 Btu/h·ft·°F (1.2 W/m·K), and have a 10°F (5.6°F) differential temperature and a flow rate of 500 gpm (32 L/s). Note the outside diameter of a nominal 6 in. (170 mm) pipe is 6.625 in. (r = 3.313 in.) (170 mm [r = 0.085 m]).

Solution

Equation 5.27 is applied to find the shape factor:

$$S_f = \frac{2\pi \times 100\ \text{ft}}{\cos h^{-1}\left(\frac{12^2 - 3.313^2 - 3.313^2}{2 \times 3.313 \times 3.313}\right)} = 262\ \text{ft} \quad \text{(I-P)}$$

$$S_f = \frac{2\pi \times 30\ \text{m}}{\cos h^{-1}\left(\frac{0.3028^2 - 0.085^2 - 0.085^2}{2 \times 0.085 \times 0.085}\right)} = 79.8\ \text{m} \quad \text{(SI)}$$

Equation 5.26 is used to find the short-circuit heat transfer between the supply and return headers:

$$q_{ss} = 0.7\ \text{Btu/h·ft·°F} \times 262\ \text{ft} \times 10°\text{F} = 1834\ \text{Btu/h} \quad \text{(I-P)}$$

$$q_{ss} = 1.2\ \text{W/m·K} \times 79.8\ \text{m} \times 5.6°\text{C} = 536\ \text{W} = 0.536\ \text{kW} \quad \text{(SI)}$$

The temperature rise in supply header water due to heat transfer from the return header (see Equation 4.2) is

$$\Delta t = \frac{q_{ss}\ (\text{Btu/h})}{500 \times Q\ (\text{gpm})} = \frac{1834\ (\text{Btu/h})}{500\ \text{Btu·min/gal·°F·h} \times 500\ \text{gal/min}} = 0.007°\text{F}$$

$$\Delta t = \frac{q_{ss}\ (\text{kW})}{4.15 \times Q\ (\text{L/s})} = \frac{0.536\ (\text{kW})}{4.15\ \text{kW·s/L·°C} \times 32\ \text{L/s}} = 0.004°\text{C}$$

Thus, the heat short-circuiting between the header pipes is small and rarely justifies additional expense to reduce it.

Steel pipe and turbulent flow are assumed to simplify the calculation since the exterior surface temperatures are very close to the fluid temperatures. This assumption results in higher transfer rates than those that would occur with HDPE pipe and nonturbulent flow. For two buried cylinders, Equation 5.27 is used to find the shape factor that is to be applied to Equation 5.26:

$$S_f = \frac{2\pi L}{\cosh^{-1}\left(\dfrac{D^2 - r_1^2 - r_2^2}{2r_1 r_2}\right)} \tag{5.27}$$

where

L = length of the pipes
D = center-to-center distance between the pipes
r_1, r_2 = radii of pipes (which will be equal in this case)

5.10 ENVIRONMENTAL IMPACT OF SURFACE-WATER HEAT PUMPS

The net environmental impact of SWHPs has been addressed in terms of overall impact to the ecosystem and bulk changes in reservoirs (Hattemer et al. 2006; Hattemer 2005). Concerns have been raised regarding the rise in reservoir temperature and potential pollution from leakage of antifreeze solutions with corrosion inhibitors. This section provides a basis for analyzing and calculating these impacts and putting them into perspective relative to activities that are largely unregulated and have far more negative thermal and pollutant impacts.

Figure 5.36 compares the hourly heat rates of a mid-size boat motor operating at cruising speed to the rates of a SWHP in an average-size lake-front home. Although the boat motor will operate far few hours than the heat pump, the total annual input into the reservoir is of the same magnitude. Also, the heat pump removes heat in the winter; the boat motor does not. Boat motors also release benzene, toluene, methyl tert-butyl ether (MTBE), ethyl benzene, xylene, and large amounts of unburned hydrocarbons (Hattemer et al. 2006). This is especially true for large high-performance two-cycle outboard motors, which remain popular in the United States. SWHPs release relatively mild fluids only when they are installed incorrectly or suffer damage. Additionally, the higher efficiencies of SWHPs (compared to conventional systems) result in lower carbon dioxide and pollutant emissions from power plants. Much of the discharge released into the air from fossil-fuel-burning power plants will eventually contribute to pollution of reservoirs and streams. Thus, regulations should be developed that recognize and minimize the relatively benign negative impact of SWHPs on reservoirs along with the positive environmental effects on the larger ecosystem.

A thorough study of the environmental, economic, and technical issues of deep-water cooling systems has been conducted for a proposed naturally chilled project for central New York (SUNY 2011). The project would incorporate and expand an existing 22 mi (36 km) by 54 in (1.37 m) diameter clear-water transmission water main and distribution network. The report concludes that there would be no harmful impact on water quality or transmission of invasive species if the intake precautions used for the Cornell University and City of Toronto systems are applied.

However, there remain issues that have not been adequately addressed for closed-loop heat pump systems. Two in particular are 1) the required reservoir sizes to ensure minimal

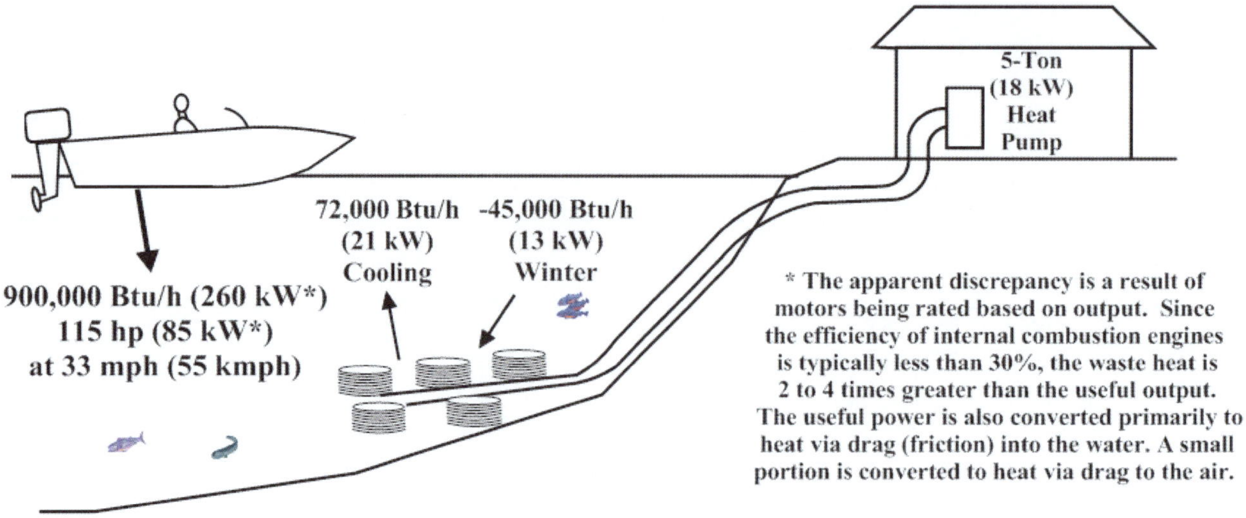

Figure 5.36 Comparative Reservoir Heat Rates for a SWHP and a Mid-Size Boat Motor

change in temperature, biological growth, impact on aquatic life, and water level and 2) the potential change in natural reservoir thermal patterns (i.e., water remaining cold in lower portions of deep reservoirs during warm months) that may result from the addition of heat from closed-loop SWHEs. This issue can be averted in open-loop SWHP systems by returning the water above the thermocline at a distance from the intake.

Hattemer et al. (2006) assessed the thermal impact of cooling 3500 homes of an average size of 3000 ft^2 (280 m^2) on a 5900 acre (2400 hectare) lake in a southern United States climate. The assumption was that 50% of the homes were cooled with coils placed in cold (50°F [10°C]), deep (50 ft [15m]) water and 50% with coils in warm (80°F [27°C]), shallow water. The analysis assumed a three month drought (no rainfall) occurred during the summer. The resulting rise in temperature was (0.5°F [0.3°C]) with a 0.12 in. (3 mm) decline in lake level due to the elevated temperature and added heat input. However, this input rise would be balanced by heat removable for winter space conditioning and domestic hot-water preheating (which is a recommended and widely used option).

The study also calculated the savings in electrical energy generation and transmission-produced pollutants per 1000 houses for 50%/50% deep water/shallow water SWHPs compared to 13 SEER air-source heat pumps. Emission offsets for 1000 homes were estimated to be 6.1×10^6 lb (2.8×10^6 kg) of carbon dioxide (CO_2), 3.5×10^4 lb (1.6×10^4 kg) of sulfur dioxide (SO_2), and 1.3×10^4 lb (0.59×10^4 kg) of nitrous oxides (NO_x). An energy analysis projected annual space-conditioning costs of $484 for deep-water SWHPs, $632 for shallow-water SWHPs, and $870 for air-source heat pumps. Water-heating cost savings generated by the water-to-air heat pumps were not included in the analysis.

The study also provides information from various sources regarding the toxicity of propylene glycol, which the U.S. Food and Drug Administration considers to be generally recognized as safe (GRAS) for use in human and animal food (except for cats) (Dow 2001). It is nontoxic to the environment and biodegrades when released in water. However, propylene glycol depletes oxygen, which has the potential to harm nearby aquatic life if released in large quantities. The study did not address the environmental impact of antifreeze corrosion inhibitors, which can be rendered unnecessary if propylene glycol and noncorrosive piping materials are applied. Corrosion inhibitors are recommended for alcohol-based antifreeze solutions since they demonstrate higher potential for problems in copper and copper-based alloys (ASHRAE 2011).

Two related areas must be considered when determining the minimum required reservoir or stream size for a SWHP system. The impact of heat extraction or rejection may result in changes to natural characteristics that affect the environment of the body of water or the performance of the SWHP itself. For example, overloading a small, shallow pond in the summer might raise the temperature of the water several degrees. Environmentally, this may negatively affect wildlife and vegetation, increase evaporation rates, and lower the water level. From a performance standpoint, the high water temperature will result in lower cooling capacity and efficiency.

Different minimum required guidelines might also result for public waters and private reservoirs built for other purposes. While the thermal impact of small SWHP systems on larger, deeper lakes is minimal, there is a point where temperatures can be noticeably altered. For public lakes, the allowable capacity per acre of surface might be much smaller than that for a pond built by a contractor that serves the dual purpose of water retention and heat pump duty. The private lake could be loaded more intensely before the temperature change significantly impacted its intended purpose. However, in an existing public lake the outcry might be huge if a small change in temperature (real or imagined) were perceived to alter the number of fish caught or the water level.

Guidelines for the minimum depth and surface area requirements for SWHPs must take into account a wide variety of conditions *and expectations*. For example, a shallow lake (<12 ft [4 m] deep) should not be expected to provide the warm winter or cool summer loop temperatures possible with vertical GCHPs or GWHPs. Nature dictates that shallow bodies of water get colder than the deep ground (and groundwater) in the winter and warmer in the summer. Designers must take this into account by limiting the capacity of SWHP equipment for a given size reservoir or stream or compensating with supplementary equipment.

As noted in Figure 5.3, there are many heat transfer modes in reservoirs. In some cases, very small ponds with high groundwater and surface-water flows will perform much better than larger, deeper stagnant lakes with dense clay (low-conductivity) lake beds. Issues of concern included the following:

- Reservoirs may have adverse level reductions from natural evaporation, leaks, or power generation. This problem will be further aggravated if excessive amounts of building heat are rejected during hot, dry periods. However, evaporation rates can be predicted for various climates if the building cooling load can be estimated by an energy balance on the water body.
- Excessive amounts of heat rejected to the bottom of a cold lake may disturb the natural thermocline and cause premature inversions. The impact on the lake ecology should be evaluated by considering the impact of the building load on deep-water temperatures.
- Although small, shallow lakes have been used to reject rather large cooling loads (100 *cooling* tons per acre) when makeup water is available, entering loop temperatures typically rise to levels (>86°F [30°C]) that may not result in suitable system efficiency.
- The heating capacities of surface bodies of water are typically much less than the cooling capacities. Winter evaporative and convective heat losses coupled with much lower solar radiation may result in freezing or near-freezing conditions. Convective heat gain from the ground to the water must be relied upon to a large extent. If lake-bottom sediments have high organic or clay content, thermal conductivity will be moderate and thermal capacity will be limited. However,

large, deep lakes will delay or avoid the onset of failure because of their large thermal capacity.

- When excessive amounts of heat are extracted from small bodies of water, the bulk water temperature will decline until the surface temperature of the coil falls below 32°F (0°C). Ice will begin to build up on the outside of the coil, which increases thermal resistance. Loop temperature will continue to decline until the heat pump shuts off, and/or the coil will float because of the ice buildup.

5.11 RECOMMENDATIONS FOR THE DESIGN OF SURFACE-WATER HEAT PUMPS

Some recommendations for the design of SWHPs are as follows:

- Conduct a thermal survey of the water body (or reference a previous survey of a similar reservoir in a similar climate) during the critical late-summer and late-winter periods. Temperatures should be taken at regular increments for the entire depth of the reservoir or stream. Information should also include if (and for how long) the surface freezes.
- Gather information about the reservoir, including depth, area, inflow, outflow, level fluctuation, and clarity. Pezent (1989) discusses the use of a Secchi disk as an indicator of clarity.
- When the heating load on a reservoir exceeds 10 tons per acre (90 kW/ha) and/or the average depth is less than 10 ft (3 m), detailed analysis that considers the above-mentioned environmental and performance issues is warranted.
- When the cooling load on a reservoir exceeds 20 tons per acre (180 kW/ha) and/or the average depth is less than 10 ft (3 m), detailed analysis that considers the above-mentioned environmental and performance issues is warranted.
- The heating and cooling loads on the building should be estimated as input for the amount of energy to be added and extracted to the surface water. This should include maximum loads and seasonal energy.
- All of this information should be linked to weather data and used to conduct an energy balance on the reservoir or stream to determine if the surface water reservoir can provide operating conditions that are acceptable from both comfort and economic perspectives.

The final report for ASHRAE RP-1385 (2009) may contain more enlightened guidance for reservoir size requirements. In the interim, examples of SWHPs attached to small reservoirs include the following:

- A manufacturing facility in Indiana with a 3 acre, 8 ft (12,000 m^2) average depth retention pond with 180 300 ft (90 m) HDPE coils (54,000 ft [2.5 m] total) connected to office heat pumps (164 tons [575 kW]) and intermittently used laboratory/plant heat pumps (259 tons [910 kW]). In peak-load winter months, the SWHEs return water to the heat pumps between 30°F and 45°F (–1°C and 7°C). In peak-load summer months, the SWHEs return water between 80°F and 100°F (27°C and 38°C).
- A 700,000 ft^2 (65,000 m^2) medical center in Illinois connected to 1500 tons (5300 kW) of heat pump equipment connected to a 15 acre (60,000 m^2) lake. After one year of operation, 180 vertical bores were added to maintain efficient performance.
- A 15,000 ft^2 (1400 m^2) community center connected to a 4 acre (16,000 m^2) lake.

5.12 REFERENCES

ASHRAE. 2009. *Development of design tools for surface water heat pump systems*. ASHRAE RP-1385. Final Report in Progress. Atlanta: ASHRAE.

ASHRAE. 2011. *ASHRAE Handbook—HVAC Applications*. Geothermal Energy, p. 34.32. Atlanta: ASHRAE.

ASHRAE. 2013a. *ASHRAE Handbook—Fundamentals*. Chapter 1, Psychrometrics. Atlanta: ASHRAE.

ASHRAE. 2013b. *ASHRAE Handbook—Fundamentals*. Chapter 14, Climatic Design Information. Appended CD-ROM. Atlanta: ASHRAE.

ASHRAE. 2013c. *ASHRAE Handbook—Fundamentals*, SI Edition. Atlanta: ASHRAE.

AWEB. 2014. Sample HVAC Project 2: Water-to-Water. Baton Rouge, LA: AWEB Supply.

CRC. 1970. *Handbook of Tables for Applied Engineering Science*. R.E. Bolt and G.L. Tuve, eds. Cleveland, OH: Chemical Rubber Company.

CUFS. 2014. Cooling Home. Ithaca, NY: Cornell University Facility Services. http://energyandsustainability.fs.cornell.edu/util/cooling/default.cfm

Degelman, L.O. 1986. *Bin and degree hour weather data for simplified energy calculations,* ASHRAE RP-385. Atlanta: ASHRAE.

Dow. 1990. *Engineering and Operating Guide for Inhibited Propylene Glycol-based Heat Transfer Fluids*. Midland, MI: The Dow Chemical Company.

Duffie, J.A., and W.A. Beckman. 1980. *Solar Engineering of Thermal Processes.* New York: John Wiley.

EIS. 2014. Surface Water Temps. Ground-Source Heat Pump Design—Keep it Simple and Solid. Northport, AL: Energy Information Services. www.geokiss.com/surwatertemps.htm

Hansen, G.M. 2011. *Experimental testing and analysis of surface water heat exchangers*. Master's thesis, Oklahoma State University, Stillwater, OK.

Hattemer, B.G. 2005. Thermal performance and environmental impact of surface water heating and cooling systems. Master's thesis, University of Alabama, Tuscaloosa, AL.

Hattemer, B.G., and S.P. Kavanaugh. 2005. Design temperature data for surface water heating and cooling systems. *ASHRAE Transactions* 111(1).

Hattemer, B.G., S.P. Kavanaugh, and D. Williamson. 2006. Environmental impacts of surface water heat pump systems. *ASHRAE Transactions* 112(1).

Heffernan, V. 2001. Toronto cools off naturally—A deep lake water cooling system. *Canadian Consulting Engineer*, Dec 1.

Holman, J.P. 1986. *Heat Transfer*, 6th ed. New York: McGraw-Hill.

InterEnergy. 1999. *BinMaker™Plus: Weather Data for Engineering*. Chicago: InterEnergy Software.

Kavanaugh, S.P., and K. Rafferty. 1997. *Ground-Source Heat Pumps: Design of Geothermal Systems for Commercial and Institutional Buildings.* Atlanta: ASHRAE.

Dow. 2001. Propylene Glycol Material Safety Data Sheet, MSDS Number P6928. Midland, MI: The Dow Chemical Company.

Peirce, L.B. 1964. *Reservoir temperatures in North-Central Alabama*, Geological Survey of Alabama, Bulletin 82. Tuscaloosa, AL: Geological Survey of Alabama.

Pezent, M.C. 1989. Thermal performance of lakes when integrated with optimized heating and cooling systems. Unpublished master's thesis, University of Alabama, Tuscaloosa, AL.

Pezent, M.C., and S.P. Kavanaugh. 1990. Development and verification of a thermal model of lakes. *ASHRAE Transactions* 96(1).

PPI. 2014. *Handbook of Polyethylene Pipe*, 2d Ed. Dallas, TX: Plastic Pipe Institute. https://plasticpipe.org/publications/pe_handbook.html

Remund, C. 2009. *Ground Source Heat Pump Residential and Light Commercial Design and Installation Guide*. Stillwater, OK: International Ground Source Heat Pump Association.

Sieder, E.N., and G.E. Tate. 1936. Heat transfer and pressure drop of liquids in tubes. *Industrial and Engineering Chemistry* (28):1429–35.

Siegel, R., and J.R. Howell. 1981. *Thermal Radiation Heat Transfer,* 2nd ed. New York: McGraw-Hill.

SUNY. 2011. Assessing the feasibility of a central New York naturally chilled water project. Final Report, USEPA Award XA-97264106-01. Albany, NY: The Research Foundation, The State University of New York. http://en.wikipedia.org/wiki/Deep_water_source_cooling

TEMA. 1978. *Standards of the Tubular Exchanger Manufacturers Association*, 6th Ed. White Plains, NY: TEMA.

USGS. 1952. *Water Loss Investigations—Lake Hefner Studies*, U.S. Geological Survey Circular 229. Washington, DC: U.S. Geological Survey.

6 Piping and Pumps for Closed-Loop Ground-Source Heat Pumps

6.1 OVERVIEW OF GCHP AND SWHP PIPING SYSTEMS AND PUMPS

The system efficiency of ground-coupled heat pump (GCHP) and closed-loop surface-water heat pump (SWHP) systems can be exceptionally high in larger buildings if

1. high-efficiency, extended-range heat pumps are used,
2. the ground and surface-water heat exchangers are of sufficient depth and length and located in mediums so that liquid temperatures entering the heat pumps are much more moderate than the outdoor air temperature,
3. the air distribution system is designed and installed so that the required fan power is small (< 15% of total system power [heat pump + fan + pump power]), and
4. the water distribution system is designed and installed so that the required pump power is small (< 10% of total system power [heat pump + fan + pump power]).

Because item 1 and especially item 2 are typically challenging to many engineers new to GSHP design, items 3 and 4 can be overlooked and not given the necessary attention to detail. Excessive air and water distribution losses with oversized and/or poorly controlled fans and pumps can nullify the efficiency made possible by a well-designed and expensive ground or surface-water heat exchanger. This chapter focuses on the design of piping and pump selection to maintain efficiency without compromising performance and installation costs.

The design of water distribution systems presents engineers with the classic challenge of optimizing the trade-off between installation costs and operating costs. Larger-diameter pipes cost more to install but require smaller pumps, result in lower energy costs, and require less maintenance. Smaller-diameter pipes are less expensive to install but more expensive to operate. Piping made of common materials, such as steel, used inside the building in many cases is less expensive because they are commonly used and supplies are readily available, but they require continuous corrosion protection. Piping materials that are resistant to corrosion, such as fibre-core polypropylene and high-density polyethylene (HDPE), can be more expensive to install inside a building because they are relatively new to the market and require more pipe hangers and flame/smoke spread wrapping when routed through plenums. However, the savings in corrosion protection costs can be significant.

Closed-loop GSHP piping systems have special characteristics that can be advantageous, while other aspects can create additional challenges. Thermally fused HDPE is the material of choice for ground and surface-water loops, as shown in Figure 6.1. Stainless steel "lake plate" heat exchangers are also available. HDPE can also be used inside the building, but it has a high degree of linear expansion, which can create problems, especially in larger-diameter pipe. Thermally fused polypropylene pipe with an inner fiberglass core has a much lower coefficient of expansion and is now being used as an alternative to steel, copper, or HDPE inside the building, as shown in Figure 6.2. However, polypropylene and HDPE are not rated to meet a flame spread index (FSI) greater than 25 or the smoke developed index (SDI) of 50 required when located in plenums and must be wrapped with materials that meet this requirement (NFPA 2015).

An additional constraint for ground and surface-water loops is providing circuits that can be purged of debris and air. Loops for larger buildings often consist of multiple parallel circuits that contain 5 to 20 U-tubes or coils that are also piped in parallel to minimize required pump head. The diameters of the sections of each circuit must be large enough to limit head loss but not so large that debris and air cannot be removed with a purge pump. Figure 6.3 demonstrates one circuit with ten U-tubes piped in parallel. Note that header diameter is reduced after the first three U-tube take-offs, again after the next several, and then until the last U-tube. In this piping arrangement the flow through the last section of the header is 1/10 of the main header flow. If the header diameters for the later sections are not reduced, the purge pump size would have to be enormous to overcome the losses in the main headers while still providing adequate purge velocity in the last section.

The benefits of thermally fused HDPE and polypropylene pipe include the following:

- Durability during field installation
- Corrosion resistance so inhibitors (that may not be allowed for piping underground or in lakes) are unnecessary
- Reduced fouling of control sensors (especially differential pressure)
- Ability to maintain smooth pipe walls and low resistance to fluid flow for life of pipe

Figure 6.1 HDPE U-Tube Loop Field and Surface-Water Loop Installations

- Limited number of joints required for ground heat exchangers
- Ease of joint fabrication
- Modest training required for fabrication proficiency compared to metal piping
- Reduced or absence of need for interior pipe insulation to prevent condensation
- Low cost compared to metal piping

Limitations of thermally fused HDPE and polypropylene pipe include the following:

- Lower pressure rating, especially at higher temperatures for HDPE
- High coefficient of expansion, especially for HDPE
- Smoke and flame spread characteristics that limit routing through plenums
- Greater number of interior piping hangers required

Figure 6.2 Equipment-Room Polypropylene Piping

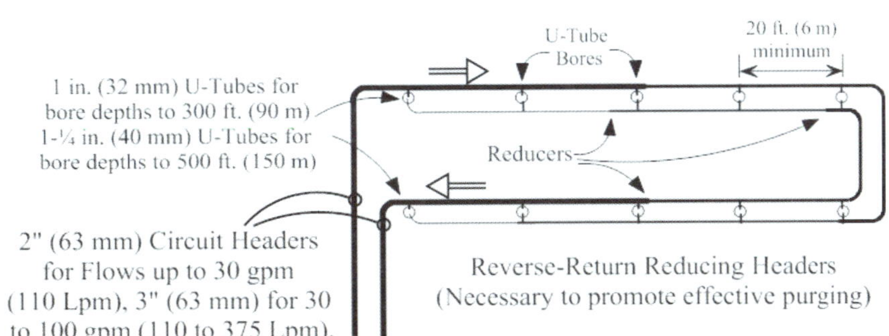

Figure 6.3 Reverse-Return Ground-Loop Circuit with Reduced Header Sections

6.2 IMPACT OF PUMP POWER

GSHPs can be very efficient systems when the power and energy of pumps and fans are optimized. The investment in efficiency of a well-designed and installed ground heat exchanger can be nullified by excessive piping losses and oversized pumps. Figure 6.4 provides four examples of systems that are otherwise properly installed but have pumps that limit the GSHP system's ability to attain full energy-saving potential.

Consider the system in Figure 6.4 with the two 385 W pumps serving a nominal 5 ton (≈60,000 Btu/h) (18 kW) heat pump. Figure 6.5 shows a screenshot of the spreadsheet *WAHPCorrector.xlsm* for a system that operates in a relatively cold climate where antifreeze solution is required. (*WAHPCorrector.xlsm* is available with this book at www.ashrae.org/GSHP.) The design entering liquid temperatures (ELTs) to the heat pumps are 80°F (27°C) in cooling and 43°F (6°C) in heating. When the 770 W for the two pumps is included, the system EER is 12.9 Btu/Wh (COP_c = 3.8) and the heating COP_h is 3.2. This represents a 16% decline for both the EER and COP when the pump power is included.

This efficiency can be substantially improved with quality design. It is possible to design a system that requires a single pump and possibly even a smaller single pump that results in much higher EER and COP.

Figure 6.4 Why Some GSHPs Use More Energy than Advertised

GLHP Rated Performance			Operating Conditions			Corrected Capacity, Power, EER and COP				
Model#	ECM-60		ELTClg	80.0	°F		Pump(s) not Included		Pump(s) Included	
TC-77°F	66.8	kBtu/h	ELTHtg	43.0	°F	TC	61.6	kBtu/h	61.6	kBtu/h
SC-77°F	0	kBtu/h	EATdbClg	75.0	°F	SC_Est	41.6	kBtu/h	41.6	kBtu/h
EER-77°F	19.5	Btu/W-h	EATwbClg	63.0	°F	SHR	0.68		0.68	
HC-32°F	43.3	kBtu/h	EATHtg	70.0	°F	EERnoCF	18.7	Btu/W-h		Btu/W-h
COP-32°F	3.9	W/W	Actgpm	15.0	gpm	kWc	4.02	kW	4.79	kW
WtrFlow	15.0	gpm	Actcfm	1800	cfm	EER	15.3	Btu/W-h	12.9	Btu/W-h
AirFlow	1800	cfm	Fan Power and Heat Correction***			HC	52.0	kBtu/h	52.0	kBtu/h
GpmPTonR	2.7		FanMotor	ECMwFCBlade		COPnoCf	4.39	W/W		W/W
CfmPTonR	323		ESP	0.5	in. water	kWh	3.95	kW	4.72	kW
GpmPTonA	2.7		FilterLoss	0.3	in. water	COP	3.85	W/W	3.23	W/W
			WAEff	30%						
			kWFan	0.51	kW		Optional Pump Power			
			FanHeat	1.7	kBtu/h		kWpump	0.77	kW	

Figure 6.5 System EER and COP Results for 5 Ton (18 kW) Heat Pump with Two Pumps

EXAMPLE 6.1— UNITARY LOOP SYSTEM DESIGN

Redesign the water distribution system that required two 385 W pumps on the 5 ton (18 kW) heat pump. The system consists of five 250 ft (76 m) nominal 3/4 in. (25 mm) U-tubes in parallel, 1 1/4 in. (40 mm) supply and return headers 75 ft (23 m) each in length, hose kits, a heat pump with a rated 10.5 ft of water (31 kPa) coil loss, and assorted fittings. The calculation is conducted in the critical mode with 20% propylene glycol-80% water fluid at 40°F (4.4°C).

Solution

Figure 6.5 shows a screenshot of the head loss calculation tool *E-PipeAlator14.xls*m (discussed later in this chapter and available with this book at www.ashrae.org/GSHP) for the original design. The pump manufacturer provides a nominal 1/6 hp (125 W) pump with a 385 W input that will deliver 26 ft of head (78 kPa) at the required 15 gpm (57 L/min) and a nominal 1/6 hp (125 W) pump with a 245 W input that will deliver 22 ft of head (66 kPa) at 15 gpm (57 L/min). The system head loss is 45.2 ft (138 kPa), which necessitates two 385 W pumps in series.

Examination of the head loss components shown in Table 6.1 indicates the primary losses are in the heat pump, the header, and the U-tubes. The two hose kits represent the fourth highest loss. Also note that the loss in the header is 4.65 ft of water per 100 ft (460 Pa/m), which is above the recommended value (see Recommendation 2 in Section 6.10). The Reynolds number in the U-tube is 3290, which indicates a transition flow regime.

A revised design with a much lower head loss that requires only a single 385 W pump can be delivered with the following adjustments:

- The header pipe diameter was increased from a nominal 1 1/4 in. (40 mm) to 1 1/2 in. (50 mm) HDPE tube. Header head loss is reduced from 12.2 to 6.7 ft of water (36 to 20 kPa).
- The U-tube diameter was increased to nominal 1 in. (32 mm), which also resulted in a 7 ft (2 m) reduction in length for each bore. The U-tube head loss is reduced from 13.5 to 3.8 ft (39 to 11 kPa). The Reynolds number indicates the flow is nonlaminar.
- The hose kits and fittings on the heat pump connections were increased to 1 1/4 in. (40 mm). The hose connection head loss is reduced from 4.3 to 1.1 ft (13 to 3 kPa).

Table 6.1 Head Loss Calculation for Original Design: Two 385 W Pumps Required

Liquid: 20% Propylene Glycol			Temperature: 40°F			Density: 64 lb/ft^3			Viscosity: 3.44 c.poise		
Heat Pump	**Flow, gpm**	**Rated Flow**		**Δh @ 60°F**							**Δh, ft**
	15	15		11.5							12.6
Coils, Valves, Other	**Flow, gpm**	**Flow Coefficient (C_v) @ 60°F**		**Quantity**							**Δh, ft**
1 in. Ball Valve	15	35		4							1.9
1 in. × 3 ft Hose Kit	15	16.4		2							4.3
Y-Strainer	15	28		1							0.7
HDPE Pipe and Fittings	**Flow, gpm**	**Nominal Diameter**	**Actual Diameter**	**Velocity, fps**	**Δh/100 ft**	**Re**	**L, ft**	**L_{eqv}**	**L_{eqv}**	**L_{eqv}**	**Δh, ft**
Main Header	15	1.25	1.36	1.3	4.65	10,404	150	4 @ 10 ft	2 @ 30 ft	2 @ 5 ft	12.2
Fitting Type								Elbow	Cls-Hdr	Red	
Vertical U-tube	3	0.75	0.86	1.7	2.57	3,290	500	1 @ 8 ft			13.5
Fitting Type								U-bend			
								Total Head Loss, ft of liquid			45.2

The total system head loss was reduced to 25.8 ft of water (77 kPa), which can be delivered by a single 385 W pump. The one-pump system EER is 14.0 Btu/Wh ($COP_c = 4.1$) and COP_h is 3.5. In both cases the improvement is 9% compared to the system with two pumps. It is likely that the savings in pump and drilling costs will be greater than the added cost of the upsized header pipe, U-tubes, antifreeze solution, hose and fittings.

One additional step involving the optimization of water flow and heat pump performance can further improve system efficiency. Liquid flow for the heat pump can be reduced from 15 to 14 gpm (57 to 53 L/min) and the total head loss becomes 22 ft of liquid (66 kPa). The two 385 W pumps are replaced with a single 245 W pump. At 14 gpm (53 L/min), the 245 W pump can deliver 23 ft of water (69 kPa). As shown in Figure 6.6, the EER is 14.4 Btu/Wh ($COP_c = 4.2$) and $COP_h = 3.85$. The cooling and heating capacities are reduced by less than 1% with the lower flow rate, and the U-tube flow remains nonlaminar.

GLHP Rated Performance			Operating Conditions			Corrected Capacity, Power, EER and COP				
Model#	ECM-60		ELTClg	80.0	°F		Pump(s) not Included		Pump(s) Included	
TC-77°F	66.8	kBtu/h	ELTHtg	43.0	°F	TC	61.6	kBtu/h	61.6	kBtu/h
SC-77°F	0	kBtu/h	EATdbClg	75.0	°F	SC_Est	41.6	kBtu/h	41.6	kBtu/h
EER-77°F	19.5	Btu/W-h	EATwbClg	63.0	°F	SHR	0.68		0.68	
HC-32°F	43.3	kBtu/h	EATHtg	70.0	°F	EERnoCF	18.7	Btu/W-h		Btu/W-h
COP-32°F	3.9	W/W	Actgpm	15.0	gpm	kWc	4.02	kW	4.26	kW
WtrFlow	15.0	gpm	Actcfm	1800	cfm	EER	15.3	Btu/W-h	14.4	Btu/W-h
AirFlow	1800	cfm	Fan Power and Heat Correction***			HC	52.0	kBtu/h	52.0	kBtu/h
GpmPTonR	2.7		FanMotor	ECMwFCBlade		COPnoCf	4.39	W/W		W/W
CfmPTonR	323		ESP	0.5	in. water	kWh	3.95	kW	4.20	kW
GpmPTonA	2.7		FilterLoss	0.3	in. water	COP	3.85	W/W	3.63	W/W
			WAEff	30%						
			kWFan	0.51	kW		Optional Pump Power			
			FanHeat	1.7	kBtu/h		kWpump	0.245	kW	

Figure 6.6 System EER and COP Results for 5 Ton (18 kW) Heat Pump with One Smaller Pump

Table 6.2 GSHP System Pump Power Benchmarks

Installed Pump Power	Power into Pump Motor	Grade	Available Head with 70% Efficient Pump at 3 gpm/ton
< 5 hp/100 tons	< 45 W/ton	A	< 46 ft of water
5 < hp/100 tons ≤ 7.5	45 < W/ton ≤ 65	B	46 to 69 ft of water
7.5 < hp/100 tons ≤ 10	65 < W/ton ≤ 85	C	69 to 92 ft of water
10 < hp/100 tons ≤ 15	85 < W/ton ≤ 125	D	92 to 138 ft of water
> 15 hp/100 tons	> 125 W/ton	F	> 138 ft of water
Installed Pump Power	**Power into Pump Motor**	**Grade**	**Available Pressure with 70% Efficient Pump at 3 L/m·kW**
$< 10.5\ W_m/kW_t$	$< 13\ W_e/kW_t$	A	< 140 kPa
$10.5 < W_m/kW_t \le 16$	$13 < W_e/kW_t \le 19$	B	140 to 210 kPa
$16 < W_m/kW_t \le 21$	$19 < W_e/kW_t \le 25$	C	210 to 280 kPa
$21 < W_m/kW_t \le 32$	$25 < W_e/kW_t \le 36$	D	280 to 420 kPa
$> 32\ W_m/kW_t$	$> 36\ W_e/kW_t$	F	> 420 kPa

W_m ≡ watts mechanical, W_e ≡ watts electrical, kW_t ≡ kilowatts thermal

Ground-Source Heat Pumps: Design of Geothermal Systems for Commercial and Institutional Buildings (Kavanaugh and Rafferty 1997) included a table that provided a benchmark grade (A, B, C, D, or F) on the rated power of the system pumps relative to the cooling or heating requirement. Table 6.2 is a reproduction of that table with both I-P and SI units. The common metric is the nominal input power to the pump (W_p) in horsepower relative the capacity in 100 tons (350 kW). This metric is easily available since pump motors have the output rating displayed on nameplates.

The use of nameplate motor power is somewhat inaccurate because motors are available in fixed increments and are almost always somewhat larger than the pump requirement of nameplate motor power. It should be recognized that the more meaningful metric is the input power to the pump motor. This value is directly related to energy use, demand, and operating cost. It is also suggested that the calculated building load in tons (or kW) be used to compute the benchmark instead of the installed equipment capacity. Details of pump and motor fundamentals are discussed in Section 6.6.

6.3 IMPACT OF PUMP ENERGY

Pump energy consumption and costs can be significant when safety factors are liberally applied or when controls are not well designed or properly functioning. This section discusses the impact of pump energy compared to heat pump consumption and provides a description of available tools to help determine when design improvements are warranted. Table 6.3 is an energy and cost calculation from *HP&PumpEnergyCalc.xls* (a spreadsheet tool available with this book at www.ashrae.org/GSHP) for the example office described in Chapter 4. The load profile shown in Figure 6.7 has been generated for the 8760 annual hours and results in the equivalent full-load hours (EFLH) for cooling (890) and heating (760) used to design the ground heat exchanger for the office building. Values calculated for the design cooling load, heating load, EER, and COP (without the pump power) are input into the spreadsheet. The corrected values for the EER and COP at full load are input along with the values at near-zero load to account for improved efficiency, as ground-loop temperature moderates at low loads.

Table 6.3 Energy Consumption and Cost for Example St. Louis Office*

Design Cooling Load	228 kBtu/h	Design Heating Load	104 kBtu/h
EER at Design Load	15.2 Btu/Wh	COP at Design Load	4.4
EER at Minimum Load	17 Btu/Wh	COP at Minimum Load	4.7
Electric Energy Cost	12.0 ¢/kWh		

%Full Load	Operating Hours		EFLH		Cooling			Heating		
	Cooling	Heating	Cooling	Heating	EER	kW	kWh	COP	kW	kWh
0%	1600	1600	0	0	17.0	13.4	0	4.7	6.5	0
10%	900	800	90	80	16.8	13.6	1220	4.7	6.5	522
20%	670	580	134	116	16.6	13.7	1836	4.6	6.6	762
30%	470	400	141	120	16.5	13.9	1953	4.6	6.6	793
40%	320	280	128	112	16.3	14.0	1793	4.6	6.7	745
50%	230	190	115	95	16.1	14.2	1629	4.6	6.7	636
60%	150	130	90	78	15.9	14.3	1289	4.5	6.7	526
70%	90	80	63	56	15.7	14.5	913	4.5	6.8	380
80%	70	60	56	48	15.6	14.7	821	4.5	6.8	328
90%	50	40	45	36	15.4	14.8	667	4.4	6.9	248
100%	30	20	30	20	15.2	15.0	450	4.4	6.9	139
Totals	**4580**	**4180**	**892**	**761**		**$1,508**	**12570**		**$610**	**5080**
Cooling and Heating Total	**8760 h**		**1653 EFLH (Cooling and Heating)**			**$2,118**	**17,649 kWh**			

**Hours of operation generated from Table 4.5 of ASHRAE RP-1120 (Carlson 2001).*

For an electric energy cost of $0.12/kWh, the annual operating energy costs for the heat pumps (not including the pumps) is $2118. Cooling-mode cost is $1508, and heating cost is $610. Note that although the EFLH are nearly the same, the cooling cost is much greater because the peak cooling load is almost twice the heating peak load.

This section analyzes three pump and pipe circuiting options to demonstrate the costs of pumping alternatives relative to heat pump operating costs. The analysis is conducted for each of the three options with an optimized pump size and then repeated for a pump that is 50% larger. Schematics of the three options are shown in Figure 6.8. The operating hours for the heat pumps were generated using Table 4.5 of ASHRAE RP-1120 (Carlson 2001) and assume no particular occupancy schedule. The specifications for the optimized pumps are as follows:

- *On-Off Pumps:* Eight 200 W pumps, 9 gpm (34 L/min·kW) each, 20% wire-to-water efficiency (these values are not good but are representative of wet-rotor pumps)
- *Constant-Speed Central Pump:* 50 ft head (150 kPa), 60 gpm (227 L/min), 51% wire-to-water efficiency
- *Variable-Speed Central Pump:* 50 ft head (150 kPa), 60 gpm (227 L/min), 51% wire-to-water efficiency, 97% variable-speed drive (VSD) efficiency, 30% minimum flow

The optimization for the central pumps is achieved by limiting head to 50 ft of water (150 kPa) and a flow rate of 3.0 gpm/ton (3.2 L/min·kW) of maximum load rather than the common practice of 3.0 gpm/ton (3.2 L/min·kW) of installed capacity.

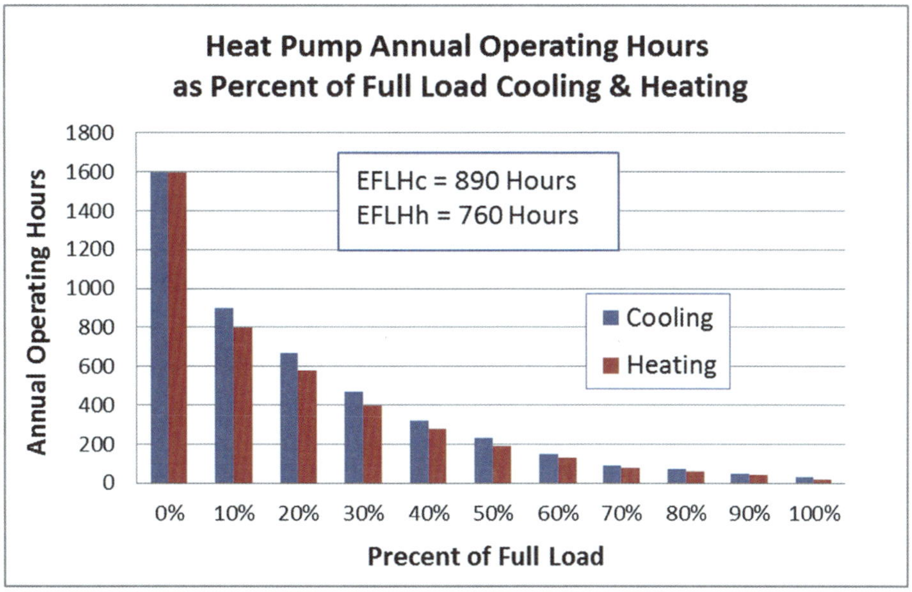

Figure 6.7 Load Profiles for St. Louis Office Building

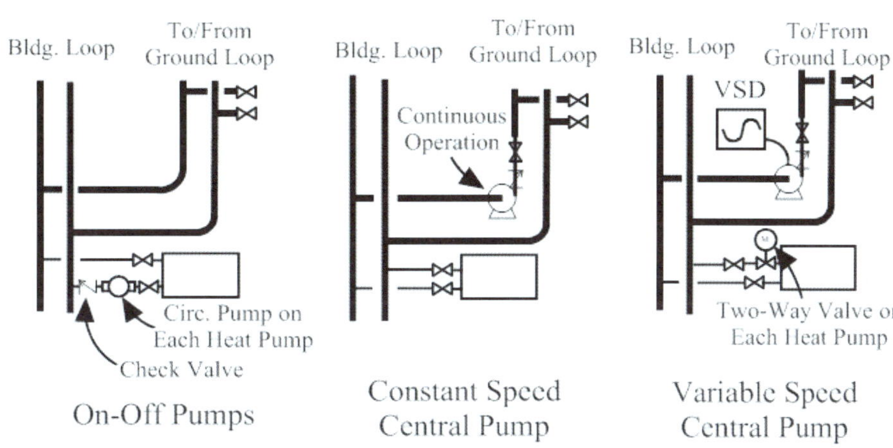

Figure 6.8 Three Pump and Piping Options for Cost Comparison

Table 6.4 indicates the optimized variable-speed pump provides the lowest cost at $251 per year, which is 12% of the heat pump cost. The on-off circulator pumps cost $310 per year, or 15% of the heat pump energy cost. Even with an optimized pump, the continuously operating pump required more than half of the entire heat pump operating cost at $1165 per year.

There is room for improvement with the best two options. Note that the variable-speed pump continues to operate when there is no load, and over half of the consumption occurs during the many hours when the pump is operating at minimum speed. VSD systems that could operate below 30% and be cycled off when no heat pumps are operating would reduce pump cost to less than $120 per year.

The wire-to-water efficiency of current on-off wet-rotor pumps is very low at 20% (ASHRAE 2003). Variable-speed wet-rotor pumps are available with much greater efficiency but currently are not economically justifiable, because the optimized system only

Table 6.4 On-Off, Constant-Speed, and Variable-Speed Pump Energy/Cost—Optimized Pump Size

	On-Off Pump kWh		Constant-Speed Pump kWh		Variable-Speed Pump kWh	
% Full Load	Cooling	Heating	Cooling	Heating	Cooling	Heating
0%	0	0	1773	1773	299	299
10%	141	125	997	887	168	149
20%	209	181	742	643	125	108
30%	220	188	521	443	88	75
40%	200	175	355	310	85	74
50%	180	148	255	211	82	68
60%	141	122	166	144	70	61
70%	98	88	100	89	54	48
80%	88	75	78	66	52	45
90%	70	56	55	44	47	37
100%	47	31	33	22	34	23
	1394	1189	5075	4632	1104	987
kWh Total	**2583**		**9707**		**2090**	
Cost Total	**$310**		**$1165**		**$251**	

Table 6.5 On-Off, Constant-Speed, and Variable-Speed Pump Energy/Cost—50% Larger Pump

	On-Off Pump kWh		Constant-Speed Pump kWh		Variable-Speed Pump kWh	
% Full Load	Cooling	Heating	Cooling	Heating	Cooling	Heating
0%	0	0	2660	2660	742	742
10%	211	188	1496	1330	417	371
20%	314	272	1114	964	311	269
30%	331	281	781	665	218	185
40%	300	263	532	465	148	130
50%	270	223	382	316	123	102
60%	211	183	249	216	105	91
70%	148	131	150	133	80	71
80%	131	113	116	100	79	67
90%	105	84	83	66	70	56
100%	70	47	50	33	51	34
	2091	1784	7613	6948	2344	2119
kWh Total	**3875**		**14561**		**4463**	
Cost Total	**$465**		**$1747**		**$536**	

costs $310 per year with the low-efficiency pumps. In this application the variable-speed pump is unnecessary, but a constant-speed pump with a higher wire-to-water efficiency with a modest cost premium would enhance economics.

The operating cost of continuously operating pumps defeats a primary benefit of GSHPs to reduce energy costs.

Table 6.5 provides the results when the analysis is repeated using pumps that are 50% larger. The cost of the on-off pump increases proportionally to $465 per year, or 22% of the heat pump operating cost. Note that the variable-speed pump is no longer the lowest-cost option. Increasing the size 50% also raises the minimum flow capacity 50%, so an increased proportion of the VSD pump operating cost is when there is no load or when the pump is operating at minimum speed. **The conventional wisdom of oversizing variable-speed pumps because they ramp down to meet the load robs the benefit of saving energy because minimum speed is almost always too high except near peak load.**

6.4 PIPING FUNDAMENTALS

The pipe pressure drop or head loss (Δp) of typical (Newtonian) fluids is determined by the Darcy-Weisbach equation (ASHRAE 2013):

$$\Delta p = f\frac{L}{D}\frac{\rho}{g_c}\left(\frac{V^2}{2}\right) \tag{6.1}$$

where

Δp = pressure loss, lb_f/ft^2 (Pa)
f = friction factor determined from Moody chart or equations
L = length of pipe, ft (m)
D = inside pipe diameter, ft (m)
V = fluid velocity, ft/s (m/s)
ρ = fluid density, lb_m/ft^3 (kg/m^3)
g_c = conversion factor for I-P, 32.2 $ft{\cdot}lb_m/lb_f{\cdot}s^2$ ($1\ kg{\cdot}m/N{\cdot}s^2$)

Equation 6.1 is modified to provide the loss in terms of fluid head as shown in Equation 6.2, which is typical practice when working with I-P units and common fluids such as water; SI practice is to use pressure loss or drop.

$$\Delta h = \frac{\Delta p}{\rho}\frac{g_c}{g} = f\frac{L}{D}\left(\frac{V^2}{2g}\right) \tag{6.2}$$

where

Δh = head loss, ft (m)
g = acceleration of gravity (32.2 ft/s^2 [9.81 m/s^2] on the surface of the earth)

While Equations 6.1 and 6.2 are relatively simple, the computation of the friction factor is more complex. The Reynolds number based on inside pipe diameter ($Re_D = \rho DV/\mu$) must be calculated using the fluid density and dynamic viscosity (μ), which varies with temperature for pure substances and with concentration for antifreeze mixtures. The calculation is further complicated by the need to have a multitude of empirically derived equations for various flow regimes (laminar, transition, and turbulent). Once Re is calculated, the relative roughness (e/D) of the inner tube wall must be determined before the friction factor can be found. This process is further complicated since the roughness (e) of pipe that has been in service for several years may be much greater than that of new pipe for which roughness data is available. Once Re_D and the relative roughness have been determined, the friction factor is found using charts such as the Moody diagram (Moody 1944) or a variety of complicated equations that typically apply to either laminar flow (Re_D < 2000 to 2300) or turbulent flow (Re_D > 4000 for rough pipes, Re_D up to 10,000 for smooth tubing). Few equations exist for transition flow (2000 to 2300 < Re_D < 4000 to 10,000), so estimates are sometimes made via interpolation between values generated using laminar-flow and turbulent-flow equations. Note that the Re_D value for the upper limit of transition flow varies significantly with pipe roughness, which creates a high degree of uncertainty. However, prudent engineering practice is to assume the conservative approach that fully turbulent flow occurs for Re ≥ 4000.

A design tool has been developed specifically for GSHP system piping design that can also be used with conventional piping such as steel, polyvinyl chloride (PVC), copper, or cross-linked polyethylene (PEX). The tool, *E-PipeAlator14.xlsm*, is available with this book at www.ashrae.org/GSHP. VisualBasic© macros have been developed for temperature-dependent fluid properties (ρ, μ) of water and common concentration mixtures of glycols and alcohols. This reduces the effort required to calculate the Reynolds number. Churchill (1977) developed a single equation for friction factor in all flow regimes that provides acceptable accuracy given the many other uncertainties in piping systems (pipe wall deterioration, fitting losses, etc.):

$$f = 8\left[\left(\frac{8}{\mathrm{Re}_D}\right)^{12} + \frac{1}{(A + B^{1.5})}\right]^{1/12} \tag{6.3}$$

where

$$A = \left\{2.457 \times \ln\left[\frac{1}{(7/\mathrm{Re}_D)^{0.9} + 0.27(e/D)}\right]\right\}$$

$$B = \left(\frac{37{,}530}{\mathrm{Re}_D}\right)^{16}$$

6.5 PIPE MATERIALS, DIMENSIONS, AND LOSS CHARACTERISTICS

An additional challenge in calculating piping loss is the variety of dimension designations. Tables 6.6 and 6.7 provide a listing of outside and inside diameters of common designations in I-P and SI units, respectively. Two traditional designations are iron pipe size (IPS) and copper tube size (CTS) in nominal inches (see Table 6.6). The term *nominal* is used since neither the outside diameter (OD) nor the inside diameter (ID) is equal to an even value. For both designations, the actual ODs are larger than the nominal values. However, the OD is equal to the nominal value for IPS pipes over 12 in. (i.e., the OD for 14 in. iron pipe is actually 14.0 in.).

Table 6.7 shows that SI pipe likewise has schedule dimension, with the nominal diameters being smaller than the actual ODs and with the IDs for Schedule 40 pipe diameters being close but not equal to the nominal diameters. However, the nominal diameter for SI dimension ratio (DR) is equal to the actual OD. While this does create some consistency, it can also cause some confusion when expressing equivalency to non-SI iron pipe sizes. For example, the equivalent DR SI pipe size to 1 in. IPS is 32 mm rather than 25 mm, because the actual OD of 1 in. IPS pipe is 1.315 in., which is near 32 mm.

The pipe wall thickness varies according to the required pressure rating of the pipe; thus, for a given pipe size the ID varies while the OD remains constant. For IPS, the designation for different thicknesses is *Schedule*, with higher numbers meaning thicker pipe walls and smaller IDs. Schedule 40 is common, with higher-pressure-rated pipe having larger numbers, such as Schedule 80, and lower-pressure-rated pipe having smaller numbers, such as Schedule 10. CTS dimensions for water service follow letter designations of K, L, and M, with K having the thickest pipe walls and smallest IDs and M having the thinnest walls and largest IDs. (Note: The nominal diameter of copper tubing for refrigeration applications is equal to the actual OD, so IDs for refrigeration tubing are less than the IDs of types K, L, and M for the same nominal diameter.)

Thermally fused HDPE pipe that is used in GSHP systems follows the IPS dimensions for OD (see Table 6.6). Like PVC pipe, the HDPE joining process requires the OD to be consistent and the ID to be varied to meet required pipe wall thickness for various pressure ratings. The ID is determined using a standard dimension ratio (SDR, or simply DR) value, which is the outside diameter divided by the pipe wall thickness (DR = OD ÷ $thkns_{wall}$). Thus, the lower the DR value, the thicker the pipe wall and the higher the pressure rating. The inside diameter is determined using

$$ID = OD \times (1 - 2/DR) \tag{6.4}$$

(Note: Thermally fused pipe dimensions are different than those of HDPE pipe joined with barbed fittings and pipe clamps. Pipe used with barbed fittings is typically consistent with Schedule 40 IPS ID to provide standard fitting sizes for this type of connection. standard inside dimension ratio [SIDR] in some cases is used to distinguish it from SDR or standard outside dimension ratio [SODR] for thermally fused pipe).

DR 11 HDPE pipe is specified for below-grade applications for pipe that has a nominal diameter of 2 in. (63 mm) and smaller (IGSPHA 2009). Because of its higher pressure rating, DR 9 is sometimes used for deep vertical bores or bores that are connected to interior piping of high-rise buildings. Because operating pressures are lower in horizontal piping, DR 13.5 or 15.5 are used for below-grade and interior header piping that is 3 in. nominal diameter (90 mm) and larger. Higher DR pipe is less expensive and has a lower pressure drop, but for the larger diameters the walls are thick enough to withstand ordinary damage during installation. Standards are available from the International Ground Source Heat Pump Association (IGSHPA) that provide additional specifications for acceptable HDPE products and installation methods. *(Note that 2 1/2 in. HDPE is not available, and 5 in. HDPE piping availability may be limited.)*

One advantage of using HDPE pipe with the DR designation is a consistent pressure rating for all pipe diameters for a particular grade of polyethylene. The only recommended method for joining this pipe is thermal fusion, which can be made with butt fusion, socket fusion (which is more common in 3/4 and 1 in. [25 and 32 mm] nominal diameter piping), or electrofusion joints. Designers should also be aware of the significant cost increase in installation equipment for tools that can fuse pipe larger than 6 in. (150 mm). Table 9.14 indicates the cost increase from $805 for a tool that can handle up to 4 in. (100 mm) pipe to $27,900 for a tool that can fuse 6 in. (150 mm) and larger pipe (RSMeans 2014). Appendix H contains recommended methods and details for these processes.

Cross-linked polyethylene (PEX) pipe is widely used in plumbing applications and in some GSHP connections. A DR designation is used, but the OD dimensions are based on copper tubing size. DR 9 is the standard for small-diameter PEX, and the thicker pipe wall combined with the smaller ODs for CTS results in IDs being significantly less than DR or Schedule pipe of the same nominal diameter. The improved flexibility (compared to HDPE) and mechanical connection method of PEX tubing typically reduce the level of effort required in making connections between the interior piping and the heat pumps.

There are a variety of approaches to determine head loss (or pressure drop) of liquid flowing through pipe based on Equations 6.1 and 6.2 and variations of Equation 6.3. These must be linked to an efficient design procedure to optimize the trade-off between using small-diameter, lower-cost pipe (that results in higher operating costs) with higher-first-cost, large-diameter pipe (that results in lower operating costs).

Table 6.6 Dimensions for Iron, HDPE, Copper, and PEX Pipe and Tubing—I-P

Pipe Diameter, in.	IPS OD, in.	ID for IPS Designated Pipe, in.					CTS OD, in.	ID for CTS Tubing, in.		
		Sch 40	Sch 80	DR 11	DR 13.5	DR 15.5		Type K	Type L	PEX DR9
3/4	1.05	0.824	0.742	0.86	NR	NR	0.875	0.745	0.785	0.68
1	1.315	1.049	1.0957	1.08	NR	NR	1.125	0.995	1.025	0.88
1 1/4	1.66	1.38	1.278	1.36	NR	NR	1.375	1.245	1.265	1.07
1 1/2	1.90	1.61	1.50	1.55	NR	NR	1.625	1.481	1.505	1.26
2	2.375	2.067	1.939	1.94	NR	NR	2.125	1.959	1.985	1.65
2 1/5	2.875	2.469	2.323	NA	NA	NA	2.625	2.435	2.465	1.89
3	3.50	3.068	2.90	2.86	2.98	3.05	3.125	2.907	2.945	
4	4.50	4.026	3.826	3.68	3.83	3.92	4.125	3.857	3.905	
5	5.563	5.047	4.813	4.55-LA	4.74-LA	8.85-LA	5.125	4.805	4.875	
6	6.625	6.065	5.761	5.42	5.64	5.77	6.125	5.741	5.845	
8	8.625	7.98	7.625	7.06	7.35	7.51	8.125	7.583	7.725	
10	10.75	10.02	9.562	8.80	9.16	9.36	10.125	9.449	9.625	
12	12.75	11.94	11.374	10.43	10.86	11.10	12.125	11.315	11.565	

NR = Not recommended for GSHPs, NA = Not available, LA = Limited availability

Table 6.7 Dimensions for Schedule and Standard Dimension Ratio Pipe—SI

Nominal Diameter, mm	Actual OD, mm	Schedule Pipe ID, mm			Actual OD, mm	DR pipe ID, mm			
		Sch 10	Sch 40	Sch 80		DR 9	DR 11	DR 13.5	DR 15.5
20	26.67	22.5	20.9	18.8	20	15.6	16.4	17.0	17.4
25	33.4	27.9	26.6	24.3	25	19.4	20.5	21.3	21.8
32	42.16	36.6	35.0	32.5	32	24.9	26.2	27.3	27.9
40	48.26	42.7	40.9	38.1	40	31.1	32.7	34.1	34.8
50	60.33	54.8	52.5	49.3	50	38.9	40.9	42.6	43.5
65	73.02	66.9	62.7	59.0	63	49.0	51.5	53.7	54.9
80	88.90	82.8	77.9	73.7	75	58.3	61.4	63.9	65.3
100	114.30	108.2	102.3	97.2	90	70.0	73.6	76.7	78.4
125	141.3	135.2	128.2	122.3	110	85.6	90.0	93.7	95.8
150	168.27	162.2	154.0	146.3	125	97.2	102.3	106.5	108.9
200	219.08	211.6	202.7	193.7	160	124.4	130.9	136.3	139.4
250	273.05	264.7	254.5	242.9	200	155.6	163.6	170.4	174.2
300	323.85	314.7	303.2	289.0	250	194.4	204.5	213.0	217.7

A traditional method of computing head loss or pressure loss is to use tables of head loss per 100 linear feet (or pressure loss per metre). The losses are found by multiplying the length of pipe by the values from Tables 6.8 or 6.9 as shown in Equations 6.5a and 6.5b. The losses through pipe fittings are found by consulting tables for equivalent lengths (L_{eqv}) of common fittings. While this method is less accurate than using K factors ($\Delta h = KV^2/2$), neither of the methods provides a high degree of accuracy given the variation and uncertainty of K factors (ASHRAE 2013).

$$\Delta h = \Delta h/100 \text{ ft} \times (L_{straight} + \Sigma L_{eqv}) \quad \text{(I-P)} \quad (6.5a)$$

$$\Delta p = \Delta p/\text{m} \times (L_{straight} + \Sigma L_{eqv}) \quad \text{(SI)} \quad (6.5b)$$

A limitation of this approach is that tables must be developed for the wide variety of pipe dimensions and water-antifreeze solutions for several different operating temperatures. Further expanding the possibilities is the fact that commonly used iron pipe wall roughness degrades, and losses increase with pipe age, especially if water treatment programs are neglected. The recommended HDPE pipe and the newly developed polypropylene pipe minimize this source of uncertainty.

Tables 6.8 and 6.9 demonstrate head and pressure loss tables for DR pipe with water at moderate temperatures. The spreadsheets used to generates these tables (*HeadLossTableIP.xlsm* and *HeadLossTableSI.xlsm*, available with this book at www.ashrae.org/GSHP) can be used to develop tables for other pipe dimensions, antifreeze solutions, operating temperatures, and pipe wall roughness.

Table 6.10 is a supplement to the head and pressure loss tables that provides a recommended maximum flow rate that results in a head loss of 3 feet of water per 100 linear feet of pipe (pressure loss ≈ 30 kPa/100 m). This assists the designer in selecting the initial flow rate through each piping section when the flow rate is known. Table values are for water and assume the system is in the cooling mode since the operating temperature is 86°F (30°C). Correction factors are provided for two common antifreeze solution fluids operating in the heating mode at 40°F (4°C).

Table 6.11 provides equivalent lengths for HDPE fittings, and Table 6.12 lists values for steel and copper fittings.

Head losses through many components such as heat pumps and water coils are given for one or more flow rates, usually at standard rating points. If the loss at some nonrated flow is desired, the following can be used:

$$\Delta h_2 = \Delta h_1 \times (Q_2/Q_1)^2 \quad (6.6)$$

Many valve manufacturers provide a flow coefficient ($C_v \equiv$ gpm) as an indicator of head loss as shown in Table 6.13. The coefficient is normally defined as the flow rate in gpm that will induce a pressure drop (Δp) of 1.0 psi. To find Δp or head loss at other flow rates, use the following:

$$\Delta p \text{ (psi)} = 1 \text{ psi} \times \left(\frac{Q \text{ (gpm)}}{C_v}\right)^2 \text{ and } \Delta h \text{ (ft of water)} = 2.31 \times \left(\frac{Q \text{ (gpm)}}{C_v}\right)^2 \quad (6.7)$$

Table 6.8 DR 11 HDPE Head Loss—Feet of Water/100 Linear Feet at 60°F*—I-P

Flow Rate, gpm	Nominal Diameter, in. 0.75	1	1.25	1.5	2	3
Inside Diameter, in.	0.86	1.08	1.36	1.55	1.94	2.86
1	0.29					
2	0.97	0.33				
3	1.98	0.67				
4	3.28	1.11	0.37			
5	4.89	1.65	0.54	0.28		
6	6.78	2.28	0.74	0.39		
8	11.42	3.81	1.23	0.64		
10	17.17	5.69	1.84	0.96	0.33	
12		7.93	2.54	1.32	0.45	
15		11.92	3.81	1.97	0.67	
20		20.27	6.43	3.32	1.12	0.17
25			9.68	4.98	1.68	0.26
30			13.55	6.96	2.33	0.36
35			18.04	9.25	3.09	0.47
40				11.84	3.94	0.60
50				17.94	5.95	0.89

Flow Rate, gpm	Nominal Diameter, in. 2	3	4	6	8
Inside Diameter, in.	1.94	2.86	3.68	5.42	7.06
60	8.33	1.24	0.37		
70	11.10	1.65	0.48		
80	14.24	2.10	0.62	0.10	
90	17.76	2.61	0.76	0.12	
100		3.17	0.93	0.14	
120		4.44	1.29	0.20	
140		5.91	1.72	0.26	
160		7.59	2.19	0.33	
180		9.46	2.73	0.41	0.11
200		11.54	3.32	0.50	0.14
250		17.58	5.03	0.75	0.21
300			7.08	1.05	0.29
350			9.46	1.39	0.38
400			12.18	1.79	0.49
450			15.23	2.22	0.61
500			18.61	2.71	0.74

Flow Rate, gpm	Nominal Diameter, in. 6	8	10	12
Inside Diameter, in.	5.42	7.06	8.80	10.43
600	3.81	1.04	0.35	0.15
700	5.09	1.38	0.47	0.20
800	6.56	1.77	0.60	0.26
900	8.20	2.21	0.74	0.32
1000	10.01	2.69	0.91	0.39
1200	14.18	3.79	1.27	0.55
1400	19.05	5.07	1.70	0.73
1600		6.54	2.18	0.94
1800		8.18	2.73	1.17
2000		10.01	3.33	1.43
2200		12.01	3.99	1.71
2400		14.19	4.70	2.01
2600			5.48	2.34
2800			6.31	2.69
3000			7.20	3.07
3500				4.11

**Tables for other pipe dimensions, fluids, and temperatures can be made with HeadLossTableIP.xlsm.*
***Head loss in tight coils (lake coils, slinky coils, etc.) is typically 3% to 4% greater than in straight pipe.*

Table 6.9 DR 11 HDPE Pressure Loss—kPa/100 Linear Metres at 20°C*—SI

Flow Rate, L/s	Outside Diameter, mm 25	32	40	50	63	75
Inside Diameter, mm	20.5	26.2	32.7	40.9	51.5	61.4
0.08	6.2	1.9				
0.17	21	6.3	2.2			
0.25	42	12.8	4.4			
0.33	71	21	7.3	2.5		
0.42	107	32	10.8	3.7		
0.50	149	44	15.0	5.1		
0.67		75	25	8.5	2.8	
0.83		113	38	12.7	4.2	
1.00		158	52	17.6	5.8	2.5
1.25			79	26	8.6	3.7
1.67			135	45	14.4	6.2
2.1				67	22	9.3
2.5				95	30	12.9
2.9				126	40	17.1
3.3				162	51	22
4.2					78	33

Flow Rate, L/s	Outside Diameter, mm 63	75	90	110	125
Inside Diameter, mm	51.5	61.4	73.6	90.0	102
5.0	109	46	19	7	4
5.8	146	61	25	9	5
6.7	188	79	32	12	6
7.5		98	40	15	8
8.3		120	48	18	10
10.0		169	68	25	13
11.7			91	34	18
13.3			117	43	23
15.0			146	54	28
16.7			179	65	35
20.0				92	49
23				124	65
27				160	84
30				200	105
33					129
37					154

Flow Rate, L/s	Outside Diameter, mm 125	160	200	250
Inside Diameter, mm	102	131	164	205
40	182	53	17.4	5.8
43	212	61	20.2	6.7
47		71	23	7.7
50		81	26	8.8
58		108	35	11.7
67		140	46	15.0
75		175	57	18.8
83			70	23
100			99	32
117			133	43
133			172	56
150				70
167				86
183				103
200				122
233				164

**Tables for other pipe dimensions, fluids, and temperatures can be made with HeadLossTableIP.xlsm.*
***Head loss in tight coils (lake coils, slinky coils, etc.) is typically 3% to 4% greater than in straight pipe.*

Table 6.10 Maximum Flow Rates for Optimum Head/Pressure Losses in GSHP Systems

	Water Flow Rate (gpm) at 3 ft of Head Loss/100 ft at 86°F						
	HDPE			Steel		PVC	Copper
	DR 11	DR 13.5	DR 15.5	Old Sch 40	New Sch 40	Sch 80	Type L
3/4	3.9	4.4	4.7	3	3.3	2.6	3.2
1	7	8	8.5	5.5	6.4	5.3	6.6
1	13	15	16	12	13	11	11.7
1	19	21	23	18	20	17	18
2	35	39	42	35	38	35	39
3	100	110	118	100	110	100	110
4	195	215	230	205	230	215	235
6	540	600	635	610	675	630	
8	1080	1200	1270	1250	1390	1325	
10	1925	2140	2275	2275	2525	2400	
12	3000	3350	3550	3600	4000	3790	

Multipliers: 20% propylene glycol at 40°F = **0.88** for nominal diameter ≤ 2 in., **0.92** for nominal diameter ≥ 3 in.

Multipliers: 20% methanol at 40°F = **0.91** for nominal diameter ≤ 2 in., **0.94** for nominal diameter ≥ 3 in.

Nominal Diameter, mm	Water Flow Rate (L/s) at 0.29 kPa/m at 30°C					
	HDPE			Steel		PVC
	DR 11	DR 13.5	DR 15.5	Old Sch 40	New Sch 40	Sch 80
25	0.25	0.28	0.30	0.19	0.21	0.16
32	0.44	0.50	0.54	0.35	0.40	0.33
40	0.82	0.95	1.01	0.76	0.82	0.69
50	1.2	1.3	1.5	1.1	1.3	1.1
63	2.2	2.5	2.6	2.2	2.4	2.2
90	6.3	6.9	7.4	6.3	6.9	6.3
125	12	14	15	13	15	14
160	34	38	40	38	43	40
200	68	76	80	79	88	84
250	121	135	144	144	159	151
300	189	211	224	227	252	239

Multipliers: 20% propylene glycol at 4°C = **0.88** for nominal diameter ≤ 63 mm, **0.92** for nominal diameter ≥ 90 mm

Multipliers: 20% methanol at 4°C = **0.91** for nominal diameter ≤ 63 mm, **0.94** for nominal diameter ≥ 90 mm

Table 6.11 Equivalent Lengths (L_{eqv}) for HDPE Pipe Fittings

	Equivalent Length, ft										
	Nominal Pipe Diameter, in.										
Fitting Type	**3/4**	**1**	**1 1/4**	**1 1/2**	**2**	**3**	**4**	**6**	**8**	**10**	**12**
Socket U-bend	12	6.4	11								
Socket U-do	9										
Socket 90 L	3.4	2.5	6	7	7						
Socket tee—Branch	4.1	5	6	10	13						
Socket tee—Straight	1.2	1.2	0.9	2	2.8						
Socket reducer (1 step)		6.1	4	3.9	4.2						
Socket reducer (2 step)			4.2		5.1						
UniCoil™	9	10									
Butt U-bend	12	22	35	43							
Butt 90 L	7	10	19	11	12	32	38	51	63	75	87
Butt tee—Branch	8	7	17	11	15	31	37	50	62	74	86
Butt tee—Straight	4.5	2.7	4	4	4	7	7	8	8	9	10
Butt reducer		4.8	6	6	7	10	13	20	26	33	39
Butt joint	2	1.2	1.3	1.3	1.2	0.8	1				
5-loop close header first take-off		17									
Last side take-off		30									
10-loop close header first take-off		20									
Last side take-off		34									

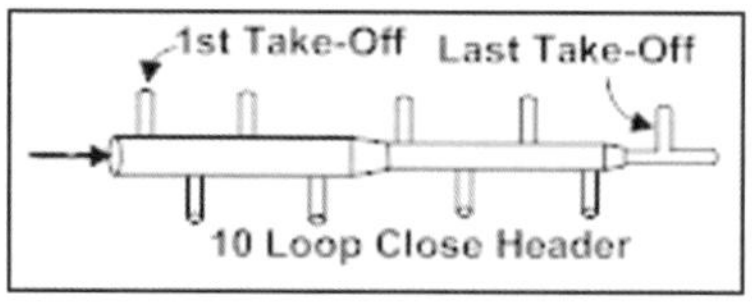

	Equivalent Length, m										
	Nominal Pipe Diameter, mm										
Fitting Type	**25**	**32**	**40**	**50**	**63**	**90**	**125**	**160**	**200**	**250**	**300**
Socket U-bend	3.7	2.0	3.4								
Socket U-do	2.6										
Socket 90 L	1.0	0.8	1.9	2.0	2.1						
Socket tee—Branch	1.2	1.6	2.0	3.0	4.0						
Socket tee—Straight	0.4	0.4	0.3	0.6	0.9						
Socket reducer (1 step)		1.9	1.2	1.2	1.3						
Socket reducer (2 step)			1.3		1.6						
UniCoil™	2.7	3.1									27
Butt U-bend	3.8	6.8	11	13							26
Butt 90 L	2.2	3.0	5.6	3.3	3.7	10	12	16	19	23	3.0
Butt tee—Branch	2.3	2.2	5.2	3.3	4.6	9.4	11	15	19	23	12
Butt tee—Straight	1.4	0.8	1.2	1.2	1.2	2.1	2.2	2.3	2.5	2.7	
Butt reducer		1.5	1.7	1.8	2.1	3.1	4.1	6.1	7.9	10	
Butt joint	0.6	0.4	0.4	0.4	0.4	0.2	0.3				
5-loop close header first take-off		5.2									
Last side take-off		9.1									
10-loop close header first take-off		6.1									
Last side take-off		10									

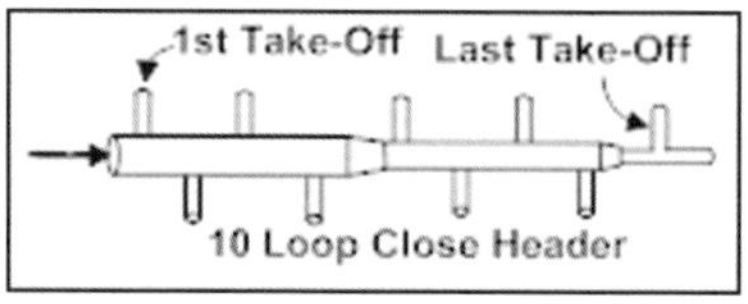

Table 6.12 Equivalent Lengths (L_{eqv}) for Iron and Copper Pipe Fittings (Kavanaugh 2006)

	Equivalent Length, ft										
	Nominal Pipe Diameter, in.										
Fitting Type	**3/4**	**1**	**1 1/4**	**1 1/2**	**2**	**3**	**4**	**5**	**6**	**8**	**10**
90° L—Screwed	2.0	2.5	3.6	4.2	5.6	9	11	14	17	22	27
90° L—Welded	1.0	1.3	1.8	2.1	2.8	4.4	5.7	7.2	8.6	11	14
45° L	1.4	1.8	2.5	2.9	3.9	6.1	8.0	10	12	15	19
Reducer	0.8	1.0	1.4	1.7	2.2	3.5	4.6	5.7	6.8	9	11
Tee—Run	1.2	1.5	2.2	2.5	3.4	5.2	6.8	8.6	10	13	16
Tee—Branch	8.0	10	14	17	22	35	46	57	68	88	108
Gate valve	1	1.3	1.8	3.1	1.4	2.2	2.9	3.6	4.3	5.5	6.8
Globe valve	24	30	43	50	34	52	68	86	103	131	162
Swing check	3.8	4.8	6.8	8	5.3	8.3	11	14	16	21	26
	Equivalent Length, m										
	Nominal Pipe Diameter, mm										
Fitting Type	**25**	**32**	**40**	**50**	**63**	**90**	**125**	**140**	**160**	**200**	**250**
90° L—Screwed	0.6	0.8	1.1	1.3	1.7	2.7	3.5	4.4	5.2	6.7	8.2
90° L—Welded	0.3	0.4	0.5	0.6	0.9	1.3	1.7	2.2	2.6	3.3	4.1
45° L	0.4	0.5	0.8	0.9	1.2	1.9	2.4	3.1	3.6	4.7	5.8
Reducer	0.2	0.3	0.4	0.5	0.7	1.1	1.4	1.7	2.1	2.7	3.3
Tee—Run	0.4	0.5	0.7	0.8	1.0	1.6	2.1	2.6	3.1	4.0	4.9
Tee—Branch	2.4	3.0	4.4	5.1	6.8	11	14	17	21	27	33
Gate valve	0.3	0.4	0.5	0.9	0.4	1	1	1	1	2	2
Globe valve	7.3	9.1	13.1	15.2	10.4	16	21	26	31	40	49
Swing check	1.2	1.5	2.1	2.4	1.6	3	3	4	5	6	8

Table 6.13 Typical Flow Coefficients (C_v) for Valves and Fittings
(C_v = Flow in gpm for Δp = 1.0 psi, Δh = 2.31 ft of water)

	Nominal Diameter, in.										
Valve/Fitting Type	**3/4**	**1**	**1 1/4**	**1 1/2**	**2**	**3**	**4**	**6**	**8**	**10**	**12**
Zone valve—Manufacturer A	23.5	37									
Zone valve—Manufacturer B	8.6	13.9		27.5	41						
Zone valve—Manufacturer C	3.5	3.5									
Hose kit—3 ft length	8	16	34	47							
Zone valve—Ball	25	35	47	81							
Ball valve—Manufacturer D	25	35	47	81	105	390	830	1250	2010	3195	
Butterfly valve					144	461	841	1850	3316	5430	
Swing check	13	21	35	45	75	195	350	990	1700	2400	
Y-strainer—IPS	18	28	43	60	95	155	250				
Y-strainer—Flange				30	70	160	260	550	920	1600	2200

6.6 PUMP FUNDAMENTALS

Different types of pumps used in closed-loop systems are shown in Figure 6.9. In-line wet-rotor circulators are commonly used in residential systems, unitary-loop commercial systems, and one-pipe loops. This design has the advantage of not requiring a seal between the pump and motor. The pumps are mounted with clamps on the suction and discharge pipes, which are connected to pump flanges. Replacement can be performed by removing the flange bolts. These pumps are limited in capacity and available head, and they have relatively poor efficiency. They are typically used in systems that have low head requirement, which offsets the poor efficiency. Newer designs have more efficient variable-speed electronically commutated motors (ECMs), which significantly lower demand and energy use. At this time, the price premium is significant and should be analyzed for economic value.

In-line circulator pumps with mechanical seals have the same mounting and service characteristics and in some cases slightly higher performance than non-ECM wet-rotor pumps. Motors can be replaced with more efficient models, but seals and couplings are necessary. Base-mounted close-coupled pumps also have seals, but the pump impeller is attached directly to the motor shaft. These pumps typically offer higher capacity and efficiency than in-line circulators.

Vertical in-line pumps and base-mounted end-suction pumps serve larger applications. Pump efficiency can be very high (over 70%), and the efficiency of motors larger than 1.0 hp (0.75 kW) is regulated with increasing efficiency as motor size increases, as shown in Table 6.8. Vertical pumps offer some advantage in terms of space requirement. Base-mounted end-suction pumps are not limited in size, are widely available, and have a history of satisfactory performance. Vertical in-line and base-mounted pumps provide seamless application of variable-speed motors. Improvements in variable-speed motors and drives have resulted in favorable economic value *when the pumps and motors are not oversized and speed controls are properly installed and maintained.*

The required input power for a pump (W_P) is computed by multiplying the volumetric flow rate (Q) by the differential pressure or head ($\Delta h = \Delta p \div \rho$) divided by the pump efficiency, as given in Equation 6.8. This value is sometimes referred to as *brake horsepower* (bhp) and includes the impact of the pump efficiency. Brake horsepower is distinguished from the pump output power, often referred to as the *water horsepower* (whp) or *hydraulic power.* Pump efficiency is the ratio of the power delivered to the water (whp) to the input power to the pump shaft (bhp).

$$W_{Pump}\,(\text{hp}) = \text{bhp} = \frac{Q \times \Delta p}{\eta_{Pump}}$$

$$= \frac{Q\,(\text{gal/min}) \times 0.1337\,(\text{ft}^3/\text{gal}) \times \Delta h\,(\text{ft of water}) \times 62.3\,(\text{lb/ft}^3)}{33{,}000\,(\text{ft}\cdot\text{lb/min}\cdot\text{hp}) \times \eta_{Pump}} \tag{6.8}$$

$$= \frac{Q\,(\text{gpm}) \times h\,(\text{ft of water})}{3960 \times \eta_{Pump}}$$

Equation 6.9 is the SI version of Equation 6.8, with the subscript m used for the unit of pump shaft or mechanical power (kW_m) to distinguish it from the motor electrical input power, for which the subscript e is used.

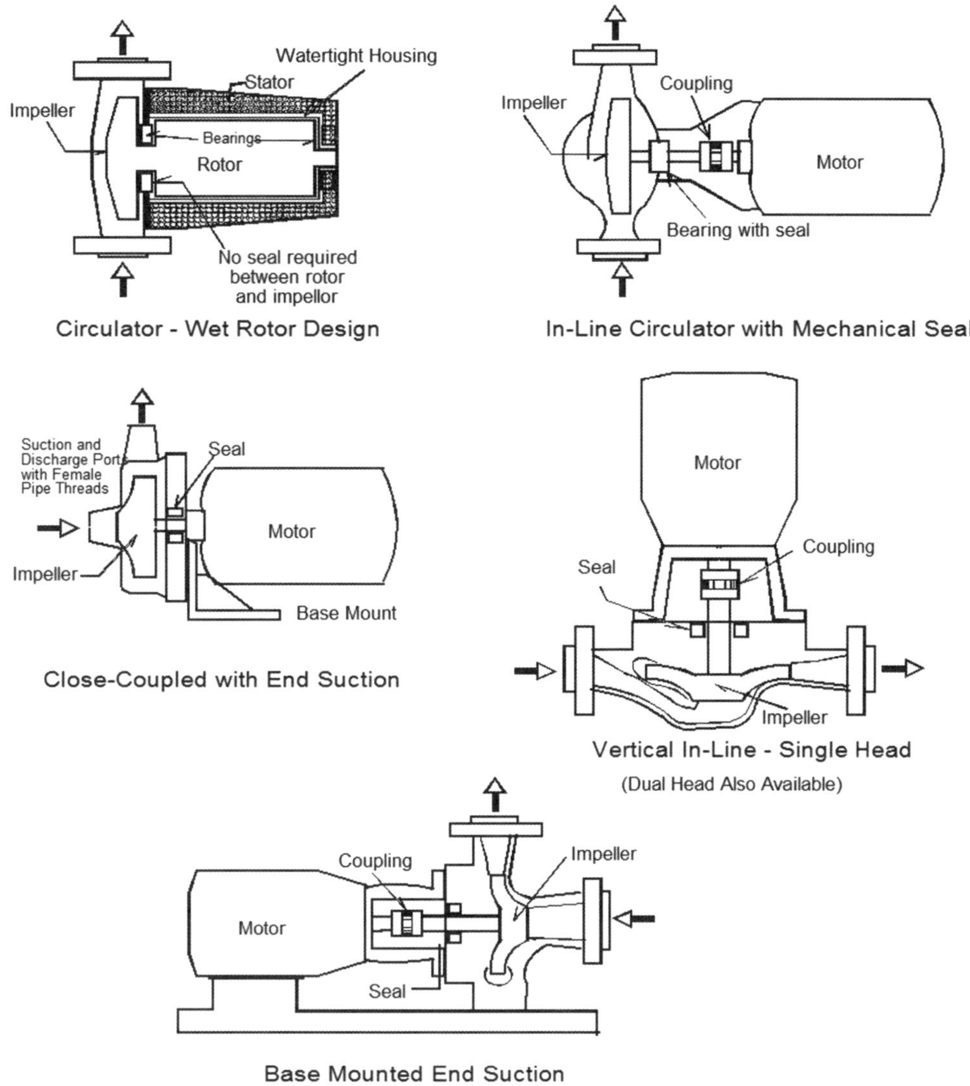

Figure 6.9 Common Pump Types Uses for Closed-Loop GSHP Applications

$$W_{Pump}\ (\mathrm{kW}_m) = \frac{Q\ (\mathrm{L/s}) \times \Delta p\ (\mathrm{kPa}) \times 1000\ (\mathrm{Pa/kPa}) \times \dfrac{\mathrm{N/m^2}}{\mathrm{Pa}} \times \dfrac{\mathrm{W{\cdot}s}}{\mathrm{N{\cdot}m}}}{1000\ (\mathrm{L/m^3}) \times 1000\ (\mathrm{W/kW}) \times \eta_{Pump}} \tag{6.9}$$

$$= \frac{Q\ (\mathrm{L/s}) \times \Delta p\ (\mathrm{kPa})}{1000 \times \eta_{Pump}} = \frac{Q\ (\mathrm{L/min}) \times \Delta p\ (\mathrm{kPa})}{60{,}000 \times \eta_{Pump}} = \frac{Q\ (\mathrm{m^3/s}) \times \Delta p\ (\mathrm{kPa})}{\eta_{Pump}}$$

It should be recognized that the more meaningful metric is the input power to the pump motor, which is directly related to energy use, demand, and operating cost. Benchmark values for both pump input power and motor input power are provided in Table 6.2. As shown in Equation 6.10, the motor input power includes motor efficiency, which declines with decreasing size and is not regulated for motors less than 1 hp (0.75 kW). The motor input power is not typically displayed on the motor nameplate and must be calculated using the motor efficiency (η_{Motor}). Full-load motor efficiencies are shown in Table 6.14 for four-pole and two-pole motors. Part-load efficiencies are nearly the same as full-load values down to 50% of full-load but decline along with power factor for lower loads. Part-load efficiencies can be found by multiplying the full-load efficiencies by the part-load multipliers (PLMs) provided in Table 6.14. If a VSD is used, its efficiency (η_{VSD}) must be included in determining motor power.

$$W_{Motor}\,(\mathrm{kW}_e) = \frac{0.746\ \mathrm{kW/hp} \times W_{Pump}\,(\mathrm{hp})}{\eta_{Motor} \times \eta_{VSD}} = \frac{W_{Pump}\,(\mathrm{W})}{\eta_{Motor} \times \eta_{VSD}} \tag{6.10}$$

Table 6.14 Minimum Motor Full-Load Efficiencies (NEMA 2009) and Part-Load Multipliers

Output Power, hp	Full-Load Efficiency ~1800 rpm (4-Pole)	Full-Load Efficiency ~3600 rpm (2-Pole)	Part-Load Multipliers (η_{PL} = PLM × η_{FL}), Percent of Full Load: 20%	40%	60%	80%
1	82.5%	74.0%	0.59	0.82	0.90	0.96
1.5	84.0%	81.5%	0.66	0.93	1.00	1.00
2	84.0%	82.5%				
3	87.5%	84.0%				
5	87.5%	86.5%				
7.5	90.2%	87.5%	0.80	0.96	1.00	1.00
10	90.2%	88.5%				
15	91.0%	89.5%	0.87	0.98	1.00	1.00
20	91.7%	89.5%				
25	92.4%	90.2%				
30	92.4%	90.2%	0.92	0.99	1.00	1.00
40	93.0%	91.0%				
50	93.6%	91.7%				

EXAMPLE 6.2— CALCULATION OF PUMP MOTOR ELECTRICAL INPUT POWER

Calculate the required four-pole motor size and power input for a pump with a 60% efficiency that delivers 50 ft of head (149.5 kPa) and 100 gpm (6.31 L/s or 378.5 L/min).

Solution

$$W_{Pump}\,(\mathrm{hp}) = \frac{Q\,(\mathrm{gpm}) \times h\,(\mathrm{ft\ of\ water})}{3960 \times \eta_{Pump}} = \frac{100\ \mathrm{gpm} \times 50\ \mathrm{ft\ of\ water}}{3960 \times 60\%} = 2.1\ \mathrm{hp} \tag{I-P}$$

$$W_{Pump}\,(\mathrm{kW}_m) = \frac{Q\,(\mathrm{L/s}) \times \Delta p\,(\mathrm{kPa})}{1000 \times \eta_{Pump}} = \frac{6.31\ \mathrm{L/s} \times 149.5\ \mathrm{kPa}}{1000 \times 60\%} = 1.57\ \mathrm{kW} \tag{SI}$$

A 3 hp (2.2 kW) pump is required, and the minimum full-load efficiency for a four-pole (~1800 rpm) motor is 87.5%. The motor will operate at 70% load (= 2.1 hp ÷ 3.0 hp), which results in a 1.0 PLM. Thus,

$$W_{Motor}\,(\mathrm{kW}_e) = \frac{746\ \mathrm{W/hp} \times 2.1\ \mathrm{hp}}{87.5\% \times 1.0} = 1790\ \mathrm{W} = 1.79\ \mathrm{kW} \tag{I-P}$$

$$W_{Motor}\,(\mathrm{kW}_e) = \frac{1.57\ \mathrm{kW}}{87.5\% \times 1.0} = 1.80\ \mathrm{kW} \tag{SI}$$

Pump curves are widely used as an alternative to the calculations demonstrated in Example 6.2. They offer a graphical visualization of the performance of pumps and permit a large amount of information to be presented in a compact form. Figure 6.10 demonstrates a typical format. The primary curves are the head (or pressure) on the vertical axis with the flow rate on the horizontal axis. Manufacturers are able to offer a much wider performance selection by providing several impellers for one pump casing. The figure shows curves for three different impeller diameters operating at a constant speed of 1750 rpm indicated with blue lines that show decreasing head with increasing flow rate.

Centrifugal pump efficiency typically is highest at flow rates greater than 50% of maximum flow capacity. Lines of constant efficiency are shown as solid green lines in Figure 6.10. For this pump the best efficiency point (BEP) occurs at a flow rate of 70 gpm (4.4 L/s) and a head of 38 ft of water (114 kPa) for the 6 in. (150 mm) diameter impeller. Efficiencies decline when smaller impellers are used, as shown in the figure. However, the power requirement will also be much lower. Lines of constant pump power are shown as dashed red lines. For this pump the lines are nearly parallel to the head versus flow rate lines, but in most cases the constant power lines have an increasingly steeper slope as flow rate increases.

Pumps should be selected to operate near the BEP. If the operating point (head and flow rate) efficiency is more than 5% below the BEP, another pump should be considered where the operating point efficiency and BEP are more closely matched.

6.7 CLOSED-LOOP WATER DISTRIBUTION SYSTEM DESIGN PROCEDURE

The suggested steps for closed-loop GSHP water distribution system design are as follows:

1. Lay out the piping network with all piping run lengths, fittings, valves, and required flows in each section.
2. Select a pipe size for each section that will result in acceptable head loss for the flow rate (see Table 6.10).

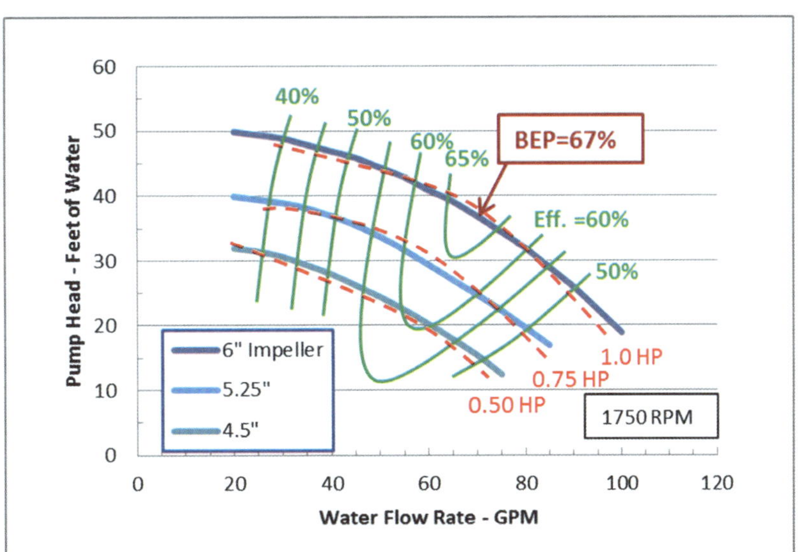

Figure 6.10 Pump Curves: Flow vs Head, Efficiency, and Power for Three Impeller Diameters

3. Include full-size purge valves (equal to or greater than circuit header diameters) in a convenient location so that individual circuits of no more than 20 vertical heat exchangers can be purged of air and debris.
4. Find the equivalent length (straight run plus equivalent length for fittings) and head/pressure loss through each section in the longest pipe run (or path that seems to have the greatest head/pressure loss). Some designers check several runs.
5. Find head loss through other components (heat pumps, control valves, etc.).
6. Locate and resize any section or component with excessive losses.
7. Sum total of losses in series flow paths and find loss through the highest head loss path.
8. Select a pump (and motor) that will result in an operating point on the pump curve that indicates the efficiency is within 5% of the BEP.
9. Calculate the required pump demand per ton (kW) of cooling capacity and redesign the system if the value is unacceptable (below a benchmark grade of A or B in Table 6.2).

The alternate central loop design for the example building that is shown in Figure 4.6 serves as an example of the design procedure in the following sections.

6.7.1 Step 1—Lay Out the Piping Network

Figure 6.11 is an expanded view of the central ground loop and building loop option shown in Figure 4.6. The ground-loop header consists of two parallel circuits, each with nine vertical U-tube heat exchangers inserted into 270 ft (82 m) deep boreholes. Each circuit has modified reverse-return headers, which minimizes the length (and therefore head loss) of the reverse-return header. In traditional designs, the reverse-return header runs parallel to the entire length of the return header, as shown in the upper right corner of Figure 1.7. In the modified design, the supply and return headers are routed in a loop so that additional length of the reverse-return header is relatively short, as shown in Figure 1.9.

The 18-bore ground loop is served by two parallel circuits with 9 bores each. Because they are in parallel, the head/pressure loss will be the same, and losses should not be added but calculated through the longer of the two circuits. The ground-loop supply and return header manifolds are in the equipment room near the purge valves. They are routed down and horizontally to outside the building wall using two 90° elbows on each header. Standard HDPE elbows are expensive and have relatively high head losses. Often necessary for interior piping, elbows in ground-loop piping can be made by bending 2 in. (63 mm) and smaller pipe in the horizontal trenches (while observing bending limitations given by Equation 6.11). Losses through these elbows are nearly equal to losses through an equivalent length of straight pipe. Larger pipe requires standard elbows or long sweep elbows fabricated from sections of coiled pipe that comply with manufacturer recommendations for minimum bending radii (Equation 6.11 for HDPE).

As shown in Figure 6.11, the supply line of the ground loop with a flow rate of 30 gpm (1.9 L/s) makes two bends before the first U-tube take-off is made. One-ninth of the flow enters the U-tube while the remaining 26.7 gpm (1.7 L/s) continues through the main supply header. As flow continues, the rate through the supply header decreases while the rate in the return header increases. The tube size is adjusted so that head losses are not excessive, while ensuring the diameters are not too large so that purging can be accomplished. To accomplish this task, the flow rate through each pipe section is noted as shown in Figure 6.11.

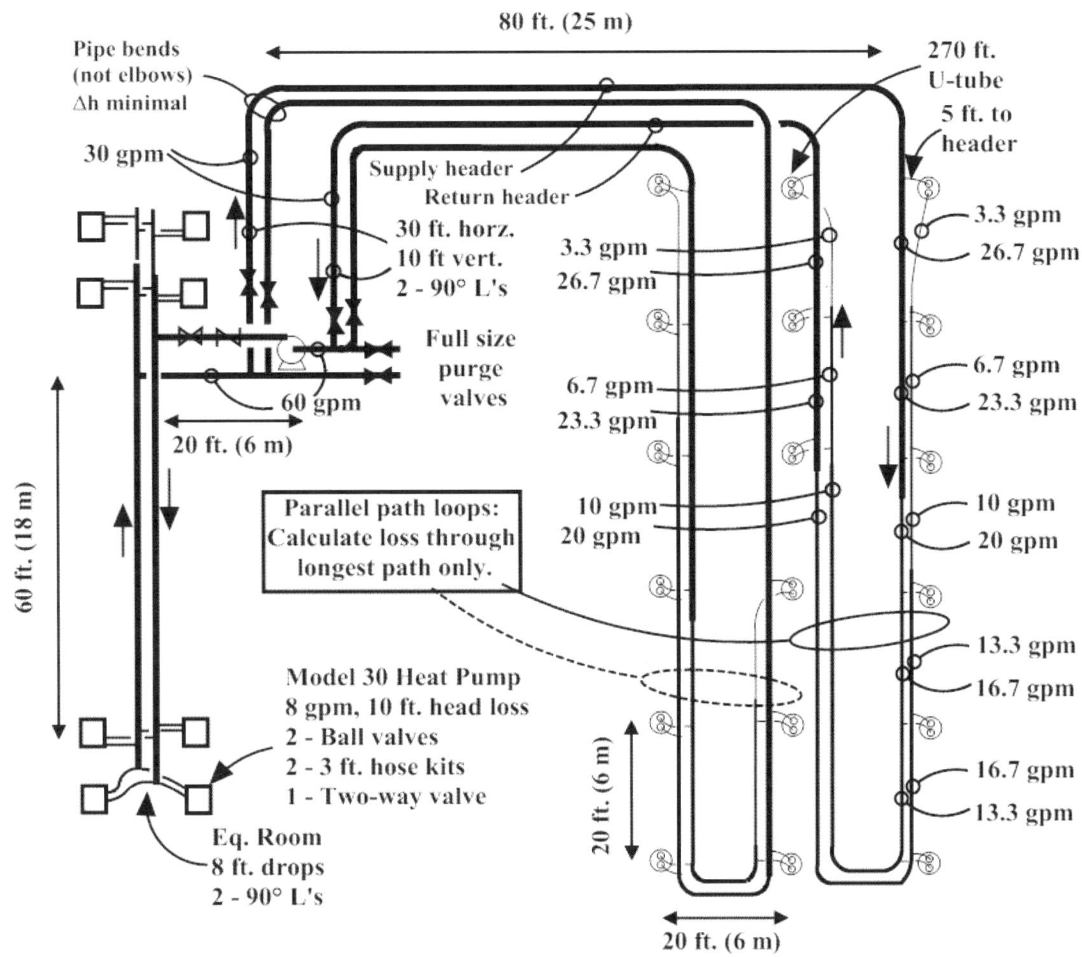

Figure 6.11 Layout of Example Pipe Network with Flow Rates for Each Section

Interior pipe routing is repeated using a similar process. In this design the heat pumps are conveniently located in two equipment closets. This arrangement permits the interior piping to be split into two parallel paths near the pump discharge then routed overhead and down into the closets. At this point hose kits are used to connect the heat pumps through two shut-off (ball) valves and a two-way control valve on each heat pump. Balancing valves and strainers at each heat pump are optional. Recall that high-efficiency water-to-air heat pumps do not require precise balancing at the expense of high-head-loss control valves and that piping systems that consist of 100% HDPE and polypropylene have limited need for heat pump strainers if systems are thoroughly purged at start-up and strainers are located on central pumps. Unitary-loop GCHPs with 100% HDPE do not typically require strainers if properly purged at start-up.

6.7.2 Step 2—Size Each Pipe Section

Table 6.15 is provided to systematize the remaining steps in the design process. The flow rate through each section of the ground-loop header is shown in column 1. Column 2 notes the piping type and dimension ratio (DR). Note that 2 in. (63 mm) and smaller pipe must be DR 11 (or possibly DR 9 for high-rise applications), while larger pipe can be DR 13.5 or 15.5 depending on operating pressures. The reason for the higher pressure rating for the smaller pipe is that surface scars that may occur during installation will have a greater relative impact on the thinner walls of smaller-diameter DR pipe than on the thicker walls of larger-diameter pipe.

Table 6.10 is used to find the appropriate pipe size shown in column 3 of Table 6.15. A maximum flow rate of 35 gpm (2.2 L/s) can be accommodated by 2 in. (63 mm) DR 11 HDPE. For the 30 gpm (0.19 L/s) design flow, the head loss is 2.33 ft of water per 100 ft of pipe (0.23 kPa/m) as indicated in Table 6.8. The supply pipe header size remains constant until after the take-off for the fourth U-tube. At this point the header size is reduced to 1 1/2 in. (50 mm) HDPE, which can accommodate flows up to 19 gpm (1.2 L/s). When the supply header flow drops to 10 gpm (0.63 L/s), the diameter is reduced to 1 1/4 in. (42 mm) and eventually to 1 in. (32 mm) pipe for the last section of the supply header. The last head/pressure loss to consider is that of the U-tube, which consists of short horizontal sections, two 270 ft (82 m) vertical tubes, and the U-bend. Recall that the head loss through only one U-tube is considered because flow through the other U-tubes is in parallel. The return header is nearly identical to the supply header except that in this design it is 20 ft (6 m) shorter than the supply. The sizing procedure is repeated for the interior pipe as shown in columns 10, 11, and 12 of Table 6.15.

6.7.3 Step 3—Locate and Size Purge Valve

The location for the purge valves is near the pump in the equipment room, which is a convenient location for the temporary connection of a purge pump required for a circuit with nine U-tubes. Note that each ground-loop circuit has isolation valves. This allows each circuit (with nine vertical loops) to be purged individually. There are no check valves or flow control valves on the ground loop, which allows installers to reverse purge flow through the ground loop. This action allows more effective air removal. The building circuit can also be purged (in one direction) by the same purge pump. This arrangement allows the entire water loop to be purged without disconnecting and reconnecting the purge pump (which reintroduces air into the system). The purge valves and connections must be a minimum of 2 in. (63 mm) nominal diameter since each circuit header is this size.

6.7.4 Step 4—Find the Equivalent Lengths and Head/Pressure Losses

The determination of equivalent lengths is shown in columns 5 through 8 of Table 6.15. Column 5 shows the length of pipe. Column 6 provides the equivalent lengths of the fittings described in column 7, which are found in Tables 6.11 and 6.12. Column 7 also indicates the quantity of each fitting type. Note that columns 6 and 7 are repeated (a and b) so that sections that contain more than one type of fitting can be accounted for in the same row. For example, note that the supply header at the fourth take-off includes the equivalent length of a straight run of a tee and a reducer. To find the equivalent length of each section shown in column 8, the straight length of pipe is added to the equivalent lengths of the fittings. The head/pressure losses for each section are determined by multiplying the values in column 4 by the values in column 8 **and dividing by 100** (since column 4 values are loss per 100 ft). This division of course is not repeated when working in SI units, as the pressure losses are provided per metre. **Note that the loss in only one U-tube is listed since they are all piped in parallel. Also note the loss in the return header between the first and last U-tube take-offs is likewise not included since it is in parallel with the supply header. To determine total system head/pressure losses, parallel-path losses are not added, only losses in series.** The calculation of interior piping equivalent lengths and losses, which include the pump suction, pump discharge, and return headers connecting the heat pumps in the equipment closet at the greatest distance from the pump, are calculated as shown in columns 13 through 17.

6.7.5 Step 5—Find Other Component Losses

The heat pump loss is provided at a nominal flow rate, which happens to be the same as the design flow. If this is not the case, the head/pressure loss can be corrected using Equation 6.6. The losses in the remaining components are based on the flow coefficient (C_v), the flow rate that results in a 1.0 psi (6.9 kPa) loss through the fitting. The losses for a flow rate of 8 gpm (0.5 L/s) through the most remote heat pump include two 3 ft (1 m) long, 3/4 in. (25 mm) nominal diameter hose kits, two ball valves, and a two-way control valve. The final loss shown in Table 6.15 is for the pump suction strainer.

Table 6.15 Head Loss Summary Table for GSHP Closed-Loop Piping Network Example—I-P

Ground Loop	**1**	**2**	**3**	**4**	**5**	**6a**	**7a**	**6b**	**7b**	**8**	**9**
Pipe Section	**Flow, gpm**	**Pipe**	**Diameter, in.**	**Δh/ 100 ft**	***L***	***L_{eqv}***	**Fitting**	***L_{eqv}***	**Fitting**	***L* Total, ft**	**Δh, ft of water**
Supply header	30	HDPE DR 11	2	2.33	130	7	2 L's at 7 ft			144	3.4
After 1st take-off	26.7	HDPE DR 11	2	1.85	20	4	Tee—straight			24	0.4
After 2nd take-off	23.3	HDPE DR 11	2	1.77	20	4	Tee—straight			24	0.4
After 3rd take-off	20	HDPE DR 11	2	1.72	20	4	Tee—straight	4	Reducer	28	0.5
After 4th take-off	16.7	HDPE DR 11	1 1/2	2.44	20	4	Tee—straight			24	0.6
After 5th take-off	13.3	HDPE DR 11	1 1/2	1.55	20	4	Tee—straight			24	0.4
After 6th take-off	10	HDPE DR 11	1 1/4	1.84	80	4	Tee—straight	4	Reducer	88	1.6
After 7th take-off	6.7	HDPE DR 11	1	2.67	20	4	Tee—straight	4	Reducer	28	0.7
U-tube	3.33	HDPE DR 11	1	0.83	565	10	U-tube	7	Tee—branch	582	4.8
Return header	30	HDPE DR 11	2	2.33	110	7	2 L's at 7 ft			124	2.9
							Ground Loop Head Loss				**15.7**
Building Loop	**10**	**11**	**12**	**13**	**14**	**15a**	**16a**	**15b**	**16b**	**17**	
Pipe Section	**Flow, gpm**	**Pipe**	**Diameter, in.**	**Δh/ 100 ft**	***L***	***L_{eqv}***	**Fitting**	***L_{eqv}***	**Fitting**	***L* Total, ft.**	**Δh, ft of water**
Ground loop to pump	30	HDPE DR 11	2	2.33	5	15	Tee—branch			20	0.5
Pump suction	60		3	1.24	5	2.2	Gate valve			7.2	0.1
Pump discharge	60		1	1.24	20	2.2	Gate valve	8.3	Swing check	30.5	0.4
Supply and return headers to 4 heat pumps	30		2	2.33	136	15	4 tees—branch	7	2 L's at 7 ft	210	4.9
Other components	18	19	20	21							
Heat pump	8										10
		C_v	**Diameter, in.**	**Quantity**							
3 ft host kits (2)	8	8	0.75	2							4.6
Ball valves	8	23.5	0.75	2							0.5
Two-way valve	8	25	0.75	1							0.2
Suction strainer	60	160	3.00	1							0.3
							Building Loop Head Loss				**21.5**
							Building and Ground Loop Head Loss				**37.3**

6.7.6 Step 6—Locate and Resize High-Loss Components

Examination of columns 4 and 13 indicates all losses are less than the recommended value of 3 ft of water per 100 feet of pipe ($\Delta p/L$ = 0.29 kPa/m). The total loss of 37.3 ft of water (112 kPa) suggests the design should merit a pump power benchmark of grade A according to Table 4.6. However, examination of column 9 indicates losses through the hose kits compose 13% of the total, so 1 in. (32 mm) nominal diameter hose kits might be advisable, especially if greater lengths are necessary.

6.7.7 Step 7—Sum Losses Through Longest Parallel Path

The losses for each section of pipe are summed in this example beginning at the point where the ground-loop supply header leaves the equipment room. There are two parallel circuits, so only the loss through the longest circuit is included. The pipe sections include the following:

- Circuit supply header main to the point of the first U-tube take-off
- Supply header through the sections for the remaining U-tube take-offs (note losses through the return side of this header between the first and last take-offs are not added because they are in parallel with supply)
- U-tube (last U-tube is used here, but any one could be used because they are in parallel)
- Circuit return header main to the pump suction header (where it joins the other circuit)
- Pump main suction and discharge to the point where flow splits to each heat pump closet
- Building interior supply header to most distant heat pump closet
- Flow though the most remote (or highest head loss) heat pump, hose connections, and valves
- Building interior return header to the point where it meets the return header from the other heat pump closet
- Building interior main header to the equipment room

As shown in Table 6.15, the total head/pressure loss is 37.3 ft of water (112 kPa).

6.7.8 Step 8—Select Pump(s)

Figure 6.12 is representation of Figure 6.10 with the curves for the smaller impeller removed. The 6 in. (152 mm) impeller would provide the necessary head of 37.3 ft of water (112 kPa pressure) for the design flow rate of 60 gpm (3.8 L/s). This point is drawn on the pump curve and indicates the pump will provide more head than required. The operating point can be determined plotting a system curve that is generated by calculating the head losses at other flow rates using Equation 6.6. Flow rates of 50 and 70 gpm (3.15 and 4.42 L/s) are used to create a curve that intersects the pump curve:

$$\Delta h_{50} = 37.3 \text{ ft} \times (50 \text{ gpm}/60 \text{ gpm})^2 = 25.9 \text{ ft of water} \qquad \text{(I-P)}$$

$$\Delta h_{70} = 37.3 \text{ ft} \times (70 \text{ gpm}/60 \text{ gpm})^2 = 50.7 \text{ ft of water} \qquad \text{(I-P)}$$

$$p = 112 \text{ kPa} \times (3.15 \text{ L/s} \div 3.8 \text{ L/s})^2 = 77.0 \text{ kPa} \qquad \text{(SI)}$$

$$p = 112 \text{ kPa} \times (4.42 \text{ L/s} \div 3.8 \text{ L/s})^2 = 152 \text{ kPa} \qquad \text{(SI)}$$

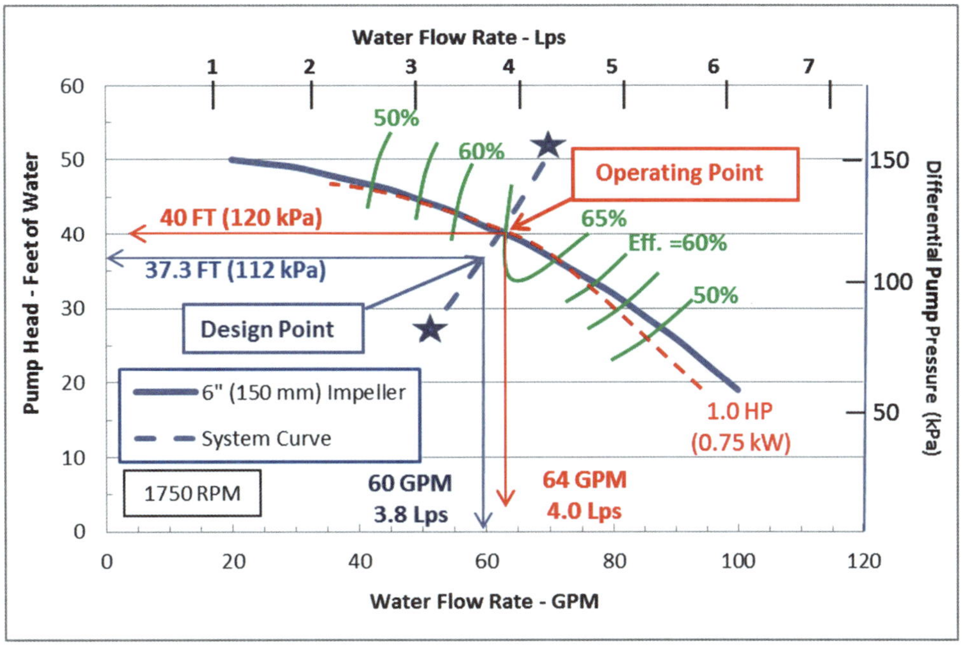

Figure 6.12 Pump Curve for Large Impeller, Showing System Curve and Operating Point

These two points are noted on Figure 6.12 with stars. The system curve is shown as a dotted blue line drawn through these two points and the design point of 37.3 ft of water (112 kPa) at 60 gpm (3.8 L/s). The operating point of this system with the pump is the point of intersection of the system curve with the pump curve, which indicates the flow rate will be approximately 64 gpm (4.0 L/s). The pump efficiency at this point will be 65%, which is only 2% less than the BEP.

The pump curve indicates a 1.0 hp motor is necessary at the operating point. This is substantiated by Equation 6.8 for this application.

A VSD could be used to lower the speed below 1750 rpm so that only 60 gpm (3.8 L/s) would be delivered at full load. The VSD could also be used to adjust flow to minimum energy use at part-load conditions. Pump flow control options are discussed in Section 6.8.

Because the motor size is above 1.0 hp for some operating points on the pump curve, a safety factor would be prudent. Options are to use a motor with a service factor of 1.25 (meaning the motor will operate 25% above rated power without overheating) or to use a 1 1/2 hp motor. The input power to the motor is determined using Equation 6.10, the minimum efficiency (Table 6.14) for a value for a four-pole motor (note rpm on pump curve), and an assumed typical full-load VSD efficiency of 97%:

$$W_{Motor}\,(\mathrm{kW}_e) = \frac{0.746\ \mathrm{kW/hp} \times W_{Pump}\,(\mathrm{hp})}{\eta_{Motor} \times \eta_{\mathrm{VSD}}} = \frac{0.746\ \mathrm{kW/hp} \times 1.0\ \mathrm{hp}}{82.5\% \times 97\%} = 0.93\ \mathrm{kW}$$

6.7.9 Step 9—Calculate Pump Power or Electrical Demand per Ton of Heat Pump Capacity

Table 4.6 indicates the building cooling load is 19 tons (67 kW), the nominal heat pump capacity is 24 tons (84 kW), and the corrected heat pump capacity is 21 tons (74 kW). While the choice is open to the standards of the individual, the middle value of corrected capacity is used here. Benchmark grades are listed in Table 6.2.

The pump power per corrected heat pump capacity is

$$\frac{W_{Pump}\,(\text{hp})}{100\text{ tons}} = \frac{1.0\text{ hp}}{21\text{ tons}/100} = 4.8\text{ hp}/100\text{ tons}(10.2\text{ W}_m/\text{kW}_t) \equiv \text{Grade A}$$

The pump motor power per corrected heat pump capacity is

$$\frac{W_{Motor}\,(\text{W})}{\text{ton}} = \frac{0.93\text{ kW} \times 1000\text{ W/kW}}{21\text{ tons}} = 44\text{ W/ton}(12.6\text{ W}_e/\text{kW}_t) \equiv \text{Grade A}$$

The design is acceptable in terms of pump and motor size.

6.8 PUMP CONTROL AND HEAT PUMP CONNECTIONS

6.8.1 Unitary Loop On-Off Control

Figure 6.13 shows the individual heat pump arrangement of a unitary GCHP system in which individual ground loops are connected to each heat pump and control is accomplished by simply turning each pump on when the compressor is activated. The connections can be made with hose kits, reinforced rubber hose with barbed fittings, HDPE with IPS adaptors, or PEX. As shown in the figure, swivel connectors are used, which makes cross-connection during system flushing convenient. Pressure/temperature (P/T) taps placed at the heat pump connections make performance verification possible. Via the P/T taps, the liquid inlet and outlet temperatures and differential pressure measurements can be made with removable probes. Flow rate can be inferred from flow versus loss data provided by the manufacturer of the heat pumps. The figure also shows three-way valves on the connections that serve the dual purpose of being isolation valves and connection ports for the purge pump.

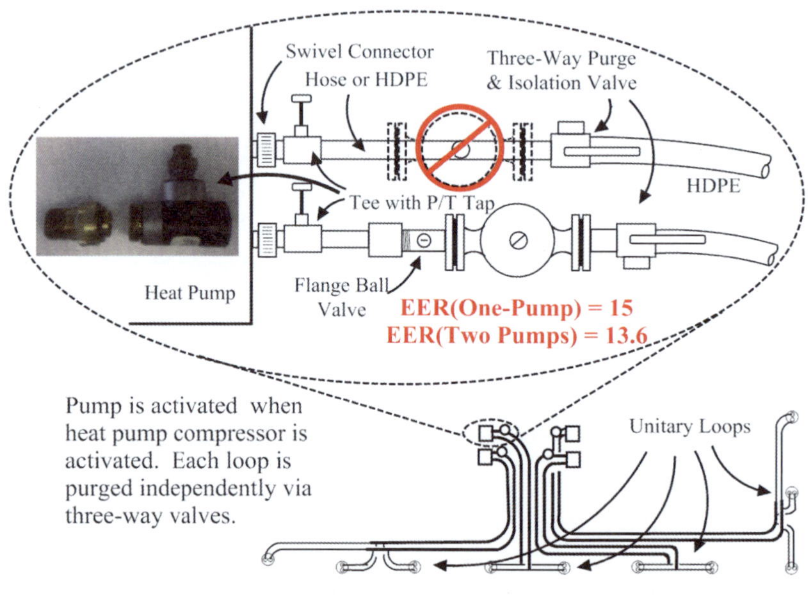

Figure 6.13 Unitary-Loop Heat Pump Connections and Pump Control

The unitary-loop option not only had the highest ENERGY STAR ratings in the survey mentioned in Section 1.6, but it also offers an excellent counter to the assertion that GCHPs are too expensive. The need for expensive controls and long runs of large-diameter building and ground-loop piping is eliminated. Another major advantage is that mechanical faults affect only one zone, unlike central-loop faults that bring down the entire building HVAC system.

This arrangement should be considered as a primary option for one- and two-story buildings with close access to ground-loop sites as shown in Figure 4.6. However, unitary loops are not universally an appropriate option. The significant cost savings for interior piping would not be realized in small-footprint high-rise buildings. The more expensive large-diameter header pipe runs for central-loop systems in tall buildings would be relatively short since they are typically vertical risers rather than long horizontal headers needed for large-footprint buildings. The value of combining zones with load diversity on a common loop is often exaggerated. There is value when load diversity is significant (i.e., when the sum of peak loads is more than 125% of the block load) and the diversified ground exchanger length is much less than the total lengths for multiple individual loop ground heat exchangers. In this situation, the cost of additional vertical bores is likely to exceed the added cost for the pipe headers and manifolds of a central loop. Another disadvantage of unitary loops is the need to measure pressure/charge level in multiple loops and provide service when pressure falls below recommended values.

A final disadvantage of unitary loops is that the relatively poor efficiency of conventional small circulator pumps will negatively affect the power input to the units, especially if two pumps are necessary. It is therefore critical to minimize friction losses to maintain high system efficiency. Consider the heat pump power (W_{HP}) input to a 36,000 Btu/h (10.6 kW) heat pump with an EER of 16.7 Btu/Wh (COP = 4.9):

$$W_{\text{HP}} = \frac{\text{TC (Btu/h)}}{\text{EER (Btu/Wh)}} = \frac{36{,}000\ \text{Btu/h}}{16.7\ \text{Btu/Wh}} = 2156\ \text{W} \qquad \text{(I-P)}$$

$$H_{HP} = \frac{\text{TC (kW)}}{\text{COP}} = \frac{10.6\ \text{kW}}{4.9} = 2.16\ \text{kW} = 2160\ \text{W} \qquad \text{(SI)}$$

With one 245 W pump the efficiency is

$$\text{EER}_{System} = \frac{\text{TC}}{W_{\text{HP}} + W_{Pump}} = \frac{36{,}000\ \text{Btu/h}}{2156\ \text{W} + 245\ \text{W}} = 15.0\ \text{Btu/Wh} \qquad \text{(I-P)}$$

$$\text{COP}_{System} = \frac{10.6\ \text{kW}}{2160\ \text{W} + 245\ \text{W}} = \frac{10{,}600\ \text{W}}{2405\ \text{W}} = 4.4 \qquad \text{(SI)}$$

while the efficiency with two 245 W pumps declines by 10%:

$$\text{EER}_{System} = \frac{\text{TC}}{W_{\text{HP}} + W_{Pump}} = \frac{36{,}000\ \text{Btu/h}}{2156\ \text{W} + 2 \times 245\ \text{W}} = 13.6\ \text{Btu/Wh} \qquad \text{(I-P)}$$

$$\text{COP}_{System} = \frac{10.6\ \text{kW}}{2160\ \text{W} + 2 \times 245\ \text{W}} = \frac{10{,}600\ \text{W}}{2650\ \text{W}} = 4.0 \qquad \text{(SI)}$$

In the future, if higher-efficiency constant-speed or variable-speed circulation pumps are available with only a modest cost premium, this issue may be resolved in terms of both electrical demand and economic value.

6.8.2 Common (Subcentral) Loop with Individual Pumps

An alternative that maintains the simple on-off control is the common (subcentral) loop with the heat pump connections arranged as shown in Figure 6.14. Figure 4.5 demonstrates the building piping layout. This option takes advantage of load diversity and minimizes the need to maintain individual loop pressures. The use of multiple common loops also reduces the need for long runs of large-diameter building and ground-loop headers and manifolds. A check valve at each heat pump is required to prevent backflow when the unit is off. A strainer may be required if the interior piping loop is steel or contains other components that are prone to corrosion. A single strainer at a central location is an option for common loops that are 100% HDPE and polypropylene.

6.8.3 One-Pipe Loop with On-Off Control (with or without VSD)

Figure 6.15 shows the heat pump connections and central-loop piping of a one-pipe GCHP system. This arrangement also achieved very high ENERGY STAR ratings in the survey mentioned in Section 1.6 and addresses the issue of cost containment through simplicity of equipment and control. Individual circulator pumps that deliver head only sufficient to overcome heat pump and connection losses are activated with the heat pump compressor. Main pumps are cycled to maintain ground-loop return temperature within a range that ensures heat pump efficiency in cooling and heating. Variable-speed pump drives can be used and are controlled by temperature sensors, which are more reliable and less expensive than controls using differential pressure transducers.

Figure 6.16 shows a vertical water-to-air heat pump and circulator pump connected to a one-pipe building loop. In this case the connections are made with a prefabricated HDPE-IPS transition fitting, an IPS-barbed hose fitting, reinforced hose, and a barbed elbow to the heat pump. This arrangement is typically less costly than hose kits. Figure 6.17 shows a polypropylene manifold for the main pumps and suction strainers of a one-pipe loop.

6.8.4 Central Loop with Variable-Speed (Frequency) Drives

In some cases central loops are a good option for ground-coupled systems in buildings with small footprints and/or significant load diversity. They are often a good option for closed-loop surface-water heat pump (SWHP) systems because there is typically a

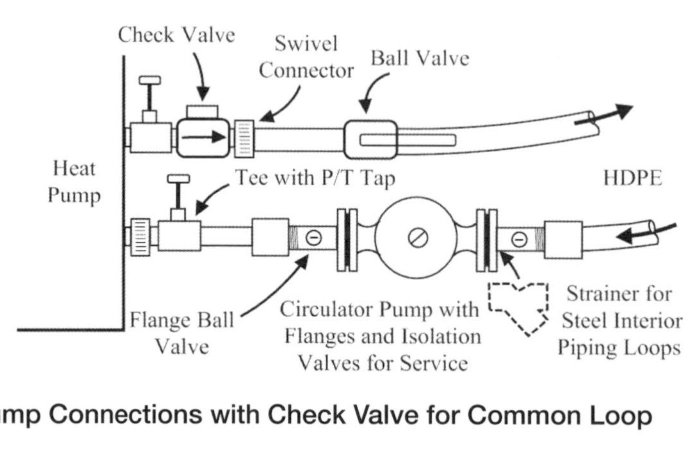

Figure 6.14 Heat Pump Connections with Check Valve for Common Loop

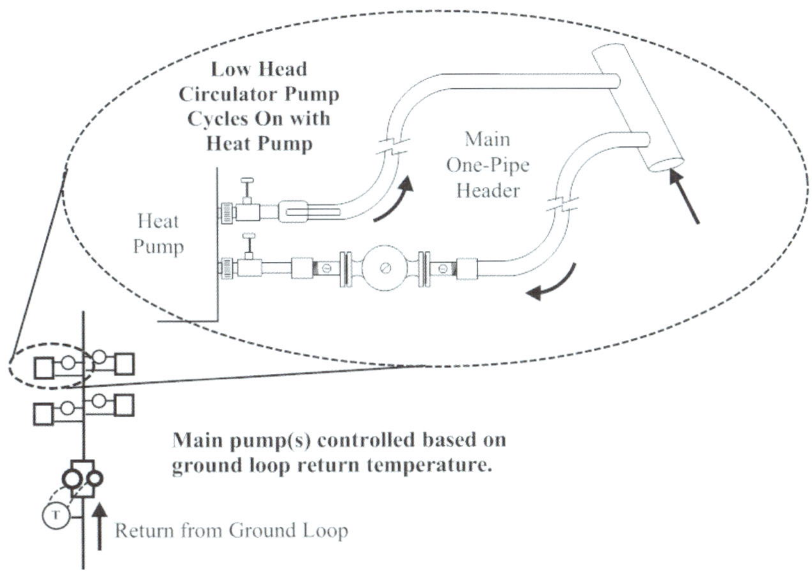

Figure 6.15 One-Pipe Loop Heat Pump Connections and Control Method

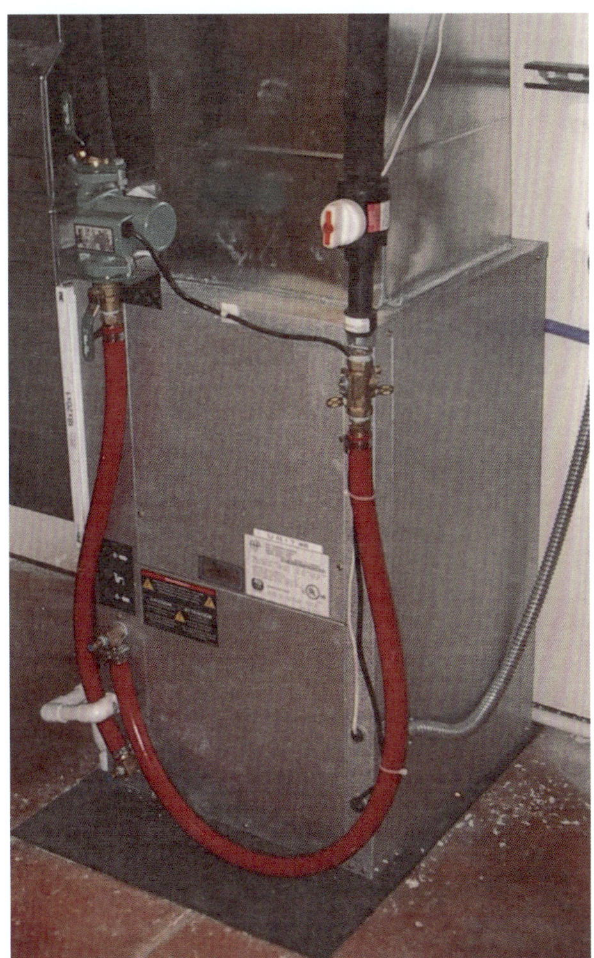

Figure 6.16 One-Pipe System Heat Pump, Circulator Pump, and Hose Connections

great distance between the building and the water reservoir. Pump energy must be minimized to capture the energy-efficient benefit of GSHPs, and VSDs (a.k.a *variable-frequency drives*, VFDs) are often used. Figure 6.18 depicts a traditional control method, which is to close two-way valves with motorized actuators on the heat pumps when units are off. The resulting reduction in system flow rate will cause pump head to increase and head loss through the piping to be lower. A differential pressure transducer is placed across the supply and return headers at a location in the pipe network remote from the pump. The differential pressure transducer signal is used to lower the operating frequency of the main pump(s) to maintain adequate differential pressure to deliver design flow rate through the most remote heat pumps. The reduction in power consumption can be significant if the pump and motor are properly sized.

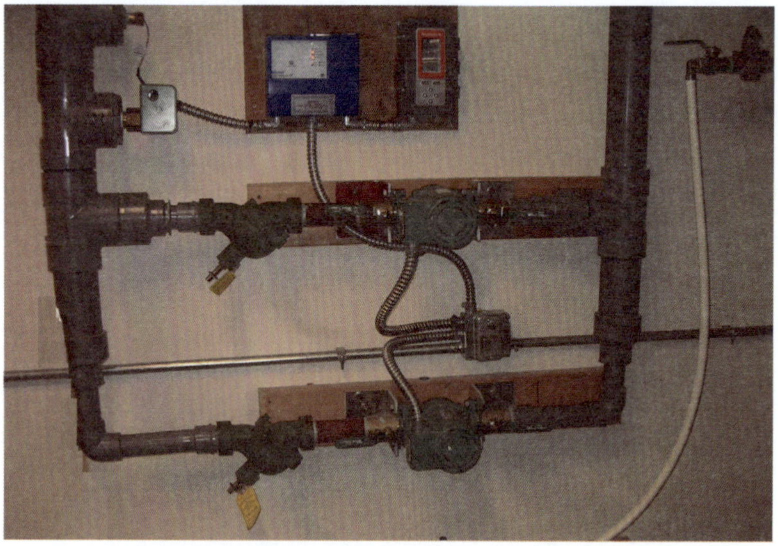

Figure 6.17 Main Pumps for One-Pipe GCHP System

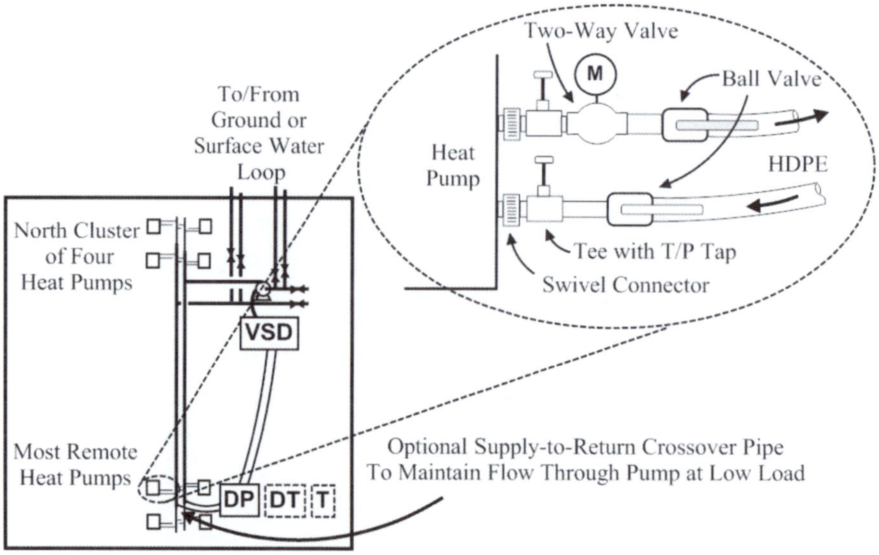

Figure 6.18 Central-Loop Heat Pump Connections and VSD Control Option

Common practice is to maintain pump operation continuously even at zero or very low part load. Thus, a crossover pipe (or a three-way valve) is installed to prevent a no-flow condition through the pump. Because recommended practice is to operate VSDs at 25% speed or more (Taco 2012), it would be prudent to deactivate the pump at no load or incorporate a much smaller constant-speed pump to operate at no or very low load in buildings that are occupied less than 50 or 60 hours per week.

Some caution is advised, because a GSHP field study indicated that less than 10% of the ground-loop VSD pumps with differential pressure transducer control were operating as intended due to faulty controls or had pumps large enough to provide near full-load flow rate at minimum motor speed (Kavanaugh and Kavanaugh 2012). Given the minimal attention to water treatment programs at these sites, it is suspected that there was a high incidence of problems at the pressure measurement locations. Suggested options include use of polyethylene or propylene interior piping or use of control schemes using temperature probes (differential temperature or ground-loop return) that are less susceptible to fouling and are less expensive to replace. However, these materials are not rated to meet a flame spread index (FSI) greater than 25 and the smoke developed index (SDI) of 50 required when they are located in plenums and must be wrapped with materials that meet this requirement (NFPA 2015).

6.8.5 Combinations of Loop Types Based on Building Layout and Load Diversity

Frequently a combination of ground loop and building loop options is optimal. Figure 6.19 shows a generalized layout of an actual 1960s-era high school in a southern location. There is little load diversity in the classrooms, offices, and library. Additionally, the offices are occupied for extended hours for 12 months per year, while the library and classroom are occupied for 40 hours per week for less than 10 months per year. It would

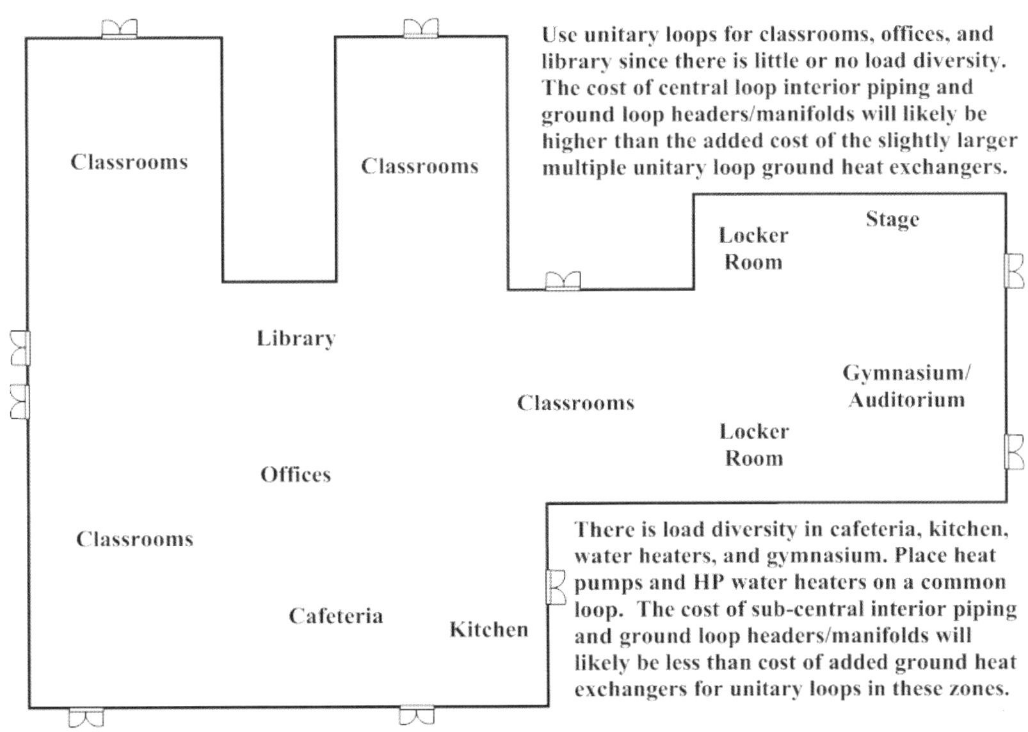

Figure 6.19 Single-Story Southeast Texas High School

be prudent to have these zones connected to unitary loops or to multiple common or one-pipe loops, one for each classroom wing and one for the office/library zones.

The cafeteria has a short but significant peak at midday, the kitchen has a high peak preceding and coincident with the cafeteria, and the gymnasium has a modest daytime load with a high peak in the evening. Also note that the kitchen and locker rooms have water heating requirements that could be satisfied or supplemented by heat pump water heaters, which extract heat from the ground loop in a climate with a high cooling load. Additionally, the peak load in the gymnasium occurs during basketball games (in the heating mode), when the kitchen and cafeteria are not occupied. Furthermore, the cafeteria, kitchen, and gymnasium are in the same area of the building and have convenient access to a potential ground-loop site. This portion of the building would be a nearly ideal candidate for a central loop. The diversity would result in a reduction in size of the ground loop. The cost of interior pipe headers would be modest since the zones are in close proximity. The heat pump capacities would be large, which would minimize the number of pipe take-offs and control valves.

6.9 GROUND-LOOP PIPING CIRCUITS

6.9.1 Ground-Loop Circuit Options

Figures 6.20 through 6.24 represent some of the more common options for ground-loop circuits. Figure 6.20 depicts the simplest unitary-loop headers, which are connected individually to a heat pump. The three- and four-U-tube circuits are direct return but are balanced by the fact that the U-tubes closest to the common take-off flow through a branch 90° elbow that has an equivalent length nearly equal to the straight runs to the more distant U-tubes. Thus, balance is attained without the reverse-return pipe.

The advantages of this option are simplicity, low cost of installation equipment, and the fact that the system can be completed reliable with less-experienced personnel. The disadvantage is the multitude of circuits that must be sustained (maintain pressure). This problem is manifest primarily for one or two years after start-up.

Figure 6.21 illustrates a 10-U-tube circuit for a very common modified reverse-return option. Flow through each U-tube is balanced by simply arranging the header in a circuitous route so that the reverse-return is very short. This eliminates the need for routing the reverse-return section for the entire length of the return header. As noted in the figure, it is critical to reduce the diameter of the main header because flow declines with each U-tube take off. If this is not done and the header diameter feeding the last U-tube take-off is equal to the diameter at the first take-off, the velocity of the liquid through the last U-tubes

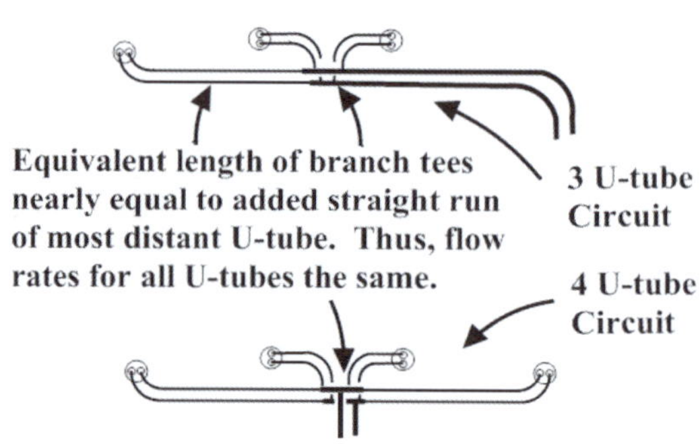

Figure 6.20 Unitary Ground-Loop Header

will likely be insufficient to remove air and construction debris during purging/flushing at start-up.

The advantage of this option is that it can accommodate a loop field with a large number of U-tubes. A disadvantage is that the circuits must be connected to manifolds with isolation valves for loops with greater than 15 to 20 U-tubes. Flow balancing is required between each circuit in most cases, but balancing each U-tube is unnecessary due to the reverse-return arrangement of the circuit. Start-up can be a challenge if the manifold for the circuits are not arranged to be individually purged through valves with diameters equal to or greater than the circuit header diameter.

Figure 6.21 indicates the elbows in the headers are long radius bends. For 2 in. nominal (60 mm) HDPE, the elbows are made by field-bending the tubing. For DR 11 and 13.5 the minimum bending radius (R_{bend}) is a function of the outside diameter (d_o) of the pipe (PP 2007):

$$R_{bend} = 25 \times d_o \tag{6.11}$$

Field-bending 3 and 4 in. (90 and 110 mm) pipe is difficult, and it is recommended that long sweep elbows be fabricated from 90° sections of coiled tubing (Elks 2005) rather than a more expensive, higher-head-loss molded fitting.

Also note that headers in Figure 6.21 would be 100 ft (30 m) in length for 20 ft (6 m) bore separation. Large-diameter tees with small-diameter take-offs for the U-tube are expensive and typically unavailable. Take-off fittings are made with side-saddle fusion, which requires a much higher level of skill and care compared to a butt or socket weld. These joints should not be made in the field, considering the poor conditions typical of loop installations even when the weather is favorable. Figure 6.22 shows a practice used to minimize side-saddle fusion joint failure. The take-off joints for the headers are made in a controlled indoor climate on sections of header pipe than can be easily shipped. More reliable butt fusion joints are made in the field to create the longer runs of headers.

Figure 6.23 shows a close header ground-loop arrangement with 10 U-tubes. Though this option is no longer popular, it remains a recommended option when the loop field is placed beneath pavement. Leaks can more easily be located, repaired, or isolated because

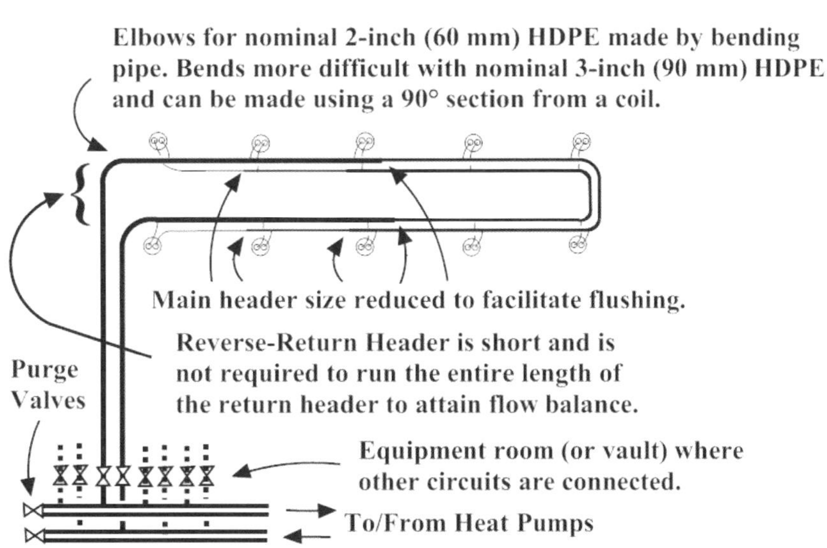

Figure 6.21 Modified Reverse-Return Ground-Loop Header

Figure 6.22 Ready-to-Ship Headers with Sidewall Take-Offs Fabricated in Controlled Conditions

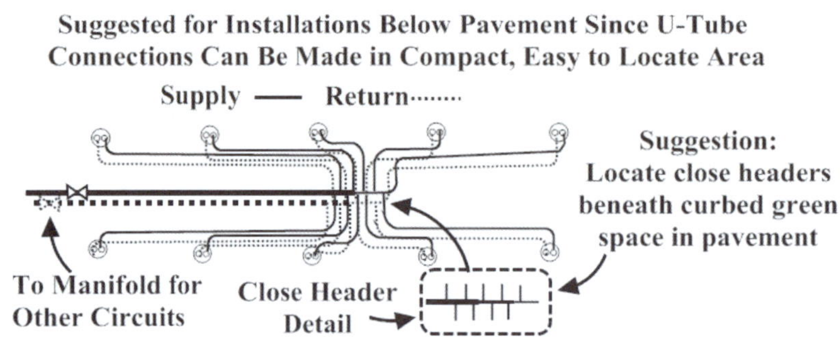

Figure 6.23 Close Headers for Ground Loops Beneath Pavement (Parking Lots)

the take-offs are in a small compact area and because the close headers are typically 4 to 8 ft (1.2 to 2.4 m) in length. It would be especially prudent to locate these headers in a curbed green space with shallow root vegetation.

The primary disadvantage of close headers is that with a large number of tubes in a confined area, care must be taken to avoid connecting U-tube supply (or return) headers together. A secondary disadvantage is the perceived need to have identical pipe lengths for each U-tube. This problem is overstated since the difference in overall length with deep bores results in minor flow imbalance, with even less imbalance in heat transfer. An example calculation is provided in Appendix I to demonstrate the needed level of concern.

Figure 6.24 depicts a standard reverse-return ground loop with three parallel circuits, each with six U-tubes. Note that the reverse-return header runs the entire length of the return header. The advantage of this arrangement is a natural balance of flow in both the individual U-tubes and the three circuits. Note that the modified reverse-return header shown in Figure 6.21 has balanced flow in the U-tubes on each circuit. However, flow

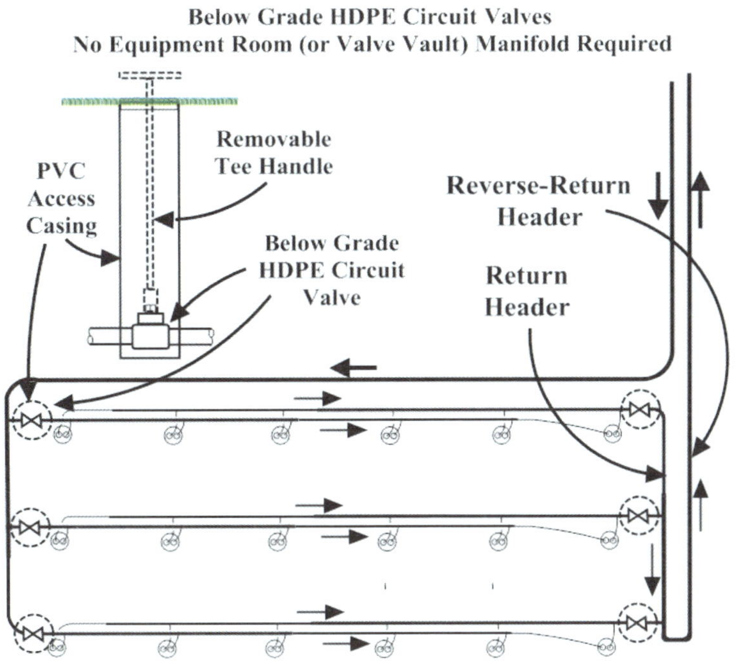

Figure 6.24 Standard Reverse-Return Ground-Loop Header with Below-Grade Circuit Valves

among circuits requires balancing, because the supply and return header lengths between the U-tubes and manifolds vary, especially if there are a large number of circuits. The disadvantage of the setup depicted in Figure 6.24 is that the reverse-return header will be longer, with increased head loss and pipe cost.

6.9.2 Manifold Options

The ground-loop design shown in Figure 6.24 has below-grade HDPE valves to isolate each circuit. This option eliminates the need for manifolds in equipment rooms or below-grade vaults and of course is not restricted to reverse-return ground loops. HDPE valves are available and are highly recommended to avoid corrosion issues, and they are connected by thermal fusion rather than with mechanical fasteners.

Figure 6.25 illustrates two equipment-room manifolds that are arranged in a manner to minimize the required floor space and provide a convenient location for purging the circuits. Figure 6.25a shows 12 parallel 2 in. (60 mm) circuits with 171 U-tubes connected to a total of 165 tons (580 kW) of water-to-air and water-to-water heat pumps. HDPE pipe is routed under the foundation and transitions to steel at the circuit isolation valves in the vertical sections shown in the figure. The building originally consisted of three stories, and the interior pipes for the eight circuits shown in Figure 6.25a were insulated to prevent condensation. Two additional floors were added later, and insulation was not used because water in the piping is operating as the condenser liquid in cooling. The installation is in a warm climate, and the water temperature never falls below the 60°F (16°C) indoor-air dew-point temperature in the winter when the liquid loop is operating as the evaporator liquid. Thus, insulation for condensation prevention was unnecessary and was likewise not used for the pipe for the added floors.

The equipment-room manifold shown in Figure 6.25b consists of nine parallel nominal 3 in. (90 mm) circuits with 144 U-tubes connected to a total of 380 tons (580 kW) of water-to-air heat pumps. The ground-loop circuit HDPE pipe for this system is also

(a) (b)

Figure 6.25 Two Equipment-Room Ground-Loop Circuit Manifolds

routed through the foundation and transitions to steel interior piping at the circuit isolation valves shown in red in the vertical sections of pipe. Both of the manifolds in Figure 6.25 take up approximately 10 ft^2 (1.0 m^2) of equipment floor area.

Figure 6.26 diagrams a below-grade valve vault, which is typically placed near large arrays of U-tubes and circuits (Kavanaugh 2009). HDPE vaults have replaced poured-in-place concrete vaults because they are less likely to fill with water. Vaults typically must be large enough to include manhole entry, lighting, and in many cases sump pumps. The circuits enter the vault through sealed connections and are tied to the main supply and return headers, which are routed to the building. Circuit flow balancing is done inside the vault. The purge valves should be routed so that connections can be made at the surface outside the vault, as shown in Figure 6.26. This eliminates the need to route the purge-pump hoses through the manhole and enhances worker safety while purging.

The primary advantage of valve-vault manifolds is that they eliminate the need to take up equipment-room space. There are several disadvantages, including cost, installation difficulty, need for electrical service, difficulty of flow balancing in a confined and inconvenient to access space, difficulty of purging if exterior connections are not available, and potential safety hazards that may result if workers are in a difficult-to-exit confined space into which a large volume of water is being pumped. The Occupational Safety and Health Administration, or the cognizant worker protection agency, would likely classify a vault as a "confined space." When this is the case, all personnel, whether entering or standing watch at the surface, must be trained and certified. All employees required to enter into confined or enclosed spaces must be instructed as to the nature of any hazards involved, about the necessary precautions to be taken, and in the use of protective and emergency equipment required (OSHA 1996).

Architects and engineers are strongly encouraged to consider the cost premium of valve vaults compared to below-grade HDPE circuit valves (Figure 6.24) or equipment-room manifolds that take up only minimal floor area if installed as shown in Figure 6.25.

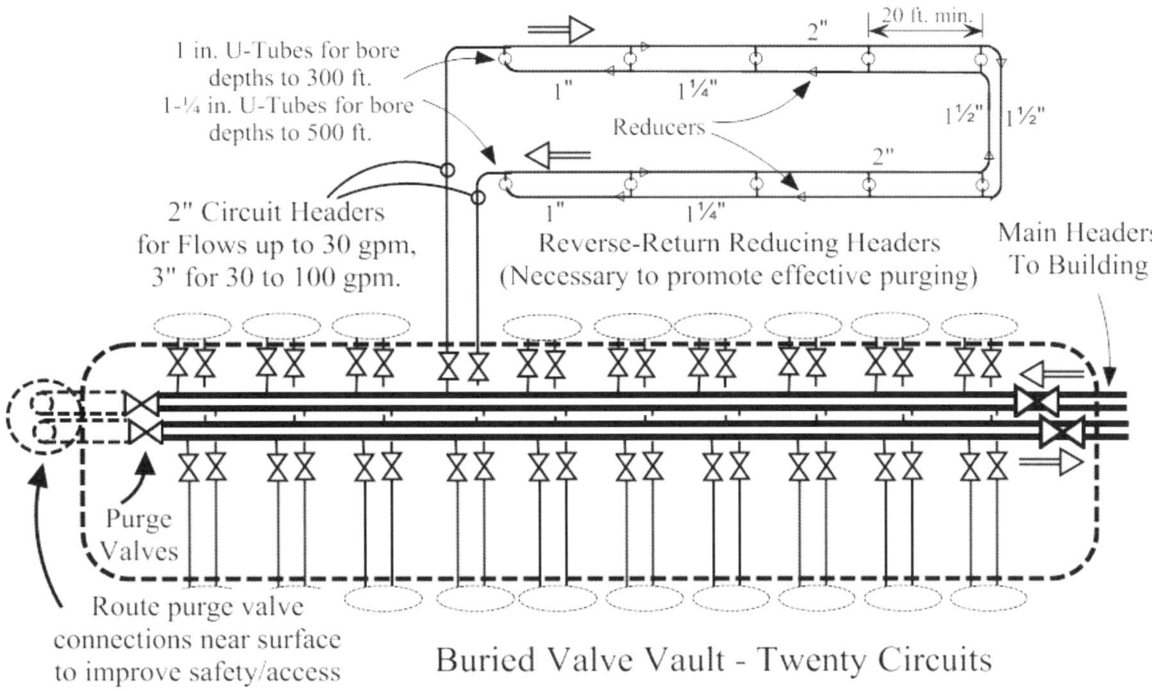

Figure 6.26 Below-Grade Valve Vault with 20 Circuits and 200 U-Tubes

The economic evaluation should compare the cost of running multiple 2 or 3 in. (60 or 90 mm) circuit headers to equipment rooms to the cost of installing a single set of larger supply and return headers between the vault and the equipment room. The cost should include the fact that 2 and 3 in. (60 and 90 mm) headers can be provided in coils and installed with devices that straighten the coils (see Figure 6.27) so that only two fusion joints are required at either end of the header. This reduces installation cost and the likelihood of poor welds. Header pipes larger than 6 in. (170 mm) must be thermally fused every 20 or 40 ft (6 or 12 m). Chapter 9 provides an example cost comparison for an HDPE below-grade manifold valve vault with an equipment-room manifold similar to those in Figure 6.25.

With all valve vaults, some degree of flooding is likely, and the relative humidity is normally near 100%. In these conditions, sweating of components will cause corrosion to any susceptible components. It is suggested that architects and engineers spend some time in a valve vault that has been in service for several years to observe the poor working environment that typically evolves.

For horizontal headers the suggested header burial depth is 4 ft (1.2 m) below grade. In warm climates 3 ft (1 m) is sufficient in terms of thermal performance, but consideration should also be given to protection from potential damage from landscaping or other potential excavation activities. One concern with on-off pump control is the possibility of low-temperature liquid entering the heat pumps at start-up. This can occur if headers are located at shallow depths, the pumps are off, and the stagnant water approaches the shallow ground temperature. This may occasionally cause low liquid temperature trip-outs.

6.9.3 Ground-Loop Purging (Flushing) and Balancing

Adequate purging of air and debris is a critical component of system start-up. The traditional rule of thumb developed by the industry for residential allocations called for a

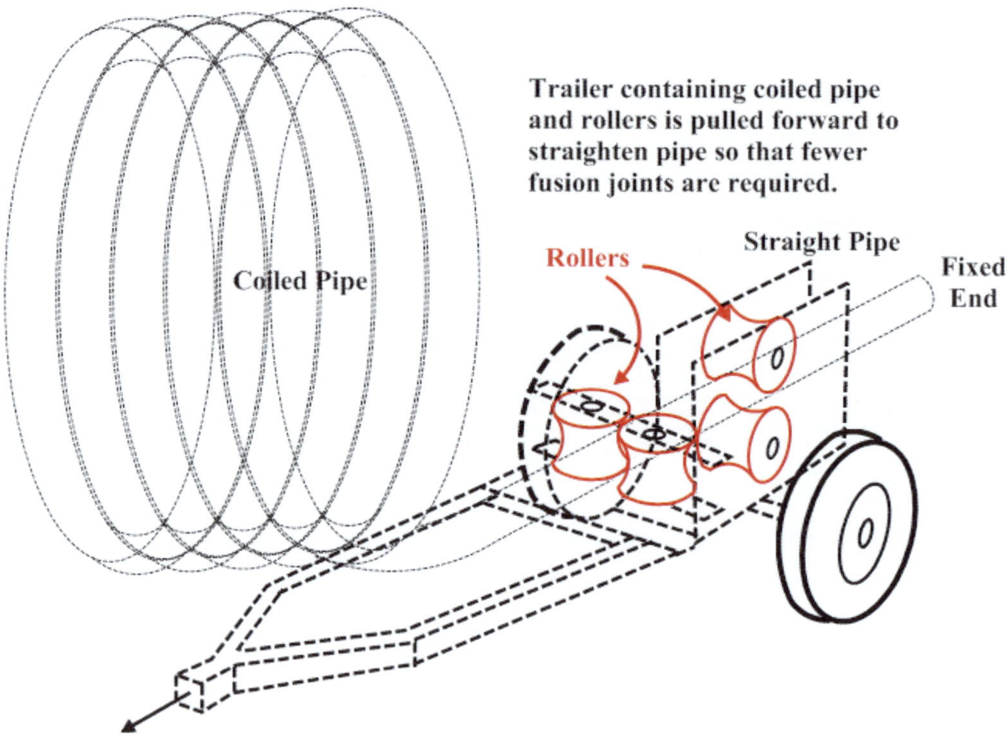

Figure 6.27 Rig to Straighten ("Tame") Coiled HDPE

purge velocity of 2 ft/s (0.6 m/s) if flow can be reversed through the system. Of course, this cannot be performed if check valves or automatic flow control valves are installed in the system. Proponents of more thorough procedures have suggested that for larger systems in which the flow cannot be reversed during purging, a velocity of 6 ft/s (1.8 m/s) may be required in some applications (PR 2014). This issue has not been adequately investigated, but it is suggested either that check valves be omitted or that bypass valves be installed in parallel with the check valves. This allows circuit balancing to be done with balancing valves that permit bidirectional flow.

Figure 6.26 displays the locations of purge valves for a valve-vault manifold. The arrangement for an equipment-room manifold would be similar to that shown in Figure 6.11. Three-way valves are typically used for unitary-loop systems, as shown in Figure 6.13.

Until independent research is conducted on this issue, the rule of thumb for purge valve sizing is that the valves be no smaller than the circuit-loop header diameters and no smaller than one-half the diameter of the main header of a central-loop system. For example, if the main header diameters shown in Figure 6.26 are 8 in. (200 mm) and the circuit header diameters are 3 in. (80 mm), the purge valve diameters should be 4 in. (100 mm).

Figures 6.28 and 6.29 display purge pumps for smaller GSHP loops with manifolds that permit reversing flow without disconnecting hose connections, which would reintroduce air into the system. Figure 6.30 demonstrates the amount of debris that can remain in a poorly managed loop field installation. Figure 6.31 shows a large trailer-mounted purge pump that may be required for very large jobs or medium-sized jobs without adequate isolation valves on circuits.

Figure 6.28 Purge Pump for 10 to 25 ton (35 to 90 kW) Circuits

Figure 6.29 Portable Truck-Mount Purge Pump for 10 to 25 ton (35 to 90 kW) Circuits

Figure 6.30 Debris Removed with Purge Pump on 300 ton (1050 kW) Ground Loop

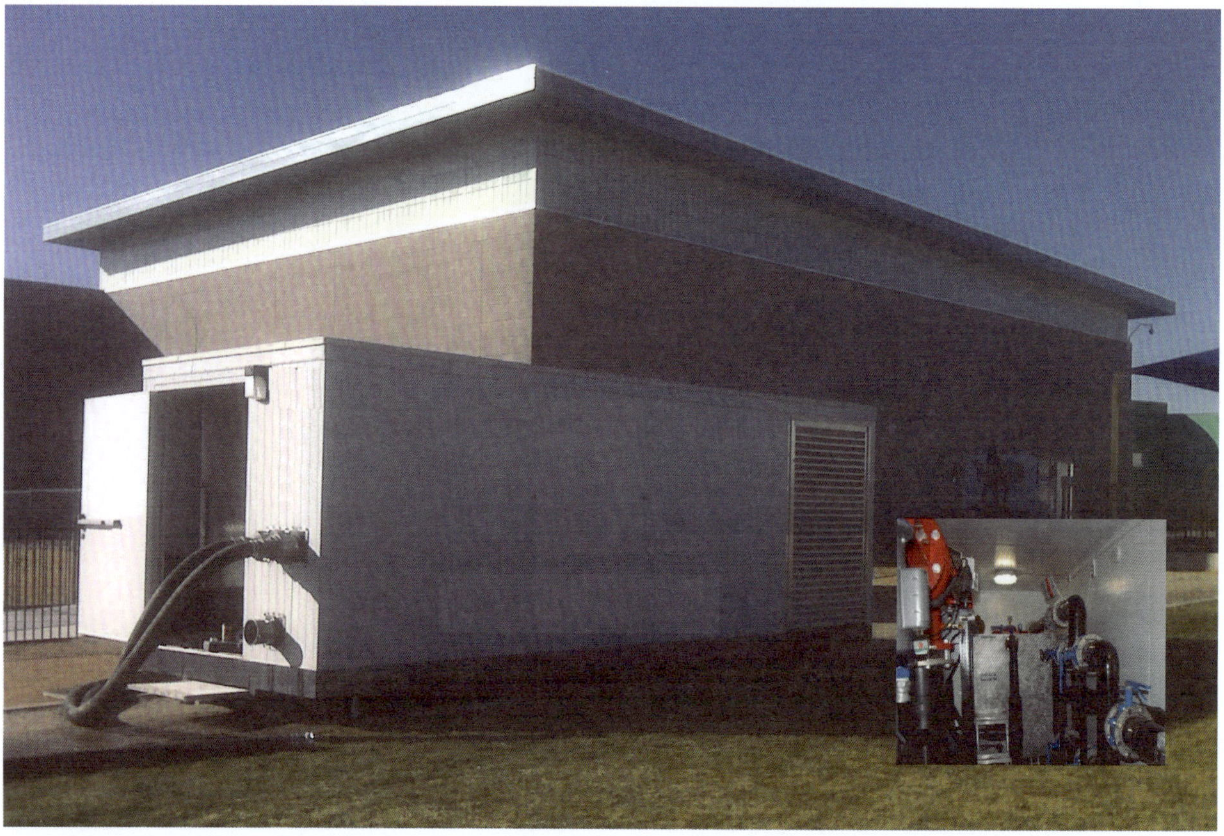

Figure 6.31 Skid-Mounted Purge Pump for Flushing Ground Loops without Circuits

6.10 SUMMARY OF PIPING AND PUMP DESIGN GUIDELINES

Recommendations for optimized piping and pump design in closed-loop GSHP systems follow:

- Use a minimum of 1 in. nominal (32 mm) U-tubes in bores up to 300 ft (90 m) in depth and 1 1/4 in. (40 mm) U-tubes in bores up to 500 ft (150 m) in depth. Avoid the use of 3/4 in. nominal (25 mm) U-tubes in bores greater than 200 ft (60 m) in depth.
- Minimize header losses to no greater than 3 ft of water per 100 ft of tubing (300 Pa/m).
- Limit closed-loop liquid flow rates to 3 gpm/ton (3.2 L/min·kW) of building block load or less. An exception is open-loop systems with high elevation heads that are typically optimized at lower flow rates, as discussed in Chapter 8.
- Specify heat pumps with head losses no greater than 12 ft of water at 3 gpm/ton (35 kPa at 3.2 L/min·kW).
- Avoid the use of circulator pumps with pump-motor efficiencies (a.k.a. wire-to-water efficiency) less than 30% for systems with head losses greater than 30 ft of water (90 kPa). (A single higher-efficiency pump is recommended rather than piping two low-efficiency circulators in series.)
- When heat pump flow balancing devices are necessary, limit head losses to no greater than 5 ft of water (15 kPa). Recall that advances in refrigerant control devices result in water-to-air and water-to-water heat pumps that are effective over a broader range of water flow rates than older equipment. Thus, precise balancing of equipment with high head-loss flow restriction devices is unnecessary.
- When using hose kits or field-fabricated hose connections, limit combined head losses to no greater than 3 ft of water (9 kPa). (For longer hoses this may require limiting losses to no greater than 3 ft of water per 100 ft of hose [300 Pa/m].)
- Install straight sections of piping near the pump inlet (especially) and discharge ports. Use suction diffusers if elbows near pump inlets are unavoidable.
- Purge-port valve diameters should be no smaller than the circuit-loop header diameters and no smaller than one-half the diameter of the main header, whichever is greater.
- The maximum number of U-tubes per circuit should be limited to 15 to ensure successful purging. Twenty U-tubes per circuit have been installed and proven possible to purge, provided flow can be reversed during purging.
- Recognize that ground exchangers have high thermal resistance (plastic pipe buried in dirt) compared to compact heat exchangers and that increasing design flow rate to affect fully turbulent flow will result in higher head losses and pumping power with minimal improvement in heat exchange.
- Recognize that laminar flow in the ground heat exchanger at low part load will have little impact on performance (Δt across the laminar boundary layer will be small because the heat rate is small) and that increasing design flow rate to affect nonlaminar flow at low part load is unnecessary and will result in higher head losses and pumping power with minimal improvement in heat exchange. *(Note: For closed-loop SWHP coils, laminar-flow coils should be avoided except at part-load operation. The thermal resistance of the interior boundary layer is typically a larger percentage of overall resistance than in ground-loop applica-*

tions. Therefore, the impact of laminar flow should be carefully considered. Calculations and design tools to address this issue are presented in Chapter 5.)

- Avoid the use of excessive amounts of antifreeze solutions, because antifreeze is costly and the increased viscosity increases pump sizes and drives up pumping energy.
- If antifreeze solutions are required but cooling is the critical design condition, perform the piping design and pump selection based on the fluid properties (i.e., viscosity) at the cooling mode liquid temperature rather than using the higher viscosity conditions at the lower heating-mode temperatures.
- Select pumps to operate near their best efficiency point (BEP).

6.11 REFERENCES

ASHRAE. 2003. *Development of guidelines for the selection and design of the pumping/piping subsystem for ground-coupled heat pump systems*. ASHRAE RP-1217 Final Report. Atlanta: ASHRAE.

ASHRAE. 2013. *ASHRAE Handbook—Fundamentals*, Pipe Sizing, p. 22.1. Atlanta: ASHRAE.

Carlson, S. 2001. *Development of equivalent full load heating and cooling hours for GCHPs applied to various building types and locations*. ASHRAE RP-1120, Final Report. Atlanta: ASHRAE.

Churchill, S.W. 1977. Friction factors equation spans all flow regimes. *Chemical Engineering* 84(24):91–92.

Elks, C. 2005. Employee at Mechanical Equipment Sales, Virginia Beach, VA. Personal communication with author.

IGSHPA. 2009. *Closed Loop/Geothermal Heat Pump Systems: Design and Installation Standards*. Stillwater, OK: International Ground Source Heat Pump Association. www.igshpa.okstate.edu/pdf_files/Standards2009s.pdf

Kavanaugh, S.P. 2006. *HVAC Simplified*. Atlanta: ASHRAE.

Kavanaugh, S.P. 2009. GSHPs: Simple is better. *ASHRAE Journal* 51(11).

Kavanaugh, S.P., and J.S. Kavanaugh. 2012. Long-term commercial GSHP performance, part 3: Ground loop temperatures. *ASHRAE Journal* 54(9).

Kavanaugh, S.P., and K. Rafferty. 1997. *Ground-Source Heat Pumps: Design of Geothermal Systems for Commercial and Institutional Buildings*. Atlanta: ASHRAE.

RSMeans. 2014. *RSMeans Mechanical Cost Data*. Norwell, MA: Reed Construction Data.

Moody, L.F. 1944. Friction factors for pipe flow. *ASME Transactions* 66:671–84.

NEMA. 2009. ANSI/NEMA MG-1-2009, *Motors and Generators*. Rosslyn, VA: National Electrical Manufacturers Association.

NFPA. 2015. NFPA 90A, *Standard for the Installation of Air-Conditioning and Ventilating Systems*. Quincy, MA: National Fire Protection Association.

OSHA. 1996. *Confined spaces*. Construction Safety and Health Outreach Program. Washington, DC: U.S. Department of Labor, Occupational Safety and Health Administration. www.osha.gov/doc/outreachtraining/htmlfiles/cspace.html

PP. 2007. *Field bending of DriscoPlex® pipe*. Technical Note PP 819-TN. Plano, TX: Performance Pipe.

PR. 2014. *Why Purge Rite?* New Waverly, TX: Purge Rite. www.purgerite.com/why.html

Taco. 2012. *Design/commissioning tips for variable speed pumping systems*. Cranston, RI: Taco, Inc.

7 Hydrology, Water Wells, and Site Evaluation

7.1 GROUNDWATER HYDROLOGY

There are many subsurface issues of common interest regardless of the system type eventually selected for a project (see Section 7.5). The presence or absence of an aquifer, aquifer type, static water level, geology, undisturbed ground temperature (or aquifer water temperature), and rig types that have worked successfully in the area are some of the issues influencing both ground-coupled heat pump (GCHP) and groundwater heat pump (GWHP) design. Though the specifics of water well design are unique to GWHP systems, many other aspects discussed in this chapter are valuable to those involved in the design of any type of GSHP system. Of particular value to both GCHP and GWHP designers are the discussions of basic hydrology and aquifer flow direction (Section 7.1) and site evaluation (Section 7.5), particularly the portion relating to interpreting water well completion reports (Section 7.5.1). Water well completion reports contain a wealth of information beneficial to GCHP design as well as GWHP designs. Additional detail on subsurface issues related to GSHP design is provided by Sachs (2002).

Production wells for access to groundwater and injection wells for returning the water to the aquifer are critical components in a GWHP system. For a successful, efficient, and cost-effective system, the engineer must be closely involved in the design of the water wells, well pumps, and associated controls. In many cases, and certainly in the most complex settings, the engineer will be working with a specialist in water well design, typically a geohydrologist, geologist, or civil engineer. While others may be responsible for the specifics of the well design, at the initial phase of the project the engineer must provide an estimate of the groundwater flow requirements in order for the well specialists to perform their job effectively. At a later stage of the project, when well flow testing is complete, data will be available to refine the design of the system to reflect actual well performance. For the engineer to participate effectively in this process, he or she must be conversant in water well terminology and basic groundwater hydrology. The goal of this section is to provide that level of background. The information in this book is not intended to provide a comprehensive treatment of water well design; this is widely available in other references (Driscoll 1986; National Water Well Association 1981; AWWA 1997; RMC 1985; BR 1995; NGWA 2014).

Precipitation falling on the surface of the earth can follow a number of pathways—it can run off directly to surface water bodies (creeks, rivers, lakes, etc.), it can evaporate into the air, or it can be absorbed into the subsurface. Water absorbed descends vertically through shallow materials, known as the *zone of aeration*, and eventually reaches what hydrologists refer to as the *zone of saturation* (Figure 7.1). Aquifers do not exist continuously in the zone of saturation, but they can exist provided certain conditions are met. For a saturated formation to be considered an aquifer, it must be characterized by passageways (pore spaces in and between the geological materials) that provide both a path through which water can flow and a volume in which water can be stored. In addition, the body must be capable of producing sufficient quantities of water to cause it to be a target for production.

Aquifers can be characterized in a number of ways, but two broad categories are confined (sometimes referred to as *artesian aquifers*) and unconfined (sometimes referred to as *water table aquifers*). When the drill rig penetrates a confined aquifer, the water level in the well bore rises above the depth where the water is first encountered. The new, higher water level is reflective of what is termed the *piezometric level* of the aquifer. This is a result of the fact that confined aquifers are under a pressure exceeding atmospheric pressure. The pressure in the aquifer is the result of it being overlain by a formation impermeable to water movement, often clay or similarly fine-grained materials. When the top of an unconfined aquifer is penetrated by the drilling operation, the water level in the well bore remains at the level at which it is initially encountered. In short, confined aquifers can be thought of as pressurized and unconfined aquifers as unpressurized. Another important issue that distinguishes confined and unconfined aquifers is how they respond to pumping of wells completed in them, which is a topic covered in more detail in Section 7.2.

Aquifers are often recharged by precipitation; this input serves to replace water withdrawn by artificial means (wells) and by natural discharge to rivers, lakes, or other aquifers. The distance between areas of recharge and areas of discharge, and thus the areal extent of aquifers, can be great, in some cases covering parts of several adjacent states.

Water present in aquifers is not a static "underground lake," but it is flowing. Flow is the result of a natural hydraulic gradient in the aquifer, with water flowing "downhill" just

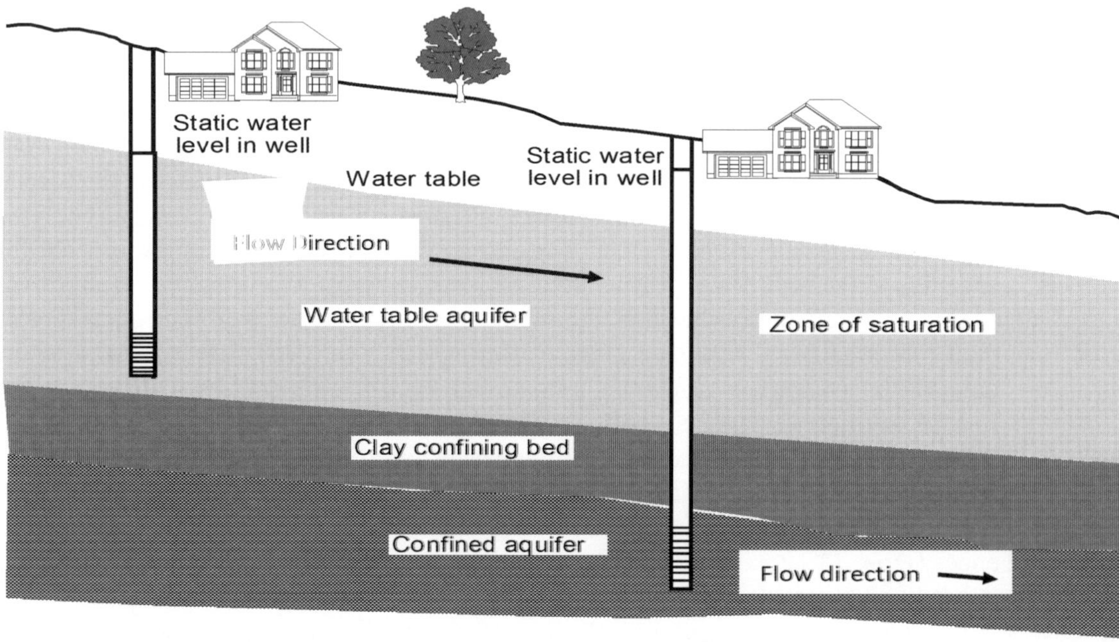

Figure 7.1 Aquifer Types—Confined (Water Table) and Unconfined (Artesian)

as it does in surface bodies, though the direction of aquifer flow may not always reflect ground surface topography. The velocity in the aquifer is a function of the available gradient and the permeability of the aquifer materials. Permeability (hydraulic conductivity), with units of gal/ft^2·day (m/day), is a measure of the quantity of water that will pass through one square foot (one square metre) of the material in one day under a gradient of 100% (a 1 ft [m] change in aquifer water level per ft [m] of horizontal distance). *Permeability* is a term associated with a specific, uniform material, and values vary widely in geological materials. Some typical values appear in Table 7.1.

Groundwater aquifer gradients are often expressed as a percentage, in a fashion similar to surface grades. For example, a difference in water level of 3 ft (0.9 m) at two points 300 ft (90 m) apart constitutes a gradient of (3/300) × 100 = 1%, or 0.01 ft/ft ([0.9/90] × 100 = 1%, or 0.01 m/m). Aquifer gradients rarely exceed 3%.

Water flow velocity can be determined by multiplying the permeability by the hydraulic gradient, in consistent units. For example, a body of medium sand (see Appendix J for grain size description) is under a hydraulic gradient of 1.5%. The velocity through the sand is

$$\text{Velocity} = P \times C \times G \qquad \text{(I-P)} \qquad (7.1a)$$

$$\text{Velocity} = P \times G \qquad \text{(SI)} \qquad (7.1b)$$

where

P = permeability, gal/ft^2·day (m/day)
G = gradient, ft/ft (m/m)
C = 0.134 ft^3/gal

For medium sand:

$$\text{Permeability} = 100 \text{ gal/day ft}^2 \qquad \text{(I-P)}$$

$$\text{Velocity} = 100 \text{ gal/day ft}^2 \times 0.134 \text{ ft}^3\text{/gal} \times 0.015 \text{ ft/ft} = 0.201 \text{ ft/day} \qquad \text{(I-P)}$$

$$\text{Permeability} = 4.1 \text{ m/day} \qquad \text{(SI)}$$

$$\text{Velocity} = 4.1 \text{ m/day} \times 0.015 \text{ m/m} = 0.062 \text{ m/day} \qquad \text{(SI)}$$

Table 7.1 Mean Permeability Values

Material	Permeability, gal/ft^2·day	Permeability, m/day
Medium gravel	10,000	400
Coarse sand	150	60
Medium sand	100	40
Silt	0.1	0.04
Shale	0.00001	0.000004
Unfractured hard rock	0.000001	0.0000004
Well-cemented sandstone	0.001	0.0004
Tuff	0.1	0.004
Friable sandstone	1.0	0.04
Fractured igneous rock	1.0	0.04
Vesicular basalt	10	0.4
Karst limestone	100	4

Note: Due the variation in materials and size ranges, permeability values can vary over a range of ±100% of the values appearing in this table.

The subsurface is not typically composed of a single, uniform material such as fine sand or coarse gravel but of a mixture of material types and sizes, and as a result permeability of homogeneous materials has limited use in practical applications. In much the same way that thermal conductivity is best determined through a test of a completed borehole, water-flow parameters in the subsurface are best determined through a test of a completed well on the site. The details of well testing are covered in Section 7.5.2, but Figure 7.2 illustrates the relationship between permeability and another item of importance—transmissivity. While *permeability* is a term more appropriate to laboratory testing of a uniform, specific material, *transmissivity* is a term reflecting the performance of an actual aquifer consisting of a mixture of materials; it is derived from analysis of the results of a well flow test. Beyond this, there is an important difference between the units of permeability and transmissivity. Permeability is a measure of the flow of water through a one square foot (square metre) cross section of material. Transmissivity is a measure of the flow through a 1 ft (1 m) wide cross section of the full thickness of the aquifer (with aquifer thickness measured in the vertical direction). The units of transmissivity are gal/ ft^2·day (m^2/day). With a known transmissivity and the storage coefficient, an index determined from a flow test, it is possible to make calculations of the impact of pumping over time and at various distances from the producing well.

Though very slow, aquifer water movement is sufficient enough to pose an important issue with respect to both open- and closed-loop heat pump applications. Because the water injected after use in an open-loop system is a few degrees warmer (in the cooling mode) or cooler (in the heating mode) than the undisturbed temperature of the aquifer

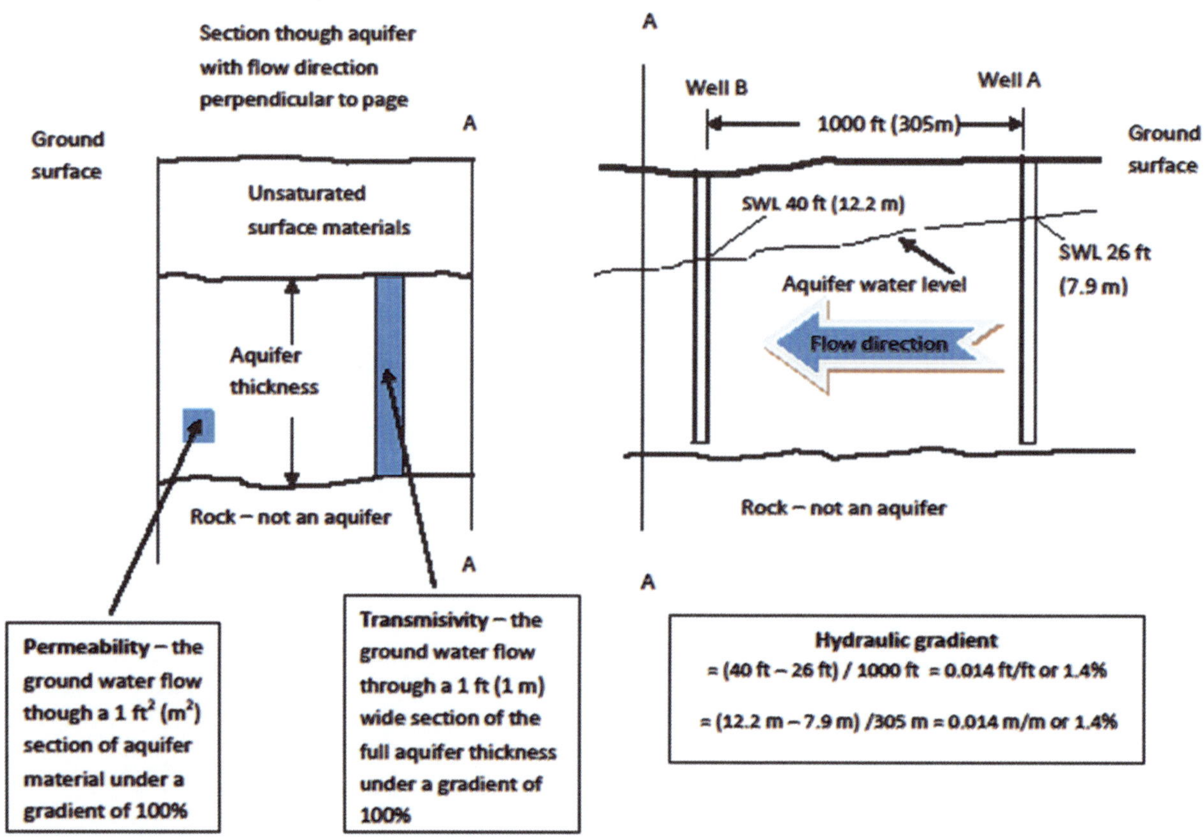

Figure 7.2 Transmissivity, Permeability, and Hydraulic Gradient

itself, there are implications for the relative placements of the production and injection wells. The injection well should always be placed down gradient, that is to say "downstream" in the context of the aquifer flow direction, from the production well. In this way, the natural flow of the aquifer helps to carry away the injected water and reduce the potential for it to migrate toward the production well. In the case of closed-loop systems that penetrate an aquifer, it is useful to orient the borefield so as to have the long dimension of the field perpendicular to the aquifer flow direction. This minimizes the number of bores potentially compromised by the impact of aquifer water thermally influenced by "upstream" boreholes.

In some cases, the aquifer flow direction has already been determined by others and this information may be discovered in the course of site evaluation research. In the event flow direction is not known, it can be determined by measuring water levels in at least three nearby wells penetrating the aquifer of interest. Figure 7.3 provides an illustration of the process. The static water level is measured in each well and converted to an elevation using the casing top elevations.

As indicated in Figure 7.3, once the elevations of the water in the three wells are established, lines can be drawn connecting the wells and then graduated in depth increments. Lines of constant groundwater elevation (dotted) can be drawn to intersect the calibrated lines connecting the wells. Groundwater flow direction is perpendicular to the lines of constant groundwater elevation. For this particular case, the production well should be located toward the upper end of the site and the injection well toward the lower end. The method described here must also consider the extent to which possible aquifer issues (aquifer thickness variation, presence of recharge areas, variation in aquifer materials, aquifer boundaries) may impact the water levels in the test wells. Details of the determination of the necessary distance between the production and injection well are covered in the next chapter.

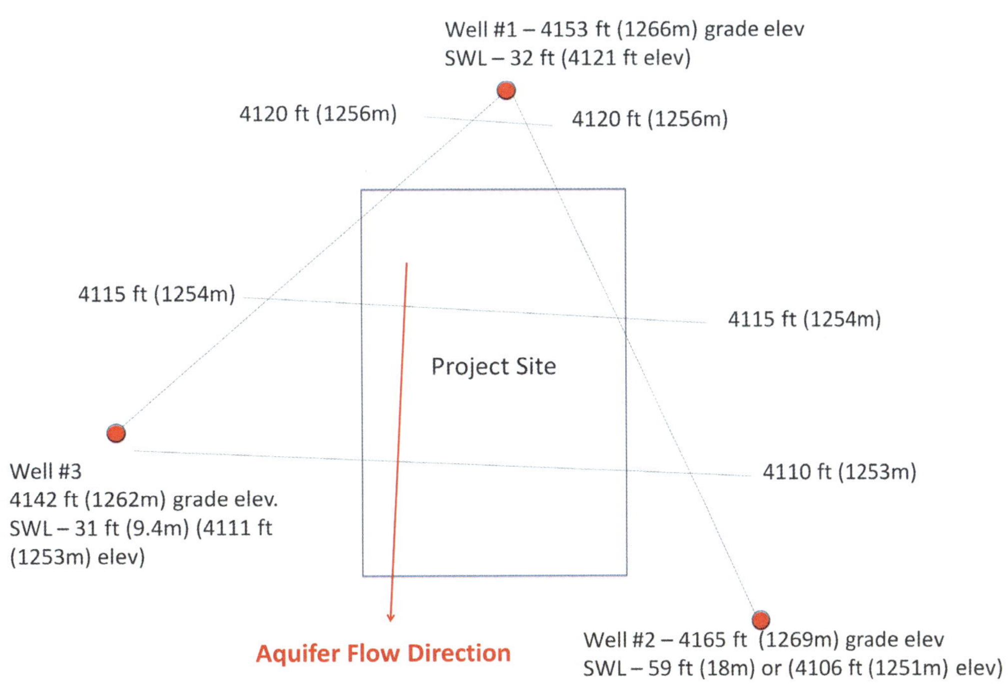

Figure 7.3 Method for Determination of Groundwater Flow Direction

7.2 WATER WELL TERMINOLOGY

Figure 7.4 shows some important terminology relating to production water wells. Static water level (SWL) is the level at which water resides in the well under nonpumping conditions and is typically measured from the ground surface (or casing top) to the water level in the well. It is reflective of the elevation of the water table in an unconfined aquifer or of the piezometric level (SWL in a well penetrating a confined aquifer). When the pump is started and water is removed from the well, there will be a drop in the water level to a lower elevation referred to as the *pumping water level* (PWL). The PWL is a function of the rate of water removal (pumping rate in gpm [L/s]), with higher pumping rates resulting in lower pumping levels. Pumping level, to be meaningful, must always be associated with a pumping rate (e.g., a 68 ft pumping level at 240 gpm [a 21 m pumping level at 15.1 L/s]) and, like SWL, is measured from ground level to the water surface in the well. The difference between the SWL and the PWL is known as *drawdown* (DD). Draw-

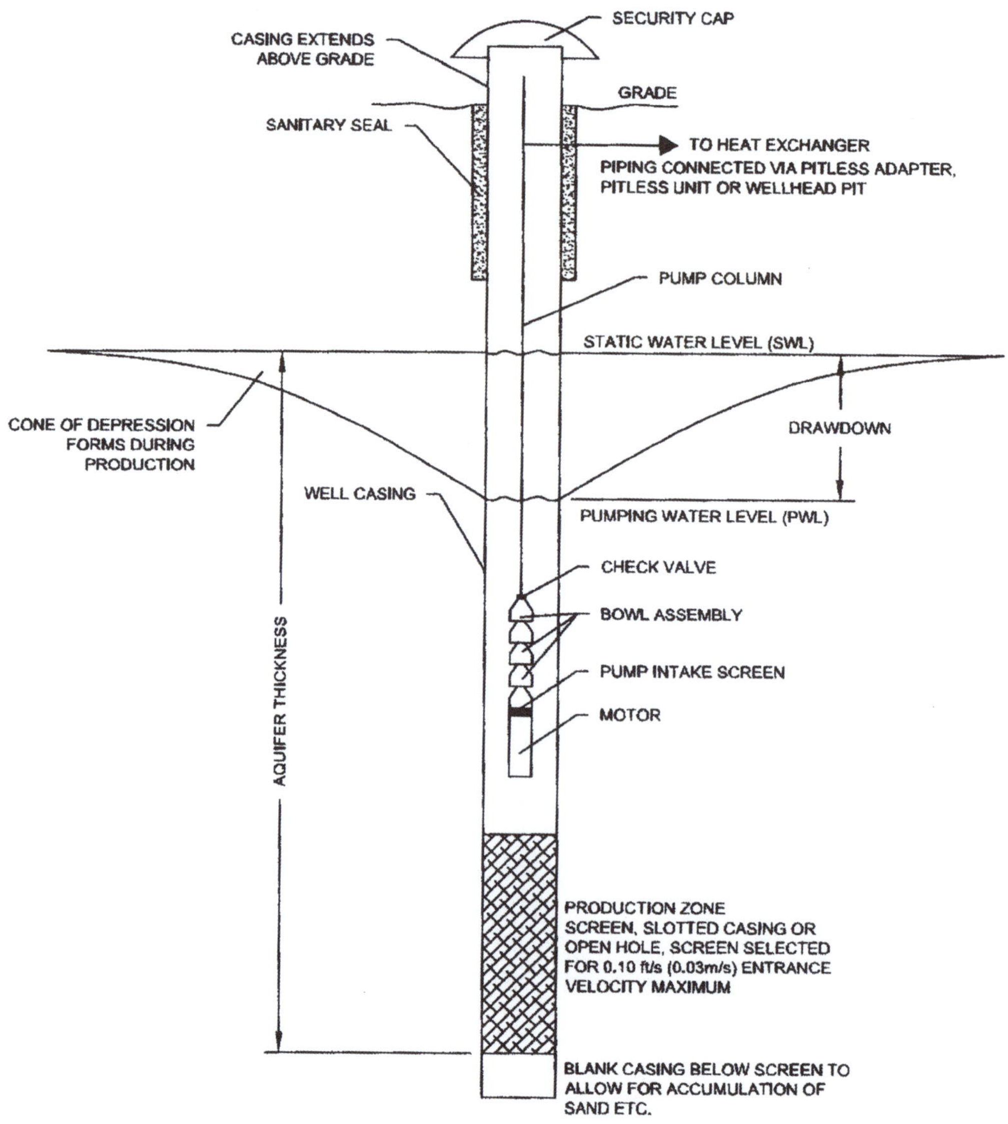

Figure 7.4 Production-Well Terminology

down, like PWL, is always associated with a pumping rate—for example, 20 ft DD at 100 gpm (6.1 m at 6.3 L/s). Specific capacity (SC) is an index of the well's ability to deliver water and is calculated by dividing pumping rate by DD. For example, a well that produces 100 gpm at a DD of 20 ft (6.1 m at 6.3 L/s) would have a SC of 100 gpm/20 ft = 5 gpm/ft (6.3 L/s/6.1 m = 1.03 L/s·m). In wells completed in confined aquifers, specific capacity is a relatively stable value over a wide range of flows (see Figure 7.5), provided the well is not drawn down below the top of the aquifer. In unconfined aquifers the specific capacity value tends to decline with increasing flow. This reaction, in an unconfined aquifer, is a result of the water passing through a smaller and smaller portion of the aquifer thickness (due to drawdown) as flow increases. The decreasing flow area results in increasing velocity and higher pressure drop. In confined aquifers, the entire aquifer thickness remains available because the drawdown occurs in the region above the aquifer.

In production-well pump head calculations, the static head (referred to as *lift* in well-pump jargon) is the sum of SWL plus DD. Thus, SC is a critical value in the context of calculation of production-well pump power requirements over a range of water flows—an issue that figures prominently in GWHP design (see Chapter 8).

Drawdown is the manifestation, at the well, of a "cone of depression" that forms around a well under pumping conditions. To cause water to flow through the aquifer toward the well, it is necessary to create an artificial pressure gradient in the aquifer. The cone reflects the pressure gradient in the zone around the well, and its shape is a function of the permeability (which is governed by the nature and size of the aquifer materials) and the manner in which the flow approaches the well. As water is drawn toward the well at a distance of, say, 50 ft (15 m), it can be thought of as passing through an imaginary cylinder 100 ft (30 m) in diameter with the well at its center. With an aquifer thickness of 30 ft (9 m), this cylinder would have a face area, the area through which the water is passing, of approximately 9400 ft^2 (873 m^2). At 10 ft (3 m) from the well, the imaginary cylinder would have a face area of 940 ft^2 (87 m^2). At 1 ft (0.3 m) from the well, the available area would be reduced to 94 ft^2 (8.7 m^2). It is apparent that with a constant flow the velocity of the water increases substantially as it approaches the well. The increase in velocity is accompanied by an increase in pressure drop as the water flows through the aquifer materials approaching the well. It is the increase in pressure drop that creates the shape of the

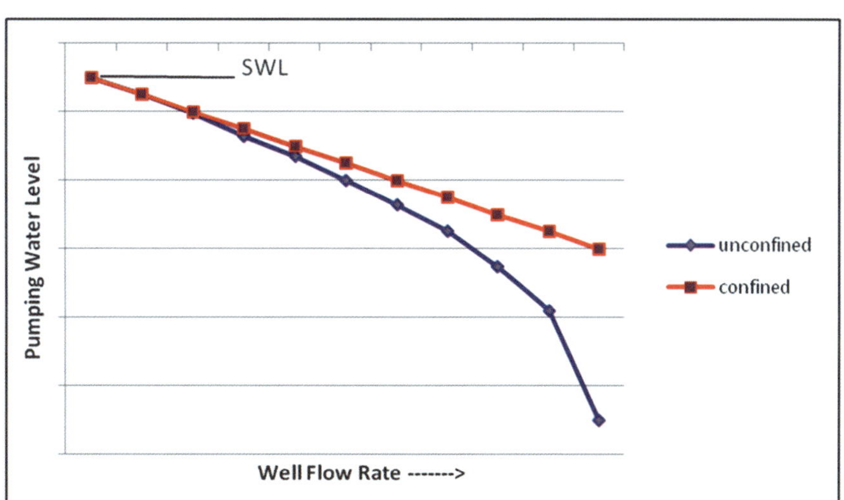

Figure 7.5 Confined and Unconfined Well Responses to Pumping

cone of depression. The high-velocity region in the near-well zone is analogous to the critical heat-transfer zone in the near-bore region of a closed-loop borehole. In some aquifers composed of particularly fine materials it is necessary to place a high-permeability material in this near-well zone to allow for reduced pressure drop and more efficient well operation. Placing high-permeability material in the near-bore zone is known as *gravel packing*, and its function, in a hydraulic sense, is very similar to the heat transfer function of high-conductivity grout in a closed-loop borehole.

The cone of depression extends away from the well for a distance determined by the nature of the aquifer materials, the production rate, and other factors. *Radius of influence* is the term applied to the distance from the well that a measurable drawdown exists. In general, aquifers characterized by high transmissivity result in cones of depression that are shallow and broad, producing a radius of influence greater than that of aquifers of low transmissivity, in which cones of depression are deep and narrow. If the cones of depression (or injection) of two wells intersect, the drawdown from one well is superimposed on the other.

Figure 7.6 illustrates some key terminology associated with injection wells. Static water level (SWL) is, as in production wells, the level at which the water resides in the

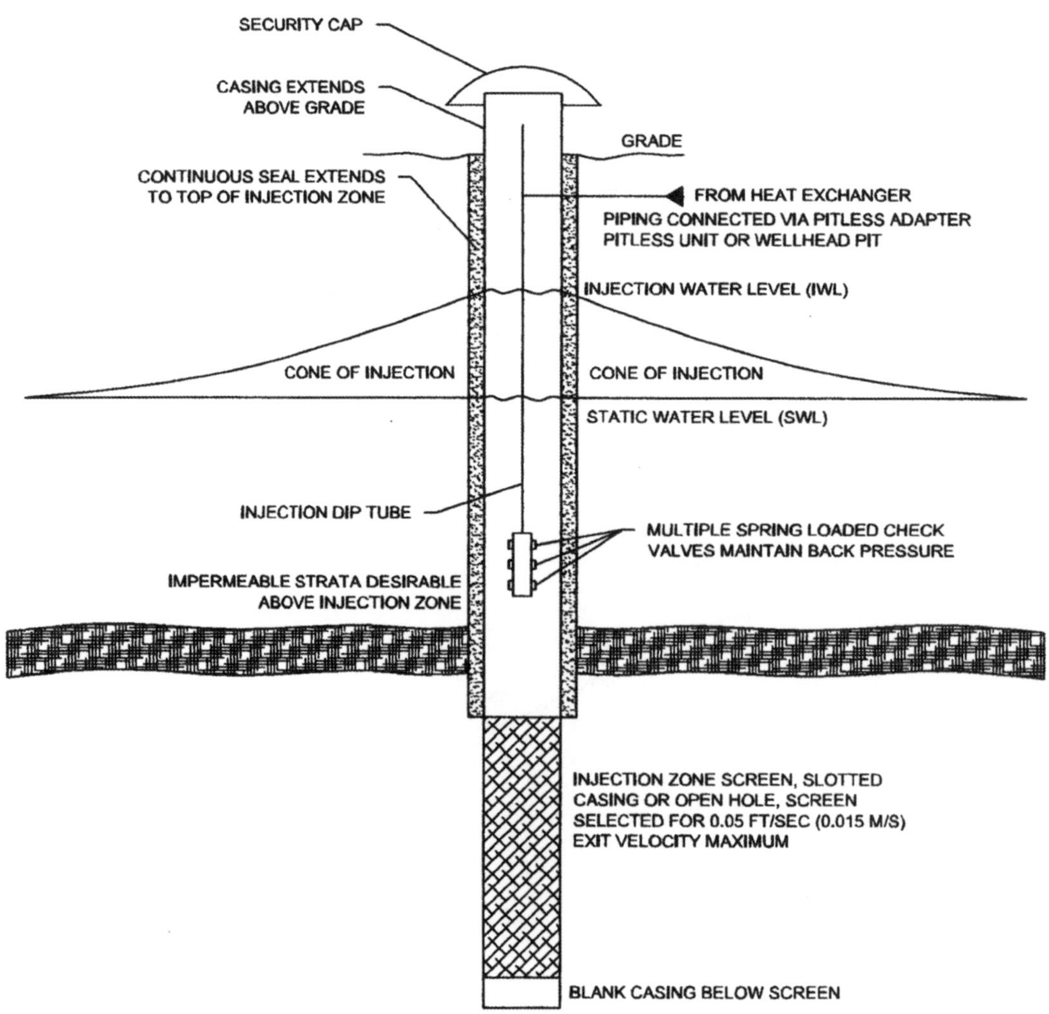

Figure 7.6 Injection-Well Terminology

well under no-flow conditions; it is measured in the same way as in production wells. When water is flowing into the well, the water level rises to a new elevation known as the *injection water level* (IWL). The IWL is measured from the ground surface to the water level in the well and is always associated with an injection rate (e.g., 15 ft at 230 gpm [4.6 m at 14.5 L/s]). Injection water level is the manifestation, at the well, of the cone of injection that forms around the well under injection conditions. In theory, for an injection well completed in the same aquifer as the production well (the usual case), the cone of injection in the injection well will be a mirror image of the cone of depression in the production well, assuming equal flows. Because the aquifer materials constitute the resistance to flow, it is logical that the pressure drop necessary to cause water to flow out of the aquifer at the production well should be the same as the pressure drop necessary to cause the same flow to reenter the aquifer at the injection well. In reality, injection wells often experience a somewhat greater cone of injection than the production cone of depression—a topic discussed in Section 7.4.6.

The difference between the SWL and the IWL is referred to as the *buildup* and is directly analogous to the DD in the production well. Injection-well specific capacity is determined by dividing the flow by the buildup, resulting in units of gpm/ft (L/s·m).

7.3 COMMON WATER WELL COMPLETION VARIATIONS

The construction details of a water well are a function of a variety of influences (desired yield, drilling method, depth, etc.), but among the most important are the nature of the geological formations the well penetrates and the nature of the aquifer in which it is completed. There are an infinite number of design variations. This section addresses three, broadly illustrating different levels of complexity and geology.

Figure 7.7 illustrates what is known as an *open-hole well*. This type of completion is characterized by the absence of any casing or screen in the production zone of the well and is used in situations where the well is completed in rock formations such as basalt, some sandstones, and limestones. Casing is used in the upper portion of the well to accommodate the surface sanitary seal (as required for all wells by most jurisdictions to a minimum of 18 ft [5.5 m]). The casing also serves as the pump housing in the well. The casing may extend down to the production zone or may be set at a shallower depth depending upon the formations encountered. Depending upon the drilling method (see Appendix K for drilling methods), a conductor casing (shown in Figure 7.7) is required in caving formations (sand, gravel, clay, etc.) to hold the hole open and facilitate placement of the sanitary seal. In some cases this casing is removed after the seal grout is placed. This type of well is relatively simple, and it may be possible for the engineer to work directly with the driller instead of using a water well design professional in sites where this type of construction is selected.

Figure 7.8 presents what is known as a *naturally developed well*. This type of completion is used in unconsolidated formations composed primarily of medium- to coarse-grained materials with some fine components in between the larger materials. Casing with a screen attached to the bottom is used in the upper portion of the hole. The length of the sanitary seal is a function of local regulations but in most cases must extend to a depth of at least 18 ft (5.5 m). The screen slot size is selected to retain a portion of the materials in the production zone and pass the fine components. The process of development (a final stage of construction described in Section 7.4.7 of this chapter) removes the fine components in the near-well zone, increasing the permeability of the materials adjacent to the

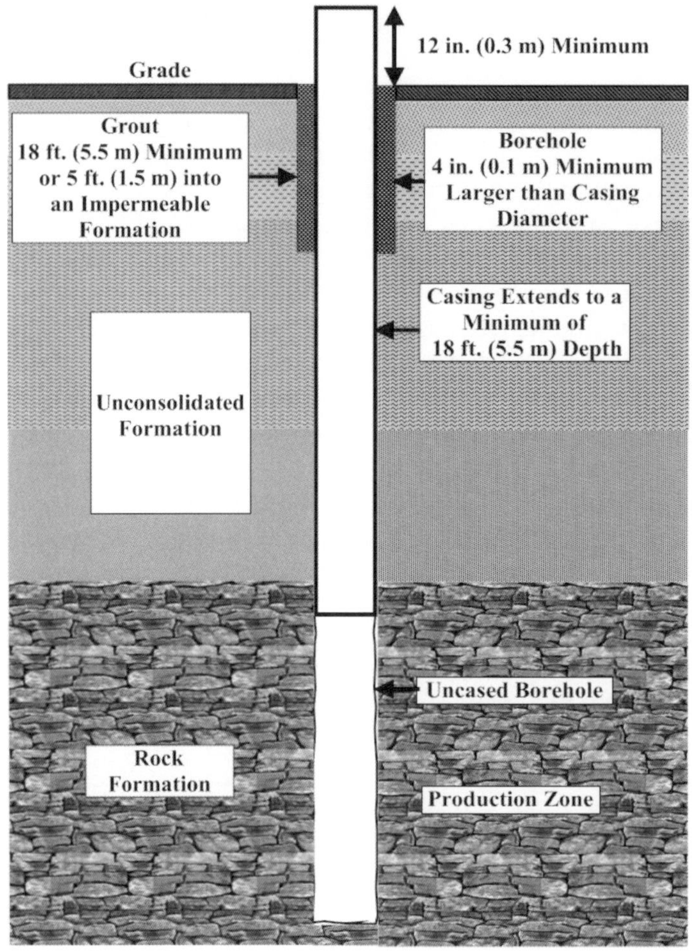

Figure 7.7 Open-Hole Well Completion

screen. Selection of the screen slot size (the size of the openings in the screen) is based on a sieve analysis of the materials produced during the drilling. The length of the screen is a function of the type of aquifer, the aquifer thickness, and the flow required from the well. Design requirements of this type of well are greater than those of open-hole wells, and engineers not experienced with water wells should work with water well design professionals in the specification of this type of well. With sufficient experience, mechanical engineers can design naturally developed wells on their own.

Figure 7.9 presents what is generally referred to as a *gravel pack* or *artificial filter* well. This is the most complex of the three wells illustrated here. It is used in settings characterized by an aquifer composed predominantly of fine-grained materials or where there are thinly stratified intervals of clay (non-water-producing) and productive zones. It is also used in some rare applications where a naturally developed well might otherwise be used. The amount of development required for a gravel pack well is normally less than that required for a naturally developed well, and in some settings the reduced development time can result in a gravel pack construction being less expensive than a naturally developed design. **There is a commonly held perception that gravel pack wells are always used in high-production applications, but this is not the case; they are required only when specific conditions dictate.** In a gravel pack well, as in a naturally developed well, casing with a screen attached to the bottom is installed in the upper por-

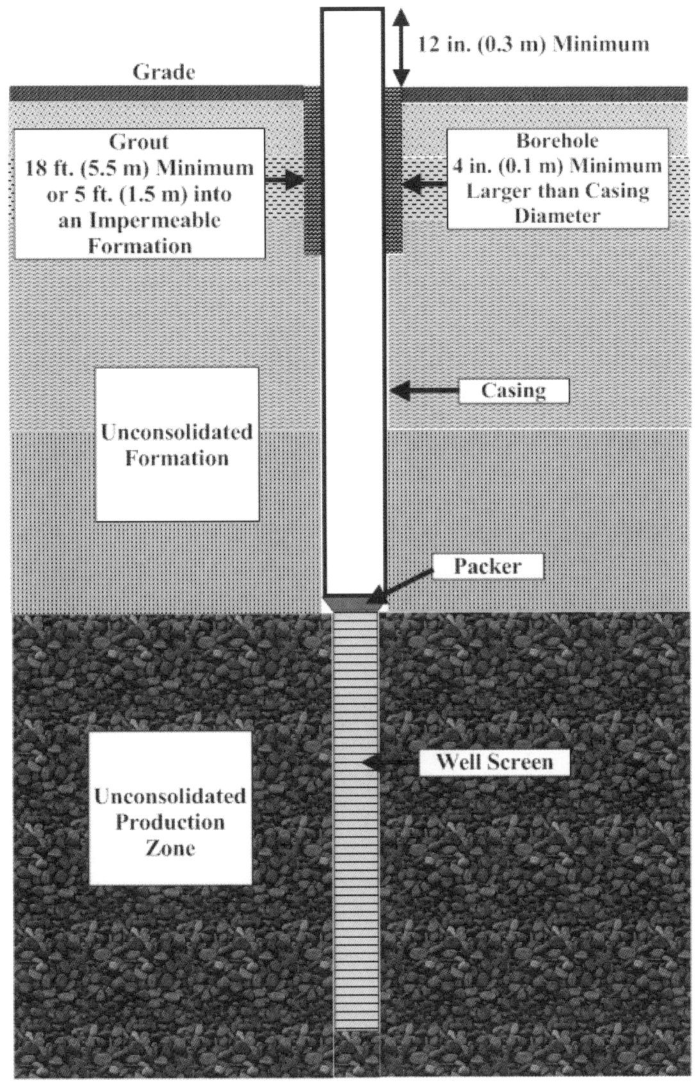

Figure 7.8 Naturally Developed Well Completion

tion of the well and placed in the production zone. Gravel pack wells are distinguished by an envelope of gravel-like material placed between the oversized well bore and the screen. This gravel performs the same function, hydraulically, as high-conductivity grout in a closed-loop borehole: it increases the conductivity in the near-bore critical zone. The gravel is selected based on a sieve analysis of the cuttings from the production zone, and the screen is selected based on the size of the gravel pack materials. The larger borehole diameter required for this construction, along with the special procedures for placing the gravel, tend to make this the most costly construction of the three well types in most applications on a per foot (metre) basis. Because of the complexity of this type of well, it is advisable for a water well design professional to be involved in a project when this design is called for.

Figures 7.7 through 7.9 illustrate very general well completion variations. The specifics of the design of a well for a particular application and site are included in the construction documents in much the same way as design details for other system components are. Often, particularly in settings appropriate to naturally developed or gravel pack wells,

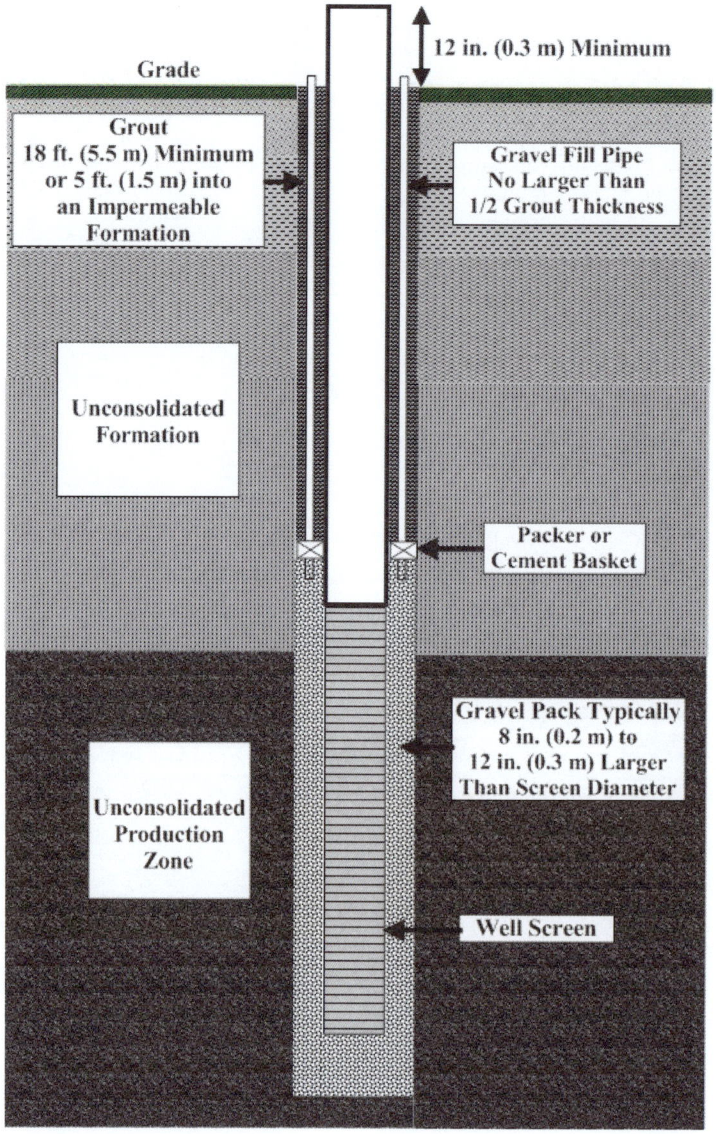

Figure 7.9 Gravel Pack Well Completion

these design details may be provided by a specialist in water well design rather than by the designer of the balance of the building mechanical system.

7.4 SELECTED TOPICS IN WATER WELL CONSTRUCTION AND DESIGN

The purpose of this section is not to provide a comprehensive treatment of the topic of water well design but to familiarize the reader with the issues involved. The following subsections discuss some of the more common issues encountered in the design, construction, and specification of water wells. Guide specifications for water wells can be found in *Water Well Specifications: A Manual of Technical Standards and General Contractual Conditions for Construction of Water Wells* (National Water Well Association 1981); *The Engineers' Manual for Water Well Design* (RMC 1985); ANSI/AWWA A100-

Table 7.2 Water Well Specification Subheadings

General conditions	Bid and contract document details, permits, use of premises, inspections, access, warranty, payment, indemnification, bonds, insurance, arbitration, clean up
Special conditions	Description of work, site subsurface information, utilities, insurance, bond, submittals, special materials, field office
Test holes and samples	Location, drilling method, logs and reports, sampling, water sampling
Well construction	Drilling methods, drilling fluid, logs and reports
Well casing and installation	Selection, size, materials, installation, joining, seating
Well grouting	Materials, installation, location, centralizers, logging, testing
Well screen	Type selection, materials, aperture size, length, installation, joining, sealing
Well filter (gravel pack)	Selection, materials, size, length, storage, disinfection, installation
Plumbness and alignment	Testing
Development	Methods, materials, sand content, records
Well testing	Type, water disposal, measurement, records, samples
Disinfection	Methods, materials, measurement, disposal
Water sampling and analysis	Type, samples, methods, laboratories
Abandonment	Sealing, grout placement, special conditions, records

97, *AWWA Standard for Water Wells* (AWWA 1997); and ANSI/NGWA-01-14, *Water Well Construction Standard* (NGWA 2014). Table 7.2 lists the subheadings included in most water well specifications.

7.4.1 Casing

The casing diameter used in shallow water wells, typical of GSHP applications, is only indirectly related to the flow required from the well. It is more directly a function of the diameter of the pump necessary to produce the flow required. Most GSHP systems use submersible-type pumps, though some lineshaft pumps have been used in the past. Table 7.3 provides general guidelines for water well casing diameters for both types of pumps. In most shallow wells, a single casing diameter is used. In deeper wells, economics or drilling method sometimes dictates a smaller casing in the lower portion of the hole (below the pump housing section).

Casing material is normally steel except in the presence of highly corrosive water, in which case nonmetallic casing (polyvinyl chloride [PVC], acrylonitrile butadiene styrene [ABS], or fiberglass) is sometimes used, though this is uncommon in GSHP applications. Caution is necessary in the use of plastic casings in larger-diameter (>6 in, [150 mm]) wells because of the substantially reduced collapse strength of plastic materials compared to steel.

7.4.2 Sample Collection

Selection of screen slot size is a function of the size of the materials in the formation. To gather the necessary information for design, samples of the cuttings from the production zone (or zones) are taken during the drilling process. The samples are typically washed on site and then placed in containers labeled with the depth interval, time, date, and well identification. In the laboratory, the samples are dried and passed through a series of sieves that separate the different grain sizes of the material. Samples produced by different drilling methods vary in accuracy in terms of their reflection of the formation interval of interest. This is particularly true for direct (mud) rotary drilling, as materials from other portions of the hole can be carried to the surface with the cuttings from the

Table 7.3 Well Casing Diameter Guidelines

Nominal Pump Bowl Diameter, in. (mm)	Suggested Casing Size, in. (mm)	Minimum Casing Size, in. (mm)	Submersible Pump Flow Range— Nominal 3600 rpm, gpm (L/s)	Lineshaft Pump Flow Range— Nominal 1800 rpm, gpm (L/s)
4 (100)	6 (150)	5 (125)	<80 (<5)	<50 (<3)
6 (150)	10 (250)	8 (200)	80–350 (5–22)	50–175 (3–11)
7 (180)	12 (300)	10 (250)	250–600 (16–38)	150–275 (9–17)
8 (200)	12 (300)	10 (250)	350–800 (22–50)	250–500 (16–32)
9 (230)	14 (360)	12 (300)	475–850 (30–54)	275–550 (17–35)
10 (250)	14 (360)	12 (300)		500–1000 (32–63)
12 (300)	16 (400)	14 (360)		900–1300 (57–82)

interval of interest. Drillers' comments as to the ease or difficulty of the drilling can offer insight into cuttings information, as well. As a result, some degree of judgment is required in interpreting the results. This portion of the specification addresses the intervals at which samples will be collected, how they are to be handled and labeled, and to whom they should be delivered.

7.4.3 Screens

Well screens are used in water wells to control the entrance of particulate (sand) into the well and to stabilize unconsolidated formations. Many types of screens are available (wire wound, louver, perforated, slotted, and bridge-slot), and the manufacturers of each make claims as to why their particular designs are superior. Selection parameters for screens involve water entrance velocity, diameter, length, material, and slot size. Diameter, length, and slot size are typically manipulated to arrive at an entrance velocity (the velocity of the water passing through the openings) of a maximum of 0.1 ft/s (0.03 m/s).

In unconfined aquifers, the screen length is typically the lower 1/3 to 1/2 of the aquifer thickness. (Aquifer thickness is the vertical distance between the top and bottom of the water-producing interval penetrated by the well. Interpretation of aquifer thickness from information in the well completion report is discussed in Section 7.5.1.) The reason for screening only the lower portion of the aquifer is that the upper portion of the aquifer will be dewatered due to drawdown. In confined aquifers, the entire aquifer thickness is typically screened unless it is unusually thick (>75 ft [23m]).

As previously stated, screen slot size is selected according to rules related to the size of the aquifer materials as determined from a sieve analysis of cuttings collected during the drilling process. In naturally developed wells, a screen slot size that retains (on the aquifer side of the screen) 40% to 50% of the aquifer materials is often chosen. This permits the finer materials to pass through the screen to be removed during the development process. Removal of the fine components opens flow passages in the formation adjacent to the well, reducing pressure drop in the near-bore zone. To make the slot size selection, a graph of the cuttings sieve analysis is necessary. Figure 7.10 provides a typical cuttings distribution curve, which can be provided by most civil engineering, geotechnology, or geohydrology laboratories. In this particular case, the 50% retained size is approximately 0.027 in. (0.68 mm) and the 40% size is 0.03 in. (0.76 mm).

Screen entrance velocity is the velocity of the water passing through the openings in the screen. As mentioned previously, a commonly used value for a maximum limit for entrance velocity is 0.1 ft/s (0.03m/s). Some references suggest this value can be as high as 0.25 ft/s (0.82 m/s), but many consultants believe and anecdotal evidence suggests that

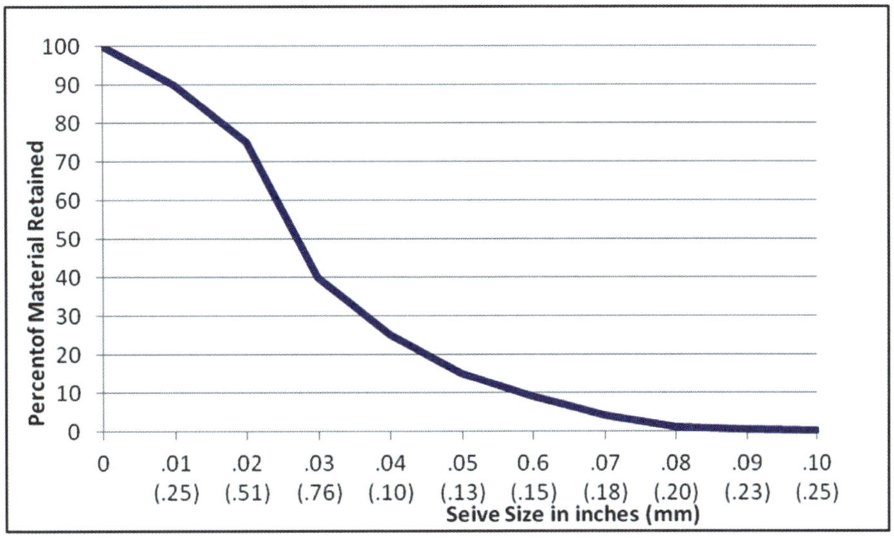

Figure 7.10 Sieve Analysis Results

EXAMPLE 7.1— SCREEN SLOT SIZE SELECTION

The screen slot size requirement is 0.40 in. (10 mm) for a naturally developed well in an unconfined aquifer of 50 ft (15 m) thickness. The well yield is to be 450 gpm (28 L/s) and a continuous slot screen is to be used. What are the length and diameter required?

Solution

Based on the flow requirement, the casing (above the screen) is likely to be in the 10 to 12 in. (250 to 300 mm) range (see Table 7.3). A screen in this same diameter range should be evaluated first.

Manufacturer data shows that a 10 in. (250 mm) screen with a 40 slot (0.040 in. [1 mm]) has an open area of 122 in.^{2}/ft or 0.847 ft^2/ft (0.26 m^2/m) of length. The flow (450 gpm [28 L/s]) is approximately 1.0 ft^3/s (0.028 m^3/s). The velocity of the water through the screen can be determined by dividing the volumetric flow of the water by the open area of the screen. For a maximum entrance velocity of 0.1 ft/s (0.03 m/s), the screen length requirement is

$$1 \text{ ft}^3/\text{s} \div 0.1 \text{ ft/s} = 10 \text{ ft}^2 \text{ ; } 10 \text{ ft}^2 \div 0.847 \text{ ft}^2/\text{ft} = 11.8 \text{ ft} \qquad \text{(I-P)}$$

$$0.028 \text{ m}^3/\text{s} \div 0.03 \text{ m/s} = 0.9 \text{ m}^2 \text{ ; } 0.9 \text{ m}^2 \div 0.26 \text{ m}^2/\text{m} = 3.5 \text{ m} \qquad \text{(SI)}$$

As this is an unconsolidated aquifer, the lower 1/3 to 1/2 is normally screened.

It appears that a smaller-diameter screen can be used in this case and still meet both the entrance velocity and length criteria. An 8 in./0.040 in. (200 mm/1 mm) slot screen has an open area of 98 in.^{2}/ft or 0.68 ft^2/ft (0.207 m^2/m).

The length required to meet the entrance velocity of 0.1 ft/s (0.03m/s) is

$$10 \text{ ft}^2 \div 0.68 = 15 \text{ ft} \qquad \text{(I-P)}$$

$$0.9 \text{ m}^2 \div 0.207 \text{ m}^2/\text{m} = 4.3 \text{ m} \qquad \text{(SI)}$$

This is a closer match to the 1/3 aquifer thickness dimension of 50 ÷ 3 = 16.7 ft (15 ÷ 3 = 5 m).

A 16 ft (5 m) section of 8in./0.040 in. (200 mm/1 mm) slot screen would meet the requirements in this application.

adhering to the 0.1 ft/s (0.03 m/s) value tends to result in a good balance between well cost and minimal maintenance (Driscoll 1986; Ralston 2000). The screen slot size is usually set by either the formation materials or the gravel pack materials; as a result, controlling entrance velocity is a function of screen length and diameter.

7.4.4 Gravel Pack

In the case of a gravel pack completion, the size of the gravel is selected based on rules relating to the size distribution of the aquifer materials in which the well is completed. One rule of thumb suggests that the 70% size of the gravel pack material be 5 times the 70% size of the aquifer materials. The screen slot size is then selected for an 85% to 100% retainage of the gravel material. To achieve maximum well performance, the gravel pack material and the screen slot size must be based on some variation of the above guidelines. Arbitrary selection from contractor's stock of these items will not provide optimum performance.

The gravel pack material is normally specified to be high silica content, well rounded with a uniformity coefficient (40% size divided by 90% size) of less than 2.5 (Driscoll 1986). The gravel is placed in the annular space between the screen and the well bore to a thickness of no greater than 8 in. (0.2 m) There are a variety of methods available for placing the gravel in the well, and the selection is a function of the well depth, the well diameter, and the driller's equipment.

7.4.5 Formation Stabilizer

Formation stabilizer is installed in wells in a similar manner as gravel pack, but it is used in most cases for support of the formations rather than for filtering or increasing permeability. Formation stabilizer is used in two situations: 1) unconsolidated alluvial and glaciofluvial sands and gravels and 2) semiconsolidated sandstones, siltstones, and sandy formations containing shells. In the former case, the stabilizer materials are selected to be close to or slightly larger than the formation materials. In the second case, the stabilizer materials are selected to be approximately 12 times the 70% size of the formation materials (Driscoll 1986).

7.4.6 Injection-Well Construction Variations

Injection wells vary from production wells in a number of ways, most notably in screen velocity and casing seal requirements. The screen velocity recommended for injection wells is 1/2 that of production wells, or 0.05 ft/s (0.015 m/s). This has been interpreted by some to mean that injection wells must be larger in diameter than production wells. This is not the case, as it is possible, particularly in unconfined aquifers, to screen more of the aquifer thickness since no allowance has to be made for drawdown as in the case of production wells. Sealing of an injection well is done in much the same way as for a well in a confined aquifer with a piezometric-level aboveground surface. The well should be cased down to the top of the injection zone and fully sealed to the ground surface (see Figure 7.6). This seal prevents the injected water from finding a path around the casing and back to the surface or to some intervening formation. Injected water should always be introduced into the well below the static water level through an injection tube (sometimes referred to as a *dip tube*). This helps reduce turbulence and air bubbles that might otherwise be formed if the water is simply discharged into the well from the surface. Air bubbles entering aquifer materials can obstruct flow just as effectively as solid particles. In addition, oxygen can promote serious scaling and fouling problems in injection wells and the near-bore zone.

EXAMPLE 7.2— PREDICTING INJECTION PRESSURE REQUIREMENTS

A building with a block load of 250 tons (88 kW) is planned for a site where the SWL is 45 ft (13.7 m). Another property 1/4 mi (400 m) from the site has an irrigation well producing 300 gpm (18.9 L/s) with a DD of 28 ft (8.5 m). The geology of the area is fairly uniform, and the target aquifer at the building site is the same one from which the irrigation well is producing.

What is the likely situation with respect to injection-well pressure requirement?

Solution

Based on the performance of the nearby well (with the understanding that data from a completed well on the actual project site is always preferable), the SC is approximately

$$300 \text{ gpm} \div 28 \text{ ft} = 10.7 \text{ gpm/ft} \qquad \text{(I-P)}$$

$$18.9 \text{ L/s} \div 8.5\text{m} = 2.22 \text{ L/s·m} \qquad \text{(SI)}$$

For a 250 ton (88 kW) load (cooling), it can be estimated that the likely flow requirement at peak will be in the range of 250 to 425 gpm (15.8 to 26.8 L/s), or roughly the same flow range as the nearby irrigation well. At the lower value, assuming the 10.7 gpm/ft (2.22 L/s·m) SC, the theoretical buildup in the injection well would be

$$250 \text{ gpm} \div 10.7 \text{ gpm/ft} = 23.4 \text{ ft} \qquad \text{(I-P)}$$

$$15.8 \text{ L/s} \div 2.22 \text{ L/s·m} = 7.1 \text{ m} \qquad \text{(SI)}$$

With a SWL of 45 ft (13.7 m), the buildup would bring the injection water level (IWL) to 45 – 23.4 = 22 ft (13.7 – 7.1 = 6.6 m)—still well below the ground surface, indicating little potential for problems with pressurization of the well.

At the higher flow, the values would be as follows: build up = 39.7 ft (12.1 m), IWL = 5.3 ft (1.6 m).

In the second case, the IWL values are more of a concern. Had the SWL in the example been 20 ft (6.1 m) instead of 45 ft (13.7 m) with the same SC, pressurization would have been indicated for both the high- and low-flow cases.

There has been some debate over the years as to the recommended screen type to be used in injection wells. As with any well, maximum open area and low velocity are desirable. Although the wire-wound (sometimes called *continuous slot* or *V-slot*) screen best addresses these issues, its design is characterized by slots that open inward (optimized for production-well flow direction), thus forming a trap for any debris that may be introduced into the well with the injected water. A distinct advantage of this type of screen, however, is that if redevelopment becomes necessary, the continuous slot design affords the most favorable arrangement for jetting through the screen and cleaning the near-bore zone. Provided the injected fluid is free of particulate matter (as is strongly recommended), the continuous-slot type screen is an effective choice pending additional research on this topic.

One of the most common concerns about injection wells is whether the well must be pressurized in order to accept the water. The reason for the concern is that it is good practice to avoid positive pressurization (meaning pressure relative to the ground surface) of injection wells if possible. Some jurisdictions prohibit positive pressurization of injection wells, but the primary reason for avoiding pressurizing injection wells is to reduce the likelihood of wells leaking at the surface or to formations shallower than the intended injection zone. When positive pressure is applied at the wellhead, the potential exists for water to find a path around the casing and out to adjacent formations or up along the casing to the surface. Ponding of water around the wellhead is a common indicator of this problem. Though it is possible to construct wells to eliminate this condition (and thereby allow safe pressurization), specific conditions must be present in the geology of the site for this to be possible. At a minimum these include an impermeable (clay) formation above the injection zone and a continuous casing seal from the top of the injection zone to the ground surface. The requirement for pressurization is relatively simple to predict. As mentioned previously, the buildup in the injection well is, in theory, a mirror image of the drawdown in the production well.

From Example 7.2, it is apparent that the SWL and SC of wells at the site can provide a clear indication of the likelihood of pressurization requirements of the injection well. The example assumes injection-well specific capacity equal to production-well specific capacity. It is not uncommon for injection wells to exhibit somewhat lower specific capacities than production wells. Such things as poor completion, fouling, scaling, and particulate accumulation can reduce injection-well performance over time. In addition, the viscosity of the water due to temperature change can also influence relative performance of an injection well compared to a production well. The relatively small temperature changes associated with GWHP systems (and the normally cooling-mode-dominant design with its favorable impact upon viscosity) tend to minimize this influence, however.

There has been no formal research into this issue in GWHP applications, but anecdotal evidence suggests that under poor conditions the specific capacity of the injection well may be 50% to 75% that of the production well. Poor conditions include one or more of the following conditions:

- Poorly controlled mud rotary drilling (can plug the injection zone with drilling fluids)
- Inadequate development of the well
- Sand-producing production well (injected sand can plug the injection zone)
- Scaling water chemistry (scale formation on the screen results in pressure drop across it)
- High potential for biofouling (particularly slime-producing bacteria can obstruct flow through screen)
- Fine-grained injection zone (more prone to plugging due to smaller flow passages)
- Low aquifer thickness (limits screen length)
- System design permitting air entry to injection line (alters water chemistry, increases scale and possibly biofouling)

On the other hand, positive conditions may permit an injection-well specific capacity of 85% to 100% of production-well specific capacity. Positive conditions include the following:

- Air drilling (eliminates drilling-mud contamination of the injection zone)
- Sand-free injection (no suspended solids in injected water to plug injection zone)
- Zero potential for air in injection line (no undesirable changes in water chemistry)
- Rock aquifer (fractures are less prone to plugging)

- Nonscaling water chemistry (reduced potential for scale formation)
- Low potential for biofouling (reduced potential for screen blockage)

In cases where it appears necessary to pressurize a single injection well and it is not possible to construct the well to accommodate pressurization, it is possible to use multiple injection wells spaced adequately apart (so as to minimize interference with each other) or to operate the system at a somewhat reduced flow at peak conditions to eliminate the positive pressure requirement. Multiple wells result in reduced flow per well, smaller buildup, and decreased pressurization requirements.

7.4.7 Well Development

Development is the process in which fine material adjacent to the screen (native material, gravel pack fine components, and clays left after drilling) is removed from the filter pack, stabilizer, or formation. The removal of the finer fraction leaves increased void space, resulting in higher permeability in the near-bore region. Development generally proceeds in two stages: initial development using the drill rig and final development by pumping the well with an engine-driven test pump. Of the many methods used for initial development, the most effective involve some variation of creating a high-velocity water flow horizontally out to the near-bore zone to dislodge the fine material and then creating a flow back toward the screen to carry away the materials. High-velocity water jetting combined with pumping the well or swabbing are two of the more effective development methods. After the initial development is complete, the final development is accomplished by pumping the well at rising flow rates followed by surging. Surging consists of stopping the pump and allowing the water in the column to backflow into the well.

It is in the development process that the sand content of the water is measured and compared to the maximum allowable sand content appearing in the well specifications. Development procedures continue until the specified sand content is achieved. Sand content is a critical component in a specification in terms of a well's cost and subsequent performance. Lower sand content (<10 ppm) typically requires more development time, which adds to construction cost. In any project in which injection is the disposal method, a sand-free well is desirable, but within realistic cost limits a maximum sand content of 5 to 10 ppm is probably acceptable. Surface separation can be used to remove the remaining particulate prior to injection to eliminate plugging of the injection well.

7.5 SITE EVALUATION FOR GWHP SYSTEMS

Investigation of a site for potential GWHP application is a multiphase process in terms of both scope and detail. As the process progresses, the scope tends to narrow and the level of detail increases. At the very earliest stages of the project the focus is on the identification of an aquifer sufficient to support the system envisioned for the site, the general regulatory setting (and agencies with jurisdiction), and any existing site infrastructure or features that may impact the development. If the results of this initial evaluation prove positive, more site-specific and detailed research is done into the local aquifer characteristics—often using well completion reports from existing nearby wells and interviews with local drillers and water well users. Contact is made with appropriate regulatory agencies, and any nearby contaminated sites are identified. This initial activity can be divided into a cursory pre-review followed by a more detailed second phase, or the entire task can be combined. For most projects in which the system type (open loop, closed loop, surface water, etc.) has been reduced to two options or possibly a single choice, the initial evaluation is often conducted as a combined effort. The final phase of site evaluation includes drilling and testing of wells. The testing typically includes flow and water level testing of the production well (or wells) and chemical and bacteriological analyses of the water produced.

In phase one of site evaluation the focus is on identifying any showstoppers in the local groundwater resource or regulatory setting. There are many resources available for clarifying the presence or absence of a groundwater aquifer in a given area. One of the best general sources of information is the *Ground Water Atlas of the United States* (USGS 1995). This publication, consisting of 13 volumes (each covering a three- to four-state area), provides regional-level information on aquifers. Information on aquifer geographic extent, geology, water quality, well yield, existing water use, and many other issues is provided for all major aquifers in a specific region. Additional detail is provided on some areas. This publication is an excellent source for the identification of local aquifers, and its bibliography provides sources to consult for additional details. The full publication is available both online and in print. Additional sources of similar information are available from most state geological surveys, state departments of geology, departments of water resources, or departments of natural resources. The particular state agency responsible for water issues varies from state to state. Most western states have a specific water resources agency. This is less common in the midwestern and eastern parts of the United States, where water responsibility is often delegated to environmental or natural resources agencies. Some states have little or no state-level agency responsible for water issues and all responsibility resides at the local or county level. Regardless of the agency responsible for enforcement, minimum regulatory requirements are set by the U.S. Environmental Protection Agency (EPA). State and local agencies are free to enact more stringent regulations, but they must meet EPA minimum requirements.

Often the states with the regulatory framework administered at the state level provide the most comprehensive information about water resources and facilitate the most favorable climate for GWHP projects. This is partly due to states' ability to employ earth-science professionals to manage and administer the groundwater regulatory system. In states where water regulation is delegated to the local authority, the resources to employ individuals with a formal education in hydrology or geology are often absent and regulatory and groundwater management rests upon those lacking the necessary scientific background.

Links to individual state agencies responsible for water resources and geology can be found on the WaterWebster™ website (WW 2011). Many of these agencies provide information on water wells, water use, water quality, and water regulatory issues and have available numerous publications on water, aquifers, well construction standards, and many other issues.

The U.S. Geological Survey (USGS) also holds vast amounts of information on water resources in individual states. Some of the more useful information available from the USGS includes continuous monitoring of water levels in aquifers with historic data plots—information that allows determination of the stability of a particular aquifer and its ability to support additional development. The USGS also monitors the water quality of aquifers. Much of this data is available online through their National Water Information System database (USGS 2014a). While most of the USGS information is available via the Internet, some of the more detailed local information references must be accessed through individual state Water Science Centers; a contact list for these centers is also available on the USGS website (USGS 2014b).

One of the hurdles facing the HVAC engineer in the course of site evaluation is developing an understanding of the terminology and unit systems of the geology and groundwater hydrology fields. These are areas unfamiliar to most engineers, and there is no

Table 7.4 Site Evaluation Issues

GCHP	GWHP
Thickness of unconsolidated materials	Aquifer presence
Nature of unconsolidated materials	Nature of aquifer materials
Thermal properties of soil and rock	Drawdown and specific capacity
Bedrock type	Well construction details
Depth to water	Static water level
Artesian aquifer presence	Aquifer type
Aquifer flow direction	Aquifer flow direction
Undisturbed ground temperature	Groundwater temperature
Rig type successful in past drilling	Rig type successful in past drilling
Contamination presence	Contamination presence
Regulatory setting	Regulatory setting

strategy more effective for becoming conversant in the terminology than learning through experience, reading, and using the many resources available. While it is possible to contract out all of the subsurface-related investigation (in the case of GWHP systems, the design of the water wells will also often be handled by others), the HVAC engineer must be familiar with these issues and fully integrate them into the overall design. To accomplish this, the engineer must be familiar with the terminology and science of geology, groundwater hydrology, and water wells.

Once an initial review concludes that a particular system type may be feasible at a site, there are several parameters of interest; these are influenced by the system type under consideration. Table 7.4 presents a summary of the key areas of interest for GWHP and GCHP systems using vertical bore heat exchangers.

7.5.1 Water Well Completion Reports

It is apparent from the topics listed in Table 7.4 that the information of interest to GWHP and GCHP system designers is similar. The best sources for providing much of this information are water well completion reports from existing wells in the area of the project site. These reports, submitted by the drillers upon completion of the construction of water wells (or borefields, in some cases), can provide data on all of the topics in Table 7.4 with the exception of the contamination and regulatory issues.

Well completion reports, sometimes referred to simply as *well logs*, are typically maintained by the water regulatory agency at the state level and in most cases are public information (except in California, where they are considered proprietary). In many states the reports reside in a database available online. The quantity of information available is substantial. To provide just one example, the Oregon water well database contains reports on 250,000 water wells in the state. Water well completion reports and other valuable subsurface information resources developed by those outside the HVAC industry have been grossly underused by those involved in GSHP projects. This section provides an introduction to reading and interpreting these documents.

Figures 7.11a and 7.11b and Figure 7.12 show examples of water well completion reports from the state of Oregon. The form shown is typical of that of many western states and somewhat longer and more detailed than that of states in the midwestern and eastern United States. While these documents are valuable information sources, it is important to consider that there are limitations on the extent to which information from nearby wells is illustrative of what can be expected at a new well drilled in the same area. Most important

LINN **RECEIVED**
52578

WELL I.D.# 0000226

STATE OF OREGON
WATER SUPPLY WELL REPORT JUL 2 7 1999
(as required by ORS 537.765)
Instructions for completing this report are on the last page of this form. WATER RESOURCES DEPT SALEM, OREGON

PAGE 1 OF 3 (START CARD) # 85423

(1) OWNER: Well Number 0000226
Name [redacted]
Address [redacted]
City [redacted]

(2) TYPE OF WORK
☒ New Well ☐ Deepening ☐ Alteration (repair/recondition) ☐ Abandonment

(3) DRILL METHOD:
☐ Rotary Air ☐ Rotary Mud ☒ Cable ☐ Auger
☐ Other

(4) PROPOSED USE:
☐ Domestic ☐ Community ☐ Industrial ☒ Irrigation
☐ Thermal ☐ Injection ☐ Livestock ☐ Other

(5) BORE HOLE CONSTRUCTION:
Special Construction approval ☐ Yes ☒ No Depth of Completed Well 202 ft.
Explosives used ☐ Yes ☒ No Type ____ Amount ____

HOLE Diameter	From	To	SEAL Material	From	To	Sacks or pounds
12	0	252	Bentonite	0	2	5 sacks
			Cement	2	152	74 sacks
			Cement Plug	202	217	8 sacks

How was seal placed: Method ☐A ☐B ☒C ☐D ☐E
☒ Other Bent Poured Dry + Probed
Backfill placed from 217 ft. to 252 ft. Material 3/4 minus Gravel
Sand Pack placed from 148 ft. to 182 ft. Size of gravel 6-9 Sand

(6) CASING/LINER:

	Diameter	From	To	Gauge	Steel	Plastic	Welded	Threaded
Casing:	8"	+1½	202	.250	☒	☐	☒	☐
					☐	☐	☐	☐
					☐	☐	☐	☐
					☐	☐	☐	☐
Liner:					☐	☐	☐	☐
					☐	☐	☐	☐

Final location of shoe(s)

(7) PERFORATIONS/SCREENS:
☐ Perforations Method
☒ Screens Type V Slot Material Stainless 304

From	To	Slot size	Number	Diameter	Tele/pipe size	Casing	Liner
169	182	.50		8"	8" PS	☐	☐
						☐	☐
						☐	☐
						☐	☐
						☐	☐

(8) WELL TESTS: Minimum testing time is 1 hour
☒ Pump ☐ Bailer ☐ Air ☐ Flowing Artesian

Yield gal/min	Drawdown	Drill stem at	Time
200	85		1 hr.
Pump set at 107			

Temperature of water 52° Depth Artesian Flow Found
Was a water analysis done? ☐ Yes By whom
Did any strata contain water not suitable for intended use? ☒ Too little
☐ Salty ☐ Muddy ☐ Odor ☐ Colored ☒ Other Upper zone
Depth of strata: 42-52, 82-92, 232-246

(9) LOCATION OF WELL by legal description:
County Linn Latitude ____ Longitude ____
Township 11S N or S Range 4W E or W. WM.
Section 25 NW 1/4 NE 1/4
Tax Lot 703 Lot ____ Block ____ Subdivision ____
Street Address of Well (or nearest address) 6500 Pacific Blvd.

(10) STATIC WATER LEVEL:
11 ft. below land surface. Date 5-25-99
Artesian pressure ____ lb. per square inch. Date ____

(11) WATER BEARING ZONES:
Depth at which water was first found 42 FT

From	To	Estimated Flow Rate	SWL
42	52	20 gpm	26½
82	92	40 gpm	25½
169	182	200 gpm	11
232	246	30 gpm	10½

(12) WELL LOG:
Ground Elevation

Material	From	To	SWL
Top Soil	0	3	
Clay Tan	3	7	
" Grey	7	17	
" Brown	17	21	
" + Gravel Small Brown	21	27	
Gravel meD w clay Brown	27	35	
Gravel Large to meD w Sand Brown	35	42	
Gravel meD + Cobbles	42	52	WB
Gravel small + Sand Coarse Brown	52	55	
Clay Grey Sandy	55	65	
Gravel meD w Sand Coarse Black	65	69	
Gravel Large + Small w some Cobbles + Black Sand	69	87	WB
Sand Black	87	92	
Cont			

Date started 4-8-99 Completed 5-25-99

(unbonded) Water Well Constructor Certification:
I certify that the work I performed on the construction, alteration, or abandonment of this well is in compliance with Oregon water supply well construction standards. Materials used and information reported above are true to the best of my knowledge and belief.
WWC Number 1454
Signed ____ Date ____

(bonded) Water Well Constructor Certification:
I accept responsibility for the construction, alteration, or abandonment work performed on this well during the construction dates reported above. All work performed during this time is in compliance with Oregon water supply well construction standards. This report is true to the best of my knowledge and belief.
WWC Number 633
Signed [signature] Date 7-26-99

ORIGINAL & FIRST COPY-WATER RESOURCES DEPARTMENT SECOND COPY-CONSTRUCTOR THIRD COPY-CUSTOMER

Figure 7.11a Water Well Completion Report Example #1

Material	From	To	SWL
Sand Coarse Grey +	92	96	
Gravel meD			
Clay Green Sandy +	96	105	
Gravel meD			
Clay Green Sandy	105	111	
Clay Green + Gravel meD	111	124	
Clay Grey	124	127	
Clay Grey + Gravel meD	127	135	
Clay Blue + Green	135	143	
Gravel Large to med	143	147	
Cemented w Dark Gree			
Sandy Clay			
Clay Brown sandy	147	152	
Clay Blue sticky	152	162	
Clay Grey Sandy	162	169	
Sand Black	169	178	WB
Sandy clay Grey w	178	182	
Packed sand Seams			
Clay Brown	182	187	
ConT			

Material	From	To	SWL
Clay Grey Sticky	187	201	
Clay Blue sticky	201	212	
" Grey Sandy	212	217	
Silt Packed	217	225	
Sand packed Fine w	225	228	
Small Gravel			
Packed Sand Fine + silty	228	232	
Clay			
Sand Black w some	232	240	WB
Small Gravel			
Sand Coarse w small	240	246	
Gravel			
Clay Blue Sticky	246	252	

Figure 7.11b Lith-Log Portion of Water Well Completion Report Example #1

are the nature of the geology and the extent to which it can be extrapolated from one site to another. This is a subject best interpreted by an earth-science professional (geologist, hydrologist, etc.). Another factor is the accuracy with which the well completion report was prepared. Drillers, who are required by law in most jurisdictions to complete the report, typically do not use conventional geological terminology in describing the material penetrated by the drilling. In addition, there may be errors in entries, well location descriptions, or other issues that impact the usefulness and accuracy of the data. Using information from existing wells to predict performance of future wells is risky and best accomplished with conservative assumptions and a realistic appreciation of the issues impacting accuracy.

The reports contain information about two different wells, well #1 (Figures 7.11a and 7.11b) and well #2 (Figure 7.12). The reports include the owner (at the time the well was constructed) in section 1; this has been blacked out for use in this publication. Section 2 describes the type of work covered in the report in terms of a new well, a modification of an existing well, or abandonment. Section 3 describes the type of rig used for the work. This is useful information for both GWHP and GCHP systems, as the type of rig in conjunction with the time required to complete the work (section 12) provides an indication of the success of that type of rig in the geologic setting present at the site.

Section 4 describes the intended use of the well. Section 5 describes the drilling in terms of depth, diameter, materials used for the seal between the casing and the borehole, and how the seal was placed. Most states allow several different methods of placing the grout seal. In both of the examples here, the drillers used Method C, which involves the use of a tremie pipe in much the same way that closed-loop boreholes are grouted. This section of the report also describes any fill material placed in the well or any open hole intervals. In the report for well #1, which was completed with a screen and gravel pack, the gravel is described and the depth interval in which it was placed is specified. It is common practice to place gravel pack material well above the top of the screened interval so as to provide material to compensate for settling of the gravel and for removal of fine components from the gravel pack during the development process. In this case the screen

51102

STATE OF OREGON
WATER SUPPLY WELL REPORT
(as required by ORS 537.765)

APR 6 1999

WATER RESOURCES DEPT
SALEM, OREGON

WELL I.D. # L 26856
START CARD # 108474

Instructions for completing this report are on the last page of this form.

(1) OWNER: Well Number ______
Name [redacted]
Address [redacted]
City [redacted]

(2) TYPE OF WORK
[X] New Well [ ] Deepening [ ] Alteration (repair/recondition) [ ] Abandonment

(3) DRILL METHOD:
[X] Rotary Air [ ] Rotary Mud [ ] Cable [ ] Auger
[ ] Other

(4) PROPOSED USE:
[X] Domestic [ ] Community [ ] Industrial [ ] Irrigation
[ ] Thermal [ ] Injection [ ] Livestock [ ] Other

(5) BORE HOLE CONSTRUCTION:
Special Construction approval [ ] Yes [X] No Depth of Completed Well 230' ft.
Explosives used [ ] Yes [X] No Type ______ Amount ______

HOLE Diameter	From	To	SEAL Material	From	To	Sacks or pounds
10	0	119	CEMT&	0		29 SKS
			BENT.		119	7 SKS
6	119	230	OPEN			

How was seal placed: Method [ ] A [ ] B [X] C [ ] D [ ] E
[ ] Other ______
Backfill placed from ______ ft. to ______ ft. Material ______
Gravel placed from ______ ft. to ______ ft. Size of gravel ______

(6) CASING/LINER:

	Diameter	From	To	Gauge	Steel	Plastic	Welded	Threaded
Casing:	6"	+1	119	.25	X	[ ]	X	[ ]
Liner:	---							

Final location of shoe(s) NONE

(7) PERFORATIONS/SCREENS:
[ ] Perforations Method NONE
[ ] Screens Type ______ Material ______

From	To	Slot size	Number	Diameter	Tele/pipe size	Casing	Liner

(8) WELL TESTS: Minimum testing time is 1 hour
[ ] Pump [ ] Bailer [X] Air [ ] Flowing Artesian

Yield gal/min	Drawdown	Drill stem at	Time
100 GPM		145 FT.	1 hr.

Temperature of water 65 F Depth Artesian Flow Found NONE
Was a water analysis done? [ ] Yes By whom ______
Did any strata contain water not suitable for intended use? [ ] Too little
[ ] Salty [ ] Muddy [ ] Odor [ ] Colored [ ] Other ______
Depth of strata: NONE

(9) LOCATION OF WELL by legal description:
County KLAMATH Latitude ______ Longitude ______
Township 39 S N or S Range 12 E E or W. WM.
Section 7 SW 1/4 NE 1/4
Tax Lot 101 Lot ______ Block ______ Subdivision ______
Street Address of Well (or nearest address) 37309 McCARTIE LN BONANZA, OR 97623

(10) STATIC WATER LEVEL:
86 FT. ft. below land surface. Date 03-30-99
Artesian pressure ______ lb. per square inch. Date ______

(11) WATER BEARING ZONES:
Depth at which water was first found 98 FT.

From	To	Estimated Flow Rate	SWL
98	230	200 GPM	86

(12) WELL LOG:
Ground Elevation 4150

Material	From	To	SWL
TOP SOIL	0	2	
YELLOW CLAY	2	10	
BROWN SANDSTONE	10	53	
SOFT BROWN ROCK	53	60	
BROWN CLAY	60	63	
BLACK ROCK	63	64	
BROWN CLAY	64	71	
YELLOW CLAY	71	83	
GRAY CLAYSTONE	83	98	
BLACK ROCK (WB)	98	137	86
BROWN ROCK (WB)	137	163	86
BLACK ROCK (WB)	163	185	86
BROWN ROCK (WB)	185	204	86
BLACK ROCK (WB)	204	226	86
BROWN ROCK	226	230	86

Date started 3-29-99 Completed 3-30-99

(unbonded) Water Well Constructor Certification:
I certify that the work I performed on the construction, alteration, or abandonment of this well is in compliance with Oregon water supply well construction standards. Materials used and information reported above are true to the best of my knowledge and belief.
WWC Number ______
Signed ______ Date ______

(bonded) Water Well Constructor Certification:
I accept responsibility for the construction, alteration, or abandonment work performed on this well during the construction dates reported above. All work performed during this time is in compliance with Oregon water supply well construction standards. This report is true to the best of my knowledge and belief.
WWC Number 777
Signed Stephen R Hughes Date 4-2-99

ORIGINAL & FIRST COPY-WATER RESOURCES DEPARTMENT SECOND COPY-CONSTRUCTOR THIRD COPY-CUSTOMER

Figure 7.12 Water Well Completion Report Example #2

extends from 167 to 182 ft (50.9 to 55.5 m) and the gravel was placed in the space between the borehole and the screen or casing from 148 to 182 ft (45.1 to 55.5 m). Well #1 was also drilled initially to a depth of 252 ft (76.8 m) and then backfilled with 3/4 minus gravel to 217 ft (66.1 m) with a cement plug between 202 and 217 ft (61.6 and 66.1 m). The interval between 202 and 182 ft (61.6 and 55.5 m) likely has 8 in. (200 mm) casing installed, though this is not clearly stated in the report. It is common to place a blank section of casing below the screened section in a well to allow for accumulation of fine material in the bottom of the well. Well #2 was completed as an open hole with 8 in. (200 mm) casing installed in the upper 199 ft (60.6 m) and open hole to the 230 ft (70.1 m) depth.

Section 6 of the report specifies the casing placed in the well in terms of depth interval, material, connection method, and wall thickness. Well #1 shows 8 in. (200 mm) steel casing to 202 ft (61.6 m) with welded connections. Note that the casing extends 1.5 ft (0.5 m) above grade level to prevent surface water from draining into the well; a requirement in many jurisdictions. Well #2 includes 6 in. (150 mm) welded steel casing to a depth of 119 ft (36.3 m) with open hole to total depth.

Section 7 describes the screen or perforated casing used in the production zone. Well #1 has a stainless steel, wire-wound screen from 167 to 182 ft (50.9 to 55.5 m). This type of screen normally has approximately 35% open area. For the 15 ft (4.6 m) screened interval this amounts to approximately 11 ft^2 (1.02 m^2) of open area. For the test flow indicated in section 8 of 200 gpm (12.6 L/s), the entrance velocity amounts to approximately 0.04 ft/s (0.012 m/s), which is well below the maximum recommended value of 0.1 ft/s (0.030 m/s). Information about a water well's screen is useful in evaluating the well's performance in terms of replicating the construction for a future well. The information in section 7 for well #2 is absent as it is an open-hole completion.

Section 8 provides information useful primarily to prospective GWHP developers, as it describes the results of the well's flow test. It should be pointed out that flow tests conducted to meet regulatory requirements are typically short (in this case only 1 h) and yield less useful information than a more formal flow test conducted for 8 to 24 h (see Section 7.5.2 of this chapter). Despite this, the information is of interest. Well #1 flow test data is the more helpful of the two examples. It shows a 200 gpm (12.6 L/s) yield at a drawdown of 85 ft (25.9 m). Adding the static water level to the drawdown suggests a pumping water level (PWL) at 200 gpm (12.6 L/s) of 85 + 11 = 96 ft (25.9 + 3.4 = 29.3 m). Specific capacity, based on the test data, would be 200 gpm ÷ 85 ft DD = 2.35 gpm/ft (12.9 L/s ÷ 25.9 m = 0.5 L/s·m). Generally, SC values of >5 gpm/ft (1.1 L/s·m) are desirable for large-capacity wells. Well #2 indicates a flow of 100 gpm (6.3 L/s) but does not provide any indication of the water level in the well. It is possible to infer something about water level from the drill stem depth (the end of the drill pipe would have had to be below the water surface), but there is considerable error associated with that assumption. The test for well #2 was conducted by airlifting, a process that involves using the rig's air compressor to inject compressed air into the water in the column pipe, causing a sufficient density decrease (resulting from the air bubbles mixed with the water) to cause the air/water mixture to flow up the column pipe. Although both reports provide some idea of flow capability, neither indicates with certainty the type of information needed for GWHP design. In neither case is there any information about whether the water level in the well has stabilized at the flow indicated—a condition that would indicate the well/aquifer could produce the flow on an extended basis.

There is information useful to both GWHP and GCHP designers on both reports in section 8: the groundwater temperature. Groundwater temperature in a given location is indicative of the undisturbed ground temperature at that location—a key design value.

This value must be judged in the context of the expected value in that area, however. This is illustrated in the case of well #2. The 65°F (18.3°C) value shown is well above the normal value (52°F to 53°F [11.1°C to 11.7°C]) for that location and is an indication of the impact of local higher-temperature geothermal resources influencing the water temperature. The extent of the influence of the geothermal resources on static water level is a complex issue and full understanding requires information beyond that available in a well completion report. This kind of high-temperature geothermal influence is limited (in shallow wells) to specific areas in the western United States and would not be encountered in the central or eastern portions of the country.

Section 9 of the report provides information on the location of the well in terms of township, range, section, subsection, and tax lot number. More recent forms in many states have added space for global positioning system (GPS) coordinates as well. In addition, some forms include space for a sketch of the well location relative to local landmarks. This information is key to accurately searching the database in which the well reports reside. While it is possible to search the database by owner name, owners may change over time, so searching by geographic location is more effective.

The static water level is identified in section 10. This information is important for both GWHP and GCHP designers as it influences drilling strategy, pumping power for GWHP systems, ground thermal property test results, and possibly ease of installation for GCHP systems. In the case of well #1, the SWL indicated is 11 ft (3.4 m). The drilling encountered four different production zones with different water levels, suggesting distinct aquifers. Three of the zones were cased off and sealed, with only one completed for production. This interval was at a depth of 167 to 182 ft (50.9 to 55.5 m). Based on the SWL of 11 ft (3.4 m) compared to the depth of the production zone, this suggests the presence of a confined aquifer. The same is true of well #2. In this case, the drilling did not encounter water until the 98 ft (29.9 m) depth, but the SWL was 86 ft (26.2 m). In this case, however, the fact that the SWL is shallower than the depth at which water was first encountered may not be indicative of a confined aquifer. Rather, it may simply be a reflection of the impact of the warmer water present in this location. If the warm-water aquifer is recharged by water of lower temperature (usually the case), the lower density of the warm water may result in a slightly elevated column of the warmer water in the well. Thus, the SWL may be a function of the density difference rather than the presence of a confined aquifer in this particular case.

Section 12 includes the description of the materials penetrated by the drilling, sometimes referred to as a *lith-log* (for *lithology log*). This information is of interest to both GWHP and GCHP designers. The material description offers some indication of the likely thermal properties and drilling ease for potential boreholes and suggests the likely permeability for water wells. It is important to note, however, that drillers rarely use conventional geological terminology to describe the materials encountered. This sometimes makes interpretation difficult. Well #1 exhibits substantial intervals of clay (see Figure 7.11b), which likely produce poor thermal conductivity, though this may be somewhat improved by the water-bearing intervals (42 to 52, 69 to 87, and 169 to 178 ft [12.8 to 15.8, 29.3 to 26.5, and 51.5 to 54.3m]). From a water well standpoint, the presence of the clay intervals above the main water-producing zone tends to support the presence of a confined aquifer with the clays acting as impermeable bodies capable of confining the aquifer pressure. Beyond this, the clays would also provide effective protection from vertical water migration in the event an injection well was operated in a pressurized condition at the site. A well-written lith-log is very helpful in determining the aquifer thickness, which is a parameter used in the calculation of required production/injection well spacing (see Chapter 8). In the case of well #1, the main production interval occurs between 169 and 178 ft (51.5

and 54.3 m) and is described as black sand. The intervals above and below this are described as sandy clay and are unlikely to produce water. As a result, the aquifer thickness in this well can be taken as 9 ft (2.7 m). Well #2 is, down to a level of 98 ft (29.9 m), largely clays and sandstone with no water. The black rock described in the interval below 98 ft (29.9 m) is likely basalt with fractures, as this is commonly encountered in the area in which this well was constructed. The aquifer thickness in this well would be interpreted as 98 to 226 ft (29.9 to 68.9 m), possibly to 230 ft (70 m) depth.

As this discussion indicates, well completion reports provide information on most of the topics of interest included in Table 7.4, and the data from these reports should be a key part of site evaluations in situations where the reports are available.

7.5.2 Well Flow Testing

Drilling and testing of the wells for a GWHP system prior to final design is strongly recommended. Flow testing of the production well is the final step in the site evaluation for a GWHP system. It provides the critical information necessary for the designer to incorporate well pumping power requirements into the design of the system and for accurate specification of the well pump and related components. Flow testing also affords an opportunity to retrieve water samples for chemical and bacteriological analyses.

Upon completion, most wells are required by regulatory authorities to be tested for performance, but these tests are often too short in duration to produce useful information (as noted in Section 7.5.1). Formal well flow testing typically requires 4 to 12 h to achieve water level equilibrium conditions in a step test. Constant-rate tests may run for 24 h or more. These are the type of tests that produce the information about well and aquifer performance necessary as input to the design of heat pump systems.

The test most commonly used in conjunction with GWHP systems is the step drawdown test. In this test, the well is pumped at several (usually three to 5) rates approximating 25%, 50%, 75%, and 100% of the expected peak requirement; data is collected at each rate until apparent equilibrium is achieved (indicated by a stable water level in the well). Appendix L includes a typical specification for a step drawdown test. Properly conducted and analyzed, this test can provide information on pumping level, drawdown, specific capacity, and well efficiency. Provided a second well is monitored for water level during the test (usually a constant-rate extension of the step test), data can be collected that allow determination of values for aquifer transmissivity and storage coefficient. The test is sometimes conducted with an engine-driven lineshaft pump to accommodate the control of the pump output. An electric submersible pump can be used provided that adequate control of the production rate can be accomplished, through either throttling or variable-speed control. In the course of the test, well water level and water flow rate are the key parameters to be monitored. Water level can be monitored automatically with a pressure transducer coupled with a data logger or manually with an electric sounder. Water flow is often measured with an orifice plate on the outlet of the pump discharge (Figure 7.13), though other types of flow meters (magnetic, paddle wheel, turbine, and ultrasonic) can be used as well. One of the key issues associated with a well test is the question of where the water will be directed for disposal. Even the shortest tests last for 4 to 6 h, and at several hundred gallons (litres) per minute, the volume of water to be disposed of is substantial. In developed areas, the operator of the local storm sewer system will have guidelines as to what is acceptable in that system. In rural areas, coordination with the county and with the local office of the environmental regulatory agency is often required. Careful coordination between the owner's representative, the contractor, and the cognizant regulatory agencies is critical to avoid delays and to ensure an uninterrupted test.

Figure 7.13 Well Flow Test with Water Flow Measurement via Orifice Plate and Ultrasonic Flowmeter

To provide the type of data useful for analysis, it is important that the flow and water level data be collected at specific time intervals with respect to the time the pump is started. For the data to be useful for analytical purposes, the flow rate must be carefully regulated to a fixed value for each segment of the test. The recommended data collection intervals are every 1 min for the initial 10 min after pump start or after an increase in flow, every 2 min for the next 10 min, every 5 min for the next 40 min, every 15 min for the next hour, and every 30 min thereafter (RMC 1985). Recovery water level readings (after the pump is stopped) are taken in the same fashion.

Figure 7.14 depicts manual water level measurement using an electric sounder (a wire with a continuity device on the end that emits a sound when the water level is encountered). The wire is calibrated with depth increments to facilitate water level determination. The test photographed was conducted with a submersible well pump and a gate valve for flow control. To some extent the test length is adjusted while testing is in progress, as the well water level must stabilize at each flow rate prior to the test at the next flow rate and the length of time required for aquifer stabilization is not predictable.

Table 7.5 provides an abbreviated example of the results from a step test. It is apparent from the results that information critical to GSHP system design calculations is readily available from the data. Most importantly, well water level and specific capacity over a range of flows can be determined from the well test data. This allows the calculation of well pump power requirements over a range of flows, values critical to the determination of optimal groundwater flow (as covered in detail in Chapter 8).

In addition to the flow and water level data, it is also important to monitor the appearance of the water. Turbidity is often encountered for short periods at water flow changes. Extended production of turbid water, however, can indicate a problem with the well.

In some cases a second test, known as a *constant-rate test*, is conducted after the step test. The purposes of this test are to confirm the ability of the well to produce at the design

Figure 7.14 Flow Test Water Level Measurement Techniques: Downhole Pressure Transducer Connected to Data Logger (Upper Left), Manual Water Level Measurement with Electric Sounder (Center), and Gate Valve for Water Flow Control

Table 7.5 Well Test Data Example

Time, min	Flow, gpm (L/s)	Water Level, ft (m)	Comments
1	90 (5.7)	72.6 (22.13)	
2	90 (5.7)	74.5 (22.71)	
3	90 (5.7)	75.0 (22.86)	
5	90 (5.7)	75.4 (22.98)	
10	90 (5.7)	75.7 (23.07)	
15	90 (5.7)	76.2 (23.22)	
30	90 (5.7)	76.9 (23.44)	
45	90 (5.7)	76.9 (23.44)	
100	140 (8.8)	80.1 (24.41)	cloudy
101	140 (8.8)	81.6 (24.87)	cloudy
102	140 (8.8)	83.0 (25.30)	
105	140 (8.8)	83.5 (25.45)	
110	140 (8.8)	84.0 (25.60)	
115	140 (8.8)	84.3 (25.69)	
130	140 (8.8)	84.8 (25.85)	
145	140 (8.8)	84.9 (25.88)	
190	180 (11.3)	95.6 (29.14)	cloudy
191	180 (11.3)	96.1 (29.29)	cloudy
192	180 (11.3)	96.7 (29.47)	cloudy
193	180 (11.3)	97.0 (29.56)	cloudy
195	180 (11.3)	97.4 (29.69)	cloudy
200	180 (11.3)	97.6 (29.75)	cloudy
210	180 (11.3)	98.5 (30.02)	cloudy
215	180 (11.3)	99.0 (30.17)	cloudy
230	180 (11.3)	99.2 (30.23)	cloudy
245	180 (11.3)	99.2 (30.23)	cloudy

rate over an extended period of time and to gather data for determination of aquifer performance parameters. The constant-rate test is typically conducted for 24 to 36 h. This type of test is only rarely used in GSHP projects, as aquifer data can sometimes be collected during the shorter-term test. In addition, GSHP production wells are not pumped at the peak rate for extended periods of time, as is the case with municipal, industrial, and irrigation wells, and injection for disposal eliminates the potential for long-term aquifer depletion.

The integration of this data into the design process is covered in detail in Chapter 8.

7.5.3 Groundwater Chemistry

Water is nature's universal solvent and, though a weak solvent, it tends to remove small amounts of various chemical constituents from the soil and rock materials through which it passes. In many cases these dissolved and suspended materials can promote scaling, corrosion, or plugging of the mechanical systems in which the water is used. Careful design and material selection can substantially reduce or eliminate the potential problems posed by the groundwater. The key strategies necessary to avoid groundwater-quality-induced problems include testing the groundwater chemistry and understanding its character, investigation of local experience with the groundwater, avoidance of designs that unnecessarily induce entrance of air to the system, isolation of the groundwater from the building loop using a heat exchanger, and understanding the implications of the way the water is used in the system.

Most water-quality-related problems occur in the injection well and the aquifer materials immediately surrounding it. The key to minimizing these problems is minimizing groundwater flow, eliminating air entry into the groundwater loop, and separating suspended material in the groundwater prior to its entry into the injection well.

Common misconceptions about groundwater quality and GSHP systems include the following:

- *Water quality is only an open-loop issue.* FALSE—Hard-water problems in desuperheaters and commercial heat pump water heating are commonly encountered in closed-loop systems.
- *Using a cupronickel heat exchanger is an effective way to deal with all water quality problems.* FALSE—Cupronickel alloys were originally developed for seawater heat exchanger applications and are effective at alleviating GSHP water quality problems. The alloy is ineffective (or only marginally better than copper) at addressing many problems commonly encountered in groundwater applications, such as hydrogen sulfide; low-pH corrosion; and iron, manganese, and carbonate scale or fouling.
- *The water meets drinking water standards, so it's acceptable to use in the heat pump.* FALSE—Drinking water standards are not designed to address corrosion and scaling. Like groundwater, you can drink tequila, too, but if you drink enough it eventually compromises your performance.
- *The water treatment guy will take care of it.* FALSE—Because of the high throughput of water in a GWHP system and regulatory limitations on chemical content of the injected water, treatment of the groundwater is not a realistic option for open-loop systems.
- *The system has been operating fine for six months, so there are no water quality issues.* FALSE—Scaling and corrosion often take years to become apparent. Initial results from the first year can be misleading with respect to long-term (5- to 20-year) performance.

The object of sampling and analysis of groundwater chemistry is not to determine if the water is of sufficient quality to use directly in the building loop of the GWHP system, because **groundwater should rarely if ever be used directly in a commercial- or institutional-scale conventional GWHP system** (as described in Chapter 8). Even water initially of apparently benign chemistry can degrade over time. In commercial- and institutional-scale GWHP systems groundwater should be isolated from the building loop with a plate heat exchanger. The goal of the groundwater analysis is to understand what problems may be encountered, how serious these problems may be, and how the system can best be configured to reduce or eliminate these problems.

Understanding water chemistry begins with a laboratory analysis of the groundwater based on a sample from the well at the site. The samples should be collected in a container provided by or approved by the laboratory performing the water analysis. Samples should not be taken during drilling or while the well is being air lifted (a procedure for producing water from the well using compressed air), because this alters the natural chemistry of the water. Tests for dissolved gases (oxygen, carbon dioxide, and hydrogen sulfide) and pH should be conducted in the field if possible, because evolution of gases can occur in a sample if it is not carefully sealed and handled.

There are four common water quality issues encountered in GWHP systems: scaling (most commonly calcium carbonate but occasionally manganese and others), iron fouling, corrosion (typically related to chloride, hydrogen sulfide, or low-pH general corrosion), and biological fouling (often related to iron-metabolizing bacteria).

Table 7.6 lists the minimum chemical constituents to be included in a water analysis for a GWHP system. If there are unusual water issues known to be a problem in the area, these should be added to the list. It is often useful to collect two samples, one from the casing before starting the pump (referred to as a *casing sample*) and the other after the pump has operated for several minutes (referred to as an *aquifer sample*). Differences in the analysis results for the two samples can provide insight into both chemical and biological reactions (see Appendix M for example well chemical and biological analysis results).

Units used for reporting water analysis results are sometimes confusing, as all constituents are not reported in consistent units. Generally, concentrations are listed in terms of parts per million (ppm) or milligrams per litre (mg/L). For the purposes of this chapter these units are considered equal. Some items (typically hardness, alkalinity, and calcium) are reported as calcium carbonate ($CaCO_3$) ppm equivalent. This practice makes the calculation of saturation and stability indices and determination of carbonate and noncarbonate hardness calculations simpler. Occasionally hardness is reported in grains per gallon or simply grains. To convert grains/gal to ppm, multiply by 17.1.

Table 7.6 Minimum Water Quality Analysis Components

pH	Chloride (Cl)
Total dissolved solids (TDS)	Carbonate (CO_3)
Iron (Fe)	Bicarbonate (HCO_3)
Total methyl orange (M) alkalinity	Hydrogen sulfide (H_2S)
Phenolpthalien (P) alkalinity	Carbon dioxide (CO_2)
Sulfate (SO_4)	Oxygen (O)
Calcium (Ca)	Manganese (Mn)
Iron bacteria	Total hardness
Slime-forming bacteria	Sulfate-reducing bacteria
Langlier saturation index (LSI)	Ryznar stability index (RSI)

The general acid/alkaline character of water is reflected in the pH or hydrogen ion concentration. A value of 7.0 is considered neutral. Most groundwater is in the range of 6.5 to 8.5. Water in building hydronic systems is typically maintained at a pH of >8.5 to ensure a minimum of corrosion. Groundwater with a pH of <7 is capable of causing general corrosion of iron alloy components; groundwater with a pH of <6.5 is capable of causing corrosion of copper alloys. Scaling tends to be associated with waters of pH >7.5 (Rafferty 2004).

Total dissolved solids (TDS) is a measure of the total amount of dissolved minerals in water. Values of >500 ppm are considered to be more prone to both potential scaling and corrosion. Most groundwater is characterized by a TDS of <750 ppm. For comparison, seawater is approximately 35,000 ppm. Generally, the electrical conductivity of the water rises as the TDS increases, and this leads to greater corrosion potential. For waters of >500 ppm TDS, it is helpful to request an analysis reporting all major anions and cations (HCO_3, SO_4, Cl, CO_3; Ca, Mg, Na, K) in the sample.

Iron can take several forms in groundwater and can combine with other elements to form more complex compounds. Reduced iron, ferrous iron (Fe_2+), is highly soluble in water, and concentrations up to 50 ppm are possible under conditions of very low oxygen and low pH. The oxidized form, ferric iron (Fe_3+), is nearly insoluble in water and rapidly comes out of solution, resulting in a red-brown film on system interior surfaces. Generally, concentrations of >0.3 ppm can result in fouling of heat exchanger and other system surfaces if oxygen is introduced into the water. The primary strategy in avoiding iron fouling problems is the rigorous avoidance of any design that allows for the entrance of air into the system. The use of open tanks for storing the groundwater is unacceptable, because, in addition to allowing oxygen to enter the system, this also lets dissolved carbon dioxide (CO_2) escape, lowering pH and exacerbating scaling. The groundwater side of the system should be maintained full and under a slight positive pressure, even when not in operation, to preclude the entrance of air in water containing ferrous iron. Designs for accomplishing this are discussed in Chapter 8. Water should enter the injection well though a dip tube (Figure 7.6) submerged below the static water level to reduce turbulence and air entrainment.

Alkalinity, pH, hardness, and calcium content are all related to scaling and can be used to calculate two indices, the Ryznar stability index (RSI) and the Langlier saturation index (LSI). Whether the indices are provided by the testing laboratory or calculated based on the water analysis results, it is important that they be based on a temperature reflective of how the water is used in the system. The saturation index and stability index were developed in the 1940s to predict the relative rate of corrosion or scaling of steel and iron alloy piping in municipal water systems. Though GWHP systems tend to have few iron and iron alloy components, the stability and saturation indices are very useful as predictors of calcium carbonate scaling potential. Phenolphthalein alkalinity (or P alkalinity), along with methyl orange alkalinity (M alkalinity) and CO_2, are helpful in the determination of the type of hardness (carbonate or noncarbonate) present in the water.

Both indices are based on the calculation of a pH of saturation (pH_s) for calcium carbonate. Though the saturation index is more commonly used, it is useful to calculate both. The stability index value is calculated according to Equation 7.2:

$$2pH_s - pH \qquad (7.2)$$

where

pH_s = pH of saturation

pH = actual pH of the groundwater

Interpretation of the RSI is based on the data in Table 7.7.

Calculation of the LSI is based on the following formula:

$$\text{Saturation index} = \text{pH} - \text{pH}_S$$

Interpretation of the saturation index is based on Table 7.8.

Both the stability and saturation indices produce a numerical value that is indicative of the relative scaling or corrosion tendency of the water. As mentioned, in the case of GWHP applications the indices are used primarily as a qualitative scaling indicator. In view of this, for GWHP applications, corrosion results may be interpreted more as non-scaling results rather than as a reliable indicator of corrosion. It is important to note, however, that the scales of both indices are logarithmic. As a result, a saturation index value of 2.0 suggests a rate of scale deposition approximately 32 times that of an index of 0.5.

The pH_S value is calculated based on the following formula:

$$\text{pH}_S = (9.3 + A + B) - (C + D)$$

where

A = (log(TDS) – 1)/10 ppm
B = (–13.12 log (((°F – 32)/1.8) + 273)) + 34.55 (I-P)
= (–13.12 log (°C + 273)) + 34.55) (SI)
C = (log (calcium hardness)) – 0.4 ppm as $CaCO_3$
D = log (M alkalinity) ppm as $CaCO_3$

In Example 7.3, the difference between the saturation index result for 85°F and 150°F (29.4°C and 65.6°C) suggests a propensity for scale formation approximately 4.5 times greater at the higher temperature. It is clear from these results that the higher temperatures encountered in the direct use of the groundwater (directly in the heat pump units) will result in a much higher propensity for scale deposition than the system using the isolation heat exchanger when operated with the same groundwater. There are also implications here for domestic hot-water heating applications. In desuperheaters and dedicated hot-water-heating heat pumps, hard water can result in scale deposition due to the high tem-

Table 7.7 Interpretation of the Ryznar Stability Index (Carrier Corp 1965)

Index Value	Interpretation
4.0 – 5.0	Heavy scale
5.0 – 6.0	Light scale
6.0 - 7.0	Balanced
7.0 – 7.5	Corrosion
7.5 – 9.0	Heavy corrosion
>9.0	Extreme corrosion

Table 7.8 Interpretation of the Langlier Saturation Index (Carrier Corp 1965)

Index Value	Interpretation
2.0	Heavy scale
0.5	Slightly scale forming
0	Balanced
–0.05	Slightly corrosive
–2.0	Serious corrosion

EXAMPLE 7.3—
EVALUATING SCALING POTENTIAL

Groundwater has the following chemistry:

- pH 8.2
- Ca hardness 165 ppm
- M Alkalinity 100 ppm
- Temperature 55°F (12.8°C)
- Total dissolved solids 500 ppm

Calculate pH_s, the saturation index, and the stability index.

Solution

A = (log(500) –1)/10 = 0.17
B = (–13.12 log(12.8 + 273)) + 34.55 = 2.33
C = log 165 – 0.4 = 1.82
D = log 100 = 2.0
pH_s = (9.3 + 0.17 + 2.33) – (1.82 + 2.0) = **7.98**

In this example, the calculated pH_s at the 55°F (12.8°C) temperature (indicative of the character of the groundwater at its undisturbed temperature) yields the following results in terms of the saturation and stability indices:

Saturation index = pH – pH_s = 8.2 – 7.98 = 0.202 (balanced)
Stability index = 2pH_s – pH = 2(7.98) – 8.2 = 7.76 (heavy corrosion)

As mentioned previously, these results in the context of a GWHP application would be considered nonscaling. The critical consideration in using the saturation and stability indices, however, is that the calculations be made based on a temperature reflective of what the water will encounter in the system. In a system with an isolation heat exchanger, the maximum surface temperature that water will encounter is approximately 85°F (29.4°C), as GWHP systems rarely operate with building loop temperatures exceeding this value (see Table 8.1). In a system in which the water is delivered directly to the heat pump units, the groundwater may encounter a temperature of approximately 150°F (65.6°C) in the hot-gas end of the refrigerant-to-water heat exchanger in cooling mode. Recalculating the results at these temperatures yields the following:

At 85°F (29.4°C):
B = (–13.12 log(29.4 + 273)) + 34.55 = 2.00
pH_s = (9.3 + 0.17 + 2.00) – (1.82 +2.0) = 7.65
Saturation index = 8.2 – 7.65 = 0.55 (slightly scale forming)
Stability index = 2(7.65) – 8.2 = 7.1 (corrosion)

At 150°F (65.6°C):
B = (–13.12 log(65.6 +273)) + 34.55 = 1.36
pH_s = (9.3 + 0.17 + 1.36) – (1.82 + 2.0) = 7.01
Saturation index = 8.2 – 7.01 = 1.19 (moderate scale)
Stability index = 2(7.01) – 8.2 = 5.82 (light scale)

peratures encountered. In space-conditioning applications, the annual quantity of operating hours in the cooling mode is also an important consideration with respect to scaling. Obviously, the greater the number of hours in cooling mode, the greater the tendency of scale deposition, as the temperatures encountered in heating-mode operation will reduce or eliminate scale formation. Removal of calcium carbonate scale can be accomplished by circulating an acid solution through the portion of the system where the deposition has occurred.

Chloride content is a contributor to corrosion of most metal alloys and is particularly injurious to 300 series stainless steel under some conditions. Under conditions of elevated temperature and chloride content, some stainless alloys are subject to pitting corrosion. Guidelines for selection of materials relative to chloride content are covered in Table 8.12 and Section 8.6.2. It is unusual for nonsaline groundwater to exhibit elevated chloride content, but it is possible in some settings. Heat exchanger plates, well screens, and potentially well pump components are the most common stainless steel components in GWHP systems.

Carbonate and bicarbonate constitute the largest portion of the alkalinity present in most groundwater. These constituents, in conjunction with pH and dissolved carbon dioxide, are also useful in checking the accuracy of a water analysis. The relative presence and concentrations of carbonate and bicarbonate are a function of the pH of the water and thus provide a check on the analytical results. Generally carbonate alkalinity exists above a pH of approximately 8.5. Bicarbonate alkalinity exists between pH 4.3 and 8.5.

Alkalinity is a measure of the ability of the water to buffer acids. It is usually reported in ppm as $CaCO_3$ equivalent. Two measures of alkalinity are commonly found in water chemistry results: M or total alkalinity, which measures all alkalinity above pH 4.3, and P alkalinity, which measures alkalinity above pH 8.3 (usually constituted by carbonate and hydroxyl alkalinity). M alkalinity is a key value in the calculation of the LSI and RSI. Three useful rules (Carrier 1965) arise from alkalinity results:

- If P alkalinity = 0, all alkalinity is caused by calcium, magnesium, and sodium bicarbonates and the water pH is < 8.5.
- If 2 × P alkalinity < M alkalinity, alkalinity is from a combination of calcium, sodium, and magnesium carbonates and bicarbonates and the pH of the water is > 8.5.
- If 2 × P alkalinity > M alkalinity, there is no bicarbonate alkalinity and all alkalinity is from calcium, sodium, and magnesium carbonates and hydroxides and the pH of the water is > 8.5.

If a water analysis reported a P alkalinity of 60 ppm and an M alkalinity of 85 ppm with a pH of 7.6 there would obviously be an error, as 2(60) > 85, so the pH should be >8.3. If erroneous results are obtained, a new sample should be collected and analyzed to determine where the error occurred. Errors in laboratory analysis results do occur, and it is important to carefully review results before system design decisions. Some consultants routinely send samples to two different laboratories to compare results.

Hardness, like alkalinity, is closely linked to scale deposition. Two types of hardness can be present in water: carbonate hardness (also known as *temporary hardness*) and noncarbonate hardness (also known as *permanent hardness*). Of these, carbonate hardness (arising from calcium and magnesium carbonates and bicarbonates) holds a far greater potential for scale deposition, as the solubility of noncarbonate hardness (from sulfates, chlorides, and nitrates) is some 70 times greater. Rules associated with hardness (Carrier 1965) include the following:

- When M alkalinity > total hardness, all hardness is caused by carbonates and bicarbonates.
- When M alkalinity < total hardness, carbonate hardness = alkalinity and noncarbonate hardness = total hardness – M alkalinity

These rules are sometimes helpful if analysis results omit total hardness or M alkalinity. With the remaining values the missing parameter can be calculated.

Hardness, and the scale it produces, is the number-one water quality problem in the United States. Water hardness is typically interpreted as indicated in Table 7.9.

Scaling problems are possible with waters of 100 ppm hardness and above, particularly at pH 7.0 and above (Rafferty 2004).

Carbon dioxide can be present in groundwater and is often a controlling factor in pH. As a dissolved gas, CO_2 is best tested in the field, but laboratory testing can be done provided samples are properly handled. If dissolved CO_2 is present and is allowed to evolve or outgas from the water (as may occur when water is stored in unpressurized piping or open tanks), the pH of the water rises and carbonate scaling may occur. One of the primary reasons for maintaining the groundwater side of systems under pressure is to prevent this occurrence. The pressure necessary to maintain the CO_2 in solution depends on the concentration. However, at concentrations less than 1000 ppm, the partial pressure of the CO_2 amounts to less than 5 psi (35 kPa).

Oxygen, like CO_2, is a dissolved gas and is associated with corrosion of iron and brass alloys if present. Generally, groundwater from depths >100 ft (>30 m) does not contain oxygen as it has been consumed through oxidation reactions with organic materials in the subsurface. Oxygen can enter an aquifer if the well drawdown is sufficient to allow water from nearby rivers or lakes to be drawn in. Mixing of oxygenated water from a surface source or shallow aquifer with iron-bearing water from another aquifer can result in serious plugging of aquifer materials and can negatively impact well production rates. As with CO_2, sample handing is critical to accurate laboratory test results and field testing is recommended.

Hydrogen sulfide (H_2S) is a dissolved gas resulting from either volcanic geologic settings or biological activity of sulfate-reducing bacteria (in water containing sulfate). When present, H_2S in concentrations greater than 0.5 ppm result in a "rotten egg" odor in the water. Copper and copper alloys are very susceptible to corrosion from H_2S at concentrations of as little as 0.5 ppm. Copper and cupronickel piping have failed in as little as five years as a result of exposure to H_2S concentrations of <1 ppm (Rafferty 1989).

Iron bacteria is a general term referring to a variety of organisms that inhabit aquifers and can colonize water wells. Contrary to popular belief, the organisms do not feed on iron components in the system; they tend to proliferate in locations where they can access dissolved iron in the water. They metabolize the iron and in the process produce thick gelatinous secretions that can seriously impair water flow. This most frequently occurs on well screens. A more complete discussion of treatment of iron bacteria infestations is presented in Appendix N.

Table 7.9 Hardness Classification

Calcium Hardness	Interpretation
<15	Very soft
15 to 50	soft
51 to 100	Medium hard
101 to 200	Hard
>200	Very hard

Biological testing for the presence of iron bacteria, slime-forming bacteria, and sulfate-reducing bacteria can be done either in the laboratory or by using a self-contained field test kit (often referred to as a *BART kit*, for bacteriological activity reaction test). In either case there are limitations. Testing for these organisms is complicated, and the interpretation of laboratory results should be done by a microbiologist or other professional familiar with the species of interest. Bacteria of all kinds are present in groundwater, and results normally confirm this fact. The mere presence of the bacteria, however, does not provide certainty that they will proliferate sufficiently to become a problem. A survey of nearby well owners regarding their experience with the water, in conjunction with the analytical results, interpreted by a professional, is key to gaining a full understanding of the microbiological character of the water. Appendix M provides an example of a biological analysis of a groundwater sample.

One common biological test for groundwater samples is the adenosine triphosphate (ATP) test, which quantifies the total viable bacteria population in the sample. A typical potable water well casing sample will yield a result of 30,000 to 65,000 cells/mL. A value in excess of 100,000 cells/mL indicates a concern for potential biofouling (Schnieders 2013).

Self-contained field test kits are available for a variety of commonly encountered organisms. Among those often used in the context of heat pump systems are tests for iron-related bacteria, slime-producing bacteria, and sulfate-reducing bacteria. These tests are accomplished by adding a small water sample to a prepackaged test kit equipped with a nutrient that stimulates growth of the specific bacteria of the test. After the water sample is added, the container is observed for several days for a visible change in appearance. The time required for the reaction to occur is an indication of the aggressiveness of the bacteria and the likelihood of future problems associated with that particular species. These test kits are manufactured for specific bacteria, and multiple kits must be used if more than one species is to be tested for. Although the tests provide a qualitative indication of future problems, they are most effective for monitoring a well on an ongoing basis. Substantial changes in the reaction time can be used as an indication of developing problems.

Sand, if present, is a suspended rather than a dissolved constituent in the water. It is typically not a problem in terms of passing through the system, where concentrations of as much as 20 ppm or more will pass through most components. The two areas where sand can be a problem are in the well pump (erosion) and the injection well (plugging). As discussed in Section 7.4.3, sand production can be minimized through careful design of the well and proper development after the well is completed. In some cases, however, it is not possible to prevent sand from entering the production stream, and the sand must be removed at the surface. When surface separation is required, strainers are the recommended device; centrifugal separators are not designed to achieve the level of removal necessary for injection, and their effectiveness is compromised by well cycling and variable-speed operation. Screen perforation size should be selected based on a sieve analysis (see Section 7.4.3) of the sand produced during the pump test of the well. In many cases two or three strainers in parallel are necessary to reduce pressure drop, particularly for removal of fine sand.

7.6 REFERENCES

AWWA. 1997. ANSI/AWWA A100-97, *AWWA Standard for Water Wells.* Denver, CO: American Water Works Association.

BR. 1995. *Ground Water Manual: A Water Resources Technical Publication*, 2d Ed. Washington, DC: U.S. Department of the Interior, Bureau of Reclamation.

Carrier Corp. 1965. *Handbook of Air Conditioning System Design*. New York: McGraw-Hill.

Driscoll, F.G. 1986. *Groundwater and Wells*, Second Edition. St. Paul, MN: Johnson Screens.

National Water Well Association. 1981. *Water Well Specifications: A Manual of Technical Standards and General Contractual Conditions for Construction of Water Wells*. Berkeley, CA: Premier Press.

NGWA. 2014. ANSI/NGWA-01-14, *Water Well Construction Standard*. Westerville, OH: National Ground Water Association.

Rafferty, K. 1989. *A materials and equipment review of selected US geothermal district heating systems*. Klamath Falls, OR: Geo-Heat Center.

Rafferty, K. 2004. Water chemistry in geothermal heat pump systems. *ASHRAE Transactions* 110(1).

Ralston, D. 2000. Design and construction of water wells for consultants. Course materials. Moscow, ID: Ralston Hydrologic Services.

RMC. 1985. *The Engineers' Manual for Water Well Design*. Los Angeles, CA: Roscoe Moss Company.

Sachs, H. 2002. *Geology and Drilling Methods for Ground-Source Heat Pump System Installation: An Introduction for Engineers*. Atlanta: ASHRAE.

Schnieders, M. 2013. Well rehabilitation: Part II, a case study. *Water Well Journal*, September.

USGS. 1995. *Ground Water Atlas of the United States*. Reston, VA: U.S. Geological Survey. http://pubs.usgs.gov/ha/ha730/gwa.html

USGS. 2014a. National Water Information System: Web Interface. Reston, VA: U.S. Geological Survey. http://waterdata.usgs.gov/nwis

USGS. 2014b. Water Science Centers Directors. Reston, VA: U.S. Geological Survey. http://water.usgs.gov/district_chief.html

WW. 2011. WaterWebster™. www.waterwebster.com/state_framebottom.htm

8 Groundwater Heat Pump System Design

8.1 INTRODUCTION

8.1.1 Background

Groundwater heat pump (GWHP), or open-loop, system design details, inside the building, are in most respects the same as those for ground-coupled heat pump (GCHP) systems. Specification and connection of heat pump units to the building loop, outdoor air strategies, and loop piping follow closely the guidelines offered in previous chapters. Loop pump guidelines (covered in Section 8.3) are based on somewhat lower pump head for GWHP systems, but loop flow rates are similar. The major difference is the ground-loop portion of the system. In open-loop systems, a small number of wells (usually one to three) provide groundwater to a plate heat exchanger that interfaces with the building loop. After it passes through the heat exchanger, all of the groundwater is returned to the ground through a similar number of injection (or reinjection) wells. It is the groundwater loop portion of the system on which this chapter focuses.

Groundwater flow in an open-loop system is analogous to loop length in a closed-loop system. The greater the loop length in a closed-loop system, the better the performance of the heat pumps as a result of more favorable operating temperatures. A key part of closed-loop design is optimizing performance versus loop cost. In open-loop systems, the greater the groundwater flow the better the performance of the heat pumps. Although there is a cost associated with increasing groundwater flow (larger wells and pumps), the more significant issue is the impact of higher well pumping power (associated with increases in groundwater flow) and its impact on system performance. A key part of open-loop design is optimizing (or at least understanding) groundwater flow with respect to system performance.

Open-loop heat pump systems were the first commercial applications of GSHP systems, with the earliest examples developed by J.D. Kroeker at sites in Oregon and Washington in the early 1950s (Knipe and Rafferty 1985; Hatten 1992). These systems, central-plant based (unitary heat pumps had not yet been developed) and using the groundwater directly in the chilled-water and heating-water loops, were, with subsequent modifications, quite successful; some remain in operation today. Properly applied, open-loop systems can offer substantial advantages in terms of capital costs while still producing system performance comparable to closed-loop systems. The most compelling advantage of open-loop systems is reduced capital cost. Figure 8.1 provides a comparison of

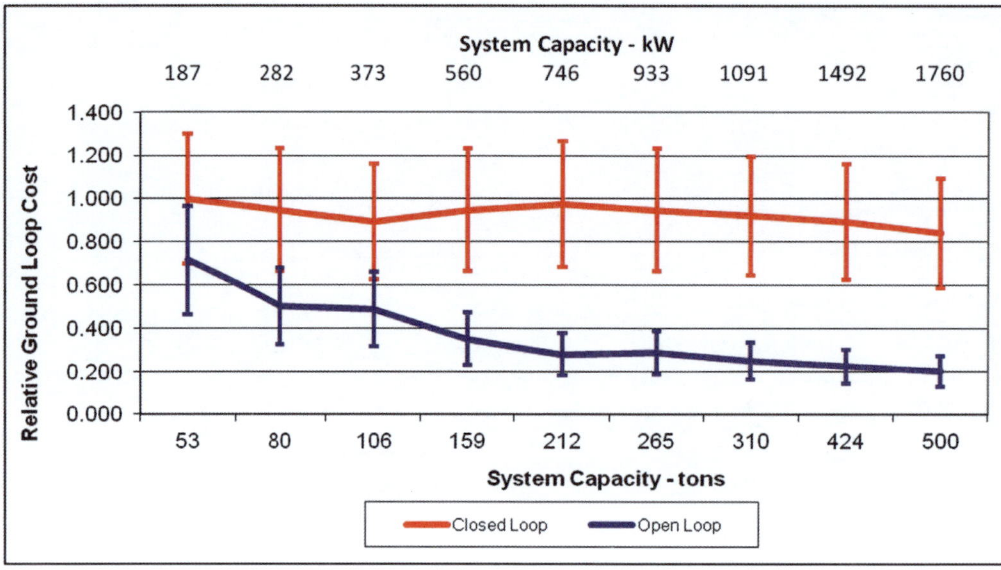

Figure 8.1 GWHP and GCHP Relative Ground-Loop Costs (Rafferty 2008)

open- and closed-loop costs for the ground-loop portion of the system. It is clear that above approximately 150 tons (528 kW), open-loop systems can offer ground-loop costs of as little as 20% of those of closed-loop systems under the most favorable conditions. While maintenance costs for open-loop systems are greater than those for closed-loop systems, the incremental maintenance cost is small in the context of the capital cost advantage (see Section 8.6.3).

Although there is a strong likelihood of reducing capital cost by using the open-loop approach in suitable applications, closed-loop designs likely will remain the most common system type in commercial and institutional settings. The reason for this is related to the necessary characteristics for a favorable open-loop application. Chief among these is an available groundwater aquifer at the site. In addition, as indicated previously, open-loop attractiveness tends to increase with system size, and large GSHP systems of any type are a small percentage of total installations. The regulatory framework must be receptive to the use of the groundwater, and the design team must be comfortable with the technology. Of these issues, design team receptiveness and aquifer availability are the most common barriers to greater GWHP system use.

8.1.2 GWHP Issues

Any discussion of GWHP systems should be prefaced by briefly addressing commonly held perceptions concerning their operation and implementation. Open-loop systems were widely applied in residential settings starting in the early 1960s and increasingly so in the 1970s. In the course of this early use, a number of problem areas were encountered. Some of these issues are unique to residential applications. Others are encountered in commercial applications, but the nature of larger system design offers effective strategies to address them that are unavailable in residential-scale systems. These issues are discussed in detail in the following list.

- ***Groundwater Quality.*** Groundwater quality is an issue in any mechanical system through which it flows. In residential open-loop systems, the groundwater is typically delivered directly to the heat pump units and any corrosion, scaling, or

Table 8.1 Approximate Heat Pump EWTs for GWHP Systems

Groundwater Temperature, °F (°C)	EWT at 1 gpm/ton (0.0179 L/s·kW)	EWT at 1.5 gpm/ton (0.0269 L/s·kW)	EWT at 2.0 gpm/ton (0.0358 L/s·kW)	EWT at 2.5 gpm/ton (0.0448 L/s·kW)
50 (10)	71 (21.7)	62 (16.7)	56 (13.3)	53 (11.7)
60 (15.6)	81 (27.2)	72 (22.2)	66 (18.9)	63 (17.2)
70 (21.2)	91 (32.8)	82 (27.8)	76 (24.4)	73 (22.8)

Basis: Building loop at 2.5 gpm/ton (0.0448 L/s·kW), heat exchanger approach 3°F (1.7°C).

fouling problems are encountered both in the heat pumps and throughout the system. In commercial open-loop systems, a plate-and-frame heat exchanger isolates the building loop from any exposure to the groundwater. The heat exchanger itself is designed to be disassembled and cleaned, and the remaining portion of the system exposed to the groundwater is limited and constructed primarily of nonmetallic piping. As a result, water quality problems and the associated maintenance costs are substantially reduced in commercial systems relative to residential installations.

- ***Thermal Impact.*** The thermal impact of a building containing any GSHP system (open or closed loop) on the ground or groundwater is a function of the building and its thermal loads only. The type of system it contains has virtually no impact on the magnitude of the thermal impact on the subsurface. An open-loop system does more directly deliver the thermal load to the groundwater. However, an aquifer penetrated by the boreholes of a closed-loop system preferentially absorbs heat relative to surrounding soil and rock. In fact, many closed-loop systems partially depend upon aquifers to reduce the local long-term thermal impact to the ground that might otherwise occur. As all aquifers are flowing (albeit very slowly), any heat signature is rapidly dissipated by heat transfer to the surrounding aquifer materials (soil, rock, clay, sand, and gravel).
- ***Pumping Power.*** Open-loop systems are characterized by higher pumping power than closed-loop systems. However, they often operate at much more favorable loop temperatures than do closed-loop systems, resulting in *system performance* (when heat pumps and well pumps are considered together) comparable to closed-loop systems. Table 8.1 provides some typical entering water temperatures (EWTs) for open-loop systems. Residential systems are characterized by much higher pump head than many commercial applications due to the use of the well pump to satisfy both the heat pump flow and high-pressure domestic needs. When coupled with the very low efficiency of fractional horsepower submersible pumps/motors (frequently in the 25% wire-to-water efficiency range) and the high water flows used in residential heat pump applications (2 to 3 gpm/ton [0.036 to 0.054 L/s·kW]—as much as twice commercial application needs), the unit well pumping power in small residential applications is far in excess of what typical commercial applications require.
- ***Aquifer Water Level Impact.*** In most jurisdictions injection of the water after use in the heat pump system is the default design. As a result, all of the flow is returned to the aquifer and there is no potential for long-term aquifer drawdown of the type that can result from surface disposal. Many older systems were designed with surface disposal of the groundwater; this constitutes a consumptive use of the water and can negatively impact aquifer water level over time. Though it is possible to use surface disposal under some conditions without neg-

ative aquifer water level impact, most modern, well-designed systems incorporate reinjection to ensure sustainability.

- ***Regulatory Framework.*** Although there are limited areas where open-loop systems are effectively prohibited, this is rare and often traceable to a reaction to poor early system designs. Limitations on production flow per well, well spacing, well function, and similar criteria can influence design, but the reality is that an open-loop system is a viable option in most jurisdictions. The U.S. Environmental Protection Agency, under the Underground Injection Control (UIC) program rules, specifically identifies allowance for geothermal injection wells (Class V) for disposal purposes (EPA 1975). In fact, in many if not most states, closed-loop borehole regulations were developed directly from existing water well administrative rules. As a result, production-well regulatory framework mirrors closed-loop borehole rules. GWHP systems do encounter a separate layer of regulatory oversight in western states with water rights systems.

8.1.3 GWHP Design Variants

Various designs have been applied to commercial open-loop heat pump systems, and each has advantages and disadvantages. In simplified form, the most common appear in Figure 8.2. In the simplest (Figure 8.2, upper left), groundwater is delivered directly to the heat pump units and groundwater flow is controlled via refrigerant pressure control valves or by motorized valves at each unit. Essentially a larger version of residential design this approach offers low capital cost but is very susceptible to water quality problems, requires high groundwater flow rates (typically 2+ gpm/ton (0.036+ L/s·kW)) often falls victim to the higher pump power problems of residential systems. This design might be considered in the smallest commercial applications (<20 tons (70 kW) but must be limited to areas of pristine water quality.

The standing column system (Figure 8.2, upper right) is a clever design developed in New England for areas of unfractured hard-rock geology (initially thought to be too costly for closed-loop drilling) that produce very little groundwater (and thus are unsuitable for conventional open systems). The systems operate with very low bore length (75 ft/ton [5.4 m/kW]) relative to closed-loop systems but relatively deep wells (1500 ft [460 m]) compared to conventional open-loop systems. Operating temperatures lie between open- and closed-loop temperatures. The major drawback to wide application of standing column systems is the direct use of the groundwater in the heat pumps, as this leaves the system open to water quality problems. A second issue is the temperature control or "bleed" rate of roughly 10% to 15% of the groundwater circulation rate to waste. Although small in terms of percentage of circulation rate, the volume can be substantial on an annual basis and has come under increasing regulatory scrutiny in recent years. Standing column systems have been widely used in New England but have made limited penetration in the balance of the country. For the region of the country in which they were developed, particularly in residential applications where the domestic water use displaces a portion of the bleed flow and where superior water quality and unfractured hard-rock geology preclude conventional open-loop systems, they can be a reasonable option. Design of standing column systems is covered in detail by Egg et al. (2013).

Conventional open-loop systems (Figure 8.2, bottom) are the focus of this chapter. In this design, the groundwater, typically at flows in the range of 1 to 2 gpm/ton (0.018 to 0.036 L/s·kW), is delivered to a plate heat exchanger that isolates the building loop from exposure to the groundwater. Building loop flow is in the same range as for closed-loop systems (2.75 to 3.0 gpm/block ton [0.05 to 0.054 L/s·kW]). In this way, the building loop

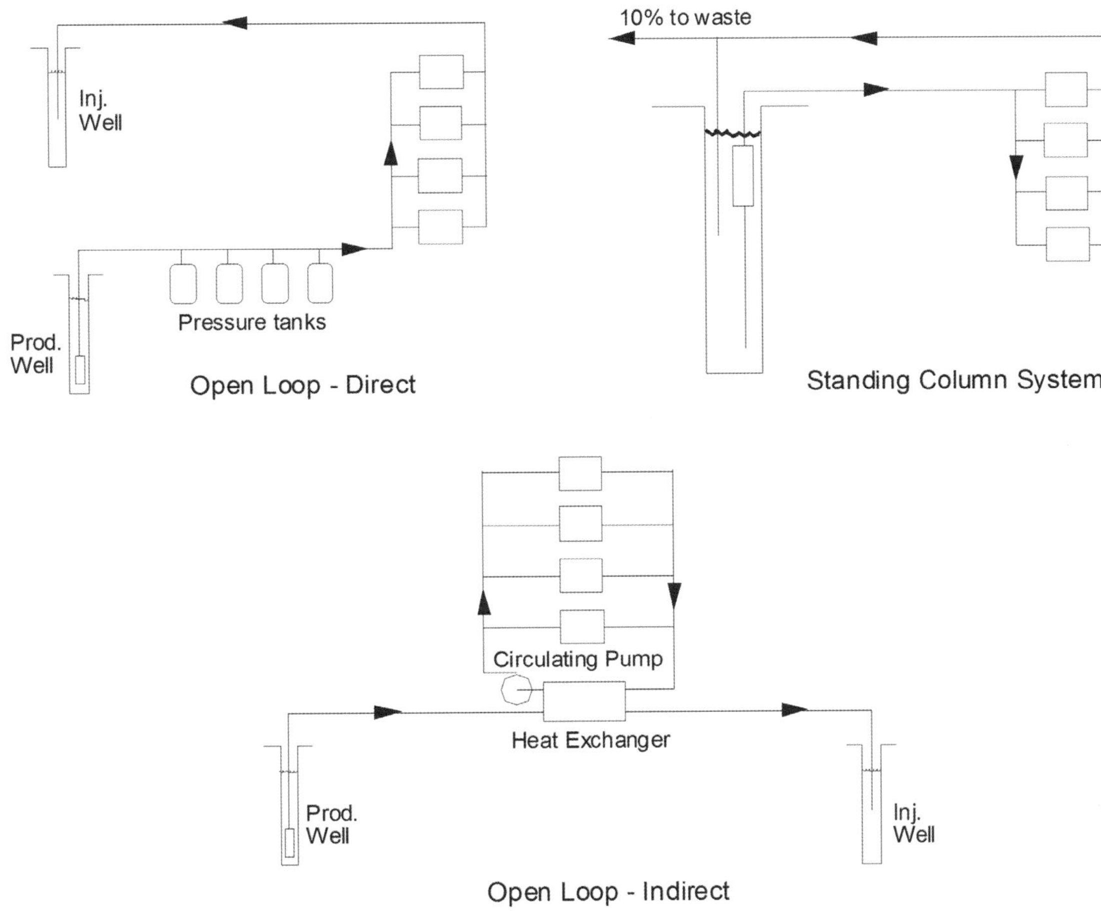

Figure 8.2 GWHP System Design Variants

is operated at a flow rate optimum for heat pump performance and the groundwater at a flow rate optimum for well pump power. The production-well pump responds to loop temperature with variable flow or cycling of the production wells or wells. Systems of this type have been installed throughout the United States in capacities up to several thousand tons.

Central-plant-based systems (not illustrated in Figure 8.2) typically consist of central chillers connected to a groundwater source via heat exchangers located in the chilled-water and condenser-water loops. The heat exchangers can be used to reject heat from the condenser loop in cooling-dominated mode and load the evaporator in heating-dominated mode. Central-plant systems, as the name implies, are composed of large, centrally located heating, cooling, and air circulation equipment. GSHP systems, regardless of the type, generate a substantial portion of the capital and operating cost savings on the basis of their use of unitary heat pumps in the zones. This eliminates much of the auxiliary energy use associated with delivering air though extensive duct systems and chilled and hot water though extensive piping loops. Coupling central-plant equipment to the ground or groundwater simply does not produce the operating cost savings that unitary heat pump designs do. In addition, the costs of central-plant GSHP systems typically far exceed unitary GSHP designs. The earliest commercial building open-loop systems (in the 1950s) used central-plant designs because unitary heat pumps were not yet available in the market. Central-plant GSHP systems are advisable only in very rare cases and only when maximum energy efficiency is not the primary goal (see Section 2.4, Tables 2.8 and 2.9).

8.2 GENERAL DESIGN APPROACH

With the exception of standing column systems, the performance of the heat pumps in GWHP systems, in terms of energy efficiency ratio (EER) or coefficient of performance (COP), increases with increasing groundwater flow. Consider a system operating in the cooling mode. For a given building cooling load, the higher the groundwater flow rate, the smaller the temperature rise on the groundwater side and the lower the leaving groundwater temperature. Regardless of whether an isolation heat exchanger is used, the lower the exit groundwater temperature, the lower the heat pump leaving water temperature and the higher the heat pump EER (COP_c) (despite a constant EWT). A similar relationship exists in the heating mode.

Although EWT is commonly used in discussing heat pump performance, it is actually the leaving water temperature (LWT) that determines unit performance. In the cooling mode, as heat is transferred from what is basically a constant-temperature process (condenser), it is the exit water temperature and the approach (the temperature difference between the exit water temperature and the refrigerant) that determines the minimum temperature at which the condensation can occur.

The effect of this is illustrated in Figure 8.3. This is a plot of the heat pump performance for a very simple system consisting of a single water-to-air heat pump unit operating in the cooling mode and supplied with groundwater from a well at a constant EWT—similar to what would occur in a residential application. As the groundwater flow to the heat pump unit is increased, the power consumption of the unit decreases (EER [COP_c] increases) due to the decreasing LWT. The figure illustrates the general trend in performance. Minimum flow in a specific heat pump is a function of the configuration of the refrigerant-to-water heat exchanger (maintaining minimum water velocity), and maximum flow is limited by head loss. The system also includes a well pump to deliver the water. Figure 8.4 illustrates well pumping power requirements assuming the case of a typical residential application in which the pump is producing to a pressure tank. As the flow for which the system is designed is increased, pumping power requirements increase.

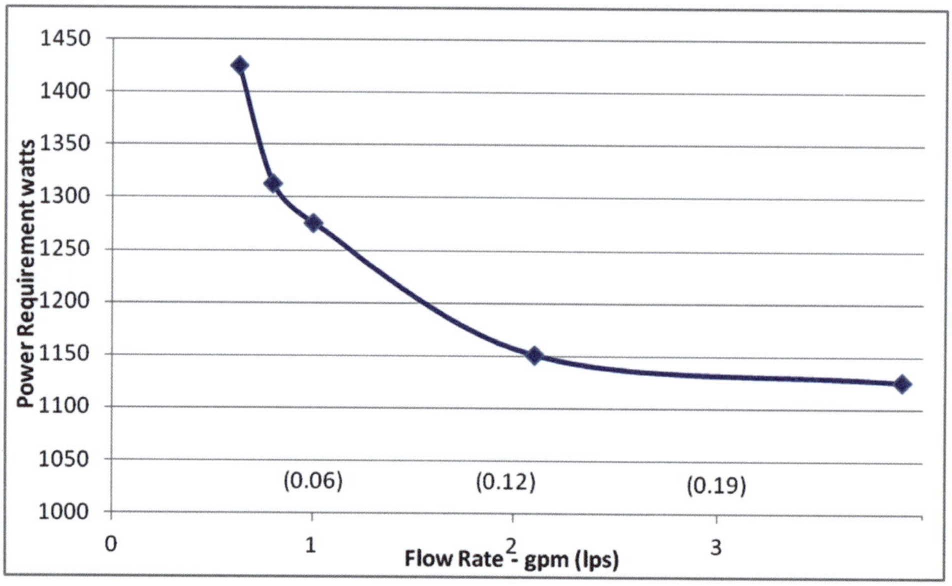

Figure 8.3 Heat Pump Performance vs Groundwater Flow

The simple system illustrated in Figures 8.3 and 8.4, in which the impact of increased groundwater flow has the opposite effect on the power consumption of the two components composing the system, constitutes a classic case of optimization. As groundwater flow is increased (Figure 8.5), total system power consumption (heat pump plus well pump) decreases to a point, reaches a minimum, and then increases. On the left side of the curve shown in Figure 8.5, the incremental gains in heat pump performance due to higher flow (and decreasing LWT) outweigh the incremental increases in pumping power requirements to provide that flow. On the right side of the curve, the increasing pump

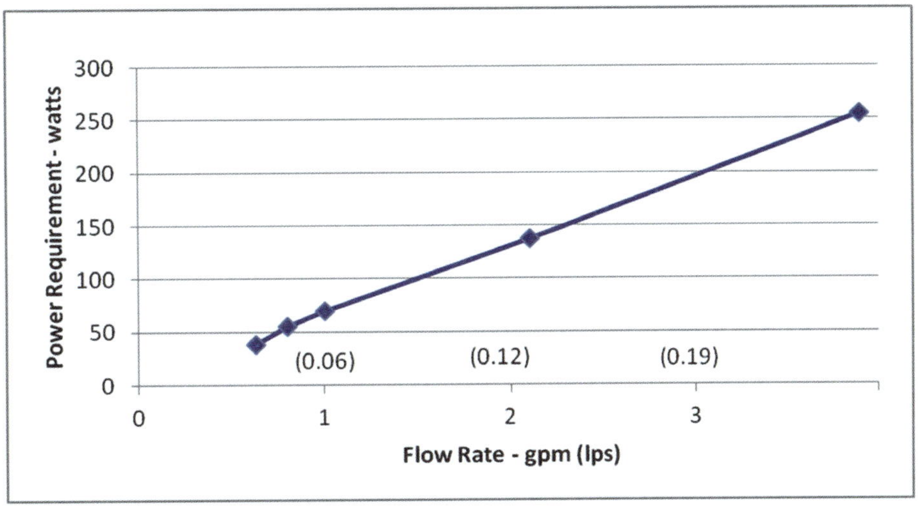

Assumes no drawdown and production to average tank pressure of 45 psi (310 kPa) and 21 ft (6.4 m) static water level (SWL) for a total head of 125 ft (373 kPa), 35% wire-to-water efficiency.

Figure 8.4 Well Pumping Power Requirement

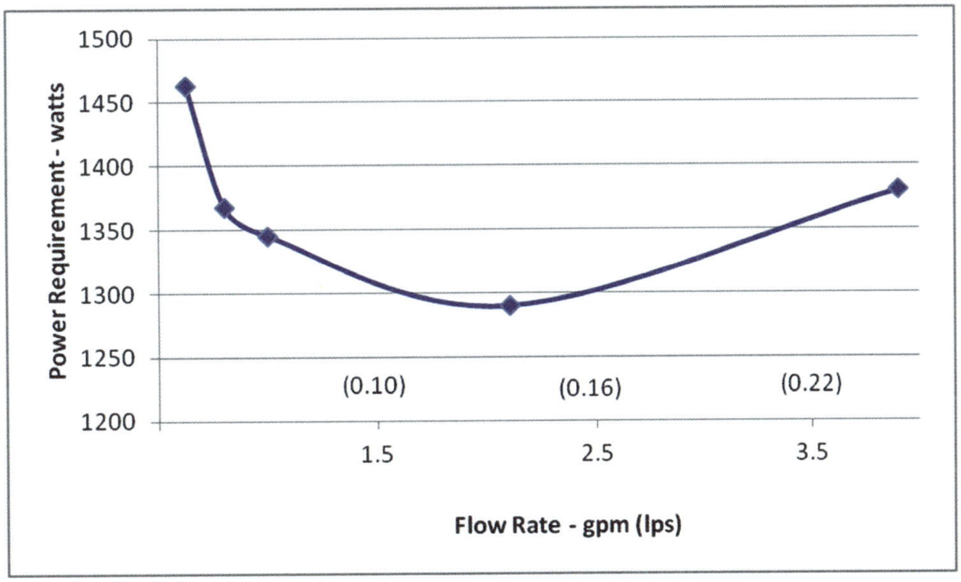

Figure 8.5 System Power Requirement vs Groundwater Flow

power outweighs incremental improvements in heat pump performance. There is a clear optimum point at which the system power consumption is minimized (system EER maximized). Groundwater flows above this point, although they result in lower heat pump power consumption, compromise overall system performance due to higher pumping power. **Every open-loop system, regardless of size and complexity, is characterized by this general relationship with an optimum flow (optimum flows are different for heating and cooling modes) for maximum *system performance.* The goal of the designer is to gather the information necessary to determine this flow at the design condition and then, as closely as possible, configure the system around it**.

Under no circumstances is it advantageous to design for flows in excess of the optimum, as this results in increased capital cost, increased well maintenance cost, and decreased system performance. There are situations (injection-well overpressurization, regulatory limitations, etc.) when it is sometimes useful to consider operation at flows somewhat less than optimal. Unfortunately, many past systems have been designed for flows well in excess of optimum values. This resulted from a focus on heat pump performance with insufficient attention paid to well pump power and system performance. It is understandable, as HVAC engineers are unaccustomed to addressing aspects of the design occurring outside the building wall, such as well pump issues. Just as in closed-loop design, where the engineer must be involved in the ground-loop design in order to produce a cost-effective and reliable system, in open-loop design the engineer must be involved in the design of the wells and well pumps as well as understand their impact on system cost and performance.

Figure 8.6 presents the results of a calculation for an office building with an 85 ton (299 kW) cooling load. To illustrate the impact of groundwater temperature and well pumping conditions on optimum groundwater flow, the calculation was run for four different cases. The blue curves represent performance with 50°F (10°C) groundwater and the red curves with 65°F (18.3°C) groundwater. The solid lines are reflective of performance at low-head well pumping conditions—75 ft (23 m) static water level (SWL) and 10 gpm/ft (2.07 L/s·m) specific capacity (SC). The dotted lines are reflective of performance at high-head well pumping conditions—300 ft (91 m) SWL and 3 gpm/ft (062 L/s·m) SC. It should be no surprise that when encountering conditions of high well pumping power requirements (low SC, deep SWL), the optimum groundwater flow tends toward lower values in terms of gpm/ton (L/s·kW) and toward higher flows at more favorable conditions (shallow SWL, high SC). The shape of the curves at low SC tends to be characterized by a more prominent peak and at high SC tends to be somewhat flatter. Clearly groundwater temperature allows for a higher system performance as a result of the more favorable temperatures. In high-pump-head cases the optimum groundwater flow amounts to values in the 1.25 to 1.3 gpm/ton (0.022 to 0.023 L/s·kW) range, and in low-pump-head cases the optimum flow is in the range of 2.1 to 2.5 gpm/ton (0.038 to 0.045 L/s·kW) in this example. The very flat nature of the curve shape for the low-pump-head cases allows for the design of a system with much lower groundwater flow than optimum (in this case 1.75 gpm/ton versus 2.1 to 2.5 gpm/ton (0.032 L/s·kW versus 0.038 to 0.045 L/s·kW)) while still preserving nearly maximum system performance. While it is not possible to arbitrarily alter the groundwater temperature or the well pumping conditions, these curves illustrate the wide variation in optimum groundwater flow resulting from groundwater conditions and the need for careful consideration of these system parameters to identify the most favorable groundwater flow rate for a particular case.

To determine the optimum flow rate for a given application, certain key information is required:

- Building block loads
- Building loop flow rate (gpm/ton [L/s·kW])

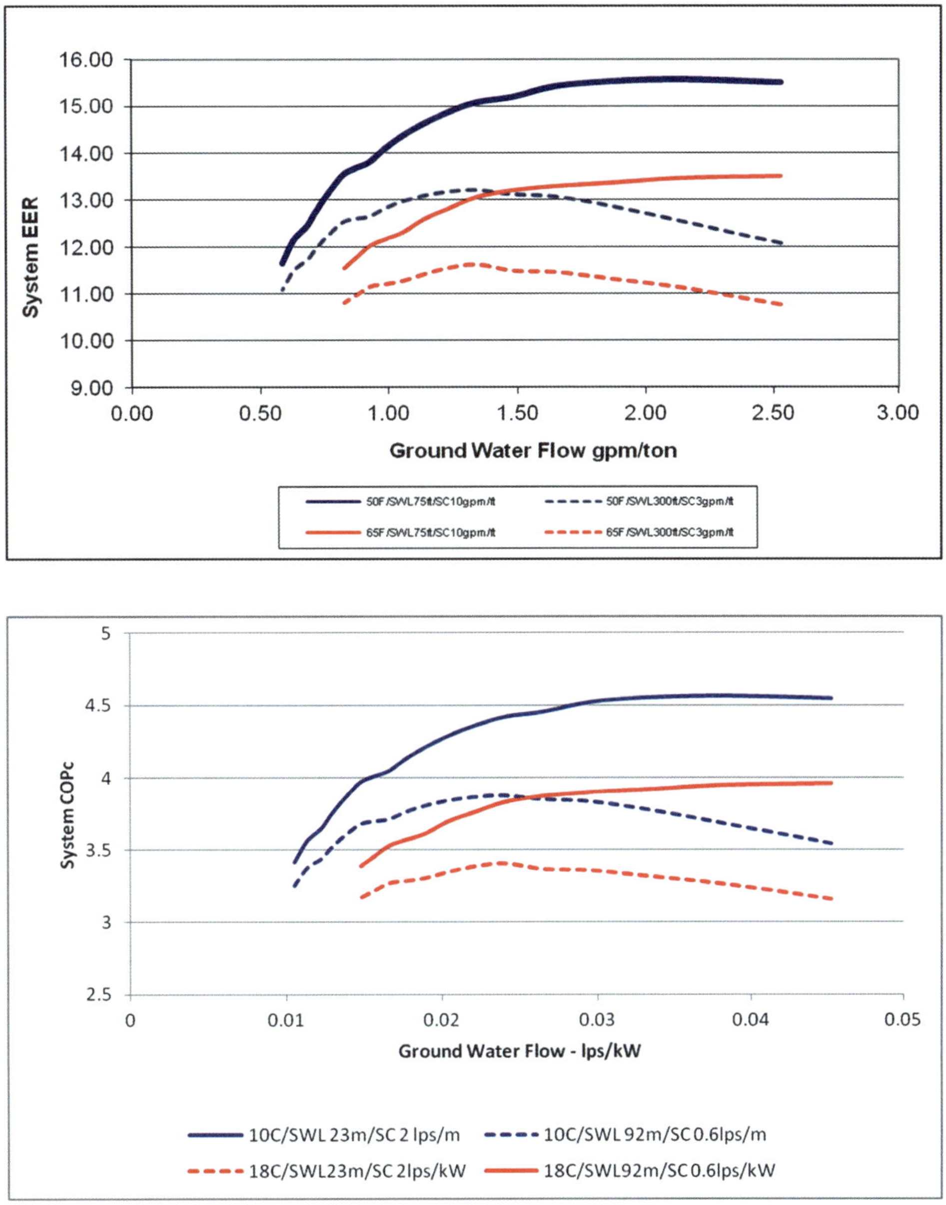

(a)

(b)

Figure 8.6 Example Optimum Groundwater Flow Rates, (a) I-P and (b) SI

- Heat pump performance (manufacturer's data) at various EWTs
- Production-well static water level
- Production-well specific capacity
- Groundwater temperature
- Platc heat exchanger approach temperature

Of these, the production-well information (SWL, SC, and groundwater temperature) obviously require information from the well itself, and this implies the necessity of drilling and testing prior to final design in order to provide the necessary information. As in the case of closed-loop design, information about the subsurface is necessary to complete the open-loop design, and the best way to accomplish this is to complete and test the well (or wells) prior to final design. Unfortunately this is not always possible and the designer must occasionally proceed on some assumptions. In many cases there is sufficient experience with the aquifer and there are a number of existing wells near the site to provide an estimate of the necessary information. Procedures for and sources of data for this purpose are discussed in Section 7.5.1.

8.2.1 Design Process—Steps

The key part of the design of any open-loop system is the system performance evaluation. This is the process of calculating the system EER (COP_c) over a range of EWTs to define the point at which the maximum *system EER* occurs. The major components (well pump, heat exchanger, piping, etc.) are selected for the duty associated with this operating point.

In brief, the approach to performance calculations associated with GWHP systems proceeds as follows, with the numbers in parentheses indicating the steps in Figure 8.7. In a system configured as in Figure 8.7 and operating in the cooling mode, a loop flow rate is established (1) along with an initial EWT for the heat pumps (2). This information, along with manufacturer's data for the heat pumps, permits the calculation of EER (COP_c) (or COP, depending on the mode to be evaluated), heat of rejection, and the LWT from the heat pumps (3). The LWT from the heat pumps is the same as the loop water temperature entering the plate heat exchanger (4). With an assumption of a heat exchanger approach temperature (5), the groundwater temperature leaving the heat exchanger can be determined (loop temperature – approach = groundwater leaving temperature) (6). With the loop heat of rejection and the groundwater ΔT (groundwater temperature – groundwater heat exchanger leaving temperature), it is possible to calculate the groundwater flow required to meet the load rejected at the heat exchanger (7). With groundwater flow and

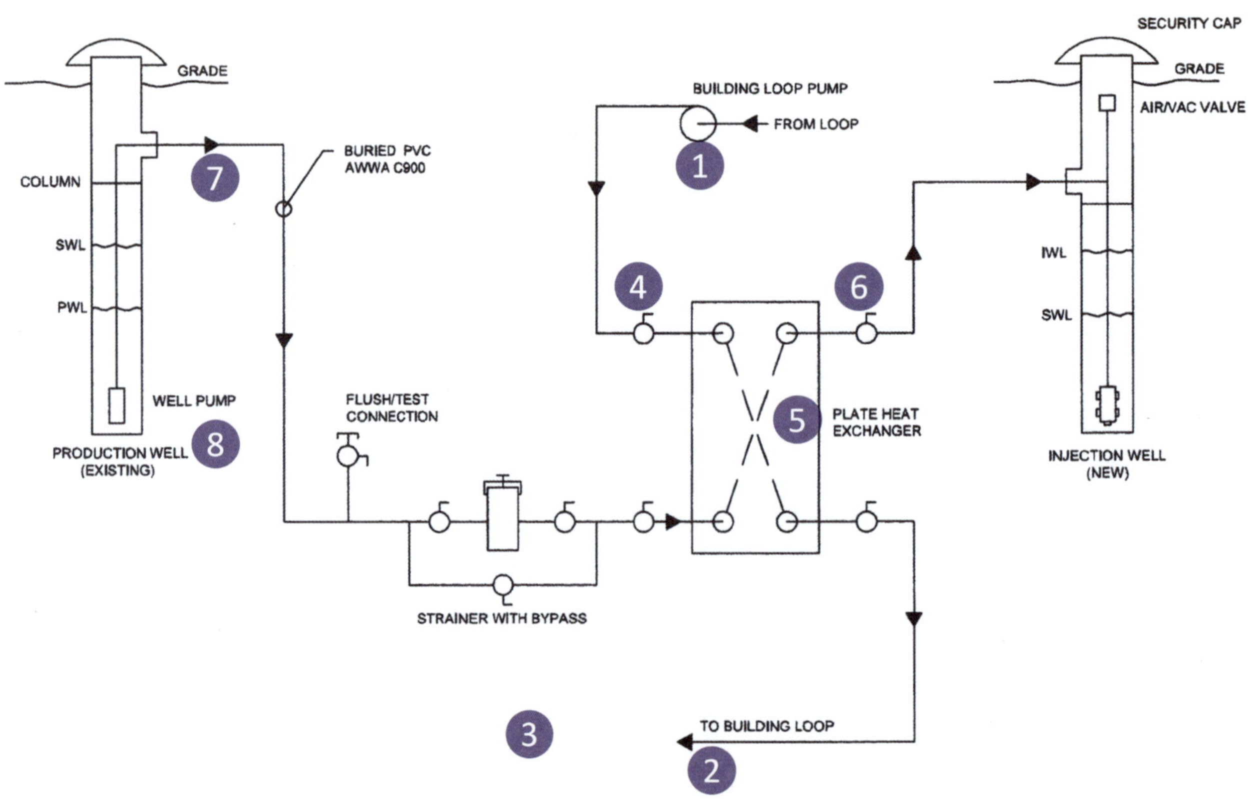

Figure 8.7 System Performance Evaluation Steps

the results of the well flow test (or information from nearby wells), the drawdown in the production well can be calculated and combined with an assumed surface head loss (for the groundwater piping and heat exchanger) to provide a calculated well pump head (8), pump horsepower, and well pump power requirement. The well pump power, the heat pump power, and the loop pump power requirements are then summed to calculate a system EER (COP_c) (or COP). This process is repeated over a range of heat pump EWTs to create a table or graph of system performance versus groundwater flow. The conditions that produce the peak system EER are the ones for which the equipment (well pump, piping, heat exchanger, etc.) is selected and for which the system is designed. Typically the cooling mode establishes the peak heat exchanger thermal duty and groundwater flow rate for commercial buildings. The lower groundwater flow requirement for the secondary mode (usually heating) can then be satisfied by variable-frequency drive (VFD) control of the well pump at a lower flow rate, cycling of the well pump, and staging of multiple well pumps or multiple wells to arrive at the necessary flow.

As suggested by this brief description, the process of making the necessary calculations to evaluate system performance over a range of groundwater flows is tedious and iterative, as is the case with closed-loop calculations, and is best accomplished with a spreadsheet or program designed for this purpose and based on the procedures outlined in this chapter. Fortunately, commercial software is available for some of the calculations necessary for open-loop design. **It is not possible by inspection, guesswork, or crystal ball to determine the optimum conditions under which a particular open-loop application should operate. It is only possible to make this determination based on the type of calculation described here.**

8.2.2 Keys to Success

Key principles in the course of open-loop design (discussed in greater detail later in the chapter) include the following:

- ***Do your homework.*** Understand the regulatory setting, and research local experience with the groundwater and wells.
- ***Test and analyze.*** Drill and test the wells early, secure well flow test results, and analyze/understand the water quality.
- ***Isolate the groundwater.*** To eliminate exposure of the building loop to the groundwater, use a plate heat exchanger with approach temperatures typically in the 2°F to 4°F (1.1°C to 2.2°C) range; note that 316 stainless steel plates and nitrile butadiene rubber (NBR) gasket materials are often satisfactory.
- ***Optimize groundwater flow.*** Flow is typically in the 1 to 2 gpm/ton (0.018 to 0.036 L/s·kW) range, but calculations should be made to verify optimum flow. The goal is to maximize *system* performance (considering power requirements of the heat pumps, loop pump, and well pump).
- ***Maintain system pressurization.*** Keep the groundwater side of the system full and pressurized to the extent possible; no open tanks should be used on the groundwater side of the system.
- ***Ensure particulate separation.*** Remove sand either with production-well completion and design or through surface separation. Strainers are the most effective means; base screen openings on a sand sieve analysis.
- ***Use injection for disposal.*** Injection should be the default disposal method, down gradient from the production well, with a 0.05 ft/s (0.015 m/s) screen velocity and an effective seal on the casing. Use a dip tube and pressure sustaining device to maintain the injection line full and pressurized.
- ***Ensure well spacing.*** Use adequate spacing between production and injection wells.

8.3 PRODUCTION/INJECTION WELL SEPARATION

As mentioned in Chapter 7, two wells located sufficiently close together will interfere with each other, and their respective drawdown or buildup impacts will be superimposed upon each other should this occur. In the case of a production well and an injection well spaced too closely, the drawdown from the production well intersecting with the buildup from the injection well will result in an artificially high gradient between the two wells, facilitating excessive water flow from the injection well to the production well. This condition can result in undesirable temperatures at the production well if sufficient flow occurs between the two wells. As a result, one of the principal questions associated with the design of a GWHP system is how far apart the production and injection wells must be to prevent or minimize this condition. It is important to understand that it is not necessary to separate the wells to such a distance that zero flow occurs between them. Water leaving the injection well must pass through hundreds of feet (metres) of soil, rock, sand, gravel, and water before reaching the production well. In the course of this, heat is exchanged with the aquifer materials, bringing the injected water temperature close to the natural aquifer temperature. Thus, some flow can be permitted between the wells, but they must be separated sufficiently to prevent excessive flow between them.

One approach is to flow-test the production well while monitoring a nearby observation well to determine aquifer transmissivity and storage coefficient values. Using this data and aquifer analysis software, it is possible to make the necessary calculations for spacing determination. This approach, under the direction of a hydrologist, should be used in all cases in which there are multiple production or injection wells required or in complex geologic/hydrologic settings. For most open-loop systems, operating with a single production well and a single injection well, the method developed by Kazmann and Whitehead (1980), with some modification, should be sufficient for establishing minimum production/injection well separation distance.

This method (Kazmann and Whitehead 1980) was developed for calculating the necessary separation distance between production and injection wells for open-loop heat pump systems. It is intended for applications characterized by geologically homogeneous settings dependent principally upon primary permeability (pore spaces between the aquifer materials rather than fractures in rock) and is based on groundwater flow, aquifer thickness, aquifer porosity, and the length of time the system operates in the dominant mode. The original method was based on the assumption that the well flows would be reversed as the system changed from heating to cooling. That is, the method assumed that at the end of the cooling season the production-well and injection-well functions would reverse, thus allowing the warm water (in the region of the cooling-mode injection well) to be delivered to the system during the heating season. The wells were then assumed to be reversed again at the beginning of the cooling season. In many regulatory jurisdictions wells must be designated as either production or injection and reversing functions is not permitted. For many if not most large commercial applications, the marginal thermal benefit derived from reversing wells seasonally may not compare favorably to the cost associated with equipping wells to perform dual duty. The cost of an additional screen to permit a well to serve as both a production and an injection well and the complex piping necessary to accommodate such operation, coupled with the fact that many commercial buildings remain in the cooling-dominated mode most of the year, tend to render this strategy of questionable value in larger commercial applications. To accommodate systems in which well function remains fixed (no switching of production and injection wells), spacing information was developed for this book that is based on the original Kazmann and Whitehead data (which accommodated up to 210-day dominant-mode operation) for the

365-day pumping period (i.e., no well reversing); this information is included in Table 8.2 in place of the original 100-, 140-, and 210-day operating modes (which assume seasonal well reversing) published by Kazmann and Whitehead.

A second issue in the original method (Kazmann and Whitehead 1980) relates to the flow rate used to calculate the separation distance. This was based on a seasonal average flow rate. In very low operating hour applications (e.g., nine-month schools), this average flow approach yields a very low value for the effective flow and appears to understate the necessary separation distance. For commercial applications it is recommended that no less than 50% of the peak groundwater flow rate used in the left column of Table 8.2. Provided the original limitations associated with the Kazmann and Whitehead method are observed (as to type of aquifer materials)—that the injection well is located down gradient of the production well and that not less than 50% of the peak groundwater flow is used to make the spacing determination—the values in Table 8.2 provide a guide for minimum production/injection well spacing.

Table 8.2 Minimum Production/Injection Well Spacing

Average Flow Rate, Q, gpm (L/s)	Aquifer Thickness, ft (m)						
	10 (3)	20 (6)	30 (9)	40 (12)	50 (15)	80 (24)	100 (30)
10 (0.6)	176 (54)	157 (48)	150 (46)	139 (42)	128 (39)	114 (35)	103 (31)
20 (1.2)	242 (74)	218 (67)	208 (63)	196 (60)	175 (53)	159 (49)	137 (42)
30 (1.9)	301 (92)	270 (82)	252 (77)	234 (71)	216 (66)	188 (57)	171 (52)
40 (2.5)	352 (107)	313 (95)	291 (89)	277 (85)	254 (77)	223 (68)	195 (59)
50 (3.2)	394 (120)	356 (109)	332 (101)	308 (94)	280 (85)	250 (76)	222 (68)
60 (3.8)	437 (133)	388 (118)	356 (109)	341 (104)	320 (98)	270 (82)	243 (74)
70 (4.4)	477 (145)	424 (129)	387 (118)	367 (112)	347 (106)	290 (88)	262 (80)
80 (5.0)	513 (156)	456 (139)	413 (126)	388 (118)	369 (113)	310 (95)	279 (85)
90 5.7)	547 (167)	483 (147)	433 (132)	412 (126)	395 (120)	331 (101)	298 (121)
100 (6.3)	582 (177)	511 (156)	462 (141)	437 (133)	398 (121)	350 (107)	316 (96)
200 (13)			652 (199)	627 (191)	573 (175)	508 (155)	456 (139)
300 (19)					680 (207)	610 (186)	550 (168)
400 (25)					790 (241)	683 (208)	626 (191)
500 (32)					897 (273)	764 (233)	690 (210)
1000 (63)					1282 (391)	1082 (330)	990 (302)

Note: Table values based on aquifer porosity value of 20%. For porosity of 10%, multiply values by 1.05. For 30%, multiply values by 0.95.

EXAMPLE 8.1— WELL SPACING

A school using a GWHP system has a peak flow rate of 167 gpm (10.5 L/s) and is producing from an aquifer of 40 ft (12 m) thickness. Determine the minimum separation distance for the injection well.

Solution

The effective flow rate for use in Table 8.2 is calculated first:

$$167 \text{ gal/min} \times 0.50 = 83.4 \text{ gpm} \qquad \text{(I-P)}$$

$$10.5 \text{ L/s} \times 0.50 = 5.25 \text{ L/s} \qquad \text{(SI)}$$

From Table 8.2, interpolating for 83 gpm (5.25 L/s) at a 40 ft (12 m) aquifer thickness results in a minimum separation distance of 395 ft (120 m).

8.4 BUILDING LOOP PUMPING FOR GWHP

The building loop piping arrangement in GWHP systems is, in most respects, the same as that used in GCHP systems. The primary difference is in the use of the plate heat exchanger (GWHP) in place of the ground loop (GCHP). As a result, the total loop pump head in GWHP systems is somewhat reduced relative to GCHP systems. A general guideline for building loop pump head in GWHP systems, assuming a plate heat exchanger pressure drop of 11.6 ft (5 psi) (3.5 m [35 kPa]), piping unit loss at 4 ft/100 ft (4 m/100 m), heat pump pressure drop of 10 ft (3 m), and a fittings allowance of 20% of the piping loss is

$$[(PL/100) \times 2 \times 1.2 \times 4.0 \text{ ft}/100 \text{ ft}] + 10 \text{ ft} + 11.6 \text{ ft} = (0.1\ PL) + 23 \text{ ft} \qquad \text{(I-P)}$$

$$[(PL/100) \times 2 \times 1.2 \times 4 \text{ m}/100 \text{ m}] + 3 \text{ m} + 3.5 \text{ m} = (0.1\ PL) + 6.5 \text{ m} \qquad \text{(SI)}$$

where PL is the piping length in ft (m) from the heat exchanger to the most distant heat pump.

For a loop flow rate of 3.0 gpm/block ton (0.05 L/s·kW), pump efficiency of 70%, motor efficiency of 85%, and a pipe length of 250 ft (76 m), this results in a value of 4.6 hp/100 tons (0.96 kW/100 kW). This value corresponds to a grade of A in terms of Table 6.2.

8.5 WELL PUMPS

Two types of well pumps are available for the range of flow rates normally required in open-loop systems: lineshaft and submersible. The lineshaft pump is characterized by an electric motor located on the surface that rotates a shaft extending down into the well and connected to the pump. Water is delivered to the surface through a pipe (known as the *column*) connected to the pump discharge. The column also houses the bearings supporting the shaft. This type of pump has only rarely been used in open-loop systems. Because of the surface electric motor and piping connections, an enclosing structure is sometimes required for protection of lubricating oil, plumbing, and electrical connections—the presence of which is avoided by the designers of office buildings and schools.

More commonly applied in GWHP systems are submersible pumps. Figure 8.8 provides an introduction to the terminology associated with this equipment. The electric motor is located at the bottom of the assembly and is connected to the pump (also sometimes referred to as the *bowl assembly*) by a short section of driveshaft. Water enters the pump at the bottom of the bowl assembly after passing through an entrance screen located between the motor and pump. Submersible motors are cooled by the water passing over the outside of the motor, and the velocity of the water passing over the motor is an important parameter. For pumps installed in very large diameter casing, or in applications in which the well's production zone is above the pump, a cooling shroud or "can" (Figure 8.9) is necessary to ensure adequate water flow past the motor. Cooling shrouds are also routinely used in the case of motors operated in conjunction with a variable-speed drive (VSD). Pumps in all cases are multiple-stage designs with additional stages added as necessary to achieve the design pump head requirement for the application. Production flow is delivered from the outlet of the pump to the surface through a pipe known as the *pump column*. It has the additional purpose of supporting the weight of the motor/pump assembly. Power is delivered to the pump through cable typically attached to the pump column at intervals of 10 ft (3 m). Installation of the motor/pump/column assembly must be done with care to avoid damage to the cable or wiring. Submersible motors of the size used in GWHP applications are normally nominal 3600 rpm designs. In contrast to lineshaft pumps, which are normally nominal 1800 rpm or less, submersible pumps are more susceptible to damage from excessive amounts of suspended sand in the production stream. For this reason and other issues discussed in Section 7.4.7, care should be exercised in the design and development of the well to ensure as low a sand content as possible in the production water.

The surface configuration of the well can consist of the column exiting the well through the top of the casing (thus anchored by a surface plate as in Figure 8.8) in a concrete pit or through a pitless adaptor (or pitless unit in larger applications) facilitating a below-grade water piping connection to the well. Pitless adaptors attach to the well casing and facilitate the connection of the production or injection line to the column or drop pipe in the well. The pitless adaptor, available in line connection sizes up to approximately 3 in. (75 mm), also includes an O-ring sealed fitting that provides the dual function of facilitating a removable connection to the buried piping and support of the pump and column. For larger piping connections, a device referred to as a *pitless unit* is used. The unit is welded to the well casing below grade and extends to just above grade, usually a total length of 4 to 5 ft (1.2 to 1.5 m). The pitless unit includes the external connection for the production or injection line and an internal O-ring sealed "spool" into which the column or dip tube is threaded. The spool supports the weight of the column pipe and pump, provides connection to the buried piping to the mechanical room, and is configured for

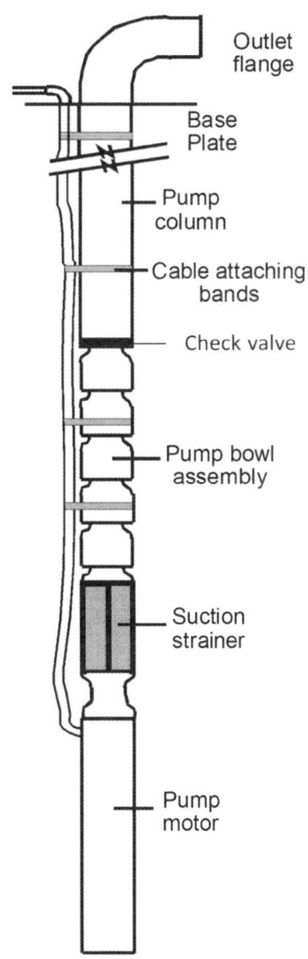

Figure 8.8 Submersible Well Pump Assembly

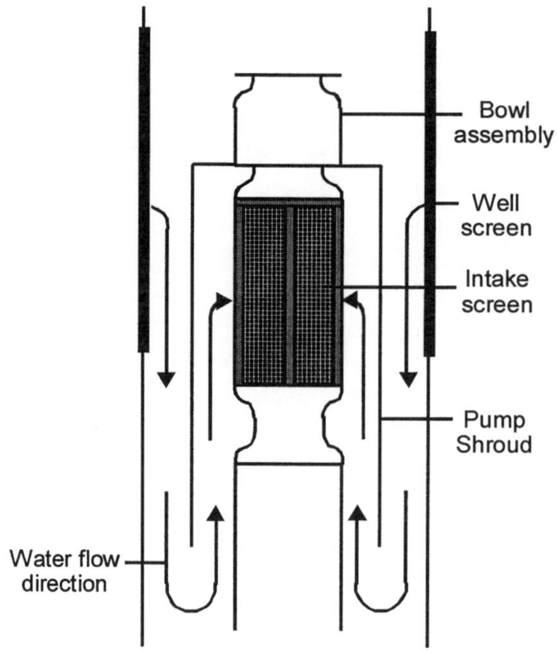

Figure 8.9 Submersible Pump Cooling Shroud

removal from the well to allow for pump replacement. For locations subject to freezing conditions, the below-grade piping connection to the well is advisable. This design also eliminates all but a small casing projection above the surface.

Most submersible motors come with a motor protection electronics package to ensure adequate protection from overload, underload, and short-cycling. Experience suggests that using the motor manufacturer's protection package is the best strategy in non-variable-speed applications. In variable-speed applications, the engineer must be certain that the factory protection is duplicated in the drive settings (discussed in Section 8.5.2). Submersible well pump motors are particularly susceptible to damage from lightning strikes and should always be equipped with surge or lightning arrestors properly grounded as per the manufacturer's instructions. In addition, starters should be quick-trip, ambient-compensated type, and overload relays should be carefully set to factory-specified current (Franklin 2007). Motors should be located at least 10 ft (3 m) off the bottom of the well to allow space for accumulation of sand and debris below the motor. Submersible pumps should always be equipped with a check valve as close to the pump discharge as possible.

Well pump selection is, like other pumps, based on flow and head, the details of which are familiar to most engineers. There is a departure from standard head loss practice, however, that arises primarily from the head loss components associated with the production well and injection well. In the course of the design of a GWHP system, the pump head is estimated first, and this value is used in the system evaluation to determine system performance versus groundwater flow (see Table 8.16). Once the optimum flow range is determined, a more detailed calculation of head loss can be made for the final design. A second issue that may depart from the experience of HVAC engineers is that most well pumps are multiple-stage devices. In many cases the manufacturer's performance curve represents a single stage and additional stages are combined to provide for the required pump head. In some cases small corrections for efficiency may be required if the number of stages required is less than five. Figure 8.10 presents a typical pump curve showing performance for two impeller diameters.

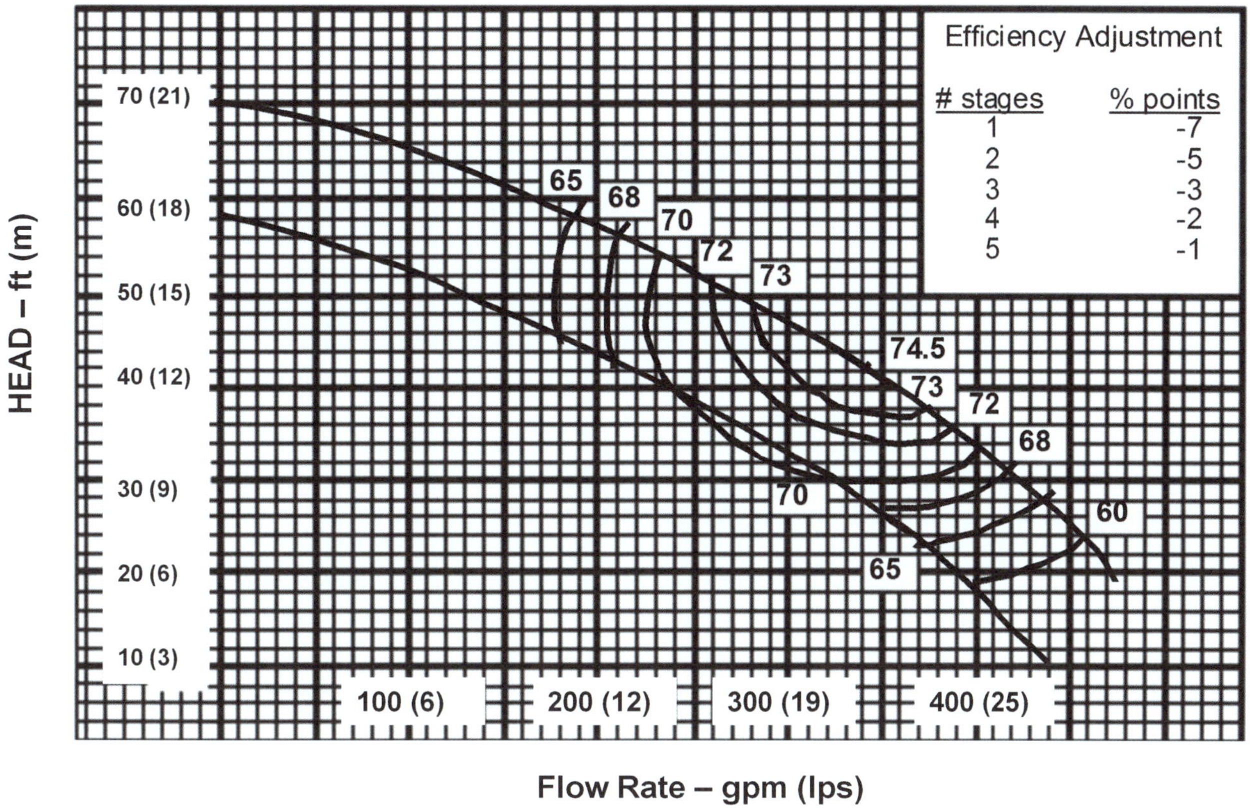

Figure 8.10 Well Pump Curve

EXAMPLE 8.2— PUMP SELECTION

Select a pump for 350 gpm at 200 ft (22 L/s at 61 m) total dynamic head (TDH):

Solution

The upper curve (for a full-size impeller) intersects the 350 gpm (22 L/s) flow rate at a head of approximately 40.2 ft/stage (12.3 m/stage). For the 200 ft (61 m) TDH requirement this results in a five-stage pump producing just over the required head at 201 ft (61.3 m). Note that the efficiency of this particular pump must be adjusted from the values appearing on the performance curve if fewer than six stages are used. Performance curve pump efficiency at the selection point amounts to 74%. In this case the efficiency penalty amounts to 1 percentage point based on the table in the upper right of the curve of Figure 8.10. As a result, the brake horsepower (bhp) requirement for the pump amounts to

$$\text{bhp} = (350 \text{ gal/min} \times 8.33 \text{ lb/gal} \times 201 \text{ ft})/(33{,}000 \text{ ft·lb/min} \times 0.73) = 24.3 \text{ hp} \qquad \text{(I-P)}$$

$$\text{bhp} = (22.1 \text{ L/s} \times 61.3 \text{ m} \times 9.8 \text{ kPa/m})/(1000 \text{ W/kW} \times 0.73) = 18.2 \text{ kW}$$

(Using Equation 6.10) (SI)

8.5.1 Well Pump Head Calculations

The components of well pump head loss include column friction loss, well static head (often referred to in pump jargon as *lift*), surface piping friction losses, and injection head (which can be positive or negative).

Column friction is the head loss in the pump column from the pump outlet to the ground surface. Because the column is straight pipe with no fittings (other than the check valve) and normally is limited in length, this component of head loss typically amounts to only a few feet (metres) barring unusual circumstances (extremely deep settings, unusually small diameter column, etc.).

The static head associated with the well is a result of the combination of SWL and drawdown (DD). It is the vertical distance water must be "lifted" by the pump to reach the surface. Lift is minimized in applications characterized by shallow SWL and high SC (low DD) and is maximized in settings with deep SWL and low SC.

Surface friction losses include all components from the production wellhead to the mechanical room and from the mechanical room to the injection well or disposal point.

Column friction is calculated in the same manner as for straight pipe in ordinary pumping applications. The length of the column is determined by the expected pumping level plus a submergence safety margin (typically 20 to 40 ft [6 to 12 m] for net positive suction head [NPSH] and seasonal water level variation) minus the length of the pump bowl assembly (normally less than 8 ft [2.4 m]). In an application with a 46 ft (14 m) SWL and a DD of 28 ft at 300 gpm (8.5 m at 1.9 L/s), the expected length of the column, assuming a 30 ft (9 m) safety margin, is

$$(46 + 28) + 30 - 8 = 96 \text{ ft} \qquad \text{(I-P)}$$

$$(14 + 8.5) + 9 - 2.4 = 29.1 \text{ m} \qquad \text{(SI)}$$

At an assumed head loss of 4 ft/100 ft (1.2 m/30 m), the estimated head loss for the column is 3.8 ft (1.2 m).

In the same application the lift portion of the pump head would amount to the sum of the SWL and the DD: 46 ft + 28 ft = 74 ft (14 + 8.5 = 22.5 m). The additional submergence for NPSH would not add to the pump head. For purposes of the system performance calculations, the lift at other groundwater flows can be calculated from the relationship

$$\text{lift} = \text{SWL} + (\text{flow/SC})$$

Although specific capacity (SC) is not a constant value, particularly in water table (unconfined) aquifers, the calculations can be rerun with a corrected value for SC once the optimal groundwater flow range is narrowed.

Surface losses include the piping, fittings, and heat exchanger losses between the production wellhead and the injection well. Since the distance between the wells may not yet be established at the time of initial calculations, it is necessary to allow a reasonable value for surface losses. Table 8.2 can provide a starting value for expected separation distance given flows under consideration and aquifer characteristics. For an effective flow rate of 150 gpm (9.5 L/s) (50% of the peak flow rate) and an aquifer thickness of 40 ft (12 m), a separation distance of approximately 532 ft (162 m) could be expected. Using a multiplier of 30% for fitting losses and routing around obstacles, an estimated total equivalent length of 692 ft (211 m) is arrived at. At 4 ft/100 ft (1.2 m/30 m) this results in a value of 27.7 ft (8.4 m) loss for piping and fittings. Adding an allowance for the heat exchanger of 7 ft (3 psi) (2.1 m [21 kPa]) results in a total surface head loss estimate of 34.7 ft (10.6 m).

Remember that in most GWHP applications the building loop side of the heat exchanger will have a higher flow rate and pressure drop than the groundwater side. Assuming that the building loop side is selected for a pressure drop maximum of 5 to 7 psi (35 to 48 kPa), the groundwater side will be lower.

Head requirements associated with the injection well can be positive, negative, or zero depending on the aquifer conditions at the site and the design of the system. In the previous example with a SWL of 46 ft (14 m) and a 28 ft (8.5 m) DD (at the production well), the theoretical injection water level (IWL) is 18 ft (5.5 m) below ground level (46 ft – 28 ft) (14 m – 8.5 m). This leaves considerable margin for avoidance of positive pressurization of the well itself assuming favorable conditions (rock geology, nonscaling water chemistry, sand-free injection water, good mud control with rotary drilling or a nonmud rotary drilling method). The pressure in the injection piping, however, is a function of the design of the dip tube. In many applications (where water chemistry suggests fouling) it is advisable to configure the dip tube to maintain a slight positive pressure on the groundwater side of the system to ensure the piping remains full and entry of air is prevented. Ideally this involves the placement of a valve at the bottom of the dip tube that modulates to maintain a positive pressure in the groundwater piping at the injection wellhead. Such valves are available, but they have rarely been used in GWHP applications to date due to cost. One design consists of a cylinder with small-diameter holes in the center section of the cylinder. A cage, also cylindrical in shape, fits into the outer cylinder and is equipped with a seal section. Moving the inner cylinder exposes or covers the holes, allowing variable flow to pass into the injection well. The position of the cage is modulated by hydraulic pressure transmitted through two small-diameter lines connected to the hydraulic power unit located on the surface. A second manufacturer uses a pneumatic design and a bellows-type valve arrangement. To date these valves, designed primarily for the aquifer storage and recovery industry, have seen little application in GWHP systems.

Four common configurations have been used for the injection dip tube (or drop pipe or injection column pipe), each resulting in a different impact on injection wellhead pressure and air entrainment:

1. Unvented dip tube with no pressure control valve (Figure 5.2).
2. Dip tube equipped with a vacuum-breaking/air-venting valve.
3. Dip tube equipped with a pressure-sustaining valve set to maintain a positive wellhead pressure at all times (Figure 8.11).
4. Dip tube equipped with a vacuum-breaking/air-venting valve and a pressure-sustaining valve set to maintain a positive pressure at the wellhead when the pump is operating (Figure 8.15).

When the potential exists for water-chemistry-related problems, the design should minimize or eliminate air entry into the system, necessitating the use of an automatic down-hole pressure control valve. Configuration 3 is preferred, but configuration 4 can be used in the event an excessive pump head penalty arises with configuration 3.

Configuration 1 can result in a negative head on the injection side of the system when the pump is in operation in applications where the IWL remains below the ground surface. As the injection flow descends in the dip tube, a vacuum is formed, effectively pulling the flow into the well. Vacuum is also sustained on pump shutdown as the water in the dip tube attempts to descend to the static water level. In most cases air eventually enters the injection piping (through fittings, etc.) to neutralize the vacuum on the pump off cycle, and this air is forced into the injection zone on pump restart. Although this configuration offers the prospect of reduced pump head under some conditions, the prospect for exposure of the piping to vacuum conditions and the likelihood of air being forced into the aquifer preclude recommendation of this design.

Configuration 2 eliminates the vacuum by placing an air/vac valve at the top of the dip tube, which promotes more stable operation by eliminating an intermittent vacuum and unventable air intrusion that would otherwise occur. This results in an injection wellhead pressure of zero (provided pressurization of the injection well itself is not necessary) under most conditions, but by admitting air to the piping this design tends to promote greater potential for water-quality-related problems in the injection well. The design also accounts for exposure to air in the dip tube as the column (above the static water level) is allowed to drain and fill with air on pump shutdown. The air is vented through the release valve at the top of the dip tube on pump start, but it is possible that a portion of it will be carried into the aquifer, as the air must rise vertically, against the direction of water flow in the dip tube to reach the vent valve. Some designers size the dip tube to compensate for any potential negative head by selecting the pipe size to result in a friction loss (as measured in feet [metres]) approximately equal to the depth of the IWL. This ensures a positive pressure at the top of the dip tube when the pump is in operation. This strategy is effective only in cases where the pump is operated at constant speed (dual setpoint loop temperature control, as discussed in Section 8.5.2). In variable-speed applications, a positive pressure is maintained in the dip tube only at higher flow rates. The vented dip tube configuration should not be used in any application in which the groundwater chemistry suggests a propensity for scaling, iron fouling, or biological fouling.

Configuration 3 is the arrangement most effective at preventing air from entering the injection side of the system under all conditions. A valve is placed at the outlet of the dip tube and set to ensure a positive pressure at the top of the dip tube, thus maintaining the injection piping full at all times. A spring-loaded check valve (or several such valves in parallel) are often selected for this duty. Spring-loaded check valves are available with adjustable crack pressure or with springs selected for a fixed crack pressure. Most manufacturers offer the valves only in smaller sizes (1 1/2 in. [37 mm] maximum), so in most applications multiple valves are required. The spring force is selected to result in a crack pressure (the pressure required to overcome the spring force and open the check valve) equivalent to the depth of the injection-well SWL. For example, if the injection-well SWL is 42 ft (12.8 m), the check valve would be selected for a crack pressure of 18 psi (124 kPa). The friction losses of the valve and the dip tube result in a slight positive pressure at the top of the dip tube when the pump is operating. The drawback is that when the pump is in operation and the water level in the injection well rises to the IWL, the pressure required to open the valve remains fixed but the differential pressure across the valve created by the columns of water inside and outside the dip tube decrease due to the rise in the well water level. This results in a pressure increase at the wellhead (when the pump is operating) that would not otherwise exist (in the absence of the valve). The magnitude of this pressure penalty is directly related to buildup (SWL minus IWL) in the injection well. For example, with a SWL of 100 ft (30.5 m) the check valve would normally be set for 45 psi (312 kPa)—100 ft SWL × 0.433 psi/ft (30.5 m SWL × 9.8 kPa/m). At design flow with an IWL at 25 ft (7.6 m), the pressure in the injection line at the wellhead would be 34.2 psi (236 kPa)—43 psi – (25 × 0.433) (312 kPa – [7.6 × 10 kPa/m])—resulting in an additional head of 79 ft (24 m) on the well pump. This configuration will result in the lowest injection-well maintenance requirements (as a result of the elimination of air entering the injection well), but the impact on well pump head (and power requirements) can be substantial, particularly in constant-speed well pump applications. The impact in a variable-speed application is the same at peak load (as for the single-speed pump), but energy requirements are reduced at part load due to operation at reduced flows (with reduced injection-well buildup and lower pressure at the wellhead). In the values appearing in Table 8.7, the pumping power penalty for this configuration is approximately 50% over that of the vented dip tube (configuration 2, Table 8.6).

Configuration 4 is an effective compromise between rigorous exclusion of air (configuration 3) and minimal impact on well pump head (configuration 2). In this case, the dip tube pressure control valve crack pressure is set for a value equivalent to the IWL. This requires the installation of an air/vac valve at the injection wellhead to allow the column of water in the dip tube to partially drain (to a level of SWL + IWL) when the pump is off. Although the air that enters the column in the off cycle is an undesirable aspect of this design, the volume of air is reduced relative to configuration 2, as the valve maintains a higher water level in the dip tube. In addition, the injection piping remains full, precluding air entry, during all operating conditions. With a high-volume air release valve at the top of the column, most of the air is vented when the pump starts. The combination of the valve set for the pressure equivalent of IWL along with an air/vac valve is a reasonable compromise in applications that would otherwise result in a large pump head penalty with the valve set for SWL (as in configuration 3). As indicated by the values of Table 8.7, the pumping power penalty associated with the configuration 4 design approximates 15% over that of the vented dip tube (configuration 2, Table 8.6).

It is important to distinguish here the difference between pressurizing the injection well and pressurizing the injection drop pipe. A positive pressure in the injection line at the surface of the injection well does not translate into a pressurization of the injection well itself, due to the air space existing between the IWL and the ground surface (Figure 8.11). As long as the IWL remains below the ground surface there is no danger of pressurizing the injection well.

In making calculations of well pump power requirements it is important to consider that submersible motors are typically somewhat less efficient than standard electric motors. Table 8.3 provides values for submersible motor efficiency compared to conventional electric motor efficiency values.

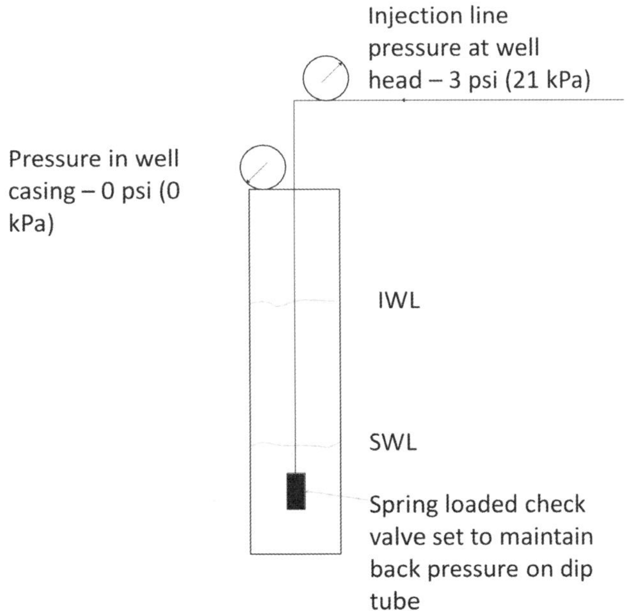

Figure 8.11 Injection Line Pressure vs Injection-Well Pressure

Table 8.3 Motor Efficiency—Submersible and Conventional Motors

hp (kW)	Submersible Motors	High-Efficiency Conventional Motors
2 (1.5)	79	84
5 (3.7)	79	85.5
7.5 (5.6)	79	87.5
10 (7.5)	80	88.5
15 (11.2)	81	89.5
20 (15)	82	90.2
25 (18.7)	83	91
30 (22.4)	83	91
40 (30)	83	91.7
50 (37.3)	83	92.4
60 (45)	84	93
75 (56)	84	93

Table 8.4 Well Pump Efficiency (Nominal 3600 rpm)

Design Flow Rate, gpm (L/s)	Pump Efficiency
25 (1.6)	60
50 (3.2)	62
75 (4.7)	70
150 (9.5)	72
200 (12.6)	75
500 (31.5)	77
750 (47.3)	78
1000 (63)	80
1250 (79)	82

Note: nominal 1800 rpm efficiency values somewhat higher.

Well pump efficiency varies with bowl diameter, flow rate, number of stages, and speed, but Table 8.4 provides approximate performance (of nominal 3600 rpm pumps) for purposes of system calculations.

Table 8.5 presents an example of production-well pump head and power requirements over a range of flow rates, illustrating the trend in head variation with flow rate. This comes from a spreadsheet constructed to make the necessary calculations previously described. In this particular spreadsheet, surface losses are calculated for a specific flow rate and pipe diameter and varied with the square of the flow. Pump column loss and injection-well dip tube head loss are included in the surface loss value. Column 3 calculates the IWL based on an injection-well SC 83% of the production-well SC. The next column displays the injection head resulting from the piping configuration. In this case it is possible, provided the injection dip tube remains full, that a portion of the negative head (minus the head loss in the dip tube) could reduce the total head on the well pump relative to the values shown in the total dynamic head (TDH) column.

Table 8.6 provides pump head values for the same application with a pressure-sustaining valve installed on the injection dip tube and selected to maintain positive pressure under all conditions (set for SWL)—corresponding to injection piping configuration 3

Table 8.5 Example Well Pump Head Values—Configuration 2

Groundwater Flow, gpm (L/s)	PWL, ft (m)	IWL, ft (m)	Injection Head, ft (m)	Surface Head, ft (m)	TDH, ft (m)	Well Pump Power, kW	Drawdown, % of Aquifer Thickness
120 (7.6)	55.0 (16.8)	−22.0 (−6.7)	0.0	8.1 (2.5)	63.1 (19.2)	2.7	16.7
166 (10.5)	60.8 (18.5)	−15.1 (−4.6)	0.0	15.4 (4.7)	76.2 (23.2)	4.5	23.1
189 (11.9)	63.6 (19.4)	−11.7 (−3.6)	0.0	20.0 (6.1)	83.6 (25.5)	5.6	26.3
212 (13.4)	66.5 (20.3)	−8.2 (−2.5)	0.0	25.2 (7.7)	91.7 (28.0)	6.9	29.4
235 (14.8)	69.4 (11.1)	−4.8 (−1.5)	0.0	30.9 (9.4)	100.3 (30.6)	8.4	32.6
258 (16.3)	72.3 (22.0)	−1.3 (−0.4)	0.0	37.3 (11.4)	109.5 (33.4)	10.1	35.8
281 (17.7)	75.1 (22.9)	2.2 (0.7)	2.2 (0.7)	44.2 (13.5)	121.5 (37.0)	12.2	39.0
304 (19.2)	78.0 (23.8)	5.6 (1.7)	5.6 (1.7)	51.8 (15.8)	135.4 (10.8)	14.7	42.2
327 (20.6)	80.9 (24.7)	9.1 (2.8)	9.1 (2.8)	59.9 (18.3)	149.8 (45.7)	17.5	45.4
350 (22.1)	83.8 (25.5)	12.5 (3.8)	12.5 (3.8)	68.6 (20.9)	164.9 (50.3)	20.6	48.6

Note: Based on SWL of 40 ft (12.2 m), production-well SC of 8 gpm/ft (1.66 L/s·m), injection-well SC of 6.7 gpm/ft (1.38 L/s·m), aquifer thickness of 90 ft (27.4 m), surface losses of 35 ft at 250 gpm (10.7 m at 15.8 L/s), pump efficiency of 70%, motor efficiency of 70%, and vented injection tube.

Table 8.6 Example Well Pumping Values—Configuration 3

Groundwater Flow, gpm (L/s)	PWL, ft (m)	IWL, ft (m)	Injection Head, ft (m)	Surface Head, ft (m)	TDH, ft (m)	Well Pump Power, kW	Drawdown, % of Aquifer Thickness
120 (7.6)	55.0 (16.8)	−22.0 (−6.7)	32.0 (9.8)	8.1 (2.5)	95.1 (28.9)	4.1	16.7
166 (10.5)	60.8 (18.5)	−15.1 (−4.6)	38.9 (11.9)	15.4 (4.7)	115.1 (35.1)	6.8	23.1
189 (11.9)	63.6 (19.4)	−11.7 (−3.6)	42.4 (12.9)	20.0 (6.1)	126.0 (38.4)	8.5	26.3
212 (13.4)	66.5 (20.3)	−8.2 (−2.5)	45.8 (14.0)	25.2 (7.7)	137.5 (41.9)	10.4	29.4
235 (14.8)	69.4 (11.1)	−4.8 (−1.5)	49.3 (15.0)	30.9 (9.4)	149.6 (45.6)	12.6	32.6
258 (16.3)	72.3 (22.0)	−1.3 (−0.4)	52.7 (16.1)	37.3 (11.4)	162.2 (49.5)	15.0	35.8
281 (17.7)	75.1 (22.9)	2.2 (0.7)	56.2 (17.1)	44.2 (13.5)	175.5 (53.5)	17.6	39.0
304 (19.2)	78.0 (23.8)	5.6 (1.7)	59.6 (18.2)	51.8 (15.8)	189.4 (57.7)	20.6	42.2
327 (20.6)	80.9 (24.7)	9.1 (2.8)	63.1 (19.2)	59.9 (18.3)	203.8 (62.1)	23.8	45.4
350 (22.1)	83.8 (25.5)	12.5 (3.8)	66.5 (20.3)	68.6 (20.9)	218.9 (66.7)	27.4	48.6

Note: Based on SWL of 40 ft (12.2 m), production-well SC of 8 gpm/ft (1.66 L/s·m), injection-well SC of 6.7 gpm/ft (1.38 L/s·m), aquifer thickness of 90 ft (27 m), surface losses of 35 ft at 250 gpm (10.7 m at 15.7 L/s), pump and motor efficiencies of 70%, an injection tube equipped with a spring-loaded check valve set for 17.3 psi (119 kPa) or SWL, and valve head loss of 14 ft (4.3 m).

previously discussed. This eliminates any potential for air to enter the injection piping or well but imposes a pump head penalty, resulting in higher pumping power requirements relative to the values in Table 8.5.

Table 8.7 illustrates the pumping values assuming injection configuration 4, in which a valve is placed on the dip tube and sized for a crack pressure equivalent to the IWL (as calculated for each specific flow in the table). Note that at the highest flow rates, as in the previous tables, the well itself becomes pressurized. While the injection piping remains under positive pressure for all conditions, when the pump is operating air will enter the dip tube on pump shutdown as the water level in the tube descends and the vacuum breaker opens. The pumping power penalty is reduced relative to the values shown in Table 8.6, however.

Table 8.7 Pumping Values—Injection Pressure Control with Configuration 4

Groundwater Flow, gpm (L/s)	PWL, ft (m)	IWL, ft (m)	Injection Head, ft (m)	Surface Head, ft (m)	TDH, ft (m)	Well Pump Power, kW	Drawdown, % of Aquifer Thickness
120 (7.6)	55.0 (16.8)	–22.0 (–6.7)	14 (4.3)	8.1 (2.5)	77.1 (23.5)	3.3	16.7
166 (10.5)	60.8 (18.5)	–15.1 (–4.6)	14 (4.3)	15.4 (4.7)	90.2 (27.5	5.4	23.1
189 (11.9)	63.6 (19.4)	–11.7 (–3.6)	14 (4.3)	20.0 (6.1)	97.6 (29.8)	6.6	26.3
212 (13.4)	66.5 (20.3)	–8.2 (–2.5)	14 (4.3)	25.2 (7.7)	105.7 (32.2)	8.0	29.4
235 (14.8)	69.4 (11.1)	–4.8 (–1.5)	14 (4.3)	30.9 (9.4)	114.3 (34.9)	9.6	32.6
258 (16.3)	72.3 (22.0)	–1.3 (–0.4)	14 (4.3)	37.3 (11.4)	123.5 (37.7)	11.4	35.8
281 (17.7)	75.1 (22.9)	2.2 (0.7)	16.2 (4.9)	44.2 (13.5)	135.5 (41.3)	13.6	39.0
304 (19.2)	78.0 (23.8)	5.6 (1.7)	19.6 (6.0)	51.8 (15.8)	149.4 (45.6)	16.2	42.2
327 (20.6)	80.9 (24.7)	9.1 (2.8)	23.1 (7.0)	59.9 (18.3)	163.9 (50.0)	19.1	45.4
350 (22.1)	83.8 (25.5)	12.5 (3.8)	26.5 (8.1)	68.6 (20.9)	178.9 (54.5)	22.4	48.6

Note: Based on SWL of 40 ft (12.2 m), production-well SC of 8 gpm/ft (1.66 L/s·m), injection-well SC of 6.7 gpm/ft (1.38 L/s·m), aquifer thickness of 90 ft (27 m), surface losses of 35 ft at 250 gpm (10.7 m at 15.7 L/s), pump efficiency of 70%, motor efficiency of 75%, an injection tube equipped with a spring-loaded check valve set for IWL, and valve head loss of 14 ft (4.3 m).

The purpose of calculations such as those summarized in Tables 8.5 to 8.7 is to provide the data to determine the power requirement of the pump over a range of groundwater flows and to incorporate these results into a calculation of system performance over the same range of flows to determine the optimum system operating point (discussed in Section 8.7.3 and shown in Table 8.16). Once that point is determined, the head loss is recalculated for that optimum flow and the pump is selected in accordance with those values.

Tables 8.5 to 8.7 also track the percentage of the available aquifer thickness that is lost to drawdown over the range of flows under consideration. Drawdown exceeding approximately 65% of the aquifer thickness (unconfined aquifers) should result in consideration of a second production well. In the event a confined aquifer is to be considered, the aquifer thickness calculations must incorporate the piezometric level.

Selection of the appropriate injection-well piping configuration for a particular application is a function of the expectation for water quality problems and the expected injection-well specific capacity. For unconsolidated aquifer settings (where the expectation of water quality problems is greater and the well maintenance requirements are higher), the selection should avoid, to the extent possible, allowing air to enter the injection piping; this would suggest using configuration 3. For aquifers in consolidated (rock) materials (where water quality problems are reduced and well maintenance requirements are low), it is possible to use configuration 2 in some cases, but configuration 4 is recommended.

8.5.2 Well Pump Control

As mentioned elsewhere in this chapter, the basis of the design of GWHP systems is to determine the minimum groundwater flow rate associated with maximum system performance in the dominant mode. The well pump and other components are then selected for that flow. The question then becomes how to control the pump to accommodate off-peak performance and flow requirements in the secondary mode (usually heating). There are several strategies available for control of well pumps. Those most commonly used include pump cycling in response to loop temperature, staging of multiple pumps in a single well or multiple wells, and variable-speed pump operation.

Table 8.8 Constant-Speed Submersible Motor Cycling Data (Franklin 2007)

Motor hp (kW)	Single Phase	Three Phase
<5 hp (3.7 kW)	100	300
7.5 to 30 hp (3.7 to 22.4 kW)	50	100
>30 hp (22.4 kW)		100

Strategies used in older systems and not recommended for future use include constant pump operation, production to a pressure tank or tanks, and production to a vented tank with secondary pump delivery to the heat exchanger. Constant pump operation is wasteful of energy (and groundwater in the event of surface disposal) in much the same way that constant uncontrolled operation of the building loop pump is. Production to pressure tanks involves, for large systems, substantial tank volume requirements and, if configured in the manner used in residential design (unnecessarily high pressure settings), can result in excessive pump power requirements. Provided the pressure settings are appropriate to system requirements and the pump is not permitted to operate constantly at other than peak conditions, this approach may be used in very small (<20 tons [70 kW]) systems, but it is not recommended in larger applications. The use of vented tanks must be avoided in any groundwater system, as the tank provides the opportunity for the escape of carbon dioxide from the water and the entry of oxygen into the system, both of which can lead to serious water quality problems, as discussed in Section 7.5.3. Open tanks were common in the first GWHP systems in the 1940s and 1950s and justified as pump control and sand removal strategies. With modern controls and more effective sand separation devices there is no justification for the use of vented tanks in modern system design.

One strategy for well pump control is cycling of the well pump in response to building loop temperature. As the loop temperature rises (cooling mode), the well pump is enabled and runs until the loop temperature is reduced sufficiently. A similar approach is used in the heating mode where the pump is enabled as the loop drops in temperature and the pump is operated until the loop rises in temperature sufficiently. This approach is often referred to as *dual setpoint control*. One of the considerations in this type of control relates to the fact that submersible pumps are limited in terms of the frequency with which they can be cycled, which is because of the need to dissipate the thermal pulse resulting from start-up. Table 8.8 provides starting frequency limitations for constant-speed submersibles.

It is apparent that for most GWHP applications the limitation amounts to 100 starts per day. A more meaningful way to express this in terms appropriate to GWHP applications is 15 minutes between starts. In the context of a GWHP system this means there must be a minimum of 15 minutes between starts as illustrated below for the cooling mode:

- Loop temperature rises to pump start setpoint, pump starts.
- Pump runs to reduce loop temperature, reaches pump off setpoint, pump stops.
- Loop temperature rises to pump start setpoint, pump starts.

The time between the two starts must be a minimum of 15 minutes. Repeated starts with shorter intervals overheat motor insulation and result in premature failure. One approach to limiting the number of starts to the specified interval is to separate the pump start and stop setpoint temperatures sufficiently to ensure adequately long pump cycle time. The cycle time for the well pump (assuming a single pump) is a function of the temperature interval between the start and stop setpoints and the thermal mass of the building loop. In a control sequence such as this, the larger the thermal mass of the building loop

EXAMPLE 8.3—
SETPOINT SELECTION

The system optimum operating temperature (as established by a calculation similar to that summarized in Table 8.16) at peak load in cooling is 81°F (27.2°C) on the building loop return. Building loop thermal mass is 10 gal/ton (9.9 L/kW). Determine the appropriate operating temperature setpoints for the cooling mode.

Solution

From Table 8.9, the required minimum controller range for cooling would be 11°F (6.1°C). Using the optimum loop temperature as the midpoint, a well pump start setting would be

$$81 + (11/2) = 86.5°F \qquad \text{(I-P)}$$

$$27.2 + (6.1/2) = 30.3°C \qquad \text{(SI)}$$

The well pump stop temperature would be

$$81 - (11/2) = 75.5°F \qquad \text{(I-P)}$$

$$27.2 - (6.1/2) = 24.2°C \qquad \text{(SI)}$$

A similar procedure would be used for the heating setpoints based on a controller range requirement of 6°F (3.3°C).

(as defined in terms of gallons [litres] of water per ton [kW] of block load) and the wider the temperature interval between start and stop settings, the longer the pump cycle time will be. Loop thermal mass is simply the sum of the volume contained in the main building loop piping and 50% of the branch piping serving the heat pump units. Past research into this issue (Rafferty 2001) has demonstrated that the shortest time between starts for the well pump occurs when the system is experiencing 50% of the peak block load (in the dominant mode, typically cooling). At peak load the pump on time is at a maximum, and at low load the pump off time is at a maximum. Table 8.9 provides initial guidelines for necessary setpoint ranges (difference between pump start and pump stop setpoint temperatures) for various system thermal values.

Based on the values in Table 8.9, it is apparent that for systems characterized by low building loop thermal mass (< 8 gal/ton [8.6 L/kW]) the required controller range becomes extremely large, resulting in the system operating at less-efficient temperatures. There are two common remedies to this situation: increasing thermal mass or staging two or more well pumps. In very small systems that will likely be served by a single production well, adding thermal mass can sometimes be easily accomplished by locating a tank or tanks near the loop circulating pump. The volume required is normally small in such applications and the tank cost and space requirements modest. (For additional information on building loop thermal mass, see Appendix G.) For larger systems, staging of multiple well pumps, particularly if multiple production wells are to be used, is a more effective strategy. Two 50% well pumps would reduce the controller range values in Table 8.10 by 50%. Thus, in provided example with the first stage well pump in operation, the required controller range would be reduced to 5.5°F (3.1°C) with a pump start temperature of 84°F (28.9°C) and a pump stop temperature of 78.5°F (25.8°C). The sec-

Table 8.9 Dual Setpoint Well Pump Control Temperature Range Requirements

Well Pump Motor hp (kW)	Loop Thermal Mass, gal/ton (L/kW)						
	2 (2.2)	4 (4.3)	6 (6.5)	8 (8.6)	10 (10.8)	12 (12.9)	14 (15.1)
	Setpoints						
	°F (°C)	°F (°C)	°F (°C)	°F (°C)	°F (°C)	°F (°C)	°F (°C)
Cooling Mode							
<5 hp (3.7 kW)	28 (15.6)	14 (7.8)	9 (5)	7 (3.9)	6 (3.3)	5 (2.8)	4 (1.4)
>5 hp (3.7 kW)	56 (31.1)	28 (7.8)	19 (10.6)	14 (7.8)	11 (6.1)	9 (5)	8 (4.4)
Heating Mode							
<5 hp (3.7 kW)	16 (8.9)	8 (4.4)	5 (2.8)	4 (1.4)	3 (1.7)	3 (1.7)	2 (1.1)
>5 hp (3.7 kW)	32 (17.8)	16 (8.9)	11 (6.1)	8 (4.4)	6 (3.3)	5 (2.8)	5 (2.8)

Notes: Basis is 15,000 Btu/h·ton (1.25 kW/kW) heat rejection in cooling, 9000 Btu/h·ton (0.75 kW/kW) heat absorption in heating; building loop thermal mass based on volume of main loop piping and 50% of heat pump branch piping. Values are based on a single well pump sized for peak block cooling load.

Table 8.10 Submersible Motor Variable-Frequency Drive Cautions (Franklin 2007)

Carrier frequency	Use low frequency for pulse width modulation drives.
Motor type	VFD operation is not recommended for single-phase submersible motors.
Voltage rise time	Limit voltage at the motor to 1000 v and rise time to <2 microseconds.
Motor overload	Overload relay must be of the quick trip type to trip at <10 s on motor stall. Ultimate trip not to exceed 115% of motor nameplate amps.
Frequency range	The frequency range is limited to 30 to 60 Hz, consult factory for operation >60 Hz.
Start/stop	Ensure 1 s maximum ramp up and ramp down between stop and 30 Hz.
Successive starts	Allow 60 s before restarting.
Filters or reactors	Filters or reactors are required if voltage is >380 and drive uses insulated gate bipolar transistor (IGBT) or bipolar gate transistor (BGT) switches and cable is longer than 50 ft (15 m). Low-pass filters preferable and must be designed for VFD operation.
Motor cooling flow	Flow at the rated nameplate frequency must be at least 0.25 ft/s (0.08 m/s) for 4 in. (100 mm) motors and 0.5 ft/s (0.15 m/s) for larger motors.

ond-stage pump would be initiated at a loop return rise above 84°F (28.9°C) (provided the first stage was already in operation) and would operate until the loop was reduced to 80°F (26.7°C).

It is important to note that staging of well pumps can be done with a single production well. Using a device known as a *Wesley tool*, it is possible to install two pumps in a single well without the need to oversize the casing. The device consists of a manifold that allows multiple pumps to be installed vertically in a coaxial fashion. This allows either multiple pumps for staging purposes or the installation of a spare pump in a single well.

Another well pump control option is variable speed. This approach offers the prospect for more stable system operating temperature, reduced aquifer sand production, and a better match to flow requirements in the secondary operating mode (usually heating). However, the primary reason for using variable speed in conventional applications—reduced energy use—is often absent in moderate to deep well pump applications. The alternative of well pump cycling and the substantial static component of the pump head (causing total head to vary not with the square of the flow but closer to directly with flow) results in little energy use incentive for using variable speed. In shallow well applications with high static water level and minimal drawdown, energy savings from variable-frequency drive (VFD) operation are greater.

A number of issues must be carefully considered in the application of variable-speed operation of a well pump, including speed limitations, motor protection, motor cooling, sand removal strategy, static head speed limitation, manufacturer coordination, and system pressurization.

Submersible motors are not commonly available in inverter duty configuration, as is the case with most conventional motors. Therefore, it is important that the engineer ensure that the design does not compromise the service life of the motor. Key cautions for submersible motor VFD applications appear in Table 8.10.

To meet the required cooling water flow rates it may be necessary to use a cooling shroud (sometimes referred to as a *flow inducer sleeve* or *can*). This shroud is an inverted can, closed at the top and open at the bottom, that fits over the pump intake and extends down over the motor in such a way as to force water to flow past the motor. Shrouds are often required in applications in which a pump is installed in unusually large-diameter casing (Figure 8.9). The inside diameter of the shroud is selected to achieve the minimum flow rate past the motor for cooling purposes.

Use of a variable-speed drive (VSD) can provide benefits in an application in which the production well produces large amounts of sand. In most wells sand production is high at pump start and decreases quickly thereafter. As a result, cycling a single-speed well pump tends to exacerbate sand production. Operation of a pump at variable flow helps to reduce sand production relative to a cycling, single-speed well pump. In addition, the flow rate necessary for GWHP systems is low for most operating hours, further reducing the sand problem. On the other hand, use of a variable-speed well pump also influences the type of sand removal device selected for the system. In any application using a variable-speed well pump, the sand removal device should be a strainer rather than a centrifugal separator. Centrifugal separators require a minimum flow to create the necessary velocities to achieve particulate separation. In variable-flow applications the flow is often below the threshold necessary for efficient operation of a centrifugal separator.

In setting the sequence of operation, no situation should be allowed in which the pump is operating at a speed insufficient to produce flow at the surface. The high static head component in well pump applications can sometimes result in a situation in which the pump speed is inadvertently reduced to a value at which it does not generate sufficient head to overcome the lift (SWL + DD) in the application. Operation in this condition can be damaging to the motor. The pump performance curve should be carefully compared to the necessary head required at all speeds to avoid this condition.

The project specifications should clearly identify that the well pump will be operated in conjunction with a VSD, and submittals must clearly confirm the fact that the manufacturer is aware of this and has provided the necessary information to the installing contractor.

System pressurization was mentioned previously in this chapter as a necessary strategy to reduce or eliminate the entrance of air into the groundwater side of the system. In variable-speed well pump applications, the engineer should carefully consider the impact of operation at off-peak flows on the pressurization control system at the injection well.

The sequence of operation for a variable-speed well pump is quite simple. The optimum operating building loop return temperature is established by the system optimization calculation (Table 8.16) for both cooling and heating modes. For example, the calculations result in a cooling-mode optimum building loop return temperature of 83°F (28.3°C) and a heating-mode temperature of 44°F (6.7°C). Well pump operation is enabled as the loop temperature approaches 83°F (28.3°C) in the cooling mode, and the pump speed is then modulated to hold the loop temperature at the designated value. Pump operation ceases when the loop falls below 81°F (27.2°C) (a small interval below the optimum temperature to avoid nuisance cycling). No well pump operation is required between 81°F and 44°F (27.2°C and 6.7°C). As the loop return approaches 44°F (6.7°C), the pump is started and allowed to modulate to maintain the 44°F (6.7°C) setpoint. On a loop temperature rise, well pump operation ceases at 46°F (7.8°C).

8.6 HEAT EXCHANGERS

Isolation of the water is the most effective strategy to avoid water quality problems in open-loop systems. Isolation using a plate heat exchanger is strongly suggested even in water of apparently benign chemistry, as it is possible for changes to occur in groundwater chemistry over the life of the system. Encroachment of salt water in coastal areas, encroachment of oxygen-saturated water at sites adjacent to rivers or lakes, mixing of aquifers of different water chemistry, and other issues that can promote water quality problems have been observed to develop after several years of operation in some existing GWHP systems—applications that indicated no particular water quality problems initially.

Frequently the use of an isolation exchanger is viewed as simply moving the problem (corrosion, scale, fouling, etc.) from the heat pumps to the heat exchanger. While this may be true to some extent, removing scale from hundreds of heat pumps and the building loop piping in a large system is a far greater task than removing it from a single heat exchanger. In addition, plate heat exchangers are manufactured to be disassembled and cleaned and are constructed of materials impervious to most groundwater. The actual heat exchange surfaces can be visually examined by removing the tie bolts and examining the individual plates (Figures 8.12 and 8.13)—something not possible with heat pump refrigerant-to-water exchangers. These considerations alone may be sufficient to justify the use of heat exchangers in most commercial applications.

An additional benefit is that the heat exchanger also facilitates the use of different flow rates on the building loop and groundwater sides of the heat exchanger. This allows the use of a building loop flow that optimizes heat pump performance (typically in the range of 2.5 to 3.0 gpm/ton [0.05 to 0.054 L/s·kW]) and a groundwater loop flow that optimizes system performance (typically in the range of 1.0 to 2.0 gpm/ton [0.018 to 0.036 L/s·kW]). Additionally, the mere presence of the heat exchanger reduces the propensity for scale simply as the result of the temperatures of the surfaces encountered by the water. In a system using the groundwater directly in the heat pump units, in the cool-

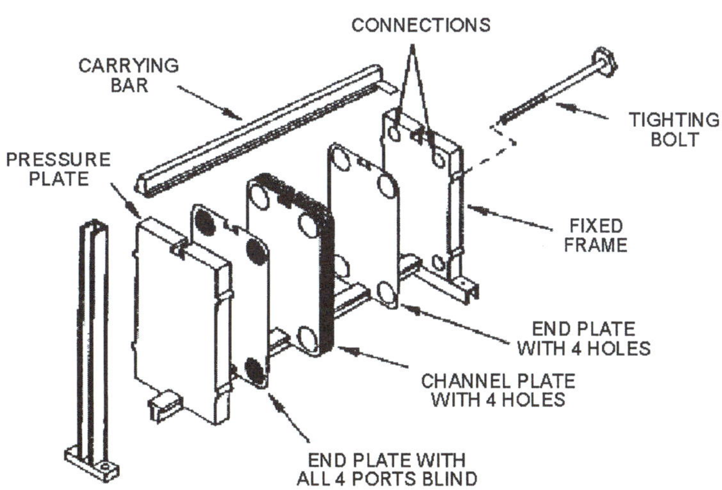

Figure 8.12 Plate Heat Exchanger

Figure 8.13 Plate Heat Exchanger Serving 115 ton (405 kW) School System

ing mode the groundwater encounters surfaces in the entering portion of the refrigerant-to-water heat exchanger of as high as 170°F (77°C). In a system with an isolation plate heat exchanger, the groundwater never encounters a surface temperature in excess of approximately 85°F (29.4°C). Formation of calcium carbonate scale, as discussed in Section 7.5.3, is a temperature-driven reaction. The higher the temperature of the surfaces encountered by the water, the greater the propensity for scale formation. As a result, the use of the isolation heat exchanger for a given water chemistry reduces the magnitude of the scale in addition to limiting it to a small portion of the mechanical system. Although the use of heat exchangers does increase capital cost (typically in the range of 100 to 300 $/ton [28 to 85 $/kW]), the benefits to the owner in terms of reduced maintenance justify its use in all but the smallest commercial groundwater systems. The cost of acid-cleaning an entire building loop to remove carbonate scale from heat pumps on a single occasion can cost as much as the plate heat exchanger for the system.

8.6.1 Approach Temperature

The groundwater flow optimization calculation (summarized in Table 8.16) for a given project tends to identify the optimum groundwater flow rate for the system. With a fixed groundwater temperature (available from the well), the flow in conjunction with the load establishes the groundwater exit temperature from the heat exchanger. As the system evaluation calculations are performed over a range of heat pump EWTs (heat exchanger leaving temperatures), the key remaining parameter in the design of the heat exchanger is the approach temperature. This is the difference between the building loop entering temperature and the groundwater leaving temperature. The approach determines how closely the loop "approaches" the groundwater temperature. Essentially, the closer the approach, the more efficient the operation of the heat pumps as a result of more favorable temperatures but the more expensive the heat exchanger as a result of increased surface area.

The focus on the leaving end of the heat exchanger arises from the fact that in most GWHP system designs, the optimum system performance results in a groundwater flow

Table 8.11 Impact of Heat Exchanger Approach

Groundwater EWT, °F	Groundwater LWT, °F	Loop LWT (Heat Pump EWT), °F	Loop EWT (Heat Pump LWT), °F	Heat Exchanger Approach, °F	Heat Exchanger LMTD, °F	Heat Exchanger U-factor, Btu/h·ft²·°F	Heat Exchanger Area, ft²	System EER	Simple Payback, years
62	86.0	84.1	95.0	9	14.5	1129	264	12.7	-
62	85.7	79.9	90.7	5	10.1	1151	369	13.2	4.4
62	85.6	77.8	88.6	3	7.7	1096	503	13.7	6.1
62	85.5	76.8	87.5	2	6.4	1120	613	13.9	13.1
Groundwater EWT, °C	**Groundwater LWT, °C**	**Loop LWT (Heat Pump EWT), °C**	**Loop EWT (Heat Pump LWT), °C**	**Heat Exchanger Approach, °C**	**Heat Exchanger LMTD, °C**	**Heat Exchanger U-factor W/m²·°C**	**Heat Exchanger Area, m²**	**System COP_c**	**Simple Payback, years**
16.7	30.0	28.9	35.0	5.0	8.1	199	24.5	3.7	-
16.7	29.8	26.6	32.6	2.8	5.6	203	34.3	3.9	4.4
16.7	29.8	25.4	31.4	1.7	4.3	193	46.7	4.0	6.1
16.7	29.7	24.9	30.8	1.1	3.6	197	56.9	4.1	13.1

LMTD = log mean temperature difference

in the range of 1.0 to 2.0 gpm/ton (0.018 to 0.036 L/s·kW) and a building loop flow in the range of 2.5 to 3.0 gpm/ton (0.045 to 0.054 L/s·kW). Given the imbalance in flows, the minimum approach temperature naturally occurs on the groundwater leaving/building loop entering side of the heat exchanger. This is in contrast to the often-held belief that the object of the design is to achieve an EWT for the heat pumps as close to the groundwater temperature as possible. In fact, for EWT to approach groundwater temperature, groundwater flow must approach building loop flow; this tends to result in a less-efficient overall design because of the higher groundwater pumping requirements.

In the course of the system design there are three primary areas in which the designer must deal with the plate heat exchanger: manipulation of the approach temperature in the course of the system performance calculation, specification of the heat exchanger based on results of the system performance calculation in the dominant mode, and evaluation of the performance of the heat exchanger in the secondary mode.

Table 8.11 provides an example of varying heat exchanger approach independently while holding other parameters constant. In this example, all of the heat exchanger selections in the table were made for the same groundwater flow and system conditions, with only the heat exchanger approach varied. The final two columns illustrate the cost and benefit of decreasing heat exchanger approach. As approach is decreased from 9°F to 2°F (5°C to 1.1°C), required heat transfer surface area increases 232% in this particular case. System performance increases from 12.7 to 13.9 EER (3.72 to 4.08 COP_c). Assuming a cost of incremental heat exchanger surface of \$45/ft² (\$484/m²), 1000 equivalent full-load hours (EFLH) cooling, and an electrical cost of 0.10 \$/kWh, the simple payback on the additional heat exchanger area appears in the final column. The example was based on a 300 ton (1056 kW) block cooling load, 75 ft (23 m) SWL, and 10 gpm/ft (2.1 L/s·m) SC.

It is apparent in Table 8.11 that the incremental cost of the 5°F (2.8°C) approach is clearly justified and the incremental cost of the 3°F (1.7°C) approach is likely justifiable for some commercial applications and most public projects, but the incremental cost of the 2°F (1.1°C) approach is unlikely to be economically justifiable. For clarity, this example maintained groundwater flow constant. In an actual design it is not possible to hold all parameters constant. As heat exchanger approach is changed, other values change in response. However, the general relationship of approach to system performance demonstrated in Table 8.11 is valid, and in many applications decreasing the approach from 10°F to 2°F (5.6°C to 1.1°C) results in approximately a 1.0 EER (0.4 COP_c) increase in system performance (Rafferty 2008). The cost of heat exchanger incremental surface area, the electrical cost, and the EFLH for the specific application exert a strong influence on the economics of heat exchanger approach. A 4°F (2.2°C) approach is justifiable in most applications, and a 2°F to 3°F (1.1°C to 1.7°C) approach is justifiable in high-electrical-cost, high-EFLH applications (Rafferty 2008).

8.6.2 Heat Exchanger Materials

Heat exchanger materials selection is influenced by several factors, including fluid temperature, fluid chemistry, and system pressure requirements. Groundwater heat pump projects typically can take advantage of the base-level (least-cost) materials offered by most manufacturers—304 stainless steel plates and NBR (medium nitrile) gaskets. The most commonly encountered water chemistry issue that requires departure from this is high chloride content (most commonly encountered in coastal areas). Table 8.12 provides plate material guidelines for high-chloride applications. For GWHP purposes, the 100°F (37.8°C) column is the most applicable.

8.6.3 Installation and Maintenance

A number of installation and maintenance issues must be carefully addressed to ensure successful plate heat exchanger applications. The first is the question of whether multiple (2 at 50%) heat exchangers are necessary. In most cases they are not necessary and a single exchanger is acceptable. The most common exceptions occur where it is not acceptable to have the system out of service at any time (corrections facilities, manufacturing processing plants, hospitals, etc.). In most all other cases it is possible to schedule maintenance and cleaning of the exchanger during unoccupied hours. Most exchangers can be opened cleaned and placed back in service in less than an eight-hour period, and cleaning is not required on more than an annual basis for most projects. Another situation in which it may be necessary to have a second exchanger is where there is extensive operation at very-low-load conditions. The performance of the exchanger should be verified at minimum load conditions to ensure stable performance. In the event performance is not acceptable at minimum load, a low-load exchanger may be necessary. For the general case, however, a single exchanger will provide acceptable performance.

Most equipment has all four piping connections on the fixed end plate of the heat exchanger. In rare cases it is necessary to locate piping connections on the movable end plate of the exchanger. In such cases it is important to ensure adequate working space to disassemble the exchanger and to use piping that is easily disassembled (grooved end connections).

Two issues related to maintenance are particularly important. One is that in the process of cleaning the exchanger only plastic brushes should be used. Most fouling is easily removed with plastic brushes, and metallic wire brushes are rarely necessary. Should it be

Table 8.12 Stainless Steel Chloride Thresholds

Material	50°F (10°C)	75°F (23.9°C)	100°F (37.8°C)
304 Stainless Steel	450	250	150
316 Stainless Steel	1000	550	375
Titanium	>1000	>550	>375

necessary to use wire brushes, only brushes of the same alloy as the plate material are acceptable. Using carbon steel wire brushes on stainless steel plates can damage the passivation of the plate material and lead to premature failure. The other issue is that when reassembling the exchanger, the manufacturer's procedure for torquing the through bolts should be strictly adhered to. Overtorquing will result in gasket failure and leaking.

It is generally not necessary to replace plate gaskets when servicing a heat exchanger; only damaged gaskets need be replaced. In some cases gaskets are glued in place (most are friction fit, however). The glue used for the gaskets can require a cure time of up to 24 hours. For this type of gasket, to minimize downtime it is useful to have on hand at least one of each type of plate (usually at least two types of plates in most exchangers) with the gaskets glued in place.

8.6.4 Heat Exchanger Connection to Loop

Figure 8.14 illustrates two approaches to the installation of a plate heat exchanger in the building loop. The most common approach, a series flow arrangement, is illustrated on the left. An alternative design, in which the exchanger is placed in a separate parallel decoupled loop, appears on the right. The series approach typically results in a heat exchanger with a higher flow (2.5 to 3.0 gpm/ton [0.045 to 0.054 L/s·kW]) on the building loop side and a lower flow (1.0 to 2.0 gpm/ton [0.018 to 0.036 L/s·kW]) on the groundwater side. This flow imbalance necessitates a greater heat transfer area in many cases due to a somewhat lower overall U-factor arising from the lower flow on the groundwater side. The parallel configuration offers the ability to operate the heat exchanger with equal flow rates on both sides, though at the expense of reduced temperature difference. The second advantage is the removal of the heat exchanger head loss from the building loop. In cases where the building loop is expected to be in heating/cooling

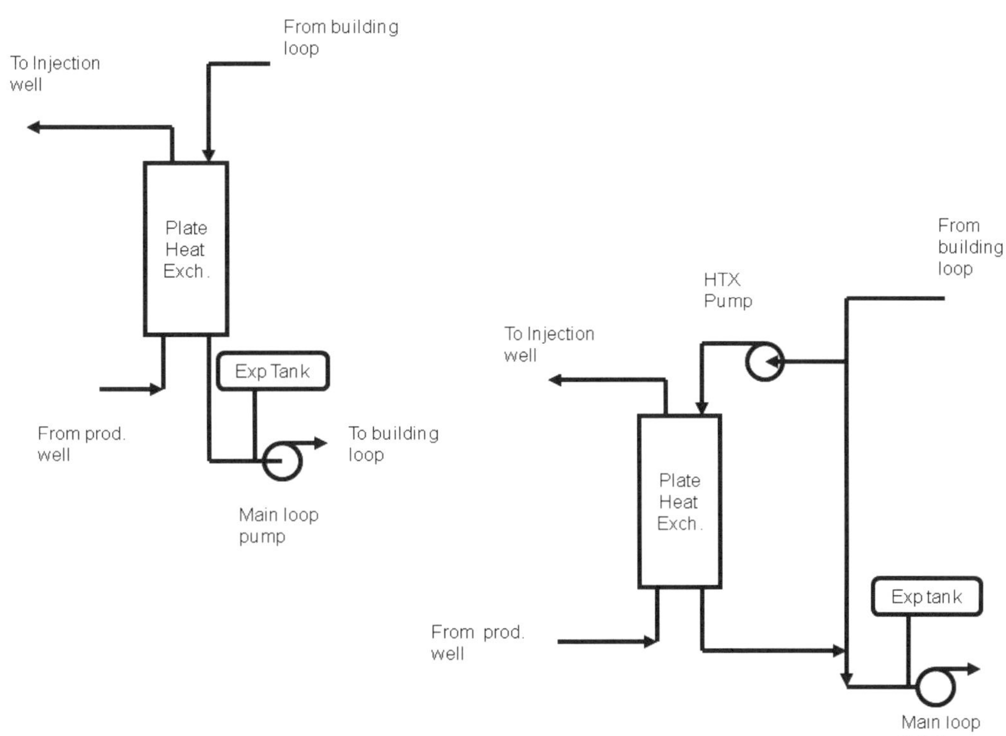

Figure 8.14 Alternative Heat Exchanger Configurations

balance for a significant portion of the year this may be a useful strategy, though this condition is rarely encountered. The savings in loop pump energy use must be balanced against the additional cost of the greater piping complexity, heat exchanger circulating pump, and associated controls, however.

8.7 SYSTEM DESIGN EXAMPLE

8.7.1 Introduction

The approach to the design of a GWHP system is similar in some respects to the practices of GCHP design, particularly in the initial phases. The consideration of building loads is the same, the design is based on the block load, and the building system is evaluated over a series of heat pump EWTs. At this point, though, the GWHP design calculations depart from the GCHP approach. In the GWHP design, at each heat pump EWT the groundwater flow necessary to achieve that EWT is calculated along with the well pump power necessary to produce it. The well pump, loop pump, and heat pump power requirements are summed to arrive at a *system* EER. The process is repeated at the next EWT. This produces a table of system performance over a range of EWTs or groundwater flows. After determination of the approximate EWT where peak system performance occurs, input values (specific capacity, groundwater loop head loss, etc.) are corrected if necessary and some final runs are made to refine accuracy. Groundwater flow is checked to ensure that the well is capable of producing that flow and that the injection well is capable of accepting it. Either additional wells are added to accommodate the flow or the system performance is evaluated at reduced flows compatible with the wells. The equipment (well pump, groundwater pipe, heat exchanger) selection is then made for the peak system performance groundwater flow compatible with site conditions.

8.7.2 Example Application Information

Consider a school with a cooling load of 90 tons (317 kW) and a heating requirement of 800,000 Btu/h (234 kW). An existing irrigation well will be used to supply the GWHP system. Table 8.13 provides information on the existing well, and selected results of a flow test on the well are provided in Table 8.14. The well produces 54°F (12.2°C) water and has not encountered any major water quality problems in the 11 years it has been in use for irrigation purposes. Disposal will be to an injection well yet to be completed. Loop pumping power requirements can be based on a flow of 2.75 gpm/block ton (0.050 L/s·kW) and a total building loop head of 62 ft (18.9 m).

Suspended solids separated during the pump test were collected and a sieve analysis produced the following results: 90% retained 0.0197 in. (0.5 mm), 80% retained 0.0232 in. (0.59 mm), 70% retained 0.0280 in. (0.71 mm), and 40% retained 0.0469 (1.2 mm).

The water chemistry results in Table 8.15 omit some of the recommended criteria listed in Table 7.6, but the information, in conjunction with the existing operating history of the well for irrigation, is sufficient to guide the system design. The scaling index is positive but on the low end of the scaling range (Table 7.8), and the combination of low pH and low hardness suggests that scaling would be minimal. In addition, iron and manganese, two common scaling/fouling sources, are both below the thresholds at which substantial problems normally occur.

The potential for biological fouling as indicated in Table 8.15 by "time to present" (the time required for the water sample to present visual reaction in the BART test tube) for both BART tests is in the low to moderate range. In the case of the slime-forming bac-

Table 8.13 Design Example Well Information

Well total depth	189 ft (57.6 m)
Well casing diameter	8 in. (203 mm)
Well screen diameter	8 in. (203 mm)
Well screen slot size	Torch cut 1/4 × 2 in., 40/ft (6 mm × 50 mm, 131/m)
Screened interval	78 to 108 ft (23.8 to 32.9 m)
Static water level	66 ft (20.1 m)
Groundwater temperature	54°F (12.2°C)

Table 8.14 Design Example Well Flow Test Information

Time, min	Flow, gpm (L/s)	Water Level, ft (m)	Comments
1	90 (5.7)	72.6 (22.1)	
2	90 (5.7)	74.5 (22.7)	
3	90 (5.7)	75.0 (22.9)	
5	90 (5.7)	75.4 (23.0)	
10	90 (5.7)	75.7 (23.1)	
15	90 (5.7)	76.2 (22.2)	
30	90 (5.7)	76.9 (23.4)	
45	90 (5.7)	76.9 (23.4)	
100	140 (8.8)	80.1 (24.4)	cloudy
101	140 (8.8)	81.6 (24.9)	cloudy
102	140 (8.8)	83.0 (25.3)	
105	140 (8.8)	83.5 (25.4)	
110	140 (8.8)	84.0 (25.6)	
115	140 (8.8)	84.3 (25.7)	
130	140 (8.8)	84.8 (25.9)	
145	140 (8.8)	84.9 (25.9)	
190	180 (11.3)	95.6 (29.1)	cloudy
191	169 (11.3)	98.1 (29.3)	cloudy
192	175 (11.3)	97.7 (21.2)	cloudy
193	178 (11.3)	99.0 (29.6)	cloudy
195	180 (11.3)	97.4 (29.7)	cloudy
200	173 (11.3)	97.6 (29.7)	cloudy
210	181 (11.3)	98.5 (30.0)	cloudy
215	180 (11.3)	99.0 (30.2)	cloudy
230	172 (11.3)	100.2 (30.2)	cloudy
245	165 (11.3)	102.2 (30.2)	cloudy

teria test, the interpretation for a three-day reaction suggests that aggressiveness is low but regular monitoring is warranted. In the case of the iron bacteria test, the reaction time of eight days is near the background level and is of less concern. The well has never required maintenance in the 11 years it has been operated. This history, along with the low level of iron and the low to moderate potential for scaling and biological fouling, suggests that rigorous exclusion of oxygen is not necessary in the design, but as minimizing air exposure is always prudent, injection piping configuration 4 is advisable.

Table 8.15 Design Example Water Chemistry

Constituent	Concentration, ppm
Chloride (Cl)	22.2
Fluorine (F)	0.73
Bicarbonate (HCO_3)	223 (as $CaCO_3$)
Sulfate (SO_4)	0.67
Dihydrogen phosphate (H_2PO_4)	0.58
Sodium (Na)	83.6
Potassium (K)	6.3
Magnesium (Mg)	3.95
Calcium (Ca)	11.8
Iron (Fe)	0.04
Manganese (Mn)	0.02
BART iron-related bacteria	8 days (time to present)
BART slime-forming bacteria	3 days (time to present)
Total hardness	56.3
Total dissolved solids (TDS)	353
Methyl orange (M) alkalinity	223 (as $CaCO_3$)
pH	7.6
Langlier saturation index (LSI) (calculated for 85°F [29.4°C])	0.38

BART = bacteriological activity reaction test

From the results of the pump test (Table 8.14), it appears that the highest flow is in excess of what the well or aquifer can produce, based on the turbid (cloudy) description of the water and the unstable flows and water levels in the test report. Information on the original construction and development of the well is not available, so it is not possible to judge whether the performance of the well is the result of insufficient development at original construction or simply the nature of the aquifer, though given the fact that well has been in operation for many years it is likely that poor development can be eliminated. The cloudy water, however, indicates that the velocity in the near-well zone is high enough to entrain fine components and that production at or above this rate is not advisable. The turbidity mentioned in the comments section of the test report (Table 8.14) for the first two readings at the 140 gpm (8.8 L/s) flow is not a concern, as wells often produce turbid water for a short period of time when the flow is suddenly increased, as it was at this point in the test.

Assuming the aquifer thickness extends from the bottom of the perforated cased interval (108 ft [33m]) to the static water level, the perforations in the casing would approximate $(108 - 78)/(108 - 66) \times 100 = 71\%$ ($[33 - 23.8]/[33 - 20.1] \times 100 = 71\%$) of the aquifer thickness, which is somewhat more than the typical 33% to 50% for a water table aquifer. The well completion report suggests that the SWL is approximately the same as the depth at which water was first encountered in drilling, suggesting a water table aquifer. Finally, the variation in specific capacity (8.2 at 90 gpm, 7.4 at 140 gpm, and 5.4 at180 gpm [1.7 at 5.7 L/s, 1.5 at 8.8 L/s, and 1.1 at 11.3 L/s]) is reflective of the performance in a water table aquifer.

The perforations in the casing result in a total inlet area of 4.08 ft^2 (0.38 m^2). At a flow of 180 gpm (0.40 ft^3/s) (11.3 L/s [0.0114 m^3/s]), the entrance velocity would amount to 0.098 ft/s (0.03 m/s), or roughly equal to the recommended maximum value of 0.1 ft/s (0.03 m/s). This assumes that all of the slots are available for water flowing into the well.

The drawdown associated with the 180 gpm (11.3 L/s) flow, however, reduces the available entrance area to only those perforations between 99 and 108 ft (30 and 33 m) depth. This results in an entrance velocity of 0.43 $ft^3/s/((9/32) \times 4.08\ ft^2) = 0.374$ ft/s (0.0114 $m^3/sec/[[9/32] \times 0.38\ m^2] = 0.114$ m/s), or nearly four times recommended entrance velocity. Poor performance at the higher flow rate could be a result of the vertical flow in the aquifer caused by the drawdown (78% of maximum drawdown) at 180 gpm (11.3 L/s). In any case, it would be prudent to limit flow to less than the 180 gpm (11.3 L/s) value for purposes of the GSHP design to ensure satisfactory performance of the well.

8.7.3 Cooling Mode

For purposes of cooling-mode performance evaluation, calculation normally begins with an EWT of 5°F (2.8°C) above groundwater temperature and works up to 25°F (13.9°C) above groundwater temperature. Based on recommendations in Section 8.6, an approach of 4°F (2.2°C) is used for this example.

At the EWT of 59°F (15°C), the heat pumps have an average EER of 17.6 (5.16 COP_c) and a total heat rejection of 1,289,000 Btu/h (378 kW) based on the 1,080,000 Btu/h (316 kW) block cooling load. Using a building loop flow rate of 2.75 gpm/ton (0.049 L/s·kW) of block load results in a temperature rise of 1,289,000 Btu/h ÷ (500 Btu·min/lb·°F·gal × 248 gpm) = 10.4°F (378 kW ÷ [15.6 L/s × 0.001163 kWh/kg·K × 3600 s/h] = 5.78°C), or a heat pump LWT of 59°F + 10.4°F = 69.4°F (15°C + 5.78°C = 20.78°C). The heat pump LWT is the same as the building loop return temperature and the heat exchanger entering temperature on the building side.

Using a heat exchanger approach temperature of 4°F (2.2°C) results in a groundwater leaving temperature of 69.4°F – 4°F = 65.4°F (20.8°C – 2.2°C = 18.6°C). At the 54°F (12.2°C) groundwater temperature available, this results in a groundwater temperature rise of 65.4°F – 54°F = 11.4°F (18.6°C – 12.2°C = 6.4°C). At the heat rejection load on the exchanger of 1,289,000 Btu/h (378 kW), the required groundwater flow rate would be 1,289,000 Btu/h ÷ (500 Btu·min/lb·°F·gal × 11.4°F) = 226 gpm (378 kW ÷ [0.001163 kWh/kg·K × 3600 s/h × 5.78°C] = 14.4 L/s). This value is far above what the existing well can produce, so it would be appropriate to begin the evaluation at a higher EWT/lower groundwater flow point. At a heat pump EWT of 66°F (18.9°C), the resulting values are as follows:

Heat pump EER = 16.0 (4.69 COP_c)
Building loop heat rejection = 1,310,175 Btu/h (384 kW)
Building loop temperature rise = 10.6°F (5.9°C)
Building loop return temperature = 76.6°F (24.8°C)
Groundwater heat exchanger leaving temperature = 72.6°F (25.6°C)
Groundwater temperature rise = 18.6°F (10.3°C)
Required groundwater flow = 141 gpm (8.9 L/s)

The 141 gpm (8.9 L/s) flow rate is within the capability of the existing well.

Continuing with the system calculations, the next few steps involve the calculation of the well pump power requirements. At the 141 gpm (8.8 L/s) flow in the well test the specific capacity (SC) was approximately 7.4 gpm/ft (1.53 L/s·m) after water level stabilization at that flow. Using this SC, the water level at the 141 gpm (8.8 L/s) flow is

$$\text{SWL} + (\text{flow} \div \text{SC}) = \text{lift}$$

$$66\ \text{ft} + (141\ \text{gpm} \div 7.4\ \text{gpm/ft}) = 85.1\ \text{ft} \qquad \text{(I-P)}$$

$$20.1\ \text{m} + (8.8\ \text{L/s} \div 1.53\ \text{L/s·m}) = 25.9\ \text{m} \qquad \text{(SI)}$$

Allowing 4 ft (1.2 m) for the head loss in the pump column brings the head loss associated with the production well (static head and column friction) to 89.1 ft (27.1 m).

Surface friction losses include the pipe to the mechanical room, the heat exchanger, and the pipe to the injection well. At this point the distances may not be known, so a placeholder value is used that can be corrected later when the expected groundwater flow range is narrowed. Assuming a heat exchanger loss of 5 psi or 11.5 ft (35 kPa or 3.5 m) and a pipe friction loss of 16 ft (400 ft at 4 ft/100 ft) (4.9 m [122 m at 1.2 m/30 m]) and a fitting adjustment of 25% of the piping loss (4 ft [1.2 m]) results in a total surface head loss of 31.5 ft (9.6 m).

As mentioned previously, the injection well is yet to be completed, so its performance is based on the performance of the existing production well. Because most injection wells demonstrate somewhat lesser performance in comparison to production wells, an "efficiency" of 80% of that of the production well is allowed for. In this particular case (given the construction of the existing production well) it may be possible to equal or better the performance of the production well with a more effective screen and development (in the injection well), so the assumed 80% performance should be sufficiently conservative. The water level in the injection well, assuming an injection-well SWL of 65 ft (19.8 m), is

$$\text{SWL} - [\text{flow} \div (\text{SC} \times \text{efficiency})]$$

$$65 - [141 \text{ gpm} \div (7.4 \times 0.8)] = 41.1 \text{ ft (below ground surface)} \qquad \text{(I-P)}$$

$$19.8 \text{ m} - [8.8 \text{ L/s} \div (1.53 \text{ L/s·m} \times 0.8)] = 12.5 \text{ m (below ground surface)} \qquad \text{(SI)}$$

This indicates that the injection-well water level will remain below ground surface, thus eliminating any concern about pressurization of the injection well. However, the "negative" 41 ft (12.5 m) of head is unlikely to be sustained consistently (see the discussion in Section 8.3.1) and would not be available at pump start. In addition, the completion of the well in unconsolidated materials (and the expectation for the same in the case of the future injection well) suggests that it would be prudent to use an injection design that reduces or eliminates the potential for air intrusion. To promote stable pressurization, eliminate vacuum potential, and reduce air infiltration, it would be wise to configure the injection-well drop pipe (dip tube) to offset some or all of this 41 ft (12.5 m). The simplest approach is to place an adjustable spring-loaded check valve on the end of the drop pipe with the setting appropriate to the head to be offset. A valve installed in the bottom of the dip tube and set for a crack pressure of 41 ft (12.5 m) would ensure a full injection pipeline under all conditions when the pump is operating. The head loss associated with the valve and the dip tube pipe friction losses would ensure a slight positive pressure at the top of the column. This eliminates the opportunity for air to enter the line and helps to reduce injection-zone water chemistry problems that might result. The only pumping penalty associated with this strategy is the lost "negative head" associated with the difference between the injection water level (IWL) and the ground surface. As a result, the total head on the well pump would amount to the following:

Lift	85.1 ft	(25.9 m)
Column friction	4 ft	(1.2 m)
Surface piping loss	31.5 ft	(9.6 m)
Injection tube/valve head loss	10 ft	(3.1 m)
Total	130.6 ft	(39.8 m)

The well pump power requirement can now be determined from the flow and head requirements:

$$\text{Theoretical horsepower} = (141 \text{ gal/min} \times 8.3 \text{ lb/gal} \times 130.6 \text{ ft}) \div 33000 \text{ ft·lb/min·hp} = 4.6 \text{ hp} \quad \text{(I-P)}$$

$$\text{Theoretical horsepower} = (8.88 \text{ L/s} \times 39.8 \text{ m} \times 9.8 \text{ kPa/m}) \div 1000 \text{ W/kW} = 3.5 \text{ kW}$$
(Using Equation 6.1) (SI)

The only remaining values necessary for the calculation are the pump and submersible motor efficiencies. From Tables 8.3 and 8.4, the expected pump efficiency for the flow range would be approximately 67% and the motor efficiency 79%.

$$\text{Well pump brake horsepower} = 4.6 \text{ hp}/0.67 = 6.9 \text{ hp} \quad \text{(I-P)}$$
$$= 3.5 \text{ kW}/0.67 = 5.3 \text{ kW} \quad \text{(SI)}$$

$$\text{Well pump power requirement} = (6.9 \text{ hp}/0.79) \times 0.746 \text{ kW/hp} = 6.5 \text{ kW} \quad \text{(I-P)}$$
$$= 5.3 \text{ kW}/0.79 = 6.5 \text{ kW} \quad \text{(SI)}$$

The loop circulating pump, assuming a pump efficiency of 65% and a motor efficiency of 87% and based on flow and head information from above, amounts to the following power requirement:

$$\text{Loop pump brake horsepower} = (2.75 \text{ gal/min· ton} \times 8.3 \text{ lb/gal} \times 90 \text{ tons} \times 62 \text{ ft}) \div (33000 \text{ ft·lb/min·hp} \times 0.65) = 5.9 \text{ hp} \quad \text{(I-P)}$$

$$\text{Loop pump brake horsepower} = (0.05 \text{ L/s·kW} \times 317 \text{ kW} \times 18.9 \text{ m} \times 9.8 \text{ kPa/m}) \div (1000 \times 0.65) = 4.4 \text{ kW}$$
(Using Equation 6.10) (SI)

$$\text{Loop pump power} = (5.9 \text{ hp} \div 0.87) \times 0.746 \text{ kW/hp} = 5.0 \text{ kW} \quad \text{(I-P)}$$
$$= 4.4 \text{ kW}/0.87 = 5.0 \text{ kW} \quad \text{(SI)}$$

In summary, for this operating condition the key results so far are as follows:

Heat Pump EWT, °F (°C)	Heat Pump EER (COP_c)	Building Loop Heat Exchanger EWT, °F (°C)	Heat Exchanger Groundwater Leaving Temperature, °F (°C)	Groundwater Flow Required, gpm (L/s)	Heat Pump Power, kW	Well Pump Power, kW	Loop Pump Power, kW	System EER (COP_c)
66 (18.9)	16 (4.69)	76.6 (24.8)	72.6 (22.5)	141 (8.9)	67.5	6.5	5.0	13.67 (4.01)
67 (19.4)								

At this point the strategy is to fill in the system performance at EWTs above and below the 66°F (18.9°C) initially calculated—obviously a time-consuming and tedious process using the manual approach outlined thus far. Fortunately there is commercially available software capable of making most of the necessary calculations. Table 8.16 provides the results of calculations for this system over a wider range of EWTs and groundwater flows.

Table 8.16 illustrates that in this case the peak system performance occurs at a heat pump EWT of 62°F or 63°F (16.7°C or 17.2°C), corresponding to a groundwater flow requirement of 179.5 gpm (11.3 L/s), or about 2.0 gpm/ton (0.036 L/s·kW). From the performance of the existing well we know that this is close to the flow at which high sand and turbid water production occurs along with excessive entrance velocity (180 gpm [11.3 L/s]). If flow is reduced to approximately 140 gpm or 1.56 gpm/ton (8.8 L/s or 0.028 L/s·kW), a flow at which the well is confirmed to perform satisfactorily, the system performance would be only slightly reduced (from 13.84 to 13.67 EER [4.06 to 4.01 COP_c]). In this particular case it seems reasonable to operate the system at slightly less than the peak performance conditions to ensure adequate well performance.

In the calculations that produced the data in Table 8.16, different SC values appropriate to each groundwater flow were used. In most spreadsheets and programs the user must enter a single SC value that the program uses for all calculations. After calculating initial results, the user then goes back and corrects the SC input for the flow that appears to produce the peak system performance. In the case of Table 8.16, a SC value appropriate to each groundwater flow requirement (based on the results of the flow test) was used to calculate drawdown and well pumping power, somewhat short-circuiting the process that would be required in most calculations.

In addition to system performance, the designer also must monitor the well pumping conditions as the system is evaluated over a range of groundwater flows. In the far right column of Table 8.16 is a listing of the calculated pumping levels in the production well based on the SC values derived from the pumping test results. As discussed in Chapter 7, a rule of thumb is that a second production well is indicated if the drawdown in the initial well approaches 66% of the available aquifer thickness. The thickness of the aquifer in this case is taken to be 42 ft (66 to 108 ft) (12.8 m [20.1 to 32.9 m]). As a result, a pumping level of greater than 66 + (0.66 × 42) = 93.7 ft (20.1 + [0.66 × 12.8] = 28.6 m) would be operating in a condition in which a second well may be advisable. In Table 8.16 the pumping level associated with the 141 gpm (8.8 L/s) groundwater flow is approximately 85 ft (25.9 m)—well within the acceptable range.

It is important to mention, however, that the design of this well and the manner in which it has evidently been operated is at variance with normally recommended practices. Drawdown of the well below the screened interval (in this case the slotted casing interval) is normally avoided in water well design and operation. When the water level is reduced to levels below the top of the screen (or perforated casing), water can cascade down into the well from the aquifer as it is dewatered, introducing air into the water. As discussed in Section 8.5.1, introduction of air is never advisable in groundwater systems. However, based on years of successful operation of this well for irrigation purposes at the approximate flow envisioned for the GSHP system, it seems reasonable to proceed with use of the well in this fashion.

A second issue is that of recommended entrance velocity in the well. The 85 ft (25.9 m) pumping level associated with the 141 gpm (8.8 L/s) flow suggests that the water will be entering through only a portion of the slotted casing installed in the well. The available open area associated with the slotted casing between 85 and 108 ft (25.9 and 32.9 m) is

$$4.08 \text{ ft}^2 \times [(108 - 85)/(108 - 78)] = 3.13 \text{ ft}^2 \qquad \text{(I-P)}$$

$$0.38 \text{ m}^2 \times [(32.9 - 25.9)/(32.9 - 23.8)] = 0.29 \text{ m}^2 \qquad \text{(SI)}$$

Table 8.16 Design Example Cooling-Mode Performance

Heat Pump EWT, °F	Heat Pump EER	Building Loop Heat Exchanger EWT, °F	Heat Exchanger Groundwater Leaving Temperature, °F	Groundwater Flow, gpm	Heat Pump Power, kW	Well Pump Power, kW	Loop Pump Power, kW	System EER	Pumping Water Level, ft
61	17.1	71.5	67.5	192.8	63.2	10.1	5.0	13.81	103.1
62	16.9	72.5	68.5	179.5	63.9	9.1	5.0	13.84	99.2
63	16.7	73.5	69.5	168.1	64.7	8.4	5.0	13.84	96.6
64	16.4	74.5	70.5	158.1	65.9	7.7	5.0	13.75	93.3
65	16.2	75.6	71.6	149.2	66.7	7.1	5.0	13.71	90.1
66	16.0	76.6	72.6	141.3	67.5	6.5	5.0	13.67	85.1
67	15.8	77.6	73.6	134.2	68.4	6.1	5.0	13.59	83.9
68	15.6	78.6	74.6	127.8	69.2	5.8	5.0	13.50	82.8
69	15.4	79.7	75.7	122.1	70.1	5.5	5.0	13.40	81.9
70	15.2	80.7	76.7	116.8	71.1	5.2	5.0	13.29	81.0
71	15.0	81.7	77.7	112.0	72.0	4.9	5.0	13.18	80.2
72	14.8	82.7	78.7	107.7	73.0	4.7	5.0	13.06	79.5
73	14.6	83.8	79.8	103.6	74.0	4.5	5.0	12.93	78.8
74	14.4	84.8	80.8	99.9	75.0	4.3	5.0	12.81	78.2
75	14.3	85.8	81.8	96.4	75.5	4.2	5.0	12.75	77.6
76	14.1	86.8	82.8	93.2	76.6	4.0	5.0	12.62	77.1
77	13.9	87.9	83.9	90.3	77.7	3.9	5.0	12.48	76.6
78	13.7	88.9	84.9	87.5	78.8	3.7	5.0	12.33	76.3
79	13.5	89.9	85.9	84.9	80.0	3.6	5.0	12.19	75.9
80	13.4	91.0	87.0	82.4	80.6	3.5	5.0	12.12	75.6
81	13.2	92.0	88.0	80.2	81.8	3.4	5.0	11.97	75.3

Heat Pump EWT, °C	Heat Pump COP_c	Building Loop Heat Exchanger EWT	Heat Exchanger Groundwater Leaving Temperature, °C	Groundwater Flow, L/s	Heat Pump Power, kW	Well Pump Power, kW	Loop Pump Power, kW	System COP_c	Pumping Water Level, m
16.1	5.01	21.9	19.7	12.1	63.2	10.1	5.0	4.05	31.4
16.7	4.96	22.5	20.3	11.3	63.9	9.1	5.0	4.06	30.2
17.2	4.90	23.1	20.8	10.6	64.7	8.4	5.0	4.06	29.4
17.8	4.81	23.6	21.4	10.0	65.9	7.7	5.0	4.03	28.4
18.3	4.75	24.2	22.0	9.4	66.7	7.1	5.0	4.02	27.5
18.9	4.69	24.8	22.5	8.9	67.5	6.5	5.0	4.01	25.9
19.4	4.63	25.3	23.1	8.5	68.4	6.1	5.0	3.99	25.6
20.0	4.57	25.9	23.7	8.1	69.2	5.8	5.0	3.96	25.2
20.6	4.52	26.5	24.3	7.7	70.1	5.5	5.0	3.93	25.0
21.1	4.46	27.0	24.8	7.4	71.1	5.2	5.0	3.90	24.7
21.7	4.40	27.6	25.4	7.1	72.0	4.9	5.0	3.86	24.4
22.2	4.34	28.2	26.0	6.8	73.0	4.7	5.0	3.83	24.2
22.8	4.28	28.8	26.5	6.5	74.0	4.5	5.0	3.79	24.0
23.3	4.22	29.3	27.1	6.3	75.0	4.3	5.0	3.76	23.8
23.9	4.19	29.9	27.7	6.1	75.5	4.2	5.0	3.74	23.7
24.4	4.13	30.5	28.2	5.9	76.6	4.0	5.0	3.70	23.5
25.0	4.08	31.0	28.8	5.7	77.7	3.9	5.0	3.66	23.3
25.6	4.02	31.6	29.4	5.5	78.8	3.7	5.0	3.62	23.3
26.1	3.96	32.2	30.0	5.3	80.0	3.6	5.0	3.57	23.1
26.7	3.93	32.8	30.5	5.2	80.6	3.5	5.0	3.55	23.0
27.2	3.87	33.3	31.1	5.0	81.8	3.4	5.0	3.51	23.0

At the flow of 141 gpm or 0.31 ft^3/s (8.8 L/s or 0.0088 m^3/s), the resulting entrance velocity amounts to 0.099 ft/s (0.03 m/s), just under the recommended 0.1 ft/s (0.031 m/s) value.

The results of Table 8.16 were based on the assumed piping head loss of 130.6 ft (39.8 m), assuming a unit loss of 4 ft/100 ft (1.2 m/30 m), a surface piping length of 400 ft (122 m), and 4 ft (1.2 m) for the production-well pump column. Given the flow rate of 141 gpm (8.8 L/s) and an aquifer thickness of 42 ft (12.8 m), Table 8.2 suggests a minimum separation distance between the production and injection wells of 367 ft (112 m). Based on a pipe size of 4 in. (100 mm) for polyvinyl chloride (PVC) AWWA C900 (2007) material at 1.4 ft/100 ft (0.42 m/30 m) and a total buried piping length of 400 ft (122 m) (allowing 50 ft [15 m] for routing around obstacles in the piping route) and a fittings allowance of 10%, the head loss for the buried piping would be

$$[(400 \text{ ft} \times 1.1)/100] \times 1.4 = 6.2 \text{ ft} \qquad \text{(I-P)}$$

$$[(122 \text{ m} \times 1.1)/30] \times 0.42 = 1.9 \text{ m} \qquad \text{(SI)}$$

Criteria for acceptable materials for the buried piping in a GWHP system include corrosion avoidance, ease of installation, contractor familiarity, and reasonable cost. Because no antifreeze or additives are involved, the issue of absolute leak avoidance is not necessary as it is in the case of closed-loop systems. Both high-density polyethylene (HDPE) and gasketed PVC are acceptable (AWWA 2007), with PVC seeing wider use as a result of the larger diameters involved in many GWHP systems. Solvent cement joined PVC is not recommended for buried piping in GWHP applications.

Allowing 75 ft (2.9 m) of piping in the mechanical room and using a 50% fittings allowance results in a head loss for the mechanical room of

$$[(75 \times 1.5)/100] \times 1.6 = 1.8 \text{ ft} + \text{heat exchanger at } 11.5 \text{ ft} = 13.3 \text{ ft} \qquad \text{(I-P)}$$

$$[(22.9 \times 1.5)/30] \times 0.42 = 0.55 \text{ m} + \text{heat exchanger at } 3.5 \text{ m} = 4.1 \text{ m} \qquad \text{(SI)}$$

The production-well column pipe would have a head loss of 1.6 ft/100 ft (0.49 m/30 m) assuming 4 in. (100 mm) steel and a length of 110 ft (33.5 m), for a total of

$$110/100 \times 1.6 = 1.8 \text{ ft} \qquad \text{(I-P)}$$

$$33.5/30 \times 0.49 = 0.55 \text{ m} \qquad \text{(SI)}$$

The adjustable spring-loaded check valves have a flow coefficient (C_v) in the 2 in. (50 mm) size of 14.5. To limit head loss, four of these valves will be used. The water flow per valve is

$$141 \text{ gpm} \div 4 = 35.2 \text{ gpm per valve} \qquad \text{(I-P)}$$

$$8.88 \text{ L/s} \div 4 = 2.22 \text{ L/s per valve} \qquad \text{(SI)}$$

The pressure drop through the valves at the flow rate for which they will be used is calculated as follows:

$$\text{Pressure drop at design flow} = (\text{Design flow rate}/C_v)^2 \times 1.0 \text{ psi}$$

$$= (35.2/14.5)^2 \times 1.0 \text{ psi} = 5.9 \text{ psi} \qquad \text{(I-P)}$$

$$5.9 \text{ psi}/0.433 = 13.6 \text{ ft} \qquad \text{(I-P)}$$

$$\text{Pressure drop at design flow} = (\text{Design flow rate}/C_v)^2 \times 6.9 \text{ kPa}$$

$$(2.22/0.092)^2 \times 6.9 \text{ kPa} = 40.7 \text{ kPa} \qquad \text{(SI)}$$

$$40.7 \text{ kPa}/0.433 = 4.2 \text{ m} \qquad \text{(SI)}$$

The injection-well dip tube would be approximately 75 ft (22.9 m) in length (SWL of 66 + 9 ft [20.1 +2.7 m] for submergence safety margin). At a 4 in. (100 mm) diameter, based on the 100 ft (30 m) production column pipe at 1.8 ft (0.55 m) loss,

$$75/110 \times 1.8 = 1.2 \text{ ft} \qquad \text{(I-P)}$$

$$23/33 \times 0.55 = 0.4 \text{ m} \qquad \text{(SI)}$$

Total well pump head = production-well column + surface loss + lift + injection valve:

$$= 1.8 + 6.2 + 13.3 + 85.1 + 13.6 + 1.2 = 121.2 \text{ ft} \qquad \text{(I-P)}$$

$$= 0.55 + 1.9 + 4.1 + 25.9 + 4.2 + 0.4 = 36.9 \text{ m} \qquad \text{(SI)}$$

The assumption in the original calculations was 130.6 ft (37.7 m). The actual head on the pump would be reduced by 9.4 ft (2.9 m), or about 7%. This would result in approximately the same percentage reduction in the well pump power requirement, thus reducing the total system power requirement approximately 0.5 kW—a difference that would change the system EER from 13.67 to 13.76 (COP_c from 4.01 to 4.04). Figure 8.15 provides a summary of the key cooling-mode values for the example.

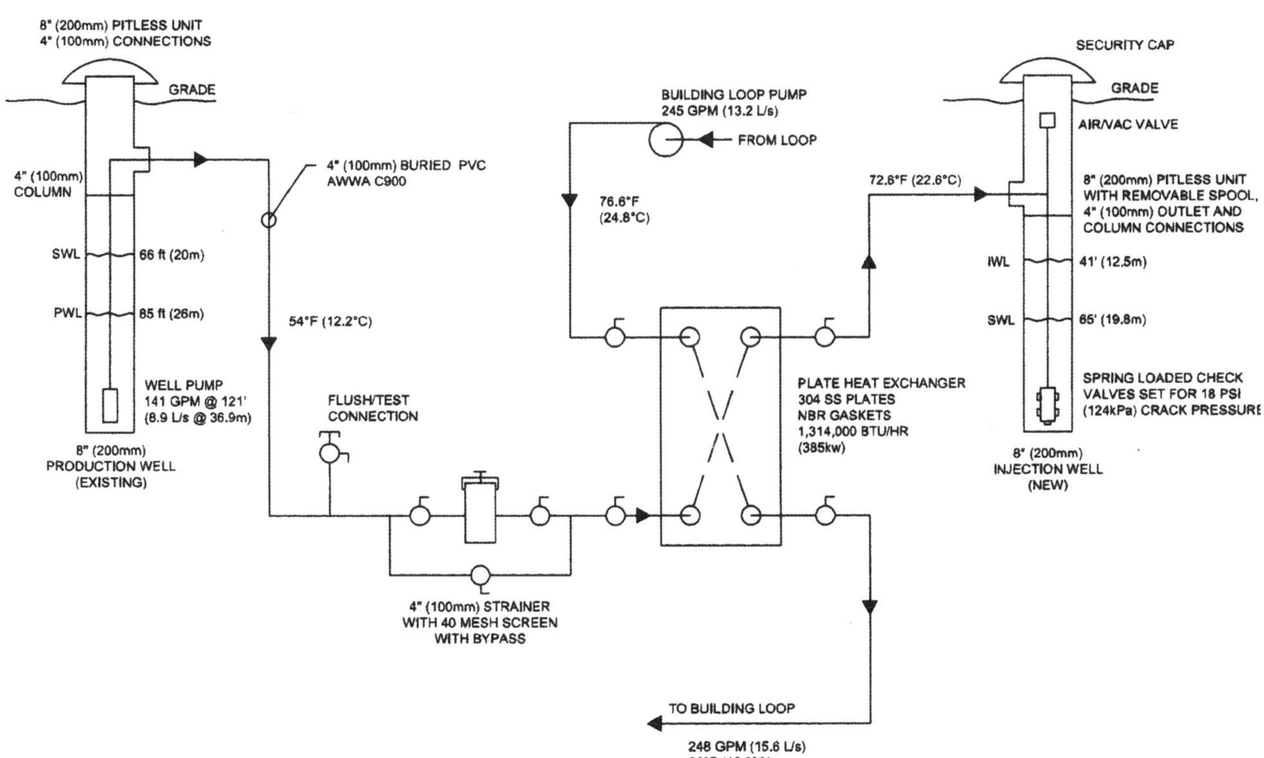

Figure 8.15 Design Example—Cooling Mode Values

8.7.4 Heating Mode

The heating mode must be evaluated with the same approach as described in the previous section to determine its performance over a range of heat pump EWT/groundwater flows.

The calculations necessary to produce the heating-mode values shown in Table 8.17 are conducted in the same manner as those for the values in Table 8.16. One difference is the assumption of a lower heat exchanger approach for the heating-mode operation. Typically the cooling mode is the dominant mode in most buildings in terms of dictating the design of the heat exchanger. As a result of the lower thermal load on the exchanger in the heating mode, excess surface area exists relative to the heating mode duty, which allows the exchanger to achieve a lower approach in heating-mode operation. The exact value for the approach is unknown until some initial calculations are made, but in most cases if the cooling mode is based on a 4°F (2.2°C) approach it is safe to conduct the heating-mode calculations on a 2°F to 3°F (1.1°C to 1.7°C) approach. In the case of Table 8.17, a 3°F (1.7°C) value was used.

In some applications, particularly those with substantial core areas, loop flow rate in the heating mode may be less than that in the cooling mode, as some core zone heat pumps may not be in operation. Most design programs, however, base the heating-mode design on the same loop flow rate used in cooling-mode operation. That is the strategy used in this example, as the school building for which the system is being designed would not have the substantial core area necessary to produce this effect. In the case of open-loop design, if substantial core areas exist and reduced heating-mode loop flow is expected, this condition may be an advantage as it may allow a somewhat larger temperature drop on the loop and hence the groundwater, thus reducing pumping power and providing greater system COP.

In the case of the example system, it appears that the heating mode could be operated at the same flow rate as the cooling mode with little impact on overall system performance. Table 8.17 values indicate a peak performance (3.32 COP) at a groundwater flow rate of 110 gpm (6.9 L/s), but there is little degradation if the likely cooling-mode flow of 140 gpm (8.8 L/s) is used (approximately 3.29 COP). Using the same flow rates for the two modes of operation could simplify pump control and potentially allow the less-expensive dual setpoint control instead of variable-speed control in this case. In the event that the application does not allow dual setpoint control and a variable-frequency drive (VFD) is used, the lower flow rate (110 gpm [6.9 L/s]) would be more appropriate, as reduced well flow rate is always more conducive to reduced well maintenance requirements.

As mentioned previously, the excess surface area issue with the heat exchanger will likely permit somewhat better performance in the heating mode than that indicated in Table 8.17. If the surface area requirements of the heat exchanger in the cooling and heating modes are compared, it is possible to infer the approximate decrease in approach arising from the surplus surface. Another method, somewhat more precise, is outlined in Appendix O; that calculation derives new exit temperatures for a specific heat exchanger configuration given information about the fluid flows, fluid specific heat, and EWTs. The calculation in Appendix O reveals that the heat exchanger in this example design would have a heating-mode performance as follows, assuming a reduced overall U-factor (from 900 to 825) due to higher-viscosity water at the heating-mode temperatures:

- Groundwater side: 141 gpm (8.8 L/s), entering at 54°F (12.2°C), leaving at 45.5°F (7.5°C)

Table 8.17 Design Example Heating-Mode Performance

Heat Pump EWT, °F	Heat Pump COP	Building Loop Heat Exchanger EWT, °F	Heat Exchanger Groundwater Leaving Temperature, °F	Groundwater Flow Required, gpm	Heat Pump Power, kW	Well Pump Power, kW	Loop Pump Power, kW	System COP
36	3.58	31.34	34.34	58.8	65.5	2.4	5.0	3.22
37	3.6	32.33	35.33	62.0	65.1	2.6	5.0	3.23
38	3.64	33.31	36.31	65.7	64.4	2.7	5.0	3.25
39	3.65	34.31	37.31	69.7	64.2	2.9	5.0	3.25
40	3.68	35.29	38.29	74.3	63.7	3.1	5.0	3.27
41	3.7	36.28	39.28	79.5	63.4	3.4	5.0	3.27
42	3.72	37.27	40.27	85.4	63.0	3.6	5.0	3.27
43	3.75	38.26	41.26	92.3	62.5	4.0	5.0	3.28
44	3.8	39.24	42.24	100.4	61.7	4.4	5.0	3.30
45	3.85	40.21	43.21	110.0	60.9	4.8	5.0	3.32
46	3.87	41.21	44.21	121.4	60.6	5.4	5.0	3.30
47	3.9	42.19	45.19	135.4	60.1	6.2	5.0	3.29
48	3.95	43.17	46.17	153.0	59.3	7.3	5.0	3.27
49	4.04	44.14	47.14	175.7	58.0	8.9	5.0	3.26
50	4.08	45.12	48.12	205.8	57.5	11.0	5.0	3.19

Heat Pump EWT, °C	Heat Pump COP	Building Loop Heat Exchanger EWT, °C	Heat Exchanger Groundwater Leaving Temperature, °C	Groundwater Flow Required, L/s	Heat Pump Power, kW	Well Pump Power, kW	Loop Pump Power, kW	System COP
2.2	3.58	-0.4	1.3	3.7	65.5	2.4	5.0	3.22
2.8	3.60	0.2	1.9	3.9	65.1	2.6	5.0	3.23
3.3	3.64	0.7	2.4	4.1	64.4	2.7	5.0	3.25
3.9	3.65	1.3	2.9	4.4	64.2	2.9	5.0	3.25
4.4	3.68	1.8	3.5	4.7	63.7	3.1	5.0	3.27
5.0	3.70	2.4	4.0	5.0	63.4	3.4	5.0	3.27
5.6	3.72	2.9	4.6	5.4	63.0	3.6	5.0	3.27
6.1	3.75	3.5	5.1	5.8	62.5	4.0	5.0	3.28
6.7	3.80	4.0	5.7	6.3	61.7	4.4	5.0	3.30
7.2	3.85	4.6	6.2	6.9	60.9	4.8	5.0	3.32
7.8	3.87	5.1	6.8	7.6	60.6	5.4	5.0	3.30
8.3	3.90	5.7	7.3	8.5	60.1	6.2	5.0	3.29
8.9	3.95	6.2	7.9	9.6	59.3	7.3	5.0	3.27
9.4	4.04	6.7	8.4	11.1	58.0	8.9	5.0	3.26
10.0	4.08	7.3	9.0	13.0	57.5	11.0	5.0	3.19

Note: Heat pump LWT below approximately 36°F (1.3°C) would require antifreeze to be used in the building loop.

- Building loop side: 248 gpm (15.6 L/s), entering at 43.4°F (6.3°C), leaving at 48.2°F (9.0°C)
- Capacity: 598,600 Btu/h (175 kW)
- Approach: 45.5 – 43.4 = 2.1°F (7.5 – 6.3 = 1.2°C)

At the higher heating-mode EWT (48.2°F vs 47.3°F at 141 gpm [9.0°C vs 8.5°C at 8.8 L/s] interpolated from Table 8.17) at which the heat exchanger would operate, the heat pumps would achieve a 3.97 COP instead of the 3.91 associated with the 141 gpm (8.8 L/s) flow in Table 8.17. Combined with the well pump power requirement at the 141 gpm (8.8 L/s) flow and the loop pump at 5.0 kW, this results in a system COP of

$$800{,}000 \text{ Btu/h} \div [(59.0 + 6.1 + 5.0) \times 3412] = 3.34 \qquad \text{(I-P)}$$

$$234 \text{ kW} \div (59 + 6.1 + 5.0) = 3.34 \qquad \text{(SI)}$$

This is slightly better than the table value of COP = 3.22, which was based on an assumed approach of 3°F (1.7°C).

The injection well for the example system has not been constructed; however, a recommendation can be made for the minimum separation distance that should be allowed between it and the existing production well. The aquifer thickness is not specifically stated in the information for the example, but based on the existing production-well construction a reasonable estimate can be made. The SWL is given as 66 ft (20.1 m) and the screened interval as 78 to 108 ft (23.7 to 32.9 m). Assuming a water table aquifer and that the lower portion of the aquifer has been screened, the aquifer thickness can be assumed to extend from 66 to 108 ft (20.1 to 32.9 m) for a total of 42 ft (12.8 m). Using a slightly more conservative value of 40 ft (12.2 m) and an effective flow rate of 50% of the peak flow, Table 8.2 suggests a minimum separation distance of 367 ft (112 m).

8.7.5 Equipment Selection Criteria, Control, and Instrumentation

The heat exchanger for the example application would be selected on the basis of the cooling-mode criteria:

- Hot side: 248 gpm (15.6 L/s), entering at 76.6°F (24.8°C), leaving at 66°F (18.9°C)
- Cold side: 141 gpm (8.8 L/s), entering at 54°F (12.2°C), leaving at 72.6°F (22.5°C)

Based on the very low chloride content of the groundwater, 304 stainless steel plates and medium nitrile rubber gaskets would be satisfactory.

The well pump would be selected for 141 gpm (8.8 L/s) at 121 ft (36.9 m).

The pump column length requirement is determined by the pumping water level (PWL) at design flow plus an allowance for required NPSH and seasonal aquifer fluctuation minus the length of the pump. The pump length is subtracted since the pump suction is at the bottom of the pump assembly.

- Pumping water level at design flow: 85 ft (25.9 m)
- Length of pump: A typical length for a seven-stage pump for 141 gpm (8.8 L/s) would be approximately 5 ft 1.5 m
- NPSH required for this pump: 8 ft (2.4 m)
- Column length required: 85 + 8 + 15 – 5 = 103 ft (25.9 + 2.4 + 4.6 – 1.5 = 31.4 m)

The actual length of column required is also influenced by the type of connection used at the wellhead—surface or subsurface (pitless adapter).

The well is currently configured for only summer operation, with a small lineshaft turbine pump discharging to a partially above-grade piping connection. To facilitate winter operation and eliminate surface piping connections, a pitless unit with 8 in. (203 mm) casing and 4 in. (102 mm) piping connections is required.

In this example design, which has the potential to operate efficiently at the same flow rate in both heating and cooling, it is possible to use the dual setpoint approach to well pump control. As mentioned previously, this type of control is influenced by the thermal mass in the building loop. Schools typically range from 4.0 to 10.0 gal/block ton (4.3 to 10.8 L/kW) in terms of building loop thermal mass. This particular school has a building loop water volume of 504 gal (1908 L), or 5.6 gal/ton (6.0 L/kW). Based on the values in Table 8.9, this would require a very substantial range (approximately 21°F [11.7°C]) on the loop temperature controller to avoid short-cycling of the well pump. Operation of the system over this large a range would result in inefficiency. Reducing the required range on the controller and bringing the loop thermal mass up to 10 gal/ton (10.8 L/kW) would require the addition of approximately 400 gal (1514 L) of additional volume to the system. The cost of adding this volume, in terms of either oversized piping or tanks, is likely to exceed the cost of using a variable-speed control on the well pump in this case.

Operation with the variable-speed well pump permits the heating-mode flow to be reduced to 110 gpm (6.9 L/s) as previously discussed. The heat exchanger, assuming an overall U-factor of 700 $Btu/h \cdot ft^2 \cdot °F$ (123 $W/m^2 \cdot °C$) due to lower water temperature and reduced flow rate, would yield a heating performance EWT for the heat pumps of approximately 46.1°F (7.8°C) at the 110 gpm (6.9 L/s) groundwater flow. This would result in a return water temperature (to the heat exchanger) of 41.3°F (5.2°C) and a system COP of 3.33.

In the cooling mode the optimum return water temperature (Table 8.16) is 76.6°F (24.8°C). The well pump would be enabled at a loop return temperature of 78°F (25.6°C) and would be modulated to maintain a loop return temperature of 77°F (25°C) in the cooling mode. At loop return temperatures below 74°F (23.3°C), the well pump would remain off. At a reduction of loop temperature to 39°F (3.9°C), the well pump would be enabled and would modulate to maintain the optimum loop return temperature of 41°F (4.4°C). At loop return temperatures above 43°F (6.1°C) in the heating mode, the well pump would remain off.

Selection of the strainer for a GWHP system is based on the results of a sieve analysis of the suspended material collected during the pump test of the production well. The slot size for the strainer screen is selected to ensure that at least 95% of the suspended material in the water is removed. In this example the sieve analysis indicated that the 90% size of the suspended material was 0.0197 in. (0.5 mm) or larger.

The 90% size from the sieve analysis suggests a requirement for a 35 mesh (Table 8.18) for complete removal, so it seems safe to specify a 40 mesh screen to ensure 95% removal of all suspended material in this case. It is sometimes necessary when selecting strainers to specify either an oversized device or two strainers in parallel to facilitate a reasonable pressure drop. In this case, however, manufacturer's data indicate that a 4 in. (100 mm) basket strainer with a 40 mesh basket will have a pressure drop of only 0.4 psi (2.8 kPa) (clean). This is acceptable and does not require oversizing or the use of dual strainers. A bypass for the strainer is used to allow for cleaning of the basket without interrupting flow (Figure 8.15).

Figure 8.16 provides a summary of suggested instrumentation for a GWHP system. Of the points shown, the following are suggested for logging on a continuous basis to aid in diagnostics:

- Production-well water level
- Injection-well water level

- Groundwater flow
- Total groundwater production (volume)
- Total heat rejection
- Total heat absorption

Well water level trends are very valuable diagnostic tools, particularly when they can be tied to specific flow rates. Changes in water levels at a specific flow, over time, can indicate fouling of the well screen, plugging of the aquifer, and other events that help to

Table 8.18 Strainer Screen Mesh Data

Mesh	Diameter, in.	Diameter, mm
20	0.0331	0.84
25	0.0280	0.71
30	0.0232	0.60
35	0.0197	0.50
40	0.0165	0.42
45	0.0138	0.35
50	0.0117	0.30
60	0.0098	0.25
70	0.0083	0.21
80	0.0070	0.18
100	0.0059	0.15

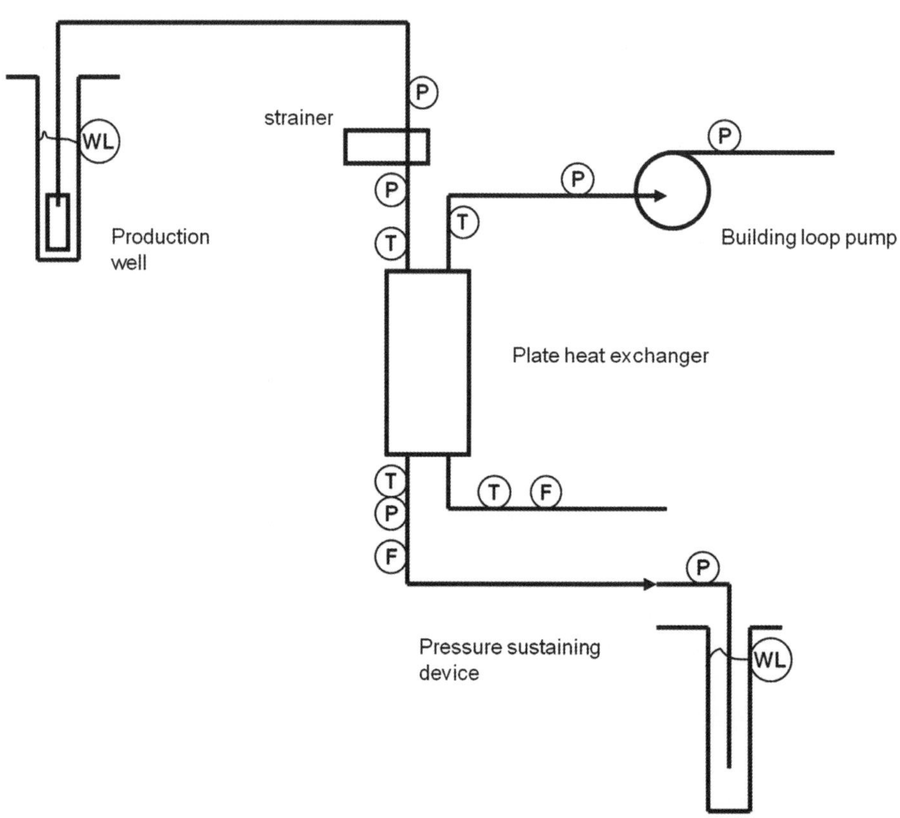

Figure 8.16 Suggested Instrumentation and Monitoring for a GWHP System

indicate when well service may be required. Some regulatory authorities require records of annual total groundwater production. Total heat rejected to and absorbed from the groundwater provides an indication of the impact of the system on the local aquifer.

Pressure drop across the groundwater strainer is a useful index of when cleaning may be required. Some maintenance personnel use heat exchanger pressure drop as an indicator of when exchanger cleaning may be required. Generally, though, the thermal performance of the exchanger will deteriorate from fouling far earlier than the same fouling will be detected through increased pressure drop. A more effective index of heat exchanger fouling is monitoring of approach (groundwater leaving temperature compared to building loop entering temperature).

8.8 GWHP ECONOMICS

8.8.1 Background

GWHP systems, under favorable conditions, can yield substantial capital cost savings compared to conventional closed-loop designs. The two systems (assuming central-loop GCHP design) are largely identical inside the building, with both using the same heat pumps, building loop piping circulating pump, and outdoor air provisions. The difference lies in the ground-loop portion of the system. The underlying reason for the open-loop cost advantage is traceable to the costs (as measured in $/ton [$/kW]) of water wells compared to closed-loop boreholes. A recent well constructed for a large open-loop system provides a useful illustration of this (Rafferty 2014). The 250 ft (76 m) deep well included a 12 in. (305 mm) casing (to 150 ft [46 m]), a 10 in. (254 mm) stainless steel continuous slot screen (100 ft [30 m]), a 20 ft (6 m) surface seal, very substantial development time (50 h), and the services of a hydrologist for design and construction management. At first glance the cost of this well, $85,000 (or $340/ft [$1115/m]) seems high, especially to those accustomed to closed-loop borehole construction costs. When the production capacity of this well is considered, however, the cost is placed in perspective. With a production of 1500 gpm (95 L/s), this well provides a capacity of 1000 tons (3520 kW) at a groundwater flow of 1.5 gpm/ton (0.027 L/s·kW). This translates into a cost of $85/ton ($24/kW) for the well, which compares favorably to equivalent borehole capacity at $18/ft and 175 ft/ton ($59/m and 15.2 m/kW), or $3150/ton ($895/kW). In both cases, however, this cost breakdown omits a number of cost items necessary to complete a system.

Just as a closed-loop system requires headers to connect the boreholes, isolation valves, vaults or manifolds, and flushing and filling, a complete GWHP ground loop includes much more than the production well to provide a complete system. The key cost items associated with the ground loop in a GWHP system include the following:

- Production well
- Well pump, drive, and electrical connection
- Piping to mechanical room
- Heat exchanger
- Piping, controls, and strainer in mechanical room
- Piping to injection well
- Injection well

Incorporating all of these GWHP costs and comparing them to the total costs of central-loop GCHP ground-loop components provides a clear picture of the relative advantages of the two system types.

Relatively little cost (capital or maintenance) data are available on open-loop systems, and most ASHRAE research has focused on closed-loop data. The cost data in this section are therefore based on 2006 to 2014 water well construction costs corrected to 2014 dollars (Rafferty 2014); to normalize the data for presentation, component parts of actual individual well construction cost results have been used to reconstruct well cost information for three different depths and three different types of completions over a range of production flow rates. Plate heat exchanger costs are based on results from recent projects as well (Rafferty 2014). The remainder of the required components (piping, controls, electrical) are based on costs in standard construction cost-estimating publications (RSMeans 2011).

8.8.2 GWHP Capital Costs

Figure 8.17 provides a comparison of the component costs for a 212 ton (723 kW) system for two cases, a 150 ft (46 m) deep open-hole well completion (red) and a 700 ft (213 m) deep gravel-pack completion (blue). In each case, one production and one injection well are included, along with the other components necessary to complete the GWHP groundwater loop (see the note at the base of the figure for details on costs). The dramatic impact of well completion type and depth on system costs is clearly demonstrated.

The 150 ft (46 m) open-hole costs represent the low end of what might be expected for well costs in general. In this case, the building mechanical costs (heat exchanger and related piping) dominate the total costs for the groundwater loop and the wells constitute less than 30% of the groundwater loop costs. The blue bars, representing costs associated with 700 ft (213 m) deep gravel pack well construction, illustrate the case of extremely high well costs. These well costs far exceed all of the other costs combined and constitute 78% of the total groundwater loop costs.

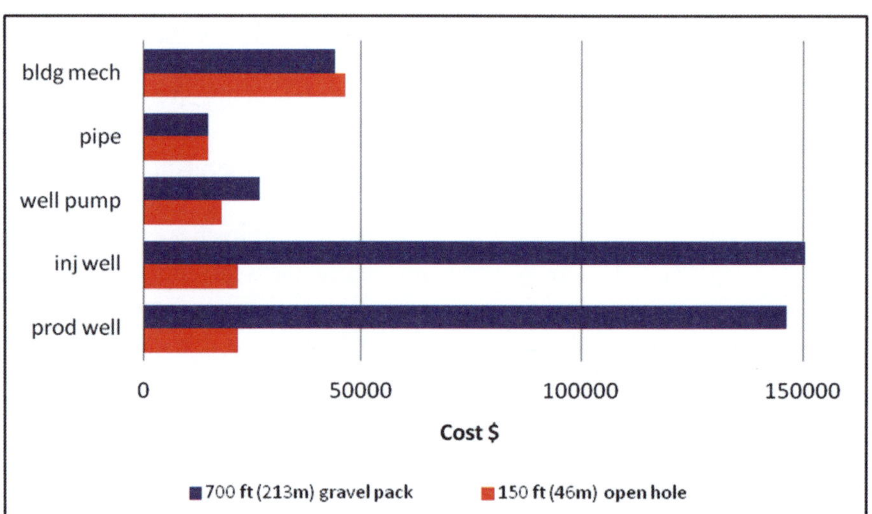

Basis is 212 ton (746 kW) system, 1.5 gpm/ton (0.027 L/s·kW). Red bars: 150 ft (46 m) deep production and injection wells, well pump (100 ft [30 m] setting) costs include VFD, electrical, and controls; building mechanical includes heat exchanger (3°F [1.7°C] approach), piping, and strainer; pipe includes PVC buried piping to and from the mechanical room. Blue bars: 700 ft (213 m) deep production and injection wells, well pump with 500 ft (152 m) setting, remainder of costs equal to 150 ft (46 m) case.

Figure 8.17 Open-Loop Component Costs—212 ton (746 kW) System

The costs of most components of GWHP ground loops are heavily influenced by the specifics of the individual design and the local aquifer and geology. In addition to the cost variations arising from different completion methods (open hole, screened, or screened and gravel packed), there are also variations caused by the type of casing and screen used. Plastic casing and screen have been used in some cases and can reduce costs substantially. These materials are limited in terms of strength and can fail if sufficient forces are imposed in grouting, cementing, or gravel packing. For very shallow wells, however, the plastic materials remain an option provided their limitations are carefully considered. A plastic well screen, installed in the well, in the 8 in. (203 mm) size, costs approximately 20% that of a stainless steel screen. Plastic well casing in the 8 in. (203 mm) size costs approximately 35% less than steel casing installed in the well. All of the cost data used in Figures 8.17 to 8.20 are based on stainless steel screens and carbon steel casing.

Well screen length, which is somewhat influenced by the aquifer type and aquifer thickness, also impacts cost. Cost data appearing here are based on screens sized for the recommended maximum entrance velocity of 0.1 ft/s (0.03 m/s) with lengths typically between 5 and 20 ft (1.5 and 6.1 m) depending on flow. The seal, especially in an injection well that will be pressurized (and where the seal must extend to the top of the injection zone), can increase costs. Seal costs for both production and injection wells are based on a depth of 40 ft (12.2 m). The cost of development, particularly in naturally developed wells, can be a major factor in total well cost. Development, the process in which fine materials in the near-well zone are removed by jetting, swabbing, and other procedures, can require significant effort in some cases, and development time can be as costly as drilling itself. Development costs shown in Figures 8.17 to 8.20 were based on a development time in hours equal to the screen length in ft (m) (i.e., 15 h for a screen of 15 ft [4.6 m] length).

Heat exchanger cost is influenced primarily by system capacity and approach temperature. The impact of approach on cost is discussed in Section 8.6.1. Very small systems incur a much higher cost per ton (kW) for the heat exchanger, as plate surface area tends to be overshadowed by the frame cost. Table 8.19 provides an example of this for two heat exchanger quotes from 2012.

Costs in Figures 8.18 to 8.20 are based on heat exchangers sized for 3 ft^2 (0.27 m^2) of surface per ton (kW) of block load (approximates 3°F [1.7°C] approach and 900 Btu/ft^2·°F [5112 W/m^2·°C). Installation is based on 25% of the exchanger cost and mechanical room piping is based on 20% of heat exchanger cost. Strainers are separately included and are based on the use of two basket strainers in parallel. The buried piping portion of the system is influenced, in terms of cost, primarily by the distances involved; this issue is typically not under the control of the designer, as well separation distance is a function primarily of system capacity and the nature of the aquifer. Distances for buried piping included here are based on separation distances of between 200 and 700 ft (61 and 213 m) depending on the groundwater flow requirement. A variety of materials for the buried piping are available, though PVC has historically been the most commonly used. Ther-

Table 8.19 Heat Exchanger Costs

Capacity, tons (kW)	Heat Transfer Area, ft^2 (m^2)	Plates and Gaskets % of Total Cost	Frame % of Total Cost	Cost of Heat Transfer Area, \$/ft^2 (\$/m^2)	Total Cost, \$/ton (\$/kW)
152 (535)	457 (42.5)	75.3	24.7	47.6 (512)	143 (40.6)
25 (88)	77.5 (7.2)	41.7	58.3	101.7 (1094)	315 (89.5)

Note: Costs include 304 stainless steel plates and NBR gaskets; designs based on 3°F (1.7°C) approach.

mally fused HDPE pipe can be used for this application, though there is no contaminant issue associated with the groundwater in the event of a leak as there is in GCHP systems. Contractors tend to be familiar with practices necessary for gasketed PVC (AWWA 2007) due to its wide use in municipal water systems; this material is the basis for piping costs used here. Table 8.20 provides a summary of the cost items included in developing Figures 8.18 to 8.20.

Figures 8.18, 8.19, and 8.20 provide a comparison of GWHP ground-loop costs for three different well depths (150, 300, and 700 ft [30, 60, and 213 m]) and three different well completions (open hole, naturally developed, and gravel pack) compared to GCHP ground-loop costs for central-loop systems. In these figures, high and low cases for GCHP costs are portrayed. The high case is based on a completed ground loop (boreholes, headers up to the building wall) at \$20/ft and 225 ft/ton (\$65.6/m and 19.5 m/kW), and the low case at \$12/ft and 175 ft/ton (\$39.4/m and 15.2 m/kW). The variation in closed-loop costs over the range of system capacities is a reflection of the initial economy of scale in borehole construction (up to approximately 100 tons [352 kW]), which is compromised by increasing horizontal loop costs (for systems up to approximately 100 to 200 tons [352 to 704 kW]), after which economy of scale again provides benefits. The higher cost curve is reflective of areas of the country where labor costs are higher, prevailing wages are in effect, experienced engineers and contractors are not available, or drilling costs are unusually high. The lower cost curve is reflective of areas where labor costs are unusually low, economical loop design (elimination of vaults, etc.) is used, experienced engineers and contractors are available, and drilling is unencumbered by difficulties.

For the case of shallow (150 ft [46 m] depth) wells, it is apparent that the GWHP costs for all well types are well below the GCHP range for all system capacities considered. For a 300 ton (1056 kW) system, the GWHP ground-loop costs would be approximately \$1,260,000 less than those for a GCHP loop in a high-cost area and \$450,000 less than those for a GCHP ground loop in a low-cost area.

Table 8.20 Summary of Costs Included in Figures 8.18 to 8.20

Production well	Drilling, casing, screen, gravel pack (where required), flow test, sanitary seal, development
Sanitary seal	40 ft (12 m) all wells
Casing	Steel—diameters 6, 8, 10, 12 in. (125, 203, 254, 305 mm) based on flow
Screen	Stainless steel, wire wound—diameters 4, 6, 8, 10 in. (100, 125, 203, 254 mm) based on flow; 0.1 ft/s (0.030 m/s) production, 0.05 ft/s (0.015 m/s) injection
Flow test	Step drawdown
Development time	Hours equal to screen length in feet
Injection well	Drilling, casing, screen, gravel pack, flow test, sanitary seal, development
Well pump	Submersible type, steel column appropriate to well depth (100, 200, 500 ft [30, 60, 152 m]), VFD, installation, wire from building, loop temperature control, 5 to 50 hp (3.7 to 37 kW) depending on flow
Consulting hydrologist	Included for all naturally developed and gravel pack wells at 8% of well cost
Buried piping	Length based on flow and required separation distance, PVC (AWWA C900 type)
Heat exchanger	304 stainless steel/NBR construction, 3°F (1.7°C) approach, 3 ft^2/ton (0.08 m^2/kW), installation at 20% of heat exchanger cost
Mechanical room piping	At 25% of heat exchanger
Strainer	Two iron-body basket strainers
Groundwater flow	1.5 gpm/ton (0.027 L/s·kW)
Contingency	15%

As well depth increases, as illustrated for 300 ft (92 m) wells in Figure 8.19, the cost-competitiveness increases between GWHP and GCHP ground loops, but only at the lower end of the capacity range and only in areas of very-low-cost GCHP construction. Only the gravel pack well construction actually crosses over into the GCHP cost range, and this only below approximately 75 tons (264 kW) system capacity under conditions of low-cost GCHP construction. Above approximately 100 tons (528 kW), GWHP construction offers substantial cost savings. In this case, a 300 ton (1056 kW) GWHP system would offer approximately $1,230,000 savings over a high-cost GCHP installation, and approximately $420,000 over the low-cost GCHP system.

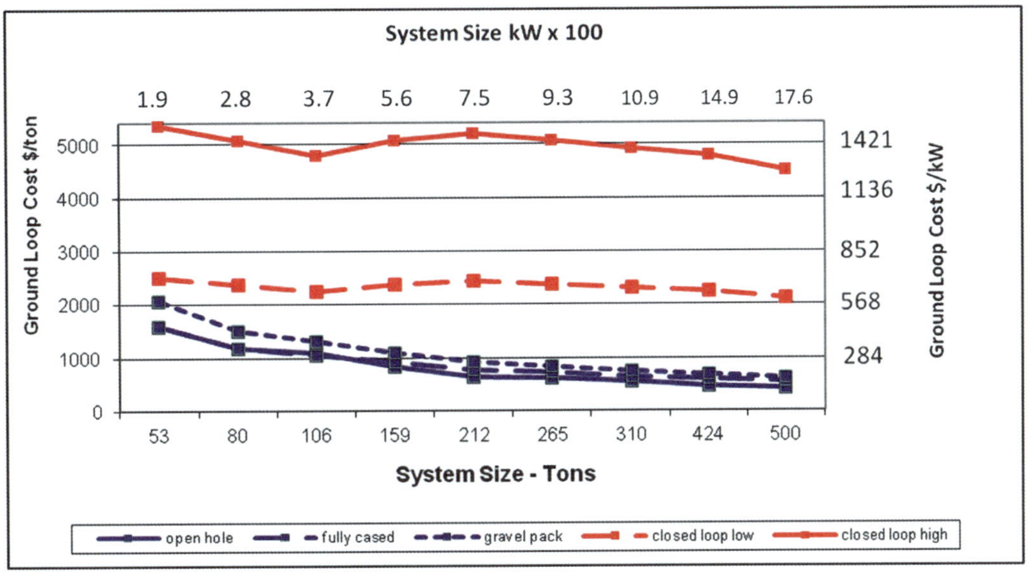

Figure 8.18 GWHP and GCHP Ground-Loop Costs—150 ft (46 m) Wells

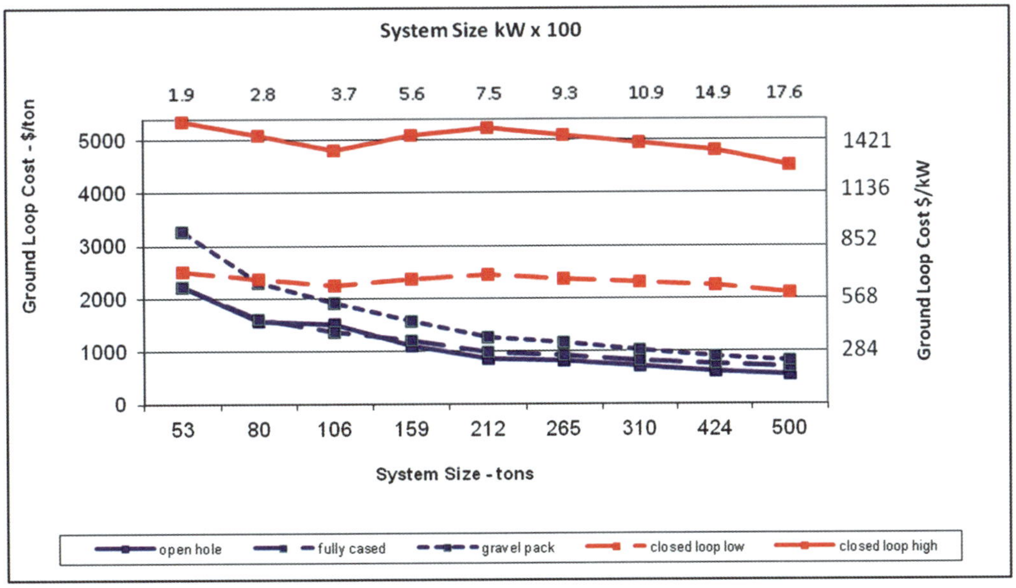

Figure 8.19 GWHP and GCHP Ground-Loop Costs—300 ft (90 m) Wells

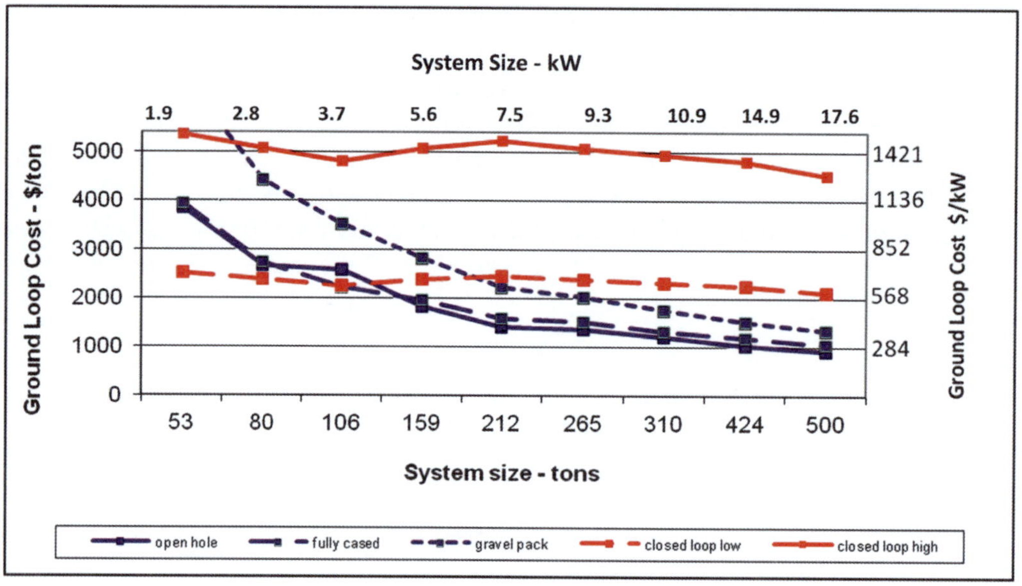

Figure 8.20 GWHP and GCHP Ground-Loop Costs—700 ft (213 m) Wells

Figure 8.20 presents the case for the highest-cost water wells considered—700 ft (213 m) depth. Here the costs are more competitive, particularly if gravel-pack type completion is required for the open-loop wells. Gravel-pack completed wells are not cost-competitive in the lowest-capacity (<75 tons [264 kW]) applications. Naturally developed and open-hole completion wells remain attractive relative to high-cost GCHP systems at capacities above approximately 80 tons (282 kW) and to low-cost GCHP systems at capacities above approximately 150 tons (528 kW). For the 300 ton (1056 kW) capacity, the GWHP ground loop would offer a savings of approximately $1,050,000, and approximately $210,000 compared to the low-cost GCHP system.

8.8.3 GWHP Maintenance Costs

While the potential savings offered by GWHP systems in some cases are attractive, they must be viewed in the context of the higher maintenance costs incurred by these systems. As in the case of capital cost data, the GWHP maintenance cost information base is sparse. An ASHRAE research project (Cane and Garnet 2000) did address maintenance costs in GWHP systems and found a median maintenance cost of $\$0.091/ft^2$ ($\$0.98/m^2$) for the seven buildings included in the study. The closed-loop systems (31 buildings) in the same report showed a median maintenance cost of $\$0.063/ft^2$ ($\$0.68/m^2$).

Given the similarity of in-building GWHP system equipment to the equipment for GCHP systems (excluding unitary or individual loops) and the very low maintenance requirements of the GCHP ground loop, it would seem reasonable to conclude that the maintenance costs of the GWHP ground-loop components would be represented by the difference between the values provided above: $\$0.028/ft^2$ ($\$0.30/m^2$). In fact, this value seems low given the likely maintenance requirements for GWHP ground-loop components.

At a minimum, regular cleaning of the plate heat exchanger, strainer blowdown, some periodic well maintenance, and periodic well pump replacements would constitute the bulk of the GWHP loop maintenance costs.

Heat exchanger cleaning is a function of the rate of fouling. In high-temperature geothermal systems handling water of several thousand parts per million, heat exchangers have frequently served for many years without cleaning (in one case eight years) (Rafferty 2014). Though there is one large GWHP installation in which the heat exchangers are cleaned monthly (Rafferty 2014), this is a result of a system design that exposes water high in iron content to aeration, resulting in severe iron fouling of the plates. In systems designed as recommended in this book, this occurrence should not be repeated. It is good practice to open heat exchangers annually, however—a procedure that even with large exchangers is possible to accomplish in a single shift with two workers.

There is little, if any, regular maintenance associated with submersible well pumps other than replacement when failure occurs. Provided the motors are not cycled excessively (see Table 8.8 for allowable cycling frequency), submersible well pumps should have a service life of approximately 15 years in low-sand-content (<10 ppm) water.

Water well maintenance requirements are, like construction costs, a strong function of the geologic setting in which they are completed. Information in Appendix N indicates that properly designed water wells, completed in the geology specified, require major maintenance (defined as 10% of well replacement cost) at the following intervals:

- Metamorphic rock (slate, schist, gneiss, marble)—15 years
- Sandstone, limestone, basalt—12 years
- Combination consolidated/unconsolidated material—8 years
- Alluvium (shallow unconsolidated material)—5 years

This study (Gass et al. n.d.) was based on municipal wells that were operated continuously; presumably wells operated on the order of 2000 EFLH per year (typical of GWHP applications) would experience service intervals somewhat to substantially beyond the values cited.

However, based on the information cited above and the well cost information collected for this book (Rafferty 2014), it is possible to calculate the predicted maintenance for different well types and capacities. For example, for a 150 ton (528 kW) system serving a 60,000 ft^2 (5580 m^2) school with two 300 ft (92 m) wells (one injection, one production) completed in sandstone, with well maintenance based on the well service intervals suggested previously and the well cost information in Figure 8.18, system maintenance costs can be calculated as follows:

- Well costs: $44,000 total, 10% = $4400; 4400/12 = $367/yr
- Heat exchanger maintenance: 8 h, 2 workers at $75/h; 16 × $75 = $1200/yr
- Strainer blowdown: 8 times per year at 0.25 h labor each—2 × $75 = $150
- Well pump replacement interval: 15 years at $6000; 6000/15 = $400/yr
- Total maintenance: $367 + $1200 + $150 + $400 = $2117/yr
  - At 60,000 ft^2: $2117/60000 = $0.035/$ft^2$, or 3.5 cents/$ft^2$
  - At 5580 m^2: $2117/5580 = $0.38/$m^2$, or 38 cents/$m^2$

This is reasonably close the results reported in the study by Cane and Garnet (2000). Adjusting for an average inflation rate of 2.5% in the interval since that study was published results in an updated incremental rate of 3.8 cents/ft^2 (41 cents/m^2). The above example, however, assumes the use of wells completed in rock—relatively low maintenance requirement wells. Substituting more maintenance-prone wells, gravel pack wells completed in alluvium with a 5 yr service interval, yields the following:

- Well costs: $138,548 total, 10% = $13855; at a 5 yr interval, $2771/yr
- Substituting the new well maintenance value into the above total = $4521/yr
  - At 60,000 ft^2: 4521/60000 = $0.075/$ft^2$, or 7.5 cents/$ft^2$
  - At 5580 m^2: 4521/5580 = $0.81/$m^2$, or 8.1 cents/$m^2$

The annual maintenance requirements calculated in these examples show good agreement with the previously published data on GWHP maintenance costs (Cane and Garnet 2000)—provided the systems include wells completed in rock geology. For higher-maintenance gravel pack wells it appears that incremental maintenance requirements for the ground-loop portion of the system could be as high as twice the amount suggested by Cane and Garnet (2000). Until such time as actual maintenance data become available, however, this issue will remain uncertain.

Comparing the incremental maintenance costs for GWHP systems to the capital cost savings available does provide some insight as to their relative impact on decision making. If the assumption is that the decision to use a closed-loop system over an open-loop system is based solely on the higher maintenance costs of the open-loop system, it is possible to construct a simple payback calculation to support that decision. The incremental capital cost of the closed-loop system over that of the open-loop system would be divided by the increased maintenance cost of the open-loop system to arrive at the simple payback. In the example above, the incremental costs of the closed-loop system over that of the open-loop system would amount to between $570,000 (high cost GCHP for 150 ton [528 kW] system) and $165,000 (low cost GCHP for 150 ton [528 kW] system). Using a value of $\$0.0365/\text{ft}^2$ (average of Cane and Garnet [2000] data and calculated maintenance cost) for the 60,000 ft^2 (5580 m^2) building incremental maintenance costs for the open-loop system over those of the closed-loop system, a simple payback of between 75 and 260 years results. The corresponding values for the higher-maintenance well case are $90,000 incremental capital cost for the low-cost GCHP and $495,000 for the high-cost GCHP with an incremental maintenance cost of $\$0.0765/\text{ft}^2\cdot\text{yr}$. This results in simple payback periods of between 20 and 108 years.

Clearly the incremental maintenance costs, when considered in the context of the incremental capital cost savings of open-loop over closed-loop systems, are not sufficient to deter decision makers from implementing GWHP systems. There may be other issues that preclude the use of open-loop systems, but it does not appear from the data available that the maintenance cost issue is a decision maker in the context of comparing open- and closed-loop systems. Figures 8.18, 8.19, and 8.20 demonstrate that open-loop systems tend to be most attractive in settings characterized by shallow wells (<700 ft [213 m]), open-hole completions, and with system capacity requirements of greater than 100 to 150 tons (350 to 530 kW), though in the case of lower-cost well construction (open hole and naturally developed), open-loop GWHPs demonstrate substantial cost advantages over closed-loop GCHPs at system capacity greater than 80 tons (280 kW) with well depth requirements of 300 ft (90 m) or less.

8.9 REFERENCES

AWWA. 2007. AWWA C900-07, *Polyvinyl Chloride (PVC) Pressure Pipe and Fabricated Fittings 4 in. through 12 in. (100 mm through 300 mm), for Water Transmission and Distribution*. Denver: American Water Works Association.

Cane, D., and J.M. Garnet. 2000. Update on maintenance and service costs of commercial building ground-source heat pump systems. *ASHRAE Transactions* 106(1).

Egg, J., G. Cuniff, and C.D. Orio. 2013. *Modern Geothermal HVAC Engineering and Control Applications.* New York: McGraw-Hill Professional.

EPA. 1975. *Manual of Water Well Construction Practices, 570/9-75-001*. Washington, DC: U.S. Environmental Protection Agency.

Franklin. 2007. *Submersible Motors: Application, Installation and Maintenance*, August 2002 Edition. Bluffton, IA: Franklin Electric.

Gass, T.E., T.W. Bennett, J. Miller, R. Miller. n.d. *Manual of Water Well Maintenance and Rehabilitation Technology*. Reprinted by the National Water Well Association from the Robert S. Kerr Environmental Research Center, USPA, Ada, Oklahoma.

Hatten, M. 1992. Ground water heat pumping lessons learned in 45 years at one building. *ASHRAE Transactions* 98(1).

Kazmann, R.G., and W.R. Whitehead. 1980. The spacing of heat pump supply and discharge wells. *Ground Water Heat Pump Journal* 1(2).

Knipe, E., and K. Rafferty. 1985. Corrosion in low temperature geothermal applications. *ASHRAE Transactions* 91(2).

Rafferty, K. 2001. Dual set point control of open-loop heat pump systems. *ASHRAE Transactions* 107(1).

Rafferty, K. 2008. Design issues in commercial open-loop heat pump systems. *ASHRAE Transactions* 114(2).

Rafferty, Kevin. 2014. Proprietary project cost and maintenance installation data collected by the author.

RSMeans. 2011. *RSMeans Mechanical Cost Data 2012*. Norwell, MA: Reed Construction Data.

9 GSHP Performance and Installation Cost

9.1 FIELD STUDY PERFORMANCE RESULTS

9.1.1 Project Overview and Loop Circuit Types

This section consists of content originally published in ASHRAE Journal *(Kavanaugh and Kavanaugh 2012a). The text has been edited to conform to the style of this book.*

Many GSHP systems have been successfully installed and operated for many years throughout the United States. However, other installations have experienced poor reliability, high energy costs, and undesirable ground-loop temperatures. Some have had significant equipment replacements, have added supplemental heating, or have installed fluid coolers. A few GSHPs have been abandoned.

A data collection and analysis project was conducted to identify common characteristics of successful GSHP systems and the incidence of unacceptable long-term temperature change (EPRI 2012). The data collection efforts were structured to gather a limited amount of the critical information for a large number of systems in the southeastern United States, Texas, and central Illinois. The approach included the following:

1. Conduct surveys (forms completed with assistance from owners, utilities, and designers)
    - Building and GSHP system description performance
    - Energy and demand from utility bills
    - Installation costs for newer sites
    - Comfort/indoor air quality/satisfaction
    - Maintenance personnel evaluation
2. Collect data including the following:
    - Temperatures: ground loop, initial ground, change with time and load
    - Loop field description: number of bores, depth, separation, U-tube size, bore grout/fill type, header arrangement and sizes, thermal property test, well logs
    - Building details: type, size, loads, occupancy, schedules, ventilation air method
    - Equipment description: heat pump type, capacity, pump system, interior piping, air distribution system, heat pump control method, pump control method

- Sufficient information to determine ENERGY STAR® rating, including energy consumption, number of occupants, occupancy hours, important internal loads, and other data depending on building application (EPA 2010)
- Installation costs for newer buildings

Figure 9.1 shows a bar graph of the surveyed commercial buildings with GSHP systems for which sufficient information was available to obtain ENERGY STAR ratings. Buildings with ratings below 75 are not officially ENERGY STAR rated because the ENERGY STAR Buildings program does not list buildings with scores below 75. A rating of 75 or higher qualifies for ENERGY STAR designation and indicates the normalized building source energy use is lower than 75% of equivalent buildings as determined from the 2003 Commercial Building Energy Consumption Survey (CBECS) published by the U.S. Energy Information Administration (EIA 2008).

Of the 36 buildings with sufficient information, 22 attained an ENERGY STAR designation. The variation in ENERGY STAR rating ranged from a low of 1 to a high of 100. While most of these systems performed well, this variation indicates GSHPs in some cases have been poorly designed and installed.

Figure 9.1 also indicates a general trend of improved ENERGY STAR ratings for newer GSHP systems. All of the surveyed sites, installed from 2005 to 2010, have ratings above 80, most of them above 90. However, the highest-rated building had been operating for ten years, and three buildings with 15 years of operation also have ENERGY STAR ratings above 90. The longest-operating system (23 years) obtained an ENERGY STAR rating of 79 in spite of the fact that it had operated in a southern climate with vertical bore

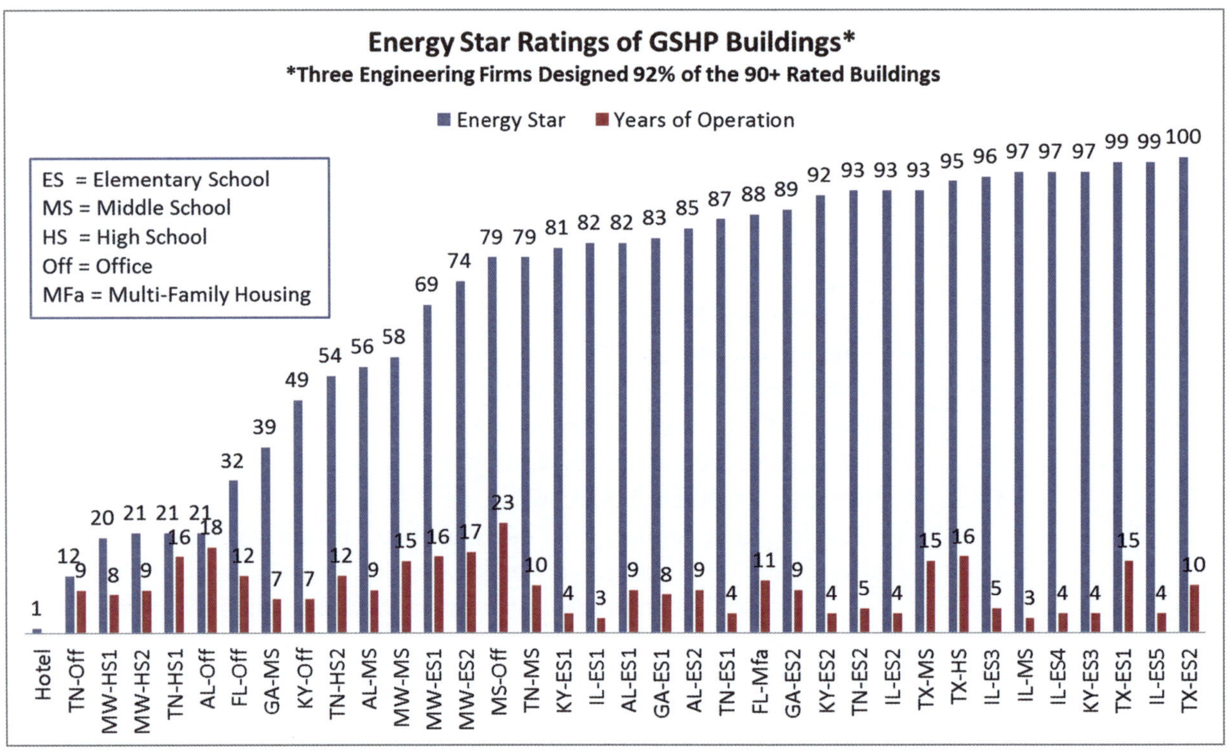

Figure 9.1 ENERGY STAR Ratings and Years of GSHP Operation for Commercial Buildings
Source: Kavanaugh and Kavanaugh (2012a)

spacing less than 15 ft (4.5 m). The lower-rated buildings have GSHP systems that had operated 9, 16, 18, and 12 years. The GSHP system in the lowest-rated building has been abandoned.

It is also interesting to note that 11 of the 12 GSHP buildings rated above 90 were designed by one of three firms. The other building rated above 90 was a result of the owner dictating to the design firm the specifications for the ground loop. On a previous project for the owner, the design firm had allowed a contractor to provide the ground loop dimensions. This system had to be supplemented by a fluid cooler after one year of operation. The owner insisted the subsequent design have a much larger ground heat exchanger with increased borehole spacing. The building achieved an ENERGY STAR rating of 92.

Figure 9.2 shows the ENERGY STAR ratings for the central loop with central pumps connected inside the building to individual heat pumps as shown in Figure 1.9. Sixteen of the 20 systems had variable-speed drives (VSDs) on the ground-loop pump motors. Two of the central systems incorporated reversible central chillers and air-handling units (AHUs) with variable-air-volume (VAV) terminals rather than individual unitary heat pumps.

Eight of the 20 central-loop systems (40%) achieved ENERGY STAR designation, with two exceeding a rating of 90. There appears to be little difference in the performance of those with VSD pump motors (6 of 16 achieving ENERGY STAR) and those with constant-speed motors (2 of 4 achieving ENERGY STAR). The central chilled-water loops with VAV air distribution systems received poor ENERGY STAR ratings.

Six systems in the survey were central one-pipe loops that were retrofits of existing schools. Five were built in the 1950s and one in 1938. As shown in Figure 9.3, all of these GSHP buildings achieved ENERGY STAR designation, with four rating 95 or higher. The lone school with a rating below 90 was built in 1938.

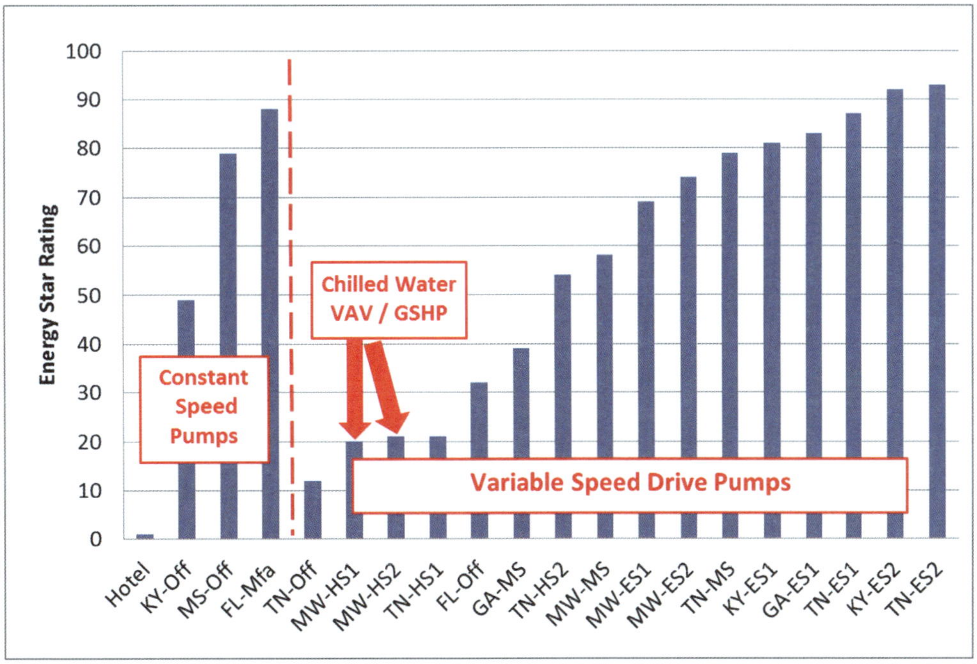

Figure 9.2 ENERGY STAR Ratings of Central-Loop GSHPs with Central Pumps
Source: Kavanaugh and Kavanaugh (2012a)

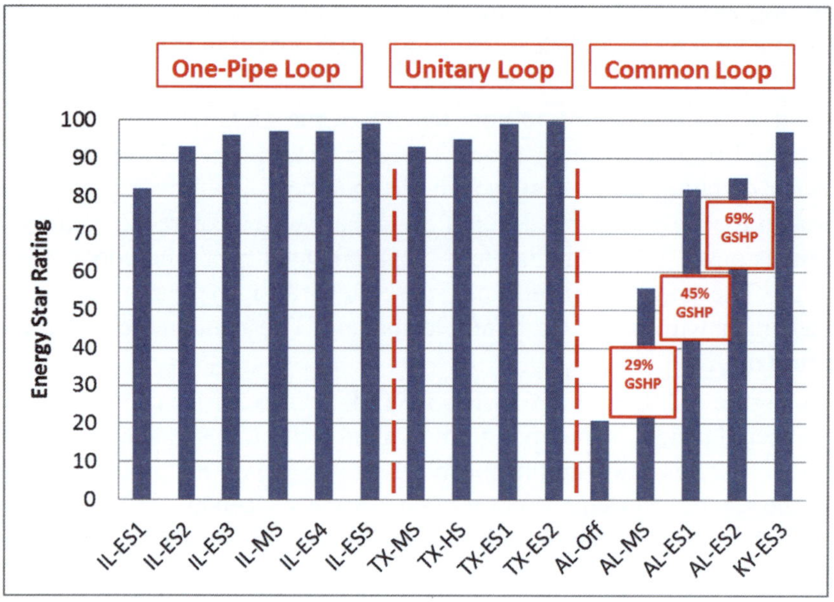

Figure 9.3 ENERGY STAR Ratings of One-Pipe, Unitary, and Common-Loop GSHPs
Source: Kavanaugh and Kavanaugh (2012a)

Four buildings in the survey were served by unitary-loop GSHPs connected to individual heat pumps with on-off circulator pumps. These buildings received ENERGY STAR ratings above 90, with one achieving a rating of 100. The two older schools are served by GSHPs in all areas. The classrooms and offices in the two newer schools are conditioned with GSHPs, while the common areas such as cafeterias, gymnasia, and kitchens are served by air-cooled equipment. All ventilation-air energy recovery units (ERUs) were supplemented by non-GSHP equipment.

Five systems in the survey were served by common loops connected to multiple heat pumps inside the building. Three of the systems were located in buildings that were also partially served by conventional unitary equipment, as noted in Figure 9.3. Four of the five systems appear to be operating effectively, while one is operating well below the average ENERGY STAR rating of 50. One building received a rating slightly above 50, but in this building only 29% of the floor space is conditioned by GSHPs. The other two buildings that are partially heated and cooled by GSHPs rated high enough to merit ENERGY STAR designation. The school building with a single common ground loop received an ENERGY STAR rating of 97.

9.1.2 Ground Loops, Pumps, Ventilation Air, and Controls

This section consists of content originally published in ASHRAE Journal *(Kavanaugh and Kavanaugh 2012b). The text has been edited to conform to the style of this book.*

Ground heat exchanger performance was found to be a critical factor in GSHP system success. Bore length (L_b) was used as a primary indicator but, as noted in previous chapters, there are several other factors that affect performance, including ground thermal properties (temperature, conductivity, and diffusivity), vertical bore separation, conductivity of the annular grout/fill, integrity of the grout/fill placement, and heat exchanger type. Some scatter in the results is expected since these characteristics vary from site to site and all these details were often not available. The impact of most of these variables is complex and often uncertain. The variation of bore length to approach temperature (dif-

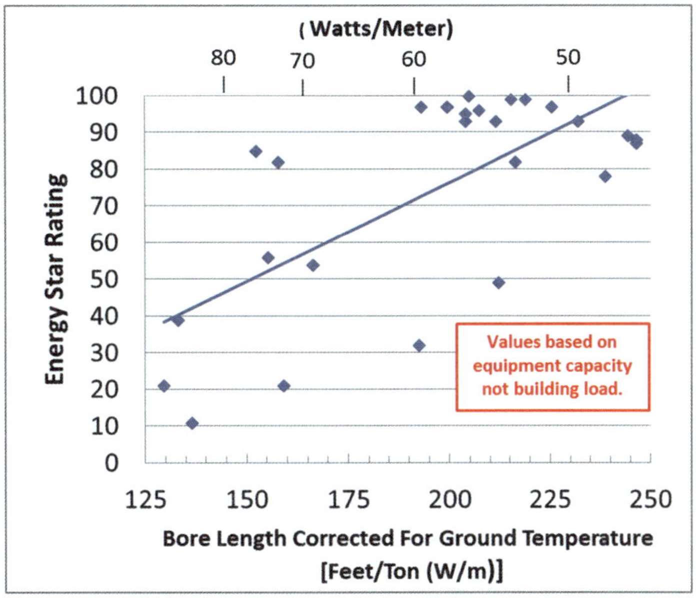

Figure 9.4 ENERGY STAR Rating vs Bore Length Normalized to 63°F (17°C) Ground Temperature
Adapted from Figure 1 of Kavanaugh and Kavanaugh (2012b)

ference between the average loop temperature and the ground temperature) is more easily normalized.

Cooling performance is a strong function of ground-loop leaving water temperature (LWT) and entering water temperature (EWT). Thus, the required cooling-mode bore length to provide high efficiency in a location with a lower ground temperature tends to be less than the required length for a warmer location. To better compare optimum ground-loop lengths for the variety of locations, the trend between installed bore length and performance is normalized for ground temperature. The adjustment is based on the average ground temperature, t_g(avg) = 63°F (17°C), and the average maximum loop temperature, (LWT + EWT)/2 ≈ 90°F (32°C), at the sites in the project survey:

$$L_b\text{/ton (Normalized)} = L_b\text{/ton} \times (90°\text{F} - t_g)/[90°\text{F} - t_g(\text{avg})]$$

A ground loop installed at 250 ft/ton (22 m/kW·ton) of bore corresponds to a normalized length of 185 ft/ton (16 m/kW·ton) for a ground temperature of 70°F (21°C), while 170 ft/ton (15 m/kW·ton) of bore results in a normalized length of 201 ft/ton (17 m/kW·ton) for a ground temperature of 58°F (14°C). The design bore lengths for the systems monitored during this project were all determined by the cooling load even though some sites had significant heating requirements.

Figure 9.4 shows the trend for ENERGY STAR rating to normalized bore length. Systems with normalized bore lengths near 150 ft/ton (13 m/kW·ton) tend to have an ENERGY STAR rating near 20, while those with normalized bore lengths of 200 ft/ton (17 m/kW·ton) are more likely to have a rating above 90. A cluster of sites with ENERGY STAR ratings above 90 have normalized bore lengths between 200 and 225 ft/ton (17 and 20 m/kW·ton). The three sites with the longest bore lengths had ENERGY STAR ratings below 90, which indicates that although bore length is important, other characteristics also affect performance results.

It is important to note the reported values are based on installed nominal cooling capacity rather than building load. The sum of the installed capacity for equipment in each zone is typically 10% to 25% greater than the load the building places on the ground loop due to load diversity and also because equipment is available in capacities of fixed increments that cannot match loads precisely.

Another important factor affecting ENERGY STAR ratings is the volumetric flow rate capacity of the ventilation air equipment. To be clear, no attempts were made to measure the actual flow rate, and only near the end of the project were carbon dioxide (CO_2) concentrations observed to estimate the amount of ventilation air. The possible correlations were for several of the newer sites for which equipment specifications were available. Figure 9.5 indicates a correlation between high ENERGY STAR rating and ventilation air equipment capacities of less than 20 cfm/person (10 L/s·person).

Figure 9.6 indicates that 81% (13 of 16) of the GSHP buildings with independent programmable thermostat control achieved ENERGY STAR designation and 56% (9 of 16) attained a rating above 90. Only 45% (9 of 20) of the GSHP buildings with a central building automation system (BAS) achieved ENERGY STAR designation, of which 15% (3 of 20) attained a rating above 90. The average ENERGY STAR rating for buildings with thermostat control was 80, and the average rating for buildings with BAS control was 61.

The reasons thermostat control provided lower energy use than BASs are likely very complex. However, one clear indication is that only 1 of the 14 variable-speed pump drives (which were controlled by a BAS) functioned properly, as indicated by differential loop temperature. Several sites had pumps large enough to provide near-full-load flow rates at minimum motor speeds. There was also minimal attention given to water treatment programs at several of these sites, and it is suspected that this may have resulted in plugging of the pressure measurement ports.

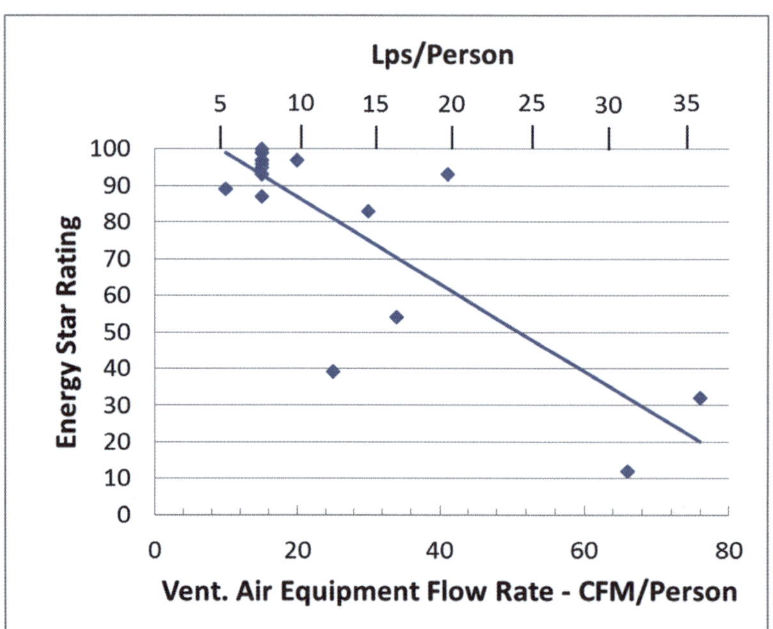

Figure 9.5 ENERGY STAR Rating vs Installed Ventilation Air Equipment Capacity
Source: Kavanaugh and Kavanaugh (2012a)

The central-loop GSHP systems had noticeably lower ENERGY STAR ratings, and most were controlled by BASs. One-pipe and individual-loop GSHPs had much higher ENERGY STAR ratings and were controlled by thermostats. A question arises: were the central-loop GSHPs less efficient because they were controlled by BASs, or were the buildings with BASs less efficient because they were used to control a central-loop GSHP?

Although these results for GSHPs were generated from a rather small data set, they are consistent with data from the 2003 CBECS (EIA 2008), as shown in Figure 9.7. Note that the buildings with unitary and packaged cooling equipment tend to use less energy than centralized systems. Additionally, the average energy consumption for all commercial buildings is less than those with energy management and control systems (EMCSs).

In summary of energy performance results, Figure 9.8 demonstrates the GSHP buildings had dramatically lower annual site energy consumption values compared to the averages in the 2003 CBECS. While most of the buildings were all electric, there were a few sites that used fossil fuel for cooking, which would add a small amount to the values shown in Figure 9.8.

As shown in Figure 9.4, vertical bore length had a strong influence on energy performance. Longer bore lengths resulted in improved ground-loop temperatures, which have a significant impact on system performance. Systems with maximum average ground-loop temperatures ([LLT + ELT]/2) below 90°F (32°C) had an average ENERGY STAR rating of 92, while those with average temperatures above 95°F (35°C) had an average rating of 53. During the field study, a large amount of data was collected; a sample of these results is presented in Figures 9.8 through 9.12.

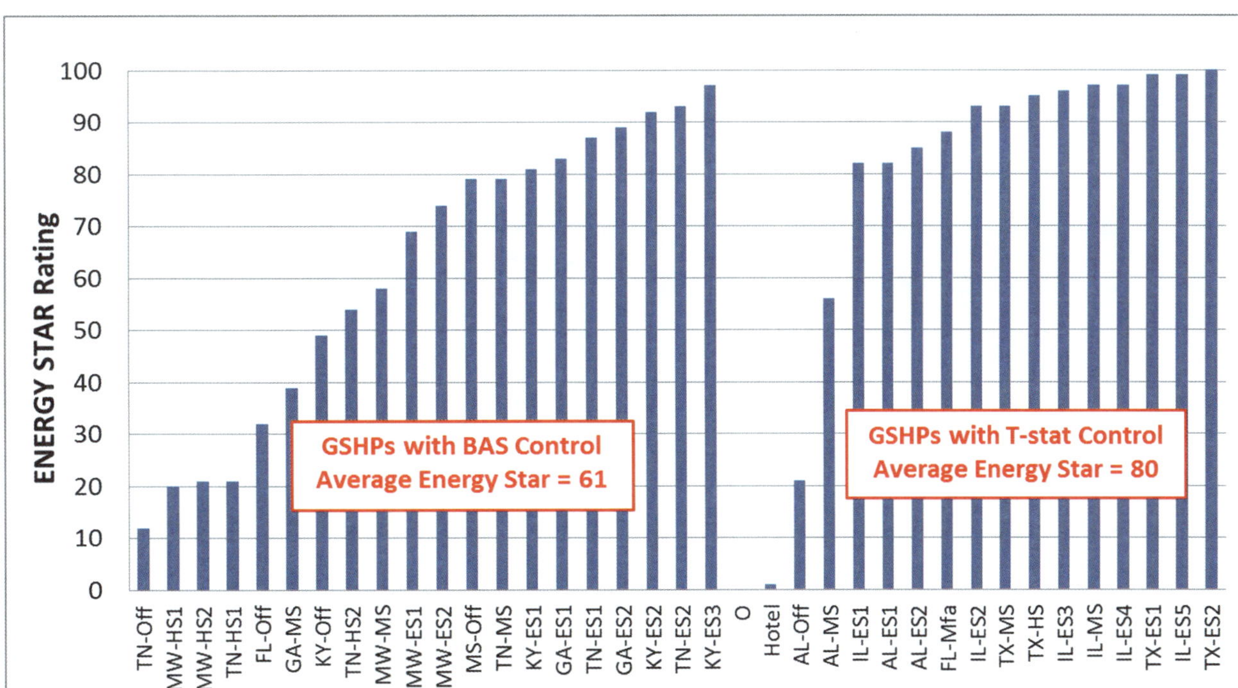

Figure 9.6 ENERGY STAR Rating and HVAC Control Type
Source: Kavanaugh and Kavanaugh (2012c)

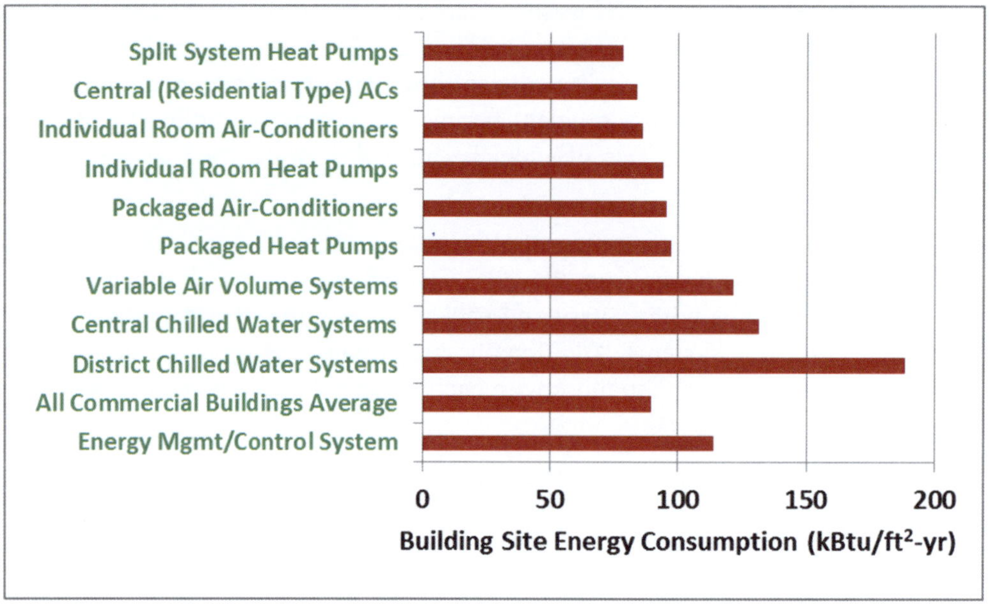

Figure 9.7 Measured Energy Consumption by Cooling System Type and EMCS
Source: Commercial Building Energy Consumption Survey (CBECS) (EIA 2008)

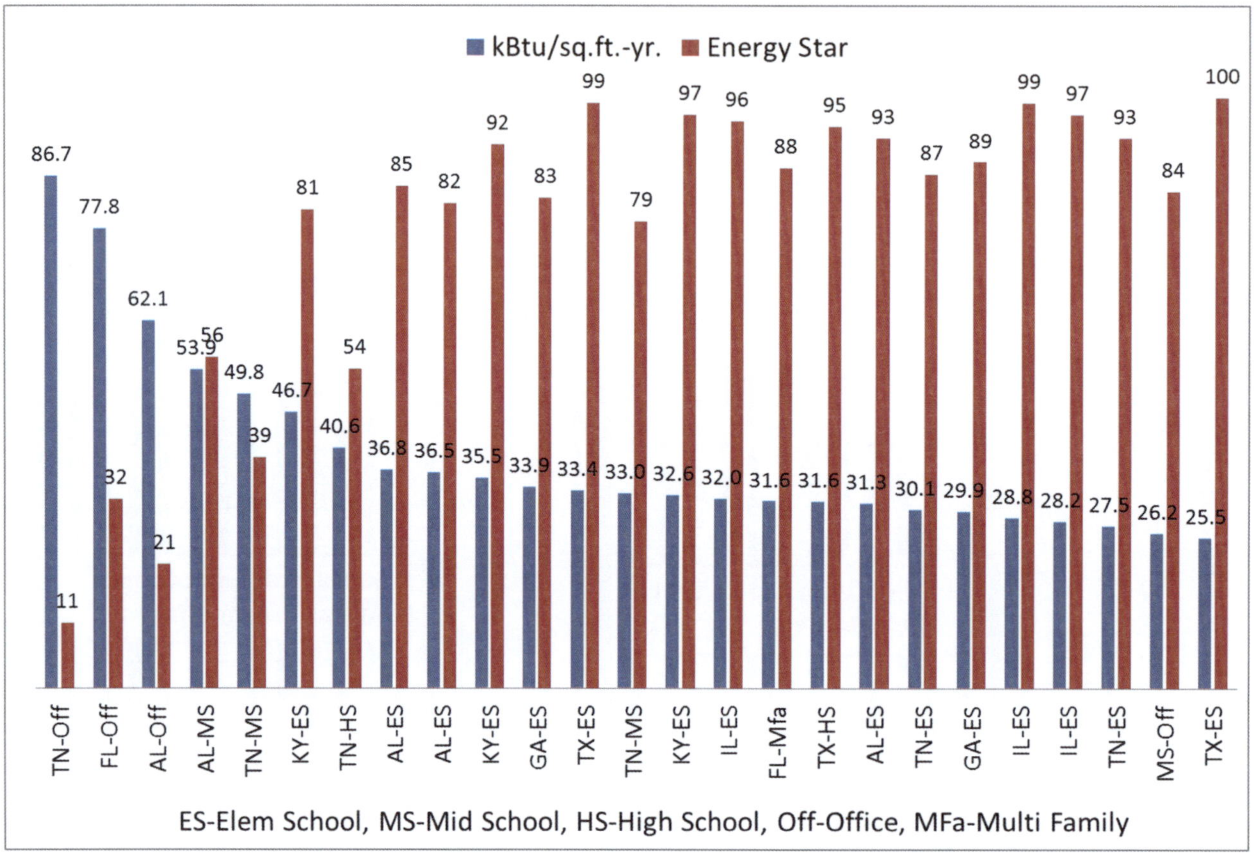

Figure 9.8 Annual Site Energy Consumption and ENERGY STAR Ratings for GSHP Buildings
Source: Kavanaugh and Kavanaugh (2012c)

9.1.3 Loop Temperatures

This section consists of content originally published in ASHRAE Journal *(Kavanaugh and Kavanaugh 2012c). The text has been edited to conform to the style of this book.*

Figure 9.9 shows recorded temperatures for a 287 ton (1000 kW) GSHP serving an 85,000 ft^2 (7900 m^2) elementary school constructed in 2003. Classrooms are served by heat pumps connected to individual loops, while a central loop with two pumps with VSDs serves other areas of the school. The figure indicates the core building ground loop is operating as intended with the ground-loop leaving temperatures remaining below 83°F (28°C) on a day when the high outdoor air temperature (OAT) was 93°F (34°C). The differential temperatures during this near-peak-load day indicate the pump is nearly the correct size, but part-load values suggest the VSD is not operating as intended. This is substantiated by the constant drive speed of 60 Hz shown in the figure.

Ground-loop temperatures recorded in a four-story 78,000 ft^2 (7200 m^2) senior apartment building are shown in Figure 9.10. A 125 ton (440 kW) GSHP system is connected to a 130-bore ground loop with 1 in. (25 mm) diameter high-density polyethylene (HDPE) U-tubes 320 ft (97 m) in depth. A total of 50 two-bedroom apartments are served by heat pumps located in interior closets placed on platforms above the water heaters. Additional heat pumps serve common areas, and two constant-speed 25 hp (19 kW) pumps are alternated to provide continuous, constant flow circulation.

Results indicate the 11-year-old system is performing well, with 83.5°F to 85.5°F (296C to 30°C) ground-loop LWTs during a day when the high OAT was 96°F (36°C). The low differential loop temperature of 4°F (2.2°C) at full load indicates the pump is delivering over twice the optimal flow rate. The local ground temperature is relatively high, but extended loop lengths resulted in good loop temperatures and high ENERGY STAR ratings.

A two-story 37,000 ft^2 (3400 m^2) office building in northwest Tennessee was constructed in 2002 with a GSHP system. Thirty-seven water-to-air heat pumps with a total nominal capacity of 106 tons (373 kW) heat and cool the building. Two 10 hp (7.5 kW)

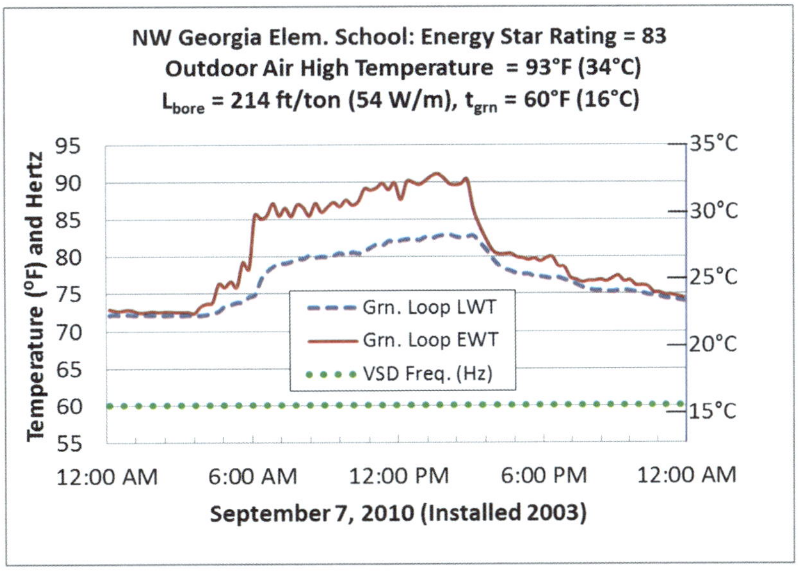

Figure 9.9 Hot-Day Loop Temperatures and VSD Speeds for 85,000 ft^2 (7900 m^2) Georgia School
Source: Kavanaugh and Kavanaugh (2012c)

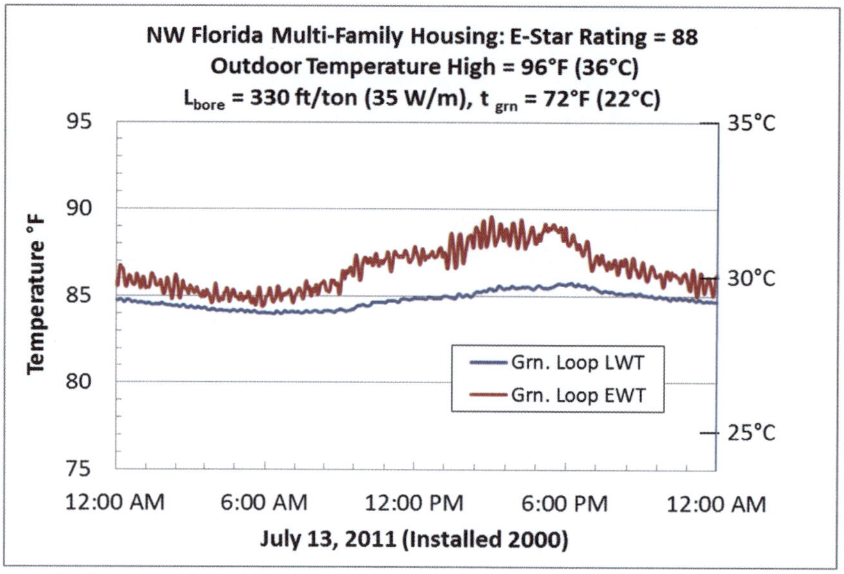

Figure 9.10 Hot-Day Loop Temperatures for 78,000 ft^2 (7200 m^2) Florida Apartment Complex
Source: Kavanaugh and Kavanaugh (2012c)

pumps with VSDs provide circulation through the interior piping, heat pumps, and loop field. Although the original design called for 93 U-tubes at a depth of 300 ft (91 m), as-built drawings indicated only 42 bores were installed. This resulted in an installed length of 119 ft/ton (97 W/m).

As shown in Figure 9.11, the ground-loop temperatures for the office were high and a likely cause for the poor ENERGY STAR rating. Peak ground-loop LWTs were 110°F (43°C), and EWTs are 117°F (47°C) on days that exceeded the local 90°F (32°C) design OAT. The 7°F (4°C) differential loop temperature at near full load indicates the pump is delivering slightly more than optimal flow. The variable-speed pump drive does not appear to be properly functioning since part-load differential temperatures are low, at 2°F (1°C).

Five schools in the field study were located in a heating-mode-dominant climate, but design ground-loop lengths are nearly the same for both heating and cooling. One of these schools is a single-story 37,400 ft^2 (3450 m^2) building constructed in 1957. An 86 ton (300 kW) one-pipe GSHP system was installed in 2007. Thirty-two vertical water-to-air heat pumps replaced the unit ventilators in the classrooms. Console units condition the offices, and ducted horizontal units serve the gymnasium, cafeteria, and kitchen. The ground loop consists of 60 nominal 1 in. (32 mm) HDPE 250 ft (76 m) vertical U-bend heat exchangers installed in a 5 × 12 grid and separated by 20 ft (6 m). A thermal property test indicated the local ground temperature was 55°F (13°C) and thermal conductivity was 1.30 Btu/h·ft·°F (2.3 W/m·K).

Figure 9.12 indicates the ground-loop leaving liquid temperature (LLT) remains between 48°F and 50°F (9°C and 10°C) on a cold day in late January. The temperature entering the ground loop (leaving the heat pumps) reached a minimum of 41°F (5°C) when the outdoor temperature was near –6°F (–21°C). The differential loop temperature is 7°F (4°C) when the loads are larger during morning start-up. However, the low differential temperatures ($\Delta t \approx$ 3°F [2°C]) indicate excess flow is being delivered at low part loads.

Concerns have been raised regarding the long-term temperatures of GSHPs that have a pronounced annual imbalance of heat transfer into or out of the ground. The maximum

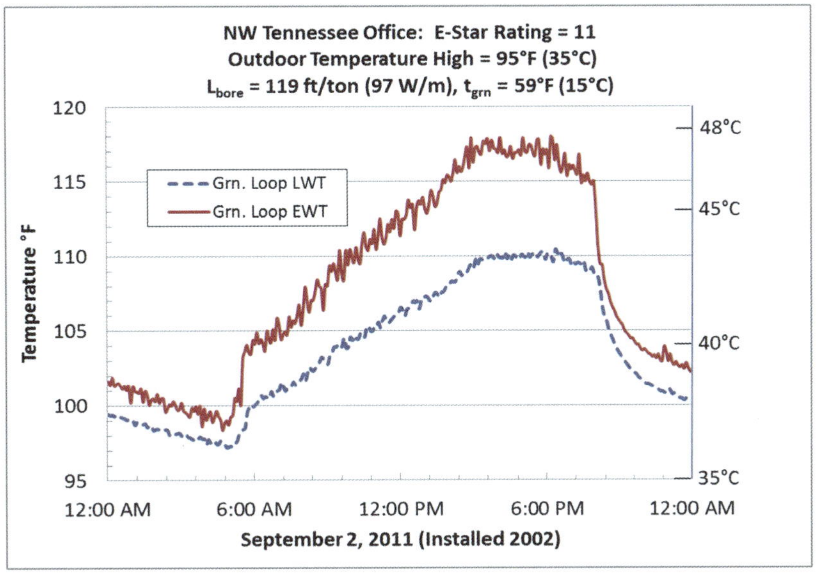

Figure 9.11 Hot-Day Loop Temperatures for 37,000 ft^2 (3400 m^2) Northwest Tennessee Office
Source: Kavanaugh and Kavanaugh (2012c)

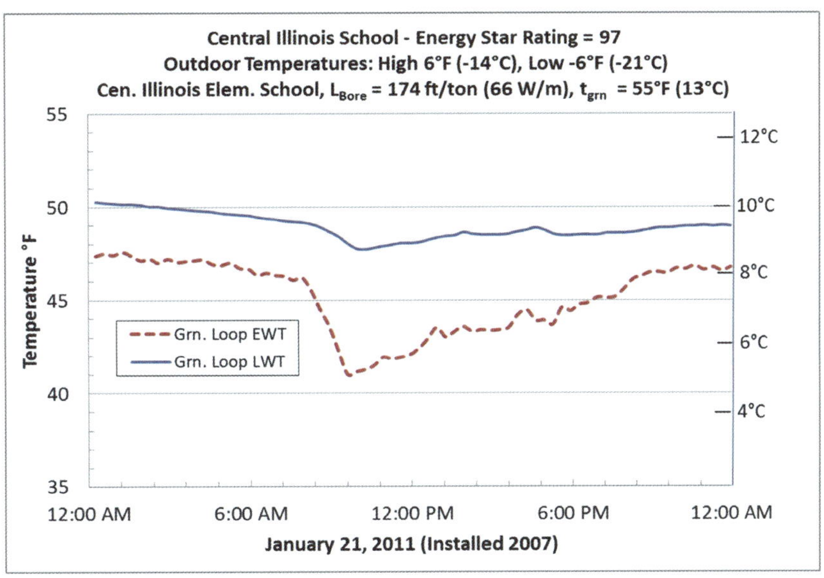

Figure 9.12 Cold-Day Loop Temperatures for 37,400 ft^2 (3450 m^2) Elementary School
Source: Kavanaugh and Kavanaugh (2012c)

approach temperatures between the ground-loop average water temperatures ([EWT + LWT]/2) and the undisturbed deep ground temperature (t_{grn}) are used as a measure of loop performance success. Figure 9.13 provides a plot of maximum approach temperature as a function of years of operation. A trend of higher approach temperatures with increased years of operation would raise concern about the expected life of ground loops with imbalanced cooling loads compared to heating loads.

Older GSHP systems appear to actually have lower approach temperatures. Results are not adjusted for many important factors such as vertical bore length, ground thermal

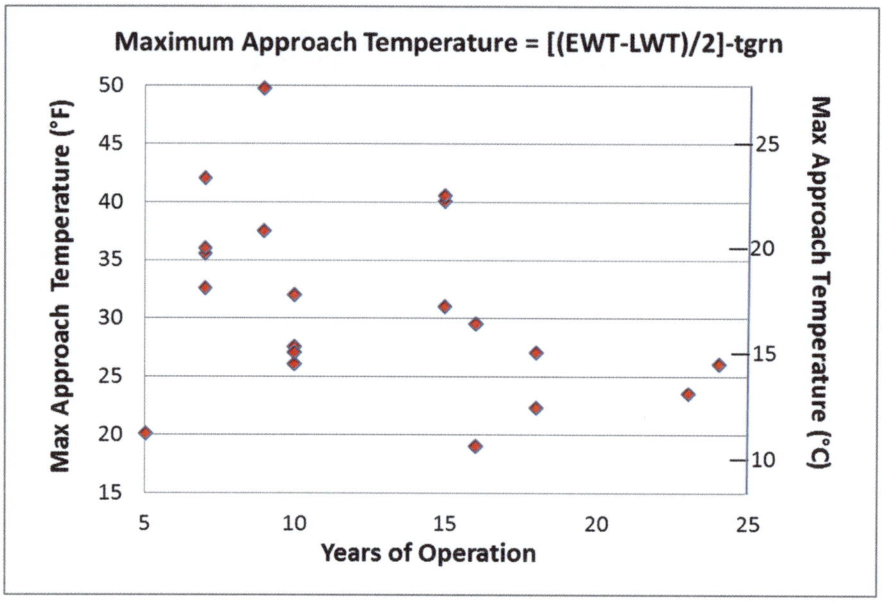

Figure 9.13 Maximum Average Ground Loop to Ground Approach Temperatures vs GSHP Age

properties, and vertical bore separation distance. The newer systems tend to have slightly shorter ground loops, but this is offset somewhat since older systems tend to have smaller vertical bore separation distances and lower-conductivity grout and fill. Three of the newer systems with high approach temperatures have vertical bore lengths less than 120 ft/ton (96 W/m), one of which is the system described in Figure 9.11.

It is recognized that this data set is small and that the presence of significant long-term temperature change cannot be excluded at this point for systems with both heating and cooling loads. Although much more field data is highly desirable, the absence of any significant trend of increased ground temperature (noted by elevated maximum approach temperature) with increased years of GSHP operation would indicate that long-term ground temperature change is not prevalent. Elevated temperatures in vertical ground loops are primarily a result of inadequate heat exchanger length. Insufficient bore separation distance, low-conductivity grout, and improper completion methods may also contribute to increase. Cooling-only or heating-only systems are problematic because long-term temperature changes are much more likely to occur.

Results from this project cannot be applied to long-term temperature decline in which the amount of heat removed from the ground in heating far exceeds the heat rejected in cooling. The transfer mechanisms are entirely different. In cooling, long-term temperature increase is mitigated by the cooling effect from reductions in moisture content (evaporation) when ground temperatures rise within the loop field. The heat rejection required to affect a 1% reduction in ground moisture is approximately the same amount needed to raise temperature 30°F (17°C) (EIS 2009). Over extended periods, the moisture content is likely to be restored to its natural condition via groundwater movement and rainwater percolation. In cold climates the heat capacity available at the freeze point of water is significant, but the impact on grout thermal and physical properties also needs further field study.

As mentioned at the beginning of Section 9.1, the study on the performance of long-term GSHP performance included results of occupant and maintenance personnel satisfaction perception (Kavanaugh and Kavanaugh 2012d; Kavanaugh and Dinse 2013).

Occupants were surveyed regarding their observations of room comfort conditions (cooling and heating), indoor air quality, lighting, acoustics, maintenance responsiveness, and ability to control on a scale from 1 (very dissatisfied) to 5 (very satisfied). In all areas the average ratings were between 3 (acceptable) and 4 (satisfied), with maintenance responsiveness and lighting the highest at 3.7 and ability to control the lowest at 3.0. The responses from maintenance personnel were limited and primarily took the form of comments and suggestions for design-related items that would enhance the maintainability of GSHPs (Kavanaugh and Dinse 2013).

9.2 PREDICTION OF THE PERFORMANCE OF GSHP DESIGN OPTIONS

The authors of a review of a large study of Leadership in Energy and Environmental Design (LEED®) buildings reported that "a large portion of the buildings are using significantly more energy than predicted" (Hinge and Winston 2009, pp. 19–20). It was also expressed that researchers "were able to obtain actual energy data for only 121 out of 585 buildings requested, and it's unclear whether that sample is representative" (p. 19). These two statements call into question the practice of depending on energy simulation to accurately predict building energy performance without sufficient validation.

Closed-loop GSHP systems (GCHPs and SWHPs) further complicate the issue since ground-loop and reservoir characteristics add additional uncertainty. While the information gathered in the GSHP field study is likely the most up to date and extensive survey of this type in the United States, it is far from being sufficient. Energy data was difficult to obtain even though electric utilities encountered in the survey had the necessary information available. The difficulty in data collection was that the building owners needed to approve access and in some cases choose not to do so. The information on installation costs was even more restricted, as discussed in the following section. Designers, contractors, and owners that are willing to share energy and cost data are likely to have completed successful GSHP projects with good energy performance and reasonable first costs.

However, the improvements in performance and cost could potentially be much higher if owners (and architects) were able to choose engineers based on quantifiable information. Publication of energy data, installation costs, and satisfaction levels allows engineers to demonstrate GSHP quality and provides owners (and possibly architects) an effective metric for selecting system options and engineering firms with proven records of success.

Until a broader base of information is available, the status of performance prediction of GSHP systems with building energy simulations is uncertain. The paragraphs that follow present an alternative system efficiency calculation that is both simple and a useful substitution for more involved simulations.

Figure 9.14 shows a comparison of the predicted versus actual energy consumption of a LEED Platinum office building with a GSHP. The building has a floor area of 78,600 ft^2 (7550 m^2) and is located in the mid-Atlantic coastal region. Local design temperatures are 91°F (33°C) in cooling and 21°F (–6°C) in heating. The actual site electrical energy use at the time of the data collection was 57.3 kBtu/ft^2 (5.3 kWh/m^2), which was 36% higher than predicted. The energy use of this building does not compare well with the GSHP systems surveyed in the field study. The energy use of this LEED Platinum building was higher than 22 of the 25 GSHP buildings shown in Figure 9.8.

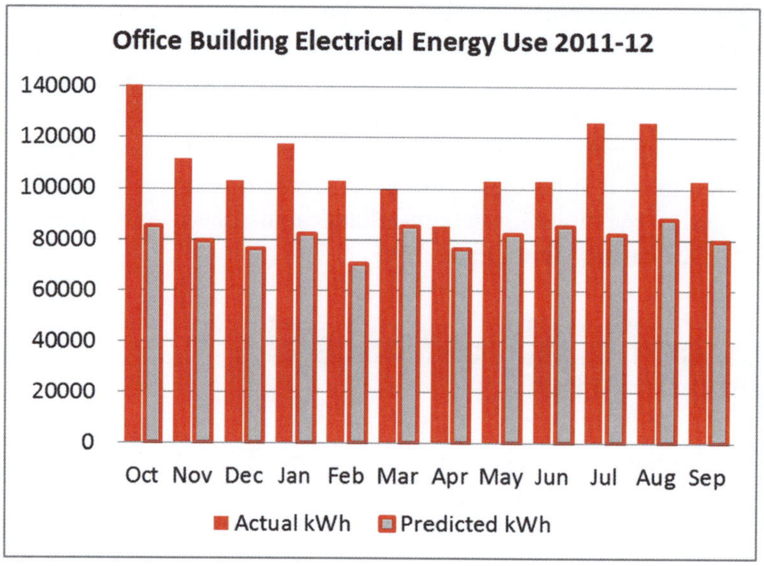

Figure 9.14 Actual and Predicted Energy Use of 78,600 ft^2 (7550 m^2) Office Building

Table 9.1 is a summary of the equipment schedule for the LEED Platinum building. The GSHP system consisted of six 29 ton (100 kW) water-to-water heat pumps with an EER of 14 Btu/Wh (COP of 4.1), which is equivalent to 0.86 kW/ton. Three 17,000 cfm (8020 L/s or 28,900 m^3/h) VAV AHUs deliver air through an underfloor air distribution (UFAD) system to the office areas. Two additional AHUs deliver flow to conference rooms. There are three return air fans with 10 hp (7.5 kW) motors and another return fan with a 1 hp (0.75 kW) motor. Water flow is provided to the ground loop by a single pump with a 20 hp (15 kW) motor. Six building loop pumps with 3 hp (2.2 kW) motors deliver flow through each heat pump. Two additional pumps with 3 hp (2.2 kW) motors are also used. Ventilation air is provided by an 11,000 cfm (5200 L/s or 18,700 m^3/h) DOAS with an ERU that has a supply fan with a 15 hp (11.2 kW) motor and an exhaust fan with a 10 hp (7.5 kW) motor.

The ground heat exchanger consists of 90 vertical bores, 400 ft (122 m) in depth, with 1.0 in. (32 mm) nominal diameter, DR 11 HDPE U-tubes, placed on 15 to 18 ft (4.6 to 5.5 m) centers. The bores were to be grouted with thermally enhanced bentonite grout with a thermal conductivity of 1.13 Btu/h·ft·°F (1.96 W/m·K). The dimensions of the ground loop appear to be adequate at 209 ft/ton (55 W/m). However, the ground loop returned water warmer than the expected 85°F (29°C) temperature in the first year of operation.

An alternative procedure for evaluating and comparing designs follows the method demonstrated in Chapter 2 to calculate system efficiency. Figure 9.15 is a screenshot of the spreadsheet tool *HVACsystemEff.xlsx* (available with this book at www.ashrae.org/GSHP) with the information from the equipment schedule entered into the appropriate rows and columns. The resulting cooling system EER is calculated to be 7.5 Btu/Wh (COP = 2.19), which is a strong indicator of why the system did not perform as well most of the GSHPs in the field study. Note that the power input of the auxiliary equipment is 93 kW, which is the sum of items 2 through 6 in the table shown in Figure 9.15. This is significant compared to the input power of the heat pumps at 147.9 kW. Furthermore, note the sum of the heat generated by the fans and chilled-water pumps is 23 tons (80 kW), which reduces the gross cooling capacity of the heat pumps by 13%. Also note

Table 9.1 Equipment Schedule for LEED Platinum GSHP Office Building

Heat Pump Unit Schedule, Water-to Water

Quantity	Cooling Mode		Ground Loop		Chilled-Water Loop	
	TC, kBtu/h (kW)	EER (COP)	EWT, °F (°C)	LWT, °F (°C)	EWT, °F (°C)	LWT, °F (°C)
6	345 (101)	14 (4.1)	85 (29)	95 (35)	65 (18)	44 (7)
	Heating Mode		**Ground Loop**		**Heating-Water Loop**	
	TC, kBtu/h (kW)	**COP**	**EWT, °F (°C)**	**LWT, °F (°C)**	**EWT, °F (°C)**	**LWT, °F (°C)**
	345 (101)	4.1	55 (13)	48 (9)	100 (38)	120 (49)

Pump Schedule (Standby Pumps Not Included)

Quantity	Service	Flow, gpm (L/s)	Head, ft (kPa)	Efficiency	Motor hp (kW)
1	Ground loop	540 (34)	92 (276)	80%	20 (15)
6	Chilled/heating water	86 (5.4)	46 (138)	76%	3 (2.2)
1	55°F (13°C) chilled water	135 (8.5)	38 (115)	68%	3 (2.2)
1	45°F (7°C) chilled water	150 (9.5)	46 (138)	64%	3 (2.2)

Air-Handling Unit Schedule

Quantity	Service	Flow, cfm (L/s)	External Static Pressure, in. (Pa)	bhp (kW)	Motor hp (kW)
3	UFAD	17000 (8020)	1.5 (375)	15.12 (11.3)	20 (15)
1	Conference	2600 (1230)	2.0 (500)	2.60 (1.9)	3 (2.2)
1	Conference	6500 (3070)	2.0 (500)	6.60 (4.9)	7.5 (5.6)

Return Air Fan Schedule (Exhaust Fans Not Included)

Quantity	Service	Flow, cfm (L/s)	External Static Pressure, in. (Pa)	bhp (kW)	Motor hp (kW)
3	UFAD	17000 (8020)	2.0 (500)	na	10 (7.5)
1	Conference	2600 (1230)	2.0 (500)	na	1 (0.75)

the 174 ton (610 kW) system is served by a 20 hp (15 kW) ground-loop pump, which results in a pump power of 11.3 hp/100 tons (24 W/kW) and garners a grade of D (see Table 6.2). Note the power of the ERU fans is not included because this type of equipment is considered a load reduction device.

The poor design EER of 7.5 (COP of 2.19) is a result of the additional 93 kW demand of the auxiliary equipment coupled with the 23 ton (80 kW) reduction in cooling capacity due to the heat generated by the fans and chilled-water pump. However, the ground-loop temperatures were warmer than anticipated and the actual EER (COP) was likely much lower.

The likely reason the ground-loop temperatures were higher than anticipated can be explained by viewing the late-winter temperature profile shown in Figure 9.16. The first obvious indication of a problem is that the ground-loop temperatures in the heating mode are higher than the normal ground temperature, which indicates the system was operating in net cooling throughout the winter. This is verified by the fact that the ground-loop EWT is almost always higher than the LWT (except for a few periods of morning start-up during cold days). This indicates the GSHP system is in the cooling mode, and the low differential temperatures reveal the variable-speed pump drive is not working as intended. If this drive is not operating correctly, the possibility arises that the VAV air distribution is

Item			Enter Chiller Data		OR		Enter AC/HP Data					
1	Chillers or Heat Pumps	Qty.	kW/ton	Tons		Qty.	EER	kBtu/h	kW-ea.	kW-tot.	kBtu/h	tons
					---	6	14	345	24.6	147.9	2070.0	172.5
					---				0.0	0.0	0.0	0.0
					---				0.0	0.0	0.0	0.0
2a	Air Handling Unit Fans		Air Flow-cfm	TP-in wtr	η fan	η mtr.	η VSD	bhp-each	---	---	---	---
		3	17000	4	71.0%	92.0%	97.0%	15.08	12.61	37.8	-129.1	-10.8
		1	2600	4	64.0%	85.0%	97.0%	2.56	2.32	2.3	-7.9	-0.7
		1	6500	4	62.0%	88.0%	97.0%	6.60	5.77	5.8	-19.7	-1.6
2b	Fan coil or VAV Fans		Air Flow-cfm	TP-in wtr	η fan	hp-ea.	η mtr. (%)		---	---	---	---
	Enter Air Flow, TP, and							0.00	0.00	0.0	0.0	0.0
	ηfan or fan hp.							0.00	0.00	0.0	0.0	0.0
								0.00	0.00	0.0	0.0	0.0
3	Return Air Fans		Air Flow-cfm	TP-in wtr	η fan	η mtr.	η VSD (%)		---	---	---	---
		3	17000	2	65.0%	92.0%	97.0%	8.24	6.89	20.7	-70.5	-5.9
		1	2600	0.75	50.0%	80.0%	97.0%	0.61	0.59	0.6	-2.0	-0.2
								0.00	0.00	0.0	0.0	0.0
4	Chilled Water Pumps		Wtr. Flow-gpm	ΔH-ft wtr	η pump	η mtr.	η VSD (%)		---	---	---	---
		6	86	46	64.0%	86.0%	97.0%	1.56	1.40	8.4	-28.6	-2.4
		1	135	38	68.0%	86.0%	97.0%	1.91	1.70	1.7	-5.8	-0.5
		1	150	46	64.0%	88.0%	97.0%	2.72	2.38	2.4	-8.1	-0.7
5	Condenser Water Pumps		Wtr. Flow-gpm	ΔH-ft wtr	η pump	η mtr.	η VSD (%)		---	---	Click Here	
	(Ground Loop For GSHPs)	1	540	92	80.0%	91.0%	97.0%	15.68	13.25	13.3	---	---
								0.00	0.00	0.0	---	---
								0.00	0.00	0.0	---	---
6	Cond./Cooling Tower Fan	0	Air Flow (cfm)	TP (in. wtr)	η fan (%)	η mtr.	η VSD	hp-each	---	---	---	---
		0	---	---	---	75.0%	100.0%		0.00	0.0	---	---
			---	---	---				0.00	0.0	---	---
			---	---	---				0.00	0.0	---	---
										kW	kBtu/h	tons
								System Totals		240.7	1798	150
				kW/ton =	1.61	EER =	7.5	Btu/W-h	COP =	2.19		

Figure 9.15 System Cooling Efficiency of Chilled-Water VAV GSHP with UFAD

also providing much greater part-load flow than intended. This suggests the fans are delivering a large percentage of the 23 tons (80 kW) of heat possible. At what should be a part-load heating condition, it appears the heat pumps are operating in cooling to overcome both the internal building loads and the excessive fan heat.

It is highly recommended that this relatively simple procedure of determining system efficiency in both heating and cooling be undertaken in all GSHP designs. The following example repeats the system efficiency calculation procedure for the GSHP system used for the example design in Chapters 4 and 6 to demonstrate an approach that makes better use of the advantages of GSHPs.

Table 9.2 is the equipment schedule for the common-loop GSHP design (and unitary system design) described in Chapter 4. The cooling capacity and EER (COP) of the eight heat pumps have been corrected for 86°F (30°C) EWT, 75°F (24°C) entering air dry-bulb temperature (EATDB), 63°F (17°C) entering air wet-bulb temperature (EATWB), and fan power based on 0.8 in. of water (200 Pa) for the external static pressure (ESP) and filter loss. The heating capacity and coefficient of performance (COP) of the eight heat pumps have been corrected for 50°F (10°C) EWT, 70°F (21°C) EATDB, and fan power based on 0.8 in. of water (200 Pa) for the ESP and filter loss. Eight nominal 1/6 hp (0.12 kW) pumps with 45% efficiency and 50% motor efficiency provide flow to each heat pump. The fan power is included in the heat pump efficiency, and no other fans are required. The ERU described in Chapter 4 is considered a load reduction device as in the previous example.

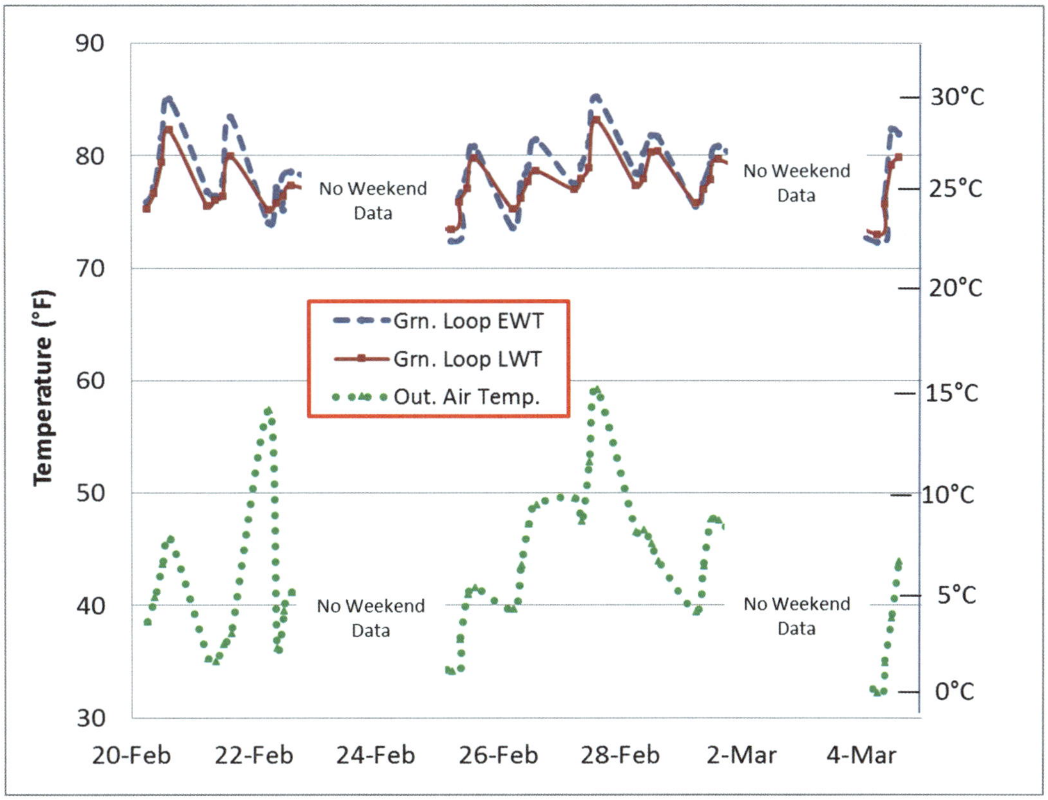

Figure 9.16 Late-Winter Temperatures for Office Building with VAV UFAD GSHP

Table 9.2 GSHP Equipment Schedule for Example 10,000 ft^2 (929 m^2) Office Building

Water-to-Air Heat Pump Schedule (EAT & Fan Heat Corrections Included)							
Quantity		Cooling Mode		Ground Loop		EAT	
	Model #	TC, kBtu/h (kW)	EER (COP)	EWT, °F (°C)	LWT, °F (°C)	DB, °F (°C)	WB, °F (°C)
3	30	26.2 (7.7)	15.1 (4.4)	86 (30)	96 (36)	75 (24)	63 (17)
2	36	32.0 (9.4)	15.3 (4.5)	86 (30)	96 (36)	75 (24)	63 (17)
3	42	37.7 (11.0)	15.1 (4.4)	86 (30)	96 (36)	75 (24)	63 (17)
		Heating Mode		Ground Loop		Heating-Water Loop	
	Model #	HC, kBtu/h (kW)	COP	EWT, °F (°C)	LWT, °F (°C)	DB, °F (°C)	WB, °F (°C)
3	30	26.6 (7.8)	4.3	50 (10)	44 (7)	70 (21)	59 (15)
2	36	31.2 (9.1)	4.4	50 (10)	44 (7)	70 (21)	59 (15)
3	42	36.0 (10.6)	4.4	50 (10)	44 (7)	70 (21)	59 (15)
Pump Schedule							
Quantity	Model #	Service		Flow, gpm (L/s)	Head, ft (kPa)	Pump, hp (kW)	Power, W
3	26-96	Model 30 heat pump		8 (30)	28 (84)	1/6 (0.12)	190
2	26-96	Model 36 heat pump		9 (34)	27 (81)	1/6 (0.12)	200
3	26-99	Model 42 heat pump		11 (42)	25 (75)	1/6 (0.12)	230

Item		Enter Chiller Data			OR	Enter AC/HP Data						
1	Chillers or Heat Pumps	Qty.	kW/ton	Tons		Qty.	EER	kBtu/h	kW-ea.	kW-tot.	kBtu/h	tons
	Clg 75 EAT & 86 EWT				---	3	15.1	26.2	1.7	5.2	78.6	6.6
	ESP+Δh(filter)=0.8" wg				---	2	15.3	32	2.1	4.2	64.0	5.3
						3	15.1	37.7	2.5	7.5	113.1	9.4
					---				0.0	0.0	0.0	0.0
2a	Air Handling Unit Fans		Air Flow-cfm	TP-in wtr	η fan	η mtr.	η VSD	bhp-each	---	---	---	---
								0.00	0.00	0.0	0.0	0.0
								0.00	0.00	0.0	0.0	0.0
								0.00	0.00	0.0	0.0	0.0
2b	Fan coil or VAV Fans		Air Flow-cfm	TP-in wtr	η fan	hp-ea.	η mtr. (%)	bhp-each	---	---	---	---
	Enter Air Flow, TP, and							0.00	0.00	0.0	0.0	0.0
	ηfan or fan hp.							0.00	0.00	0.0	0.0	0.0
								0.00	0.00	0.0	0.0	0.0
3	Return Air Fans		Air Flow-cfm	TP-in wtr	η fan	η mtr.	η VSD (%)	bhp-each	---	---	---	---
								0.00	0.00	0.0	0.0	0.0
								0.00	0.00	0.0	0.0	0.0
								0.00	0.00	0.0	0.0	0.0
4	Chilled Water Pumps		Wtr. Flow-gpm	ΔH-ft wtr	η pump	η mtr.	η VSD (%)	bhp-each	---	---	---	---
								0.00	0.00	0.0	0.0	0.0
								0.00	0.00	0.0	0.0	0.0
								0.00	0.00	0.0	0.0	0.0
5	Condenser Water Pumps		Wtr. Flow-gpm	ΔH-ft wtr	η pump	η mtr.	η VSD (%)	bhp-each	---	---	Click Here	
	(Ground Loop For GSHPs)	3	8	28	45.0%	50.0%	100.0%	0.13	0.19	0.6	---	---
		2	9	27	45.0%	50.0%	100.0%	0.14	0.20	0.4	---	---
		3	11	25	45.0%	50.0%	100.0%	0.15	0.23	0.7	---	---
6	Cond./Cooling Tower Fan	0	Air Flow (cfm)	TP (in. wtr)	η fan (%)	η mtr.	η VSD	hp-each	---	---	---	---
		0	---	---	---				0.00	0.0	---	---
			---	---	---				0.00	0.0	---	---
			---	---	---				0.00	0.0	---	---
										kW	kBtu/h	tons
								System Totals		18.5	256	21
				kW/ton =	0.87	EER =	13.8	Btu/W-h	COP =	4.04		

Figure 9.17 System Cooling Efficiency for Unitary GSHP in Example Building

Figure 9.17 presents the cooling calculation efficiency results for the common-loop/unitary-loop GSHP. The absence of a significant amount of auxiliary equipment and accompanying heat produces a design load EER of 13.8 Btu/Wh (COP of 4.04), which is a significant improvement compared to the chilled-water VAV UFAD GSHP in the previous example. Figure 9.18 presents the heating results indicating the full-load design COP is 3.97. In both cases the rows for entering values for auxiliary equipment contain a large number of blanks.

Perhaps of equal importance, it should be noted that the absence of auxiliary equipment is accompanied by an absence of cost. Thus, simple GSHPs have three significant advantages over traditional central-air-and-water-distribution HVAC systems attached to GSHP loops: they cost less to install, require much less input power, and can have simpler control.

9.3 FIELD STUDY INSTALLATION COST RESULTS

This section consists of content originally published in ASHRAE Journal *(Kavanagh et al. 2012). The text has been edited to conform to the style of this book.*

Performance and cost surveys were collected and site visits were performed for 40 locations. Although the survey included 23 building owners, only 4 building owners or engineers completed the installation cost portion of the survey. Fortunately, they provided

Item			Enter AC/HP Data		OR							
1	Rev.Chillers or Heat Pum	Qty.	COP	kBtu/h					kW-ea.	kW-total	kBtu/h	Eqv.Tons
	Htg. 70 EAT & 50EWT	3	4.3	26.6	---				1.8	5.4	79.8	6.7
	ESP+Δh(filter)=0.8" wg	2	4.4	31.2	---				2.1	4.2	62.4	5.2
		3	4.4	36					2.4	7.2	108.0	9.0
					---				0.0	0.0	0.0	0.0
2a	Air Handling Unit Fans		Air Flow (cfm)	TP (in. wtr)	η fan (%)	η mtr. (%)	η VSD (%)	bhp-ea.	---	---	---	---
2b	Fan coil or VAV Fans		Air Flow (cfm)	TP (in. wtr)	wire-air (OR	hp-eac	η mtr. (%)	bhp-ea.	---	---	---	---
	Enter Air Flow, TP, and											
	ηfan or fan hp.											
3	Return Air Fans		Air Flow (cfm)	TP (in. wtr)	η fan (%)	η mtr. (%)	η VSD (%)	bhp-ea.	---	---	---	---
4	Chilled Water Pumps		Wtr. Flow (gpm)	ΔH (ft. wtr)	η pump (%	η mtr. (%)	η VSD (%)	bhp-ea.	---	---	---	---
5	Condenser Water Pumps		Wtr.Flow (gpm)	ΔH (ft. wtr)	η pump (%	η mtr. (%)	η VSD (%)	bhp-ea.	---	---	Click Here	
	(Ground Loop For GSHPs)	3	8	28	45.0%	50.0%	100.0%	0.13	0.19	0.6	---	---
		2	9	27	45.0%	50.0%	100.0%	0.14	0.20	0.4	---	---
		3	11	25	45.0%	50.0%	100.0%	0.15	0.23	0.7	---	---
										kW	kBtu/h	Eqv.Tons
								System Totals		18.4	250.2	20.9
						kW/ton :	0.88		COP =	3.97		

Figure 9.18 System Heating Efficiency for Unitary GSHP in Example Building

cost data for multiple buildings, some of which were monitored and several that were too new for performance rating or were still under construction. The results are heavily weighted toward the two system types that achieved the highest ENERGY STAR ratings. Costs were available for seven systems with a one-pipe central loop in the building with small pumps that circulate liquid from a common supply and return pipe through the heat pumps. Data were collected for seven unitary-loop GSHPs in which each a heat pump is connected to an individual loop and circulation is provided by a small on-off pump. Data for three central loop systems were also included, along with results from a previous Electric Power Research Institute (EPRI) and Tennessee Valley Authority (TVA) project (Zimmerman 2000) and an ASHRAE research project (Caneta Research 1995).

The increase in the HVAC component costs of GSHP systems since the 1995 study is 177%, while the increase in the ground-loop portion was only 52%. In this recent study, the ground-loop portion of GSHP systems accounted for 26% of the total while the HVAC component composed 74% of the total. Thus, attempts to reduce GSHP cost by focusing primarily on the ground loop seem illogical. The lack of responses to the cost component of the surveys is disappointing given commercial GSHPs are often avoided because of high installation cost. Emphasis should be placed on gathering additional detailed cost information to expand the results and further develop the conclusions of this study.

Figure 9.19 shows the costs for the complete GSHP system and the ground-loop portion based on floor area. The Illinois (IL) systems are one-pipe loops, the Texas (TX) systems are unitary loops, and the Tennessee and Georgia (TN/GA) systems are central loops. The ground-loop costs for the IL and TN/GA loops include the vertical bore and exterior header costs, while the TX systems also include the interior building piping and pump costs.

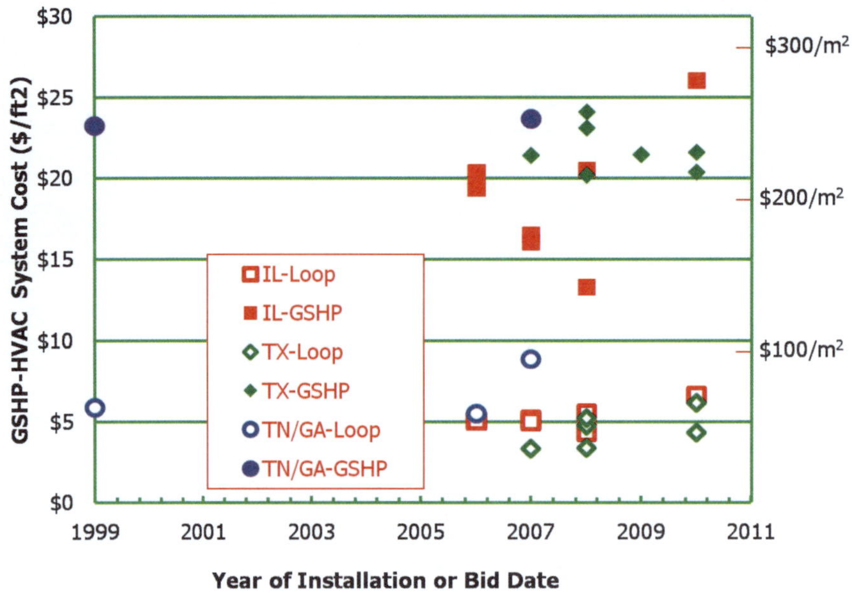

Figure 9.19 GSHP System and Ground-Loop Cost Based on Building Floor Area

The average system cost including the ground loop was $20.75/ft^2 ($223/m^2) with a high of $26.10/ft^2 ($281/m^2) and a low of $13.34/ft^2 ($144/m^2). The average ground-loop cost was $5.29/ft^2 ($57/m^2) with a high of $8.89/ft^2 ($96/m^2) and a low of $3.35/ft^2 ($36/m^2). The average cost of the ground loop was 25.5% of the total GSHP system cost based on floor area. Costs for the TX systems include non-GSHP equipment that served the common areas.

Figure 9.20 shows the costs for the total GSHP system and the ground-loop cost based on the rated capacity of the heat pumps. Again the ground-loop costs for the IL and TN/GA loops include the vertical bore and exterior header costs while the TX systems also include the interior building piping and pump costs. The system cost for the TX systems based on equipment capacity are not included because the common areas are heated and cooled by non-GSHP equipment. One of the TN/GA sites only included the ground-loop cost.

The average GSHP system cost was $7694/ton ($2190/kW) with a high of $9206/ton ($2620/kW) and a low of $6291/ton ($1790/kW). These values include the cost of the ground loop. The average ground-loop cost was $2483/ton ($706/kW) with a high of $4076/ton ($1160/kW) and a low of $1209/ton ($344/kW). As shown in Figure 9.20, the low value was for the system installed in 1999, which also had a relatively short loop length for the rated capacity of the installed equipment. The average cost of the ground loop was 32.3% of the total GSHP system cost based on rated equipment capacity.

Figure 9.21 provides the costs for the ground loop based on the length of the vertical bore. The average ground-loop cost was $11.77/ft ($38.62/m) with a high of $15.00/ft ($49.20/m) and a low of $6.76/ft ($22.18/m). These values include the cost of exterior headers for the IL and TN/GA systems, and the TX systems also include interior piping and pumps.

Figure 9.22 shows the costs for the total GSHP system and the ground-loop cost based on floor area from two previous studies. A condensed publication (Caneta Research 1998) of a large survey from an ASHRAE-sponsored research project (Caneta Research 1995) studied systems located in colder climates, including Canadian buildings. The sys-

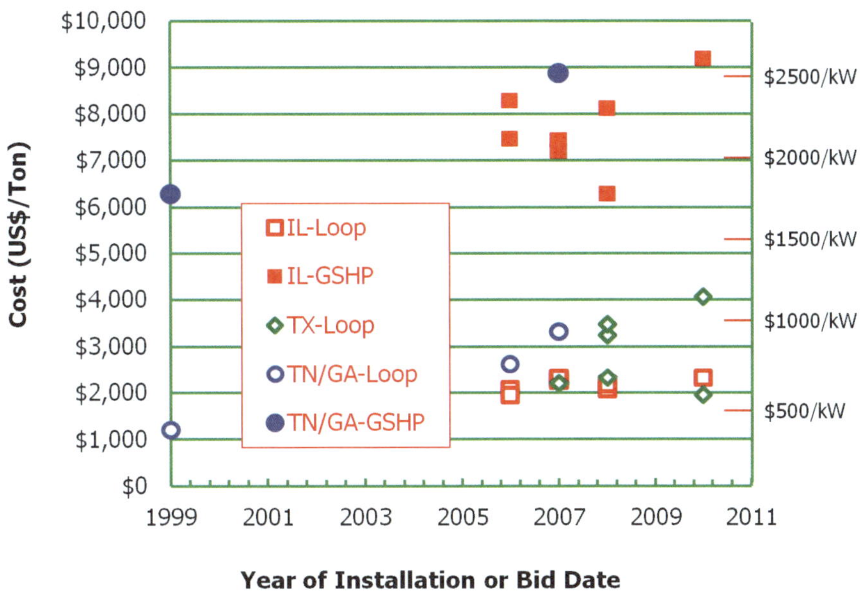

Figure 9.20 GSHP System and Ground-Loop Cost Based on Cooling Capacity

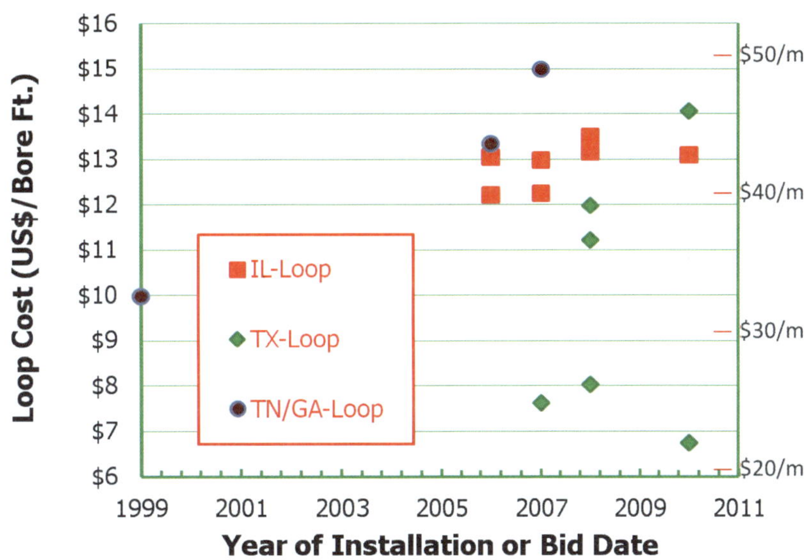

Figure 9.21 Ground-Loop Cost Based on Vertical Bore Length

tem installed in 1990 is a unitary loop, while the other five sites have central loops. Horizontal closed-loop and open-loop groundwater systems were also surveyed, but only the vertical closed-loop systems are shown in Figure 9.22. An EPRI cost and maintenance survey in the Tennessee Valley was conducted on several GSHP schools (Zimmerman 2000); all of these systems are vertical central-loop GSHPs.

The average GSHP system cost for the 1995 survey was \$9.07/ft^2 (\$98/m^2), with a very high variation in cost between the maximum of \$14.34/ft^2 (\$154/m^2) and minimum of \$2.67/ft^2 (\$29/m^2). The average ground-loop cost was \$3.49/ft^2 (\$38/m^2), with an even more pronounced variation between the maximum of \$7.38/ft^2 (\$79/m^2) and minimum of

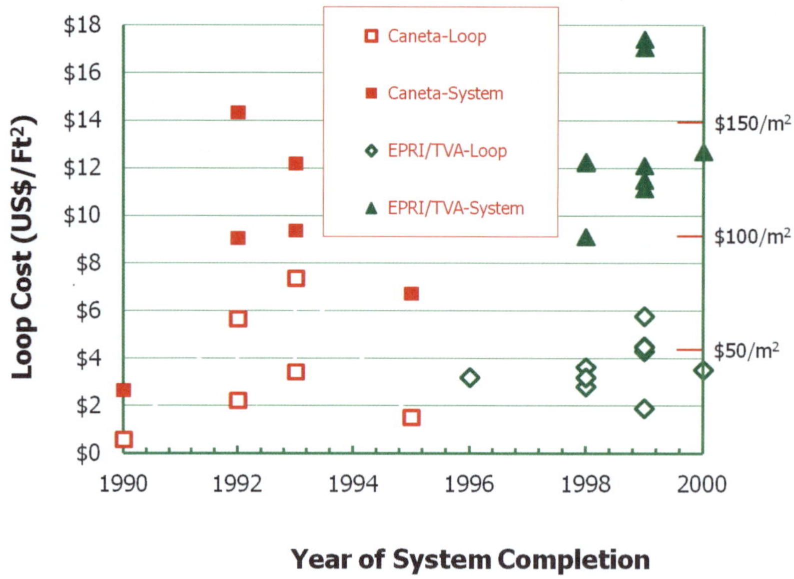

Figure 9.22 Previous GSHP System and Loop Cost Studies (Caneta Research 1995; Zimmerman 2000)

$0.60/ft^2 ($6.46/m^2). The average cost of the ground loop was 38.5% of the total GSHP system cost based on floor area, which is notably higher than the value for the more recent survey (25.5%).

The average GSHP system cost for the 2000 survey was $13.08/ft^2 ($138/m^2) with a maximum of $17.41/ft^2 ($187/m^2) and a minimum of 9.10/ft^2 ($98/m^2). The average ground-loop cost was $3.76/ft^2 ($40/m^2) with a maximum of $5.80/ft^2 ($62/m^2) and a minimum of $1.93/ft^2 ($21/m^2). The average cost of the ground loop was 30.1% of the total GSHP system cost based on floor area, which is greater than the value for the more recent survey (25.5%). It may be of interest that results were not influenced by LEED-related costs since no buildings were rated.

Table 9.3 provides a detailed listing of system costs for seven elementary schools in central Illinois. Ground-loop costs are fairly consistent based on bore length (±5%) and equipment capacity (±9%).

Ground-loop costs range between $1957 and $2344 per ton ($556 and $666 per kilowatt) and between $12.23 and $13.50 per bore foot ($40.11 and $44.28 per bore metre). The HVAC cost per unit area for the lowest-cost building is $8.92/ft^2 ($96/m^2). The highest-system-cost school, at $26.10/ft^2 ($281 m^2), was a new building that unexpectedly had a low floor area per unit cooling capacity of 353 ft^2/ton (9.33 m^2/kW), a low ENERGY STAR rating (75), and a high HVAC cost of $19.45/ft^2 ($209/m^2).

More detail for the lowest-cost system listed in Table 9.3 is provided in Table 9.4. The ground-loop cost represents 33% of the total GSHP system cost at $13.34/ft^2 ($144/m^2). The most significant interior HVAC items were piping (18.9%), heat pump equipment (21.6%), and controls (8%). Itemized costs for the ground loop were not provided beyond what is shown in Table 9.4.

Table 9.5 summarizes the HVAC and ground-loop costs for seven schools in central Texas. The common areas in these schools are conditioned with conventional HVAC systems, while GSHPs serve primarily the classrooms. The rated capacity of the GSHP equipment is considerably less than that of the conventional equipment, although the per-

Table 9.3 Specification and Cost Details for Illinois Elementary School One-Pipe Loop GSHPs

Installation Type		Retrofit	Retrofit	Retrofit	Retrofit	Retrofit	Retrofit	New
GSHP Installation Date		2006	2006	2007	2007	2008	2008	2010
Building Construction Date		1954	1954	1957	1954	1938	1956	2010
Building Size	**ft^2**	23,700	43,200	37,400	31,000	19,000	55,150	76,900
	m^2	2200	4000	3500	2900	1770	5130	7150
Equipment Capacity	**tons**	59	115	86	67	48	117	218
	kW	208	405	300	235	170	410	770
GSHP System Cost	**$**	490,000	859,000	621,000	499,000	390,000	736,000	2,007,000
GSHP System	**$/ton**	8305	7470	7221	7448	8125	6291	9206
	$/kW	2360	2125	2050	2120	2310	1790	2620
GSHP System	**$/ft^2**	20.68	19.88	16.60	16.10	20.53	13.35	26.10
	$/m^2	222.46	213.95	178.66	173.20	220.86	143.60	280.82
Vertical Bore Cost	**$**	82,000	NA	129,000	98,000	72,000	144,000	NA
Vertical Bore Length	**ft**	10,000	18,400	15,000	12,000	8,000	18,000	39,000
	m	3050	5610	44,570	3660	2440	5490	11,900
Vertical Bore	**ft/ton**	169	160	174	179	167	154	179
	W/m	68	72	66	64	69	75	64
Ground-Loop Cost	**$**	123,000	225,000	195,000	156,000	105,000	243,000	511,000
Ground Loop	**$/ton**	2085	1957	2267	2328	2188	2077	2344
	$/kW	593	556	645	662	622	591	666
Ground Loop	**$/ft**	12.30	12.23	13.00	13.00	13.13	13.50	13.10
	$/m	40.34	40.11	42.64	42.64	43.05	44.28	42.98
Exterior Header and Purge	**$**	40,000	NA	66,000	59,000	33,000	99,000	NA
HVAC System Cost	**$**	367,000	634,000	426,000	342,000	289,000	492,000	1,496,000
HVAC System	**$/ft^2**	15.49	14.68	11.39	11.03	15.21	8.92	19.45
	$/m^2	166.62	157.91	122.56	118.71	163.67	95.99	209.32
Ground Loop Percent of Total		25.1%	26.2%	31.4%	31.3%	26.9%	33.0%	25.5%
HVAC Percent of Total		74.9%	73.8%	68.6%	68.5%	74.1%	66.8%	74.5%

centage of area served by GSHPs is typically larger. For example, 61.8% of the floor area of the school built in 2007 is served by GSHPs but the GSHP capacity is only 36% of the total system capacity.

The average total system costs were \$5190/ton (\$1475/kW) and \$21.75/ft^2 (\$234/m^2). The average ground-loop length was 282 ft/ton (24 m/kW) with a cost of \$10.19/ft (\$33.43/m) of bore and \$2924/ton (\$831/kW). The variation in cost per unit length is significant, which may be a result of slow construction activity at the time that resulted in lower-than-normal ground-loop costs. However, the average cost of the ground loop was 33% of the total, which is higher than the average for systems in this survey. This is expected, because the loop lengths are significantly longer for this hot climate and high ground temperature.

Table 9.6 provides cost information for several GSHP systems installed between 1990 and 1995 and listed in an ASHRAE-sponsored research project (Caneta 1995). The aver-

Table 9.4 Itemized Component Retrofit Costs for Illinois Elementary School, 55,150 ft^2 (5125 m^2) with One-Pipe GSHP

	117 ton (410 kW) One-Pipe GSHP: 90 bores at 200 ft (61 m)					
Item	**$/ft²**	**$/m²**	**$/ton**	**$/kW**	**Total $**	**%**
One-pipe loop	2.52	27.12	1188	338	138,951	18.9%
Insulation	0.44	4.73	207	59	24,258	3.3%
Equipment	2.50	26.90	1180	336	138,075	18.8%
Equipment mark-up	0.38	4.09	177	50	20,700	2.8%
Pumps	0.16	1.72	74	21	8600	1.2%
Expansion tank	0.05	0.54	26	7	3000	0.4%
Air venting	0.01	0.11	4	1	450	0.1%
Equipment installation	0.25	2.69	118	34	13,800	1.9%
Electric/controls	1.07	11.51	506	144	59,189	8.0%
Sheet metal	0.46	4.95	216	61	25,305	3.4%
General work	0.72	7.75	340	97	39,780	5.4%
Condensate drainage	0.11	1.18	51	15	6000	0.8%
Balance	0.09	0.97	43	12	5040	0.7%
Chemical	0.02	0.22	12	3	1375	0.2%
Glycol	0.14	1.51	68	19	7900	1.1%
HVAC total	8.93	96.09	4209	1197	492,423	66.9%
Ground loop total	4.41	47.45	2078	591	243,117	33.1%
GSHP total	13.34	143.54	6287	1788	735,540	100.0%

Table 9.5 Specification and Cost Details for Central Texas Unitary-Loop GSHPs

School Building Type		Elementary	Elementary	Elementary	Middle	High	Middle	Elementary
Bid Date		2007	2008	2008	2008	2009	2010	2010
Building Size	**ft²**	112,300	112,300	111,600	177,300	411,800	177,700	112,300
	m²	10,400	10,400	10,400	16,500	38,300	16,500	10,400
Percent GSHP Floor Area		61.8%	61.8%	61.4%	59.4%	37.1%	62.9%	61.4%
Heat Pump Capacity	**tons**	163	163	163	262	345	308.5	163
	kW	573	573	573	921	1213	1085	573
Total System Capacity	**tons**	459	463	459	804	1574	838	463
	kW	1614	1628	1614	2828	5536	2947	1628
HVAC/GSHP System Cost	**$**	2.41E+06	2.71E+06	2.58E+06	3.54E+06	8.86E+06	3.63E+06	2.43E+06
HVAC/GSHP System	**$/ton**	5244	5844	5625	4405	5630	4332	5246
	$/kW	1491	1662	1599	1253	1601	1232	1492
HVAC/GSHP System	**$/ft²**	21.43	24.10	23.14	19.98	21.52	20.43	21.63
	$/m²	231	259	249	215	232	220	233
Vertical Bore Length	**ft**	47,850	47,560	47,560	71,050	91,930	83,230	47,560
	m	14,585	14,496	14,496	21,656	28,020	25,369	14,496
Vertical Bore	**ft/ton**	294	292	292	271	266	270	292
	W/m	39	40	40	43	43	43	40
Ground-Loop Cost	**$**	376,000	591,000	553,000	606,000	NA	510,000	693,000
Ground Loop	**$/ton**	2307	3626	3393	2313	NA	1653	4252
	$/kW	656	1031	965	658	NA	470	1209
Ground Loop	**$/ft**	7.86	12.43	11.63	8.53	NA	6.13	14.57
	$/m	25.78	40.77	38.15	27.98	NA	20.10	47.81

Table 9.6 Cost Details for ASHRAE RP-863 Study (Caneta 1995)

Building Type		Golf Clubhouse	Secondary School	Elementary School	Office	Education Center	Hotel
Installation Date		1990	1992	1992	1993	1993	1995
Location		Pennsylvania	Ontario	Minnesota	Virginia	New York	Pennsylvania
Building Size	**ft^2**	15,000	181,000	78,000	26,700	8000	39,900
	m^2	1400	16,800	7250	2480	745	3700
Heat Pump Capacity	**tons**	25.5	410	193	100	24	97
	kW	90	1442	679	352	84	341
GSHP System Cost	**$**	40,000	2,595,000	706,100	325,800	75,000	269,380
GSHP System	**$/ton**	1569	6329	3659	3258	3125	2777
	$/kW	446	1800	1040	926	889	790
GSHP System	**$\$/ft^2$**	2.67	14.34	9.05	12.20	9.38	6.75
	$\$/m^2$	29	154	97	131	101	73
Vertical Bore Length	**ft**	3000	72000	28000	15840	4000	9000
	m	914	21946	8534	4828	1219	2743
Vertical Bore	**ft/ton**	118	176	145	158	167	93
	W/m	98	66	80	73	69	124
Ground-Loop Cost	**$**	9000	1,030,200	176,500	92,030	59,040	61,950
Ground Loop	**$/ton**	353	2513	915	920	2460	639
	$/kW	100	714	260	262	699	182
Ground Loop	**$/ft**	3.00	14.31	6.30	5.81	14.76	6.88
	$/m	9.84	46.93	20.68	19.06	48.41	22.58

age GSHP system cost was $9.06/ft^2 ($98/m^2), with a wide variation between $2.67/ft^2 and $14.34/ft^2 ($29/m^2 and $154/m^2). Average ground-loop cost was $3.49/ft^2 ($38/m^2) and ranged from $0.60/ft^2 to $7.38/ft^2 ($6.50/m^2 to $79/m^2). Average cost per unit capacity was $3453/ton ($982/kW) and varied from $1569/ton to $6329/ton ($446/kW to $1800/kW), and the average cost based on vertical bore length was $8.51/ft ($28/m) with a $3.00/ft to $14.76/ft ($10/m to $48/m) range. The average bore length was 143 ft/ton (12.4 m/kW), and variation was also notable from a low of 93 ft/ton (8.1 m/kW) to a high of 176 ft/ton (15.3 m/kW).

Table 9.7 lists costs for the three GSHP systems in the EPRI/TVA study (Zimmerman 2000) that contained the most complete detail. The GSHP system costs for the three buildings were $11.47/ft^2, $14.92/ft^2, and $17.06/ft^2 ($123/m^2, $161/m^2, and $184/m^2). Costs for the ground loop were $4.31/ft^2, $4.63/ft^2, and $5.79/ft^2 ($46/m^2, $50/m^2, and $62/m^2). The average GSHP system costs for the nine buildings in the survey with complete data were $13.08/ft^2 ($141/m^2) and $4190/ton ($1190/kW). The average bore length was 148 ft/ton (12.8 m/kW) and the typical floor area per unit of cooling capacity was 330 ft^2/ton (8.7 m^2/kW). This study also contained building energy consumption, operating cost, and maintenance information.

Caution is advised in applying the results of this survey directly when estimating costs for GSHP projects. Reasons for uncertainty include the following items:

- This is a limited data set and should be considered a step toward greater transparency in publishing HVAC and GSHP system costs.

Table 9.7 Itemized Cost per Unit Floor Area for EPRI/TVA Study (Zimmerman 2000)

Item	Cost, $/ft²			Cost, $/m²		
	Low	Mid	High	Low	Mid	High
Total GSHP system cost	11.47	14.92	17.06	123	161	184
Major equipment	2.51	2.59	3.11	27.01	27.87	33.46
Piping/valves	1.91	1.46	2.99	20.55	15.71	32.17
Pumps/controls	0.24	0.24	0.12	2.58	2.58	1.29
Ductwork	2.01	2.39	3.16	21.63	25.72	34.00
HVAC controls	0.08	1.89	1.33	0.86	20.34	14.31
Other	0.41	1.72	0.56	4.41	18.51	6.03
Total interior cost	7.16	10.29	11.27	77	111	121
Drilling (and casing)	1.90	2.30	3.39	20.44	24.75	36.48
Pipe and U-tubes	0.32	1.00	0.28	3.44	10.76	3.01
Grouting	0.31	0.42	0.23	3.34	4.52	2.47
Trenching/headers	0.88	0.74	1.18	9.47	7.96	12.70
Compaction	0.66		0.43	7.10		4.63
Other	0.24	0.17	0.28	2.58	1.83	3.01
Total exterior cost	4.31	4.63	5.79	46	50	62
Exterior cost percent of total	37.6%	31.0%	33.9%	37.6%	31.0%	33.9%

- There was a high degree of reluctance to share itemized costs, which is to be expected for higher-cost GSHPs. Therefore, it is suspected that the averages in the recent survey may be lower than the actual national average.
- Optimum vertical bore lengths in colder climates tend to be shorter than lengths required in hot climates.
- Drilling conditions, local code requirements, and labor rates vary considerably from region to region and can have dramatic effects on costs.
- Very large projects can create variations in costs due to a "feast or famine" effect for the ground-loop contractor industry. Contractors who dedicate their entire capacity for months or years to a single project at a distant location endanger losing their local steady clientele.
- The Texas systems benefit from an infrastructure that has developed over 25 years. Consistent opportunities for loop contractors are available, contractor travel distances are short, equipment has been optimized for local conditions, the local geology is less uncertain, and engineers have adjusted designs to optimize installation efficiency. Therefore, it is not uncommon for a single rig to install in excess of 1500 vertical feet (460 m) of heat exchanger per day.

Observations of the results of the information gathered in the studies conducted by Caneta Research (1995), Zimmerman (2000), and EPRI (2012), can be summarized with the following statements.

- The average costs in the 2012 study were \$20.75/ft^2 (\$223/m^2) for the GSHP system, which included \$15.46 (\$166/m^2) for the HVAC portion and \$5.29/ft^2 the ground loop (\$57/m^2).

- The average costs in the 2000 study were \$13.08/ft^2 (\$141/m^2) for the system, \$9.32 (\$100/m^2) for the HVAC portion, and \$3.76/ft^2 (\$40/m^2) for the ground loop.
- The average costs in the 1995 study were \$9.07/ft^2 (\$141/m^2) for the system, \$5.58 (\$100/m^2) for the HVAC portion, and \$3.49/ft^2 (\$40/m^2) for the ground loop.
- In the sixteen years since the 1995 study, the cost of the interior portion of GSHP systems has increased by 177% while the cost of the ground-loop portion has increased only 52%.
- The percentage of ground-loop costs to total GSHP system cost declined from 38.5% in 1995 to 30.1% in 2000 to 25.5% in 2011.
- The focus of future cost containment efforts in commercial GSHPs should concentrate on the HVAC systems while not neglecting efforts to improve efficiency and expand opportunities for ground-loop installations.
- Greater emphasis should be placed on gathering detailed cost information to expand and improve the results and conclusions of the three studies discussed in this section.

9.4 ESTIMATION OF THE COST OF GSHP DESIGN OPTIONS

Cost estimation of any type of HVAC system has become increasingly difficult and variable. System complexity has increased and, in spite of easy access to information via the Internet, the most common listing on vendor websites is "Call for Price Quotation." The chapter on costs in *ASHRAE Handbook—HVAC Applications* has almost no information (ASHRAE 2011). GSHP costs are likewise variable and uncertain, as noted by the results presented in the previous section.

This section relies primarily on *RSMeans Mechanical Cost Data 2014* (RSMeans 2014) coupled with a few HVAC and GSHP industry sources (GPI 2014; TCI 2011), including an online vendor that provides updated HVAC and "geothermal heat pumps" prices (IWA 2014). The information focuses on GCHP costs and interior HVAC costs. Chapter 8 provides information for groundwater heat pump (GWHP) costs. Information for SWHPs is limited in this book, but a study is in progress (ASHRAE 2009) and additional information will be available once a final project report is submitted.

Table 9.8 provides the 2014 price information available for equipment costs. These costs include material, labor, overhead, and profit. The information is limited to the equipment itself and does not include the associated distribution systems unless noted. It is highly recommended that a summary cost analysis for the various GSHP equipment options be performed with this table at the earliest possible stage in the design process. Example 9.1 provides a compelling illustration of the importance of this suggestion if cost optimization is an important factor to the GSHP design team.

Table 9.9 is a supplement to Table 9.8 in that it provides costs for a variety of water-to-air and water-to-water heat pumps rated for extended-range GSHP applications. It also includes efficiency information and the cost of optional heat recovery coils for preheating domestic water. Shipping costs are not included in these prices as they are in Table 9.8.

Table 9.8 HVAC Equipment Installation Costs (Material, Labor, and Profit) (RSMeans 2014)

Unitary Equipment tons	kW	Split Air Heat Pump with Auxiliary	Packaged Air Heat Pump with Auxiliary	WSHP	Packaged Cooling Electric Heat	Packaged Cooling VAV Electric Heat	RTU-CAV Gas Heat
		Cost in US $ (2014)—Equipment and Installation Only (No Accessories)					
3	11	4125	4875	3350			6050
5	18	6525	6775	4750	10,600		8075
10	35	12,700	15,400	12,300	16,900	19,100	15,800
20	70	24,400	20,900	21,600	30,600	34,300	30,100
50	175		41,700	53,000	79,500	88,000	56,500

Chillers (Nonreversing) tons	kW	Packaged Reciprocating Air-Cooled	Reciprocating Water-Cooled	Screw Air-Cooled	Screw Water-Cooled	Centrifugal Water-Cooled	Direct Absorption Gas—Duplex
		Cost in US $ (2014)—Equipment and Installation (Pipe Network Not Included)					
25	90		28,500				
50	175		49,800				
100	350	93,500	83,500		80,500		170,000
200	700	159,000		170,000	111,000		225,500
300	1050	212,000		224,500	150,000		266,500
400	1400	292,000	-			170,500	340,500
1000	3500		-			524,000	840,000

Cooling Towers, Coolers, and Boilers tons	kW	Forced Centrifugal Fan	Induced Axial Fan	Fiberglass Axial Fan	Fluid Cooler	Plate Heat Exchanger	Boiler—Heating Water Gas Fired
		Cost in US $ (2014)—Equipment and Installation (Pipe Network Not Included)					
50	175	13,800	12,500		30,900		15,900
100	350	21,300	15,600	15,800	62,000	36,000	24,600
200	700	33,400	32,500	30,000	123,000	58,000	37,100
300	1050	46,300	43,500			80,000	50,100
400	1400	55,500	54,000			102,000	90,000
1000	3500	146,000	118,000				243,000

Air Handling and Outdoor Air ft³/min	m³/h	CAV with Heating and Cooling Coils	VAV-CW with Coils	Central AHU VAV	Field-Fabricated VAV	Make-Up Air Chilled Water or Direct Expansion	Energy Recovery (Wheel Type)
		Cost in US $ (2014)—Equipment and Installation (Ductwork Not Included)					
2,000	3,400	13,000	17,900			27,400	9825
5,000	8,500	21,300	25,900	39,325		32,600	13,200
10,000	17,000	31,500	49,000	67,800		66,000	17,500
20,000	34,000	78,500	90,500	116,700		130,500	30,400
40,000	68,000				94,800		56,500
100,000	170,000				252,500		

Terminal Units ft³/min	m³/h	Fan Coil with Electric Heat	Fan Coil Four Pipe	VAV Hot Water	Fan-Powered VAV Hot Water		
		Cost in US $ (2014)—Equipment and Installation (No Zone Duct Included)					
400	680	2150	2490	4425	5025	*Add $1450 for zone duct	
800	1360	3000	3512	5350	5975	*Add $2900 for zone duct	
1200	2040	5575	4095	6550	7400	*Add $4350 for zone duct	
2000	3400	9275	5219	5875	11,825	*Add $5800 for zone duct	
Pumps—Bronze		1/12 hp (60 W)	1/3 hp (125 W)	1 hp (0.75 kW)	5 hp (3.7 kW)	10 hp (7.5 kW)	20 hp (15 kW)
Circulator in-line		$810	$1350	$2000			
Base-mounted					$10,600	$16,800	$20,200

CAV = constant-air volume; CW = condenser water; RTU = rooftop unit; VAV = variable air volume

Table 9.9 Heat Pump Online Catalog Prices (IWA 2014)

Single-Speed Water-to-Air and Water-to-Water Heat Pumps—Cost and EER at 77°F (25°C) ELT									
		Manufacturer A Water-to-Air		Manufacturer B Water-to-Air		Manufacturer A Water-to-Water		Manufacturer B Water-to-Water	
tons	kW	Cost	EER	Cost	EER	Cost	EER	Cost	EER
2	7	$2190	19.7	$2635	19.2	—	NR	—	NR
2.5	9	$2230	17.6	$3130	19.3	—	NR	—	NR
3	10.5	$2310	17.9	$3160	16.6	$2550	NR	$5650	NR
3.5	12	$2365	17.0	$3290	18.9	—	NR	—	NR
4	14	$2710	20.0	$3380	18.2	$2980	NR	$5670	NR
5	17.5	$3000	17.2	$3470	18.0	$3270	NR	$6430	NR
Heat recovery unit (HRU) water pre-heater (desuperheater): Add $289 to $350									

Dual-Capacity and Variable-Speed Water-to-Air Heat Pumps—Cost and EER at 77°F (25°C) ELT									
		Manufacturer B Water-to-Air Heat Pump, Dual		Manufacturer B Water-to-Air Heat Pump, Variable Speed		Manufacturer C Water-to-Air Heat Pump, Dual (a)		Manufacturer C Water-to-Air Heat Pump, Dual (b)	
tons	kW	Cost	EER	Cost	EER	Cost	EER	Cost	EER
3	10.5	$3870	20.1	$6570	20.4	$6940	20.3	$5160	18.2
4	14	$4090	19.2	$7110	20.2	$7400	19.3	$5760	17.9
5	17.5	$4480	19.7	—	—	$7900	18.8	$5780	17.5
6	21	$4700	18.0	—	—	$7930	16.9	—	—
HRU (desuperheater): Add $289						HRU (desuperheater) Included in cost			

Commercial Single-Speed Water-to-Air Heat Pumps—Cost and EER at 77°F (25°C) ELT			
tons	kW	Cost	EER
6	21	$5610	15.2
7.5	26	$6290	16.1
10	35	$6970	16.1
12	42	$8600	NR
15	53	$9400	NR
20	70	$10,300	NR
HRU (desuperheater): Add $425			

EXAMPLE 9.1— GSHP EQUIPMENT COST

Compare the primary equipment cost for a 400 ton (1400 kW) system for the two GSHP options shown in Figure 2.16 (multiple common loops) and Figure 2.17 (chilled-water VAV). The common-loop system consists of ten loops with ten heat pumps each (three 3 ton [11 kW], four 4 ton [14 kW], and three 5 ton [18 kW]). Each unit has a single 1/6 hp (0.12 kW) pump.

The chilled-water GSHP VAV system consists of two 200 ton (700 kW) reversible water-cooled screw compressor chillers (assume the cost is the same as for the nonreversible chiller), eight 20,000 cfm (34,000 m^3/h) AHUs each connected to 15 fan-powered VAV terminals (five 800 cfm [1360 m^3/h], five 1200 cfm [2040 m^3/h], and five 2000 cfm [3400 m^3/h]). Water flow to the ground loop is provided by two 20 hp (15 kW) base-mounted pumps, and the chilled-water loop is supplied by three 10 hp (7.5 kW) pumps.

Solution

Table 9.10 shows the tabular results of the comparison, with the equipment for the common-loop heat pump system being less than 25% of that for the chilled-water VAV system. These costs include material and labor to install the units but do not include the costs of connection to the air, water, and electrical systems. The heat pump systems consists of 100 heat pumps and 100 circulator pumps for a total of \$504,000, or \$1260/ton (\$360/kW).

The chilled-water GSHP VAV system includes 2 chillers, 8 AHUs, 120 fan-powered VAV terminals, and 5 pumps, for a total of \$2,254,400, or \$5636/ton (\$1610/kW).

Table 9.10 GSHP Equipment Cost: Common-Loop Heat Pump vs Chilled-Water VAV

Option 1—Ten Common-Loop GSHPs—40 tons (140 kW) Each			
Quantity		**Unit Cost**	**Total Cost**
30	3 ton (11 kW) heat pump	\$3350	\$100,500
40	4 ton (14 kW) heat pump*	\$4050	\$162,000
30	5 ton (14 kW) heat pump	\$4750	\$142,500
100	1/6 hp (0.12 kW) in-line circulator pumps*	\$990	\$99,000
		Total	\$504,000
		Cost/ton	\$1260
	Interpolated values	**Cost/kW**	\$360

Option 2—Chilled-Water VAV GSHP—Two 200 ton (700 kW) Chillers—Central Loop			
Quantity	**Components**	**Unit Cost**	**Total Cost**
2	200 ton (700 kW) water-cooled screw chillers	\$111,000	\$222,000
8	20,000 cfm (34,000 m^3/h) VAV AHUs	\$116,700	\$933,600
40	800 cfm (1360 m^3/h) fan-powered VAV terminals	\$5975	\$239,000
40	1200 cfm (2040 m^3/h) fan-powered VAV terminals	\$7400	\$296,000
40	2000 cfm (3400 m^3/h) fan-powered VAV terminals	\$11,825	\$473,000
2	20 hp (15 kW) base-mounted pumps	\$20,200	\$40,400
3	10 hp (7.5 kW) base-mounted pumps	\$16,800	\$50,400
		Total	\$2,254,400
		Cost/ton	\$5636
		Cost/kW	\$1610

Table 9.11 summarizes the interior pipe and fitting costs for three common materials used in GSHP applications. The table does not include the cost of pipe insulation because HDPE and polypropylene typically do not require insulation for GSHP applications, except in colder climates where the pipe outside surface temperature may occasionally fall below the room air dew point. The cost for steel pipe is for grooved joint fittings. Welded steel pipe is slightly higher in cost; values are provided in *RSMeans Mechanical Cost Data 2014* (RSMeans 2014). The HDPE pipe values reflect the cost of butt fusion joints and fittings for all sizes. The polypropylene pipe values assume socket fusion joints and fittings up to 4 in. (100 mm) and butt fusion for larger pipe and fittings.

Table 9.12 provides the costs of underground HDPE pipe installation based on a 4 ft (1.2 m) burial depth. The source of the data in the table does not include the labor costs for fusion joints or the trenching and backfilling costs, so these must be added into the

Table 9.11 Interior Pipe and Fitting Installation Costs (Material, Labor, and Profit) (RSMeans 2014)

Pipe Material	Nominal Diameter, in. (mm)	Straight $/ft	Straight $/m	90°L $/fitting	45°L $/fitting	Coupling $/fitting	Tee $/fitting	Red $/fitting
	1 (32)	15.9	52	47	47	32.5	72.5	
Steel—Black (Schedule 40)	1.25 (40)	18.3	60	50.5	50.5	32.5	77.5	
	1.5 (50)	21	69	54	54	36	83	
	2 (60)	26	85	61	61	46.5	92.5	67.5
Grooved joint hangers at 10 ft (3 m) centers	3 (80)	43	141	96.5	96.5	61.5	127	82.5
	4 (100)	53	174	114	114	85	181	99
	6 (150)	98.5	323	256	256	141	400	165
Piping 10 ft (3 m) above floor	8 (200)	137	449	465	465	208	770	315
	10 (250)	177	581	755	755	289	1400	555
	12 (300)	200	656	1150	1150	320	1950	935
	1 (32)	1.9	6	13.5		13.5	17.7	
HDPE DR 11	1.5 (50)	2.4	8	17.0		21.0	24.8	
	2 (60)	4.0	13	17.0	13.5	28.7	21.2	20.3
Butt fusion fittings hangers at 3 to 4 ft (1 to 1.2 m) centers	3 (80)	4.9	16	34.0	34.0	36.9	39.0	20.3
	4 (100)	8	27	47	47	48	57	30
	6 (150)	20	66	109	109	73	142	71
Piping 10 ft (3 m) above floor	8 (200)	34	110	268	268	96	350	109
	10 (250)	53	173	1000	1000	115	1044	187
	12 (300)	77	252	1055	1055	135	1400	306
	3/4 (25)	15.85	52	19	19	21	32	21
	1 (32)	17.5	57	26	26	25.5	38	23
Polypropylene DR 11	1.25 (40)	20.5	67	29	29	27	42.5	27
	1.5 (50)	24.5	80	36.5	36.5	32.5	56	38.5
Hangers 3 per 10 ft (1 per m)	2 (60)	30	98	42	41.5	38.5	66.5	62
	3 (80)	40.5	133	87.5	92	63.5	116	108
Piping 10 ft (3 m) above floor	4 (100)	54.5	179	153	163	99	189	148
	6 (150)	60.5	198	320	335		450	221
	8 (200)	90.5	297	630	550		705	286
	10 (250)	122	400	895	735		970	

**Cost of insulation not included because interior HDPE and polypropylene often do not require insulation.*

total cost. Trenching and backfilling costs for other burial depths are directly proportional to burial depth (a 20% deeper trench costs 20% more) since the source cost is based on the volume of the excavation. The table includes the cost of sleeves used for wall or floor penetrations. The assumed length of the sleeve is 12 in. (250 mm) and the sizes are based on the pipe diameter, not the outer diameter of the sleeve. Unfortunately, *RSMeans Mechanical Cost Data 2014* (RSMeans 2014) does not list the cost for side-saddle tees commonly used for U-tube take-offs. Field connection of these fittings should be avoided and shop fabrication is highly recommended (as shown in Figure 6.22). It is suggested that the installation cost of a HDPE tee given in Table 9.9 be used as a substitute for the side-saddle tee.

Figure 9.23 provides the costs of underground valve vaults based on bids submitted to a contractor in the Midwest (TCI 2011). The costs are for 3 in. (80 mm) DR 11 circuit

Table 9.12 Ground-Loop Header Installation Costs (Material, Labor, and Profit) (RSMeans 2014)

HDPE DR 11 Diameter,	Straight Pipe	Butt Fusion	Trench/Backfill 4 ft Depth*			Pipe Flange	Pipe Sleeve	Fusion Tool	
			12 in.	24 in.	36 in.			Rental	Cost
in.	$/ft	$/weld	$/ft	$/ft	$/ft	$/flange	$/sleeve	$/day	$
1	0.87	8.6	1.67				156	44.5	805
1.5	1.1	13.4	1.67				190	44.5	805
2	1.84	18.3	1.67	3.33		19.35	211	44.5	805
3	2.21	23.5	1.67	3.33		22.5	238	50.5	805
4	3.7	30.5	1.67	3.33		30.5	276	50.5	805
6	9.2	46.5		3.33		43.5	420	113	27,900
8	15.3	61		3.33	5.00	63	525	113	27,900
10	24	73.5			5.00	100	560	196	27,900
12	35	86			5.00	133	620	196	27,900
HDPE DR 11 Diameter,	**Straight Pipe**	**Butt Fusion**	**Trench/Backfill 1 m Depth***			**Pipe Flange**	**Pipe Sleeve**	**Fusion Tool**	
			0.25 m	**0.5 m**	**0.75**			**Rental**	**Cost**
mm	**$/m**	**$/weld**	**$/m**	**$/m**	**$/m**	**$/flange**	**$/sleeve**	**$/day**	**$**
32	2.9	8.6	4.48				156	44.5	805
50	3.6	13.4	4.48				190	44.5	805
60	6.0	18.3	4.48	8.96		19.35	211	44.5	805
80	7.2	23.5	4.48	8.96		22.5	238	50.5	805
100	12	30.5	4.48	8.96		30.5	276	50.5	805
150	30	46.5		8.96		43.5	420	113	27,900
200	50	61		8.96	13.44	63	525	113	27,900
250	79	73.5			13.44	100	560	196	27,900
300	115	86			13.44	133	620	196	27,900

**Common earth, 1/2 yd^3 (0.4 m^3) excavator, vibrating roller compaction*

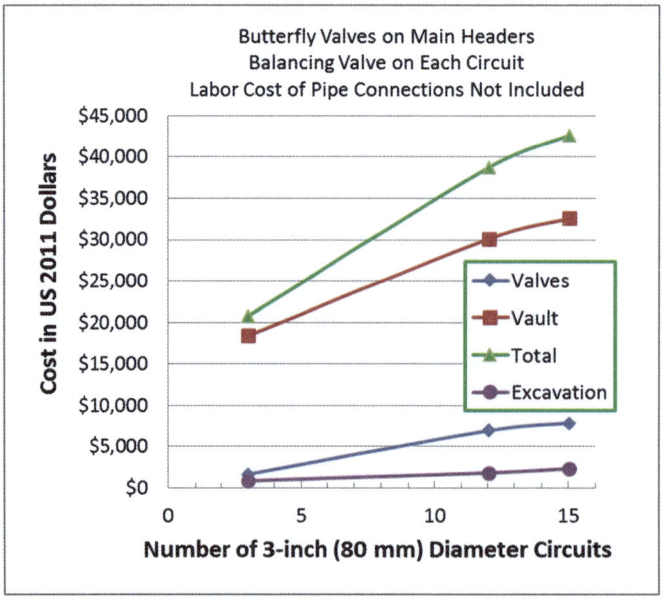

Figure 9.23 Underground Valve Vault Costs (TCI 2011)

headers, which typically can support up to 35 tons (120 kW) of heat pump capacity. (DR 13.5 and 15.5 HDPE can support slightly more capacity.) Excavation costs were adapted from *RSMeans Mechanical Cost Data 2014* (RSMeans 2014) for underground storage tank installation, excavation, and backfill costs. The figure does not include the cost of fusing the main header (two joints) and circuits (two per circuit), but these values can be approximated by inserting the labor cost for each butt fusion joint listed in Table 9.12. Costs for the valves assume they are installed by the vault manufacturer prior to shipment. Vaults are also available with 2 in. (60 mm) diameter circuits (up to 12 tons [40 kW]) and 4 in. (100 mm) circuits (up to 80 tons [280 kW]). Vaults with these larger circuit pipes require an extremely large purge pump that may elevate start-up cost if one is not locally available.

EXAMPLE 9.2— TO VAULT OR NOT TO VAULT

For the 400 ton (1400 kW) central ground loop described in Example 9.1, compare the cost of using an underground valve vault with the cost of routing all ten circuit headers into the equipment room. A schematic of these two options is shown in Figure 1.9. The distance between the vault and the equipment room is 200 ft (60 m) and the straight sections of HDPE are shipped in 40 ft (12 m) lengths. Assume the excavating cost for the single headers and the multiple circuit headers are the same.

Solution

Option 1 is the vault with the manifold and valves. The vault has ten 3 in. (80 mm) circuits and 8 in. (200 mm) main headers. It is assumed the vault is also equipped with two 4 in. (100 mm) butterfly valves for purging. The main headers are routed from the vault to 90° elbows below the equipment room and connected to 5 ft (1.5 m) risers, pass through sleeves in the floor, and are terminated into a flange. The total length of the two headers is 410 ft (125 m), which requires 16 butt fusion welds (10 on the straight pipe, 4 on the elbows, and 2 on the flanges). Twenty 3 in. (80 mm) butt fusion welds are required to connect the circuits to the valve vault.

Option 2 is to locate the manifold and valves in the equipment room and route the 10 circuits (20 pipes) directly from the loop field. Figure 6.25 contains photographs of this arrangement in equipment rooms. Circuit headers are routed below the equipment room and are bent 90° upward, pass through sleeves, and are terminated in flanges and circuit balancing valves. The total length of the twenty headers is 4100 ft (1250 m), which requires 160 butt fusion welds (120 on the straight pipe and 40 on the flanges). Twenty 3 in. (80 mm) side-saddle fusion welds are required to connect the circuits to the 8 in. (200 mm) main header in the equipment. One end of each main header is terminated into a flange and butterfly valve. The other end is terminated into a reducer, flange, and two 4 in. (100 mm) butterfly valves for purging.

Table 9.13 shows the results of the comparison of the two options and indicates the equipment room option costs 38% less than the valve vault option. It should be recognized that as header lengths increase, the difference between vault option and equipment room option decreases. However, the header lengths in this case would have to increase from 200 to 1500 ft (60 to 460 m) for the costs of the options to be the same.

Table 9.13 Costs of Ground-Loop Manifold and Valves in Vault vs Equipment Room

Quantity	Option 1—Vault with Manifold and Valves	Unit Cost	Total Cost
1	HDPE vault with valves—8 in. (200 mm) mains, ten 3 in. (80 mm) circuits	$35,000.00	$35,000
410	8 in. (200 mm) HDPE DR 11 pipe	$15.30	$6,273
2	8 in. (200 mm) 90° elbows	$268.00	$536
2	8 in. (200 mm) flanges	$63.00	$126
2	8 in. (200 mm) pipe sleeves	$525.00	$1,050
20	3 in. (80 mm) butt fusion welds	$23.50	$470
16	8 in. (200 mm) butt fusion welds	$61.00	$976
			$44,431
		Cost/ton	$111.08
		Cost/kW	$31.74
Quantity	**Option 2—Ten Circuits to Equipment Room Manifold and Valves**	**Unit Cost**	**Total Cost**
4100	3 in. (80 mm) HDPE DR 11 pipe	$2.21	$9,061
20	8 in. (200 mm) HDPE DR 11 pipe	$15.30	$306
40	3 in. (80 mm) flanges	$22.50	$900
20	3 in. (80 mm) pipe sleeves	$238.00	$4,760
20	3 in. (80 mm) pipe saddle fitting to 8 in. (200 mm) main	$39.00	$780
160	3 in. (80 mm) butt fusion welds	$23.50	$3,760
20	3 in. (80 mm) saddle fusion welds	$70.00	$1,400
4	8 in. (200 mm) butt fusion welds	$30.50	$122
2	4 in. (100 mm) butt fusion welds	$23.50	$47
1	Valve set—two 8 in. (200 mm) butterfly valves, 20 3 in. (80 mm) balancing valves, two 4 in. (100 mm) butterfly valves (purge)	$6,000.00	$6,000
2	8 in. (200 mm) flanges	$63.00	$126
2	8 × 4 in. (200 × 100 mm) reducers	$109.00	$218
2	4 in. (100 mm) flanges	$30.50	$61
			$27,541
		Cost/ton	$68.85
		Cost/kW	$19.67

Figure 9.24 is an example of a vertical-loop cost calculator provided by a product manufacturer (GPI 2014). The intended use for this calculator is to apply the results of ground-loop calculations as described in Chapters 3 and 4 to determine the costs of various loop options. The results shown in the table are based on alternatives discussed in the example design in Chapter 4. The loop options were eighteen 270 ft (82 m) vertical bores when using a grout with a 0.90 Btu/h·ft·°F (1.56 W/m·K) thermal conductivity or eighteen 332 ft (101 m) vertical bores when using a grout with 0.42 Btu/h·ft·°F (0.73 W/m·K) conductivity. The output provides the net cost of the vertical heat exchangers and the amount of materials required. The base case was using the conventional lower-conductivity bentonite grout. The second option was to raise the grout conductivity by either mixing large amounts of silica with the bentonite or adding high-performance graphite (HPG). In this example the thermally enhanced (TE) grout options were the lower-cost alternatives because of the reduced loop length (compared to the pure bentonite), and the grout with the HPG had lower material-handling costs (compared to the silica sand TE grout mixture).

INPUT INFORMATION AREA (Yellow Cells Only)	TE Grout		TE Grout		Conventional	
Grout Type	w / Silica Sand		with HPG		Non-TE Grout	
Thermal Conductivity or Solids Content	0.88	TC	0.88	TC	20	%
Grout Cost / Bag	9.50	$	9.50	$	8.00	$
Sand or Graphite Cost / Bag	3.40	$	11.00	$		
Grout Freight Cost / Truckload	2050	$	2050	$	2050	$
Grout Bags / Truckload	918	#	918	#	918	#
Sand or Graphite Freight Cost / Truckload	1550	$	3580	$		
Sand or Graphite Bags / Truckload	910	#	1456	#		
Bore Diameter	5.25	in	5.25	in	5.25	in
U-Bend Diameter	1	in	1	in	1	in
Bore Depth	270	ft	270	ft	332	ft
Number of Bores	18	#	18	#	18	#
Loss to Formation	5	%	5	%	5	%
Grout Mixing Time / Unit (Bag)	8	min	8	min	5	min
Grout Pumping Rate	8.5	gpm	10	gpm	10	gpm
Labor Rate / Hour	30.00	$	15.00	$	15.00	$
U-Bend Cost / Bore Foot	0.75	$	0.75	$	0.75	$
Drilling Cost / Foot	8.00	$	8.00	$	8.00	$
Other Equipment or Material Cost / Foot	0.115	$	0.115	$		
Header Cost per Bore	975.00	$	975.00	$	975.00	$
TOTAL GROUT VOLUME CALCULATIONS						
Bore Volume / Foot (0.0408*(BD^2))	1.125	gal	1.125	gal	1.125	gal
Pipe Volume / Foot (0.0408*((UD^2)*2))	0.141	gal	0.141	gal	0.141	gal
Bore - Pipe Volume / Foot	0.984	gal	0.984	gal	0.984	gal
Volume / Bore (V/F*BD)	278.96	gal	278.96	gal	343.02	gal
Total Grout Volume Required (V/B*NB)	5,021	gal	5,021	gal	6,174	gal
GROUT YIELD AND BAGS NEEDED						
Required Sand or Graphite	200	lbs	7.5	lbs		
Grout Yield / Bentonite Bag	30.1	gal	18.5	gal	26.8	gal
otal Bags of Bentonite, based on Full Pallets (TGV/GY)	167	#	272	#	231	#
Truckloads of Bentonite (TBN/GBpTL)	1	#	1	#	1	#
Total Bags of Sand or Graphite Needed	668	#	136	#		
Truckloads of Sand (TSN/SBpTL)	1	#	1	#		
GROUT BATCHES PER BORE	9.3		7.5		12.8	
DELIVERED GROUT COST PER GALLON	1.484	$	1.930	$	0.630	$
TOTAL COST PER BORE						
Grout Cost	413.88	$	538.30	$	215.98	$
Cost to Mix Grout	37.07	$	30.16	$	16.00	$
Cost to Pump Grout	16.41	$	6.97	$	8.58	$
Pipe Cost	202.50	$	202.50	$	249.00	$
Drilling Cost	2160.00	$	2160.00	$	2656.00	$
Other Equipment Cost	31.05	$	31.05	$		
Header Cost per Bore	975.00	$	975.00	$	975.00	$
Total Cost per Bore	3,835.91	$	3,943.99	$	4,120.56	$
Savings per Bore		$	108.08	$	284.65	$
TOTAL COST FOR PROJECT	**69,046.33**	$	**70,991.74**	$	**74,170.02**	$
Additional Cost		$	1,945.41	$	5,123.69	$
Additional Cost		%	2.74	%	6.91	%
Total Tons of Dry Material on Site	20.9		7.8		5.8	

Figure 9.24 Vertical Ground-Loop Cost Calculator with Grout Conductivity Comparison
Printed with permission of GeoPro, Inc.

9.5 CHARACTERISTICS OF QUALITY GSHPs

The last article in the *ASHRAE Journal* series on long-term commercial GSHP performance (Kavanaugh and Meline 2013) suggested a format for presenting a summary of characteristics of completed projects that could be used to gauge quality. This format includes the following:

- A short summary, description of building (type, floor area, date of construction, etc.)
- ENERGY STAR rating
- Cost of mechanical system and total cost of construction
- Annual electrical and fossil-fuel energy use
- Electrical demand
- Energy expenditures
- HVAC equipment summary (heat pumps, pumps, ERUs, boilers, etc.)
- Ground-loop description
- Occupant satisfaction levels
- Maintenance staff satisfaction levels

Though these items are critical to determination of quality, satisfaction, and economic value, it appears they are rarely quantified or published. However, they should be. Owners and architects should request this information when interviewing engineering firms, and engineering firms should be proud to provide this information if they have done their work well.

The article concluded with a listing of characteristics of successful GSHPs and successful GSHP engineering design firms. These characteristics are repeated here.

Characteristics of successful GSHPs include the following:

- The ENERGY STAR rating of the building exceeds 90.
- Maximum loop temperatures returning from the ground tend to be below 90°F (32°C) for systems in which the cooling mode determines the loop length.
- The systems surveyed during this project were primarily ten-month schools and 8:00 a.m. to 5:00 p.m. offices located in areas where the measured ground thermal conductivity was between 1.0 and 1.5 Btu/h·ft·°F (1.7 and 2.6 W/m·°C). Under these circumstances, the successful vertical ground loops tend to be in the range of 200 to 240 ft of vertical bore per installed ton (17 to 21 m/kW) of cooling capacity for a ground temperature of 63°F (17°C). This corresponds to a range of 155 to 185 ft/ton (13 to 16 m/kW) for 55°F (13°C) ground temperature and 270 to 320 ft/ton (23 to 28 m/kW) for 70°F (21°C) ground temperature.
- The ground-loop lengths of systems in this survey were all dictated by the cooling-mode requirements. This resulted in advantageous heating-mode ground-loop temperatures, even at the coldest sites in central Illinois. At the one site that was monitored continuously, the ground-loop return temperature remained above 46°F (8°C) when the outdoor temperature was –6°F (–21°C).
- The primary equipment type tends to be water-to-air heat pumps.
- Installed outdoor ventilation air equipment capacity tends to be 20 cfm/person (9.4 L/s·person) or less.
- Systems with heat pumps circuited to individual ground loops, small central loops, or multiple common loops outperformed systems with large central loops by a significant margin.
- Pump control tends to be on-off for these smaller loops rather than variable speed.

- Ground-loop pump power tends not to exceed 10 hp/100 tons (kW_P/kW_{HP}). This value is deemed to be average (grade = C) using recommended guidelines.
- Due to the selection of piping materials and the pH level of the fill water, piping systems tend not to require chemical treatment. However, caution is advised against using polyvinyl chloride (PVC) pipe. It is not recommended for service in GSHP systems.
- Control is provided by individual thermostats or a building automation system (BAS) that is simple with a clear and concise sequence of operation so program adjustments (or retrocommissioning) can be performed by the maintenance staff.
- When surveyed, occupants rate indoor comfort, indoor air quality, acoustics, lighting, maintenance responsiveness, and system controllability as satisfactory.
- When surveyed, the maintenance staff rates system serviceability, quality of design, and quality of installation as satisfactory.
- Owners and designers are satisfied with utility cost, and they openly share results (and permit ENERGY STAR rating).
- Owners and designers are satisfied with the installation costs, will openly share itemized results, and are confident the project provides positive economic value.

Characteristics of successful GSHP engineering design firms include the following:

1. The engineering firm performs all system design (including the ground-loop and HVAC controls) and is open to feedback for suggested modifications that benefit the owner and building occupants.
2. In situations where a firm has designed several buildings for an owner, the engineers regularly communicate with maintenance supervisors and staff (and are not afraid to enter their break room at lunch).
3. In situations where a firm has designed a *single* building for an owner, the owner regularly refers the engineers because of the quality of the work product.
4. The engineering firm is familiar with the capabilities of the local ground-loop and mechanical contractors and the corresponding level of monitoring to ensure systems are installed as designed.
5. The engineering firm has a basic understanding of local geology, groundwater regulations, drilling techniques, and ground-circuit header assembly. The firm provides designs that are sensitive to the resulting constraints and are therefore respected by ground-loop contractors (who typically do not hold engineers in high regard).
6. Because ENERGY STAR rating (unlike LEED rating) is based on measured energy performance, requires minimal paperwork, uses information routinely provided by the utilities, and is relatively simple and inexpensive, the engineering firm maintains a listing of ratings for completed projects.
7. The engineering firm tracks, maintains, and openly shares records of mechanical and ground-loop costs. Contractors and subcontractors are encouraged (or possibly required) to submit itemized bids to identify components or designs that are not good value.
8. The owners and engineering firm allow occupant and maintenance satisfaction surveys to be conducted and review results and comments in order to improve quality.
9. The engineering firm oversees the design through construction and performs system commissioning as an included service rather than a separate line item in their fee proposals that might be eliminated by a client.

9.6 REFERENCES

ASHRAE. 2009. *Development of design tools for surface water heat pump systems*. ASHRAE RP-1385, Final Report in Progress. Atlanta: ASHRAE.

ASHRAE. 2011. *ASHRAE Handbook—HVAC Applications*, Chapter 37, Owning and Operating Costs. Atlanta: ASHRAE.

Caneta Research. 1995. *Operating experiences with commercial ground-source heat pump systems*. ASHRAE RP-863 Final Report. Atlanta: ASHRAE.

Caneta Research. 1998. *Operating Experiences with Commercial Ground-Source Heat Pump Systems*. Atlanta: ASHRAE.

EIA. 2008. Detailed tables, 2003 CBECS survey data. Commercial Building Energy Consumption Survey (CBECS). Washington, DC: U.S. Energy Information Administration. www.eia.gov/consumption/commercial/data/2003/

EIS. 2009. Ground heat exchanger model uncertainty. *Instructional Manual—GshpCalc 5.0, GSHP Design Software*. Northport, AL: Energy Information Services.

EPA. 2010. How the rating system works. ENERGY STAR Portfolio Manager Overview. Washington, DC: U.S. Environmental Protection Agency. www.energystar.gov/index.cfm?c=evaluate_performance.pt_neprs_learn

EPRI. 2012. *Long-term performance of commercial ground source heat pumps*. Final Report Draft, EP-P40851/EP-P40852. Palo Alto, CA: Electric Power Research Institute.

GPI. 2014. Grout cost calculator. Elkton, SD: GeoPro, Inc.

Hinge, A.W., and D.J. Winston. 2009. Documenting performance. *High Performing Buildings*, Winter.

IWA. 2014. Heating & Air Conditioning / Geothermal Heat Pump. Ingram's Water and Air Equipment, Paducah, KY. http://ingramswaterandair.com/heating-conditioning-geothermal-heat-pump-c-45_82.html

Kavanaugh, S.P., and D.R. Dinse. 2013. Long-term commercial GSHP performance, part 6: Maintenance and controls. *ASHRAE Journal* 55(1).

Kavanaugh, S.P., and J.S. Kavanaugh. 2012a. Long-term commercial GSHP performance, part 1: Project overview and loop circuit types. *ASHRAE Journal* 54(6).

Kavanaugh, S.P., and J.S. Kavanaugh. 2012b. Long-term commercial GSHP performance, part 2: Ground loops, pumps, ventilation air, controls. *ASHRAE Journal* 54(7).

Kavanaugh, S.P., and J.S. Kavanaugh. 2012c. Long-term commercial GSHP performance, part 3: Loop temperatures. *ASHRAE Journal* 54(9).

Kavanaugh, S.P., and J.S. Kavanaugh. 2012d. Long-term commercial GSHP performance, part 5: Comfort and satisfaction. *ASHRAE Journal* 54(12).

Kavanaugh, S.P., and L. Meline. 2013. Long-term commercial GSHP performance, part 7: Achieving quality. *ASHRAE Journal* 55(2).

Kavanaugh, S.P., M. Green, and K. Mescher. 2012. Long-term commercial GSHP performance, part 4: Installation costs. *ASHRAE Journal* 54(10).

RSMeans. 2014. *RSMeans Mechanical Cost Data 2014*. Norwell, MA: Reed Construction Data.

TCI. 2011. HDPE vault quotation, TR2057. Submitted to Tri-County Irrigation, Goodfield, IL.

Zimmerman, D.R. 2000. *Documentation of operation, maintenance & construction cost of geothermal heat pump systems in schools*. Final Report, EP-P3128/C1476. Palo Alto, CA: Electric Power Research Institute.

A

Conversion Factors

Figure A.1 is a screenshot of UnitsConverter.xlsx, which is available with this book at www.ashrae.org/GSHP. The spreadsheet enables quick conversion of units from I-P to SI units and vice versa. There is instruction on how to use the spreadsheet available in the Excel file. Figure A.1 is useful for manual conversion of units.

Input Values in Yellow Cells - Read Converted Units in Blue Cells								
	I-P to SI				SI to I-P			
Temperature	68	°F =	20.0	°C	0	°C =	32.0	°F
Length	100	ft =	30.5	m	100	m =	328.1	ft
Length	1.00	in =	2.54	cm	1.00	cm =	0.394	in
Distance	1.00	mi =	1.61	km	1.00	km =	0.62	mi
Mass	1.00	lb =	0.45	kg	1.00	kg =	2.20	I2
Area	1,000	ft^2 =	93	m^2	1000	m^2 =	10764	ft^2
Volume	1000.0	ft^3 =	28.3	m^3	100	m^2 =	3531.0	ft^2
Volume	100.0	gal =	378.5	litres	100	litres =	26.4	gal
Power	1	hp =	0.75	kW	1	kW =	1.34	hp
Density	1.00	lb/ft^3 =	16.02	kg/m^3	1000.00	kg/m^3 =	62.44	lb/ft^3
Velocity	1000	fpm =	5.080	m/s	1	m/s =	197	fpm
Pressure	100	psi =	689.5	kPa	100	kPa	14.5	psi
External Static Presure	1	in. of water =	249.0	Pa	1000	Pa =	4.02	in. of water
Liquid head	100	ft of water =	30.5	m of water	100	m of water =	328.1	ft of water
Head-Pressure	100	ft. water =	299.9	kPa	100	kPa	33.3	ft. water
Head-Pressure	3	ft. water/100'	0.295	kPa/m	1	kPa/m	0.33	ft. water/100'
Specific Heat	1.00	Btu/lb-°F	4.19	kJ/kg-K	1.00	kJ/kg-K	0.24	Btu/lb-°F
Therm Cond	1.000	Btu/h-ft-°F =	1.731	W/m-K	1.000	W/m-K =	0.578	Btu/h-ft-°F
Thermal Resistance	1.000	h-ft-°F/Btu =	0.578	m-K/W	1.000	m-K/W =	1.731	h-ft-°F/Btu
Heat Transfer Coefficient	100.0	Btu/h-ft^2-°F =	567.8	W/m^2-K	100.0	W/m^2-K =	17.6	Btu/h-ft^2-°F
Hydraulic Conductivity (permeability)	1.000	gal/ft^2-day	0.041	m/day	0.010	m/day	0.246	gal/ft^2-day
Air Flow	1000	cfm =	472.0	Lps	1000	Lps =	2118.6	cfm
Air Flow	1000	cfm =	1699.2	m^3/hr	1000	m^3/hr	588.5	cfm
Water Flow	1	gpm =	3.79	Lpm	1	Lpm =	0.264	gpm
Water Flow	1000	gpm =	63.09	Lps	1	Lps =	15.9	gpm
Water Flow	100	gpm =	22.7	m^3/hr	100	m^3/hr =	440.2	gpm
Specific capacity (water well)	1.000	gpm/ft =	0.021	Lps/m	0.0100	Lps/m =	0.483	gpm/ft
Loop thermal mass	1.00	gal/ton =	1.07	L/kW	1.00	L/kW =	0.93	gal/ton
Acerage to sq. ft.	1	acres	43560	ft^2	43560	ft^2	1.0	acres
Acerage	1	acres	4047	m^2	1000	m^2	0.247	acres
Acerage	1000	acres	405	ha (10^4 m^2)	1000	ha =	2470	acres
Cooling load/Capacity	1	Ton =	3.52	kW	1	kW =	0.284	Ton
Cooling load/Capacity	1000	kBtu/h =	293.1	kW	1000.0	kW =	3412.0	kBtu/h
Cooling load/Lake area	10	Ton/acre =	86.9	kW/ha	10	kW/ha =	1.2	Ton/acre
Cooling load/Face Area	10.0	kBtu/h-ft^2	31.5	kW/m^2	100.0	kW/m^2	31.7	kBtu/h-ft^2
Cooling Eff. EER←→COP	10.00	Btu/W-h =	2.93	W/W	1.00	W/W =	3.41	Btu/W-h
Cooling Eff. kW/ton←→COP	1.00	kW/Ton =	3.52	W/W	1.00	W/W =	3.52	kW/Ton
Bore Length/hp cap.	200	ft/ton =	17.3	m/kW	10	m/kW =	115.4	ft/ton
Feet per ton to W/m	200	ft/ton =	57.7	W/m	10	W/m	1153.9	ft/ton
Bore Length/floor area	1.00	ft/ft^2 =	10.76	m/m^2	10.00	m/m^2 =	0.93	ft/ft^2
Pump power/hp cap.	10	hp/100 ton =	2.12	kW$_m$/100 kW$_t$	1	kW$_m$/100 kW$_t$ =	4.7	hp/100 ton
Pump power/hp cap.	10	hp/100 ton =	21.2	W$_m$/kW$_t$	1	kW$_m$/100 kW$_t$ =	4.7	hp/100 ton
Floor Area/capacity	100	ft^2/ton =	2.64	m^2/kW	10.00	m^2/kW =	378	ft^2/ton
Water Flow/capacity	1	gpm/ton =	1.08	Lpm/kW	1	Lpm/kW	0.93	gpm/ton
Demand/Floor area	10.00	W/ft^2 =	107.60	W/m^2	10.00	W/m^2 =	0.93	W/ft^2
Air Flow/capacity	400	cfm/ton =	53.7	Lps/kW	100	Lps/kW	745.1	cfm
Solar Radiation	100	Btu/hr-ft^2=	0.315	kW/m^2	1.00	kW/m^2 =	317	Btu/hr-ft^2
Solar Radiation	500	Btu/day-ft^2=	5.7	MJ/day-m^2	10.0	MJ/day-m^2	881	Btu/day-ft^2=
Lake Flow	10000	ft^3/s	283.21	m^3/s	10	m^3/s =	353.1	ft^3/s
Header Cfactor Ht Loss	0.002	gpm/ft	0.00041	Lps/m	1	Lps/m	4.83243	gpm/ft
Fin pitch - fin spacing	12	fins/inch =	2.1	mm	1	mm =	25.4	fins/inch
Cost/Area	$10.00	$/ft^2 =	$107.64	$/m^2	$1.00	$/m^2 =	$0.09	$/ft
Cost/Length	$1.00	$/ft =	$3.28	$/m	$10.00	$/m =	$3.05	$/Ton
Cost/Cooling load	$10,000	$/Ton =	$2,843	$/kW	$1,000	$/kW	$284	
Cost/Volume	$10.00	$/gal =	$2.64	$/litre	$1.00	$/litre	$3.79	$/gal

Figure A.1 HVAC and GSHP Units Converter (UnitsConverter.xlsx)

B Standards and Recommendations for GSHP Components and Procedures

This appendix provides references to several publications and standards for procedures and specifications that are specific to the GSHP industry.

Ground Formation Thermal Property Test

- ASHRAE. 2011. *ASHRAE Handbook—HVAC Applications*. Chapter 34, Geothermal Energy, pp. 34.13–34.14. Atlanta: ASHRAE.
- IGSHPA. 2013. *Closed-Loop/Geothermal Heat Pump Systems: Design and Installation Standards (1B.3.1)*. Stillwater, OK: International Ground Source Heat Pump Association.

Ground Heat Exchanger Materials—Closed Loop

- IGSHPA. 2013. *Closed-Loop/Geothermal Heat Pump Systems: Design and Installation Standards (1C)*. Stillwater, OK: International Ground Source Heat Pump Association.
- PPI. 2011. *Handbook of Polyethylene Pipe*, 2d Ed. Irving, TX: Plastics Pipe Institute.

Ground Heat Exchanger Pipe Flushing, Purging, Pressure and Flow Testing—Closed Loop

- IGSHPA. 2013. *Closed-Loop/Geothermal Heat Pump Systems: Design and Installation Standards (1E)*. Stillwater, OK: International Ground Source Heat Pump Association.
- PPI. 2011. *Handbook of Polyethylene Pipe*, 2d Ed. Irving, TX: Plastics Pipe Institute.

Ground Heat Exchanger Pipe Joining Methods—Closed Loop

- IGSHPA. 2013. *Closed-Loop/Geothermal Heat Pump Systems: Design and Installation Standards (1D)*. Stillwater, OK: International Ground Source Heat Pump Association.
- PPI. 2011. *Handbook of Polyethylene Pipe*, 2d Ed. Irving, TX: Plastics Pipe Institute.

Ground Heat Exchanger Vertical Borehole Placement and Backfilling—Closed Loop

- IGSHPA. 2013. *Closed-Loop/Geothermal Heat Pump Systems: Design and Installation Standards (2B.1)*. Stillwater, OK: International Ground Source Heat Pump Association.

Ground-Source Heat Pumps

- Table 2.8, Recommended Minimum Allowable Heat Pump Efficiencies
- ASHRAE. 2012. ANSI/AHRI/ASHRAE ISO Standard 13256-1: 1998 (RA 2012), *Water-Source Heat Pumps-Testing and Rating for Performance—Part 1: Water-to-Air and Brine-to-Air Heat Pumps*. Atlanta: ASHRAE.
- ASHRAE. 2012. ANSI/AHRI/ASHRAE ISO Standard 13256-2: 1998 (RA 2012), *Water-Source Heat Pumps Testing and Rating for Performance—Part 2: Water-to-Water and Brine-to-Water Heat Pumps.* Atlanta: ASHRAE.

Groundwater Piping

- AWWA. 2007. AWWA C900-07, *Polyvinyl Chloride (PVC) Pressure Pipe and Fabricated Fittings 4 in. through 12 in. (100 mm through 300 mm), for Water Transmission and Distribution*. Denver: American Water Works Association.

Water Wells

- NGWA. 2014. ANSI/NGWA 01-14, *Water Well Construction Standard*. Westerville, OH: National Ground Water Association.
- AWWA. 1997. AWWA A100-97, *Standard for Water Wells*. Denver: American Water Works Association.

C Pressure Ratings and Collapse Depths for Thermoplastic Pipe

C.1 HIGH-DENSITY POLYETHYLENE PIPE PRESSURE RATINGS

High-density polyethylene (HDPE) pipe with designation PE 4710 is currently the customary pipe for ground heat exchangers. It has higher pressure ratings than PE 3406/3408, which previously was the more commonly used pipe. Table C.1 provides pressure ratings for HDPE PE 4710 pipe at various dimension ratios (DRs) and temperatures, and Table C.2 provides pressure ratings for HDPE PE 3406/3408 pipe at various DRs and temperatures.

C.2 FIBERGLASS-CORE POLYPROPYLENE PIPE PRESSURE RATINGS

Polypropylene pipe with a fiberglass core is currently available for the interior piping of GSHP systems. The pipe has much lower expansion coefficients than HDPE pipe and will not deform to the extent of HDPE pipe runs. Like HDPE, polypropylene has good chemical and corrosion resistance, so inhibitor requirements are substantially reduced compared to metal piping. This provides an advantage in jurisdictions that impose limits on chemical treatments for below-grade piping applications. Also, the recommended joining practice is thermal fusion, which offers substantial integrity in applications with GSHP temperature swings compared to glued or screwed plastic pipe. Fusion joints for pipe diameters of 4 in. (110 mm) or less are typically made by socket fusion and larger diameters are made by butt fusion. Note also the pressure ratings at higher temperatures are improved compared to those of HDPE. Installation cost is comparable to groove-joint steel, higher than HDPE, and lower than welded steel piping. Table C.3 provides pressure ratings for fiber-core polypropylene pipe at various DRs and temperatures.

C.3 HDPE PIPE COLLAPSE DEPTHS

The collapse of HDPE piping due to external pressures of high-density grouts has not been an issue with optimum bore depths less than 300 ft (90 m). But bore depths have increased at sites with higher loads to available land area. Equations are available to predict external pressures (or bore depths) that would cause pipe collapse when the density

Table C.1 Pressure Ratings for HDPE PE 4710 Pipe

Temperature, °F	Pressure Rating, psig					
	Dimension Ratio (DR)					
	17	15.5	13.5	11	9	7
30	220	222	258	323	403	538
40	176	195	226	282	353	470
73.4	126	139	161	202	252	336
110	95	104	121	151	189	252
140	63	70	81	101	126	168
Temperature, °C	**Pressure Rating, kPa (gage)**					
	Dimension Ratio (DR)					
	17	**15.5**	**13.5**	**11**	**9**	**7**
–1.1	1517	1531	1779	2227	2779	3710
4.4	1214	1345	1558	1944	2434	3241
23.0	869	958	1110	1393	1738	2317
43.3	655	717	834	1041	1303	1738
60.0	434	483	558	696	869	1158

Table C.2 Pressure Ratings for HDPE PE 3406/3408 Pipe

Temperature, °F	Pressure Rating, psig					
	Dimension Ratio (DR)					
	17	15.5	13.5	11	9	7
30	160	177	205	256	320	427
40	140	154	179	224	280	373
73.4	100	110	128	160	200	267
110	75	83	96	120	150	200
140	50	55	64	80	100	133
Temperature, °C	**Pressure Rating, kPa (gage)**					
	Dimension Ratio (DR)					
	17	**15.5**	**13.5**	**11**	**9**	**7**
–1.1	1103	1220	1413	1765	2206	2944
4.4	965	1062	1234	1544	1931	2572
23.0	690	758	883	1103	1379	1841
43.3	517	572	662	827	1034	1379
60.0	345	379	441	552	690	917

of the external fill material is greater than that of the fluid inside the pipe. However, these equations apply to liquids, and most grouts typically set up after a short period of time to a consistency of peanut butter. It has not been well researched, but it is likely the equations may be somewhat conservative. They are presented here since caution is advised when installing loops to depths greater than 300 ft/90 m. (This caution is coupled with the recommendation to install bores with greater separation distance to minimize cross-drilling when boring greater than 400 ft [120 m]. A 25 ft [7.5 m] separation is suggested, but 20 ft [6 m] is the absolute minimum.)

In rare cases, manufacturing defects at extrusion facilities have resulted in U-tube coils being supplied with thinner-than-specified pipe walls. It is highly recommended

Table C.3 Pressure Ratings* for Fiber-Core Polypropylene Pipe

Temperature, °F	Pressure Rating, psig			
	Dimension Ratio (Available Diameters)			
	17 (6 to 10 in.)	11 (1 1/4 to 10 in.)	9 (1 to 4 in.)	7.4 (1/2 to 3/4 in.)
73	139	220	280	350
140	71	115	145	183
180	50	78	100	120
	50-Year Pressure Rating, 1.5 Safety Factor			
Temperature, °C	**Pressure Rating, kPa (gage)**			
	Dimension Ratio (Available Diameters)			
	17 (160 to 250 mm)	**11 (40 to 250 mm)**	**9 (32 to 110 mm)**	**7.4 (20 to 25 mm)**
22.8	958	1517	1931	2413
60.0	490	793	1000	1262
82.2	345	538	690	827
	50-Year Pressure Rating, 1.5 Safety Factor			

Ratings may vary for different manufacturers.

Table C.4 Apparent HDPE Elastic Modulus at 73.4°F (23°C) (PPI 2011)

Load Duration	PE 3xxx		PE4xxx		Temperature, °F (°C)	Compensation Factor C_T
	psi	MPa	psi	MPa		
1 h	74000	510	78000	538	40 (4)	1.49
10 h	62000	428	65000	448	60 (16)	1.18
100 h	52000	359	55000	379	73.4 (23)	1
1000 h	44000	303	46000	317	80 (27)	0.93
1 year	38000	262	40000	276	100 (38)	0.73
10 years	32000	221	34000	234	120 (49)	0.58

that coils being installed into deep boreholes be checked at the site for proper wall thickness and ovality (a.k.a. "out-of round") before installation.

The allowable unconstrained pipe wall buckling pressure (P_{wu}) is

$$P_{wu} = \frac{f_o \times 2E}{N_s \times (1 - \mu^2)} \times \left(\frac{1}{\mathrm{DR} - 1}\right)^3 \times C_T \qquad \text{(C.1)}$$

where

f_o = ovality factor
E = apparent modulus of elasticity at 73.4°F (23°C) from Table C.4
N_s = safety factor
μ = Poisson's ratio (0.45 for HDPE)
DR = dimension ratio
C_T = temperature compensation factor

The values in Table C.4 are used in Equation C.1 to find the unconstrained pipe wall buckling pressure (P_{wu}).

The ovality factor (f_o) is a function of the deflection percentage (DP):

$$\text{DP (\%)} = (d_i - d_{Min})/d_i \tag{C.2}$$

where

d_i = inside diameter of round pipe

d_{Min} = minimum inside diameter of out-of-round pipe (with no internal or external pressure)

The following are values for f_o based on DP values:

$f_o = 1.0$ for DP = 0%
$f_o = 0.85$ for DP = 2%
$f_o = 0.70$ for DP = 4%
$f_o = 0.54$ for DP = 6%
$f_o = 0.42$ for DP = 8%
$f_o = 0.36$ for DP = 10%

Pipe buckling is possible when the external pressure on the pipe resulting from the grout ($P_{ext} = \rho_{grout} \times$ depth) exceeds the pressure inside the pipe resulting from the fill water ($P_{int} = \rho_{water} \times$ depth). Thus, the buckling depth is as follows if the pipe is not pressurized:

$$\text{Buckling depth} = P_{wu} / (\rho_{grout} - \rho_{water}) \tag{C.3}$$

Equation C.3 can be corrected for additional pipe pressure at the surface (P_{add}) as shown in Equation C.4:

$$\text{Buckling depth} = (P_{wu} + P_{add}) / (\rho_{grout} - \rho_{water}) \tag{C.4}$$

EXAMPLE C.1— CALCULATION OF PIPE BUCKLING DEPTH

Find the buckling depth of a DR 11, PE 4710 HDPE pipe at 80°F (27°C) for a thermally enhanced grout with a density of 12.5 lb/gal (1496 kg/m^3) (see Table 3.2). Use a safety factor of 1.5, no additional pressure, and 2% ovality, and assume the grout stays in a liquid form for 1 hour.

Solution

$$P_{wu} = \frac{f_o \times 2E}{N_s \times (1 - \mu^2)} \times \left(\frac{1}{\text{DR} - 1}\right)^3 \times C_T$$

$$= \frac{0.85 \times 2 \times 78000 \text{ psi}}{1.5 \times (1 - 0.45^2)} \times \left(\frac{1}{11 - 1}\right)^3 \times 0.93 \qquad \text{(I-P)}$$

$$= 103 \text{ psi}$$

$$P_{wu} = \frac{f_o \times 2E}{N_s \times (1 - \mu^2)} \times \left(\frac{1}{\text{DR} - 1}\right)^3 \times C_T$$

$$= \frac{0.85 \times 2 \times 538{,}000 \text{ kPa}}{1.5 \times (1 - 0.45^2)} \times \left(\frac{1}{11 - 1}\right)^3 \times 0.93 \qquad \text{(SI)}$$

$$= 710 \text{ kPa} = 710{,}000 \text{ kg/m·s}^2$$

$$P_{wu} = \frac{f_o \times 2E}{N_s \times (1 - \mu^2)} \times \left(\frac{1}{\mathrm{DR} - 1}\right)^3 \times C_T$$

$$= \frac{0.85 \times 2 \times 538{,}000 \text{ kPa}}{1.5 \times (1 - 0.45^2)} \times \left(\frac{1}{11 - 1}\right)^3 \times 0.93 \quad \text{(SI)}$$

$$= 710 \text{ kPa} = 710{,}000 \text{ kg/m·s}^2$$

The densities of the water at 80°F (27°C) and of the grout are found in Table 3.2:

$$\rho_{water} = 62.2 \text{ lb/ft}^3 \text{ (994 kg/m}^3\text{)}$$

$$\rho_{grout} = 12.5 \text{ lb/gal} \times 7.48 \text{ gal/ft}^3 = 93.5 \text{ lb/ft}^3 \text{ (1496 kg/m}^3\text{)}$$

Thus,

$$\text{Buckling depth} = P_{wu} / (\rho_{grout} - \rho_{water})$$
$$= 103 \text{ lb/in.}^2 \times 144 \text{ in.}^2\text{/ft}^2 \div (93.5 \text{ lb/ft}^3 - 62.2 \text{ lb/ft}^3)$$
$$= 474 \text{ ft} \quad \text{(I-P)}$$

(Purists would include the terms $g/g_c = 32.2 \text{ ft/s}^2 \div 32.2 \text{ lb}_m\text{·ft/lb}_f\text{·s}^2$.)

$$\text{Buckling depth} = P_{wu} / (\rho_{grout} - \rho_{water})$$
$$= 710{,}000 \text{ kg/m·s}^2 \div [(9.81 \text{ m/s}^2) \times (1496 \text{ kg/m}^3 - 994 \text{ kg/m}^3)]$$
$$= 144 \text{ m} \quad \text{(SI)}$$

C.4 REFERENCES

Carda, R. 2014. Construction docs for closed loop ground heat exchanger: System installation meet design intent. Presented in Seminar 13 at the ASHRAE Annual Conference, Seattle, WA, June 28–July 2.

PPI. 2011. *Handbook of Polyethylene Pipe*, 2d Ed. Irving, TX: Plastics Pipe Institute.

D Vertical-Loop Installation Equipment and Procedures

D.1 VERTICAL-LOOP DRILLING METHODS

Figure D.1 shows the typical details of a small rotary drilling rig that is well suited to vertical ground-coupled heat pump (GCHP) loop installation. Rigs must be powerful enough to drill through difficult formations and yet small and flexible enough to quickly move from borehole to borehole. Recall that recommended borehole sizes range between 3 1/2 to 5 1/4 in. (9 to 13 mm) in diameter; therefore, a very powerful top-head drive is not necessarily required. However, a powerful mud pump (5 × 6 in. [13 × 15 mm] minimum) is suggested since it will enhance the drill bit's cutting capability and rapidly remove the cuttings. Most loop contractors prefer a minimum drill stem length of 20 ft (6 m) since the time spent adding (and removing) drill stems is often a very large percentage of the total drilling time.

As noted in Appendix J, for many unconsolidated formations, drilling mud must be added into the pit to prevent borehole collapse. If drilling mud is added, the drilling fluid will be pumped into the formation and will not return to the mud pit. Thus, circulation will be lost. Drilling mud is normally a bentonite clay that forms a thin, temporary clay wall at the outside surface of the bore.

Mud rotary is the preferred drilling method in clay, sands, and some soft rocks. In harder formations many drillers prefer to use air rotary drilling. In this method, the mud pump is replaced with a large compressor to remove the cuttings. An advantage of rotary drilling is that the mud pit is no longer required. Cuttings are blown to the top of the borehole and form a large "ant bed" type of pile. Soft- and medium-hardness rock is typically drilled with a roller-cone bit as shown in Figure D.1 (although drag bits have also been used). Harder formations often require downhole hammers (also shown in Figure D.1) to attain acceptable penetration rates.

Many loop contractors prefer rock to softer unconsolidated formations because of the absence of the "mess" of drilling mud. However, one of the more undesirable formations is the combination of the two. For example, an unconsolidated formation that requires drilling mud may occur to a depth of 50 ft (15 m) with hard rock beneath that is difficult to drill with mud rotary. If the driller switches to air at 50 ft (15 m), the bore wall may collapse on top of the drill string before it can be removed, and the hole will fill back in before the U-tube can be inserted. In some cases foams can be used with air to create the same effect as drilling mud. However, in many cases temporary or permanent casing must be inserted to prevent collapse. This is not only costly, but it also slows loop installation speed.

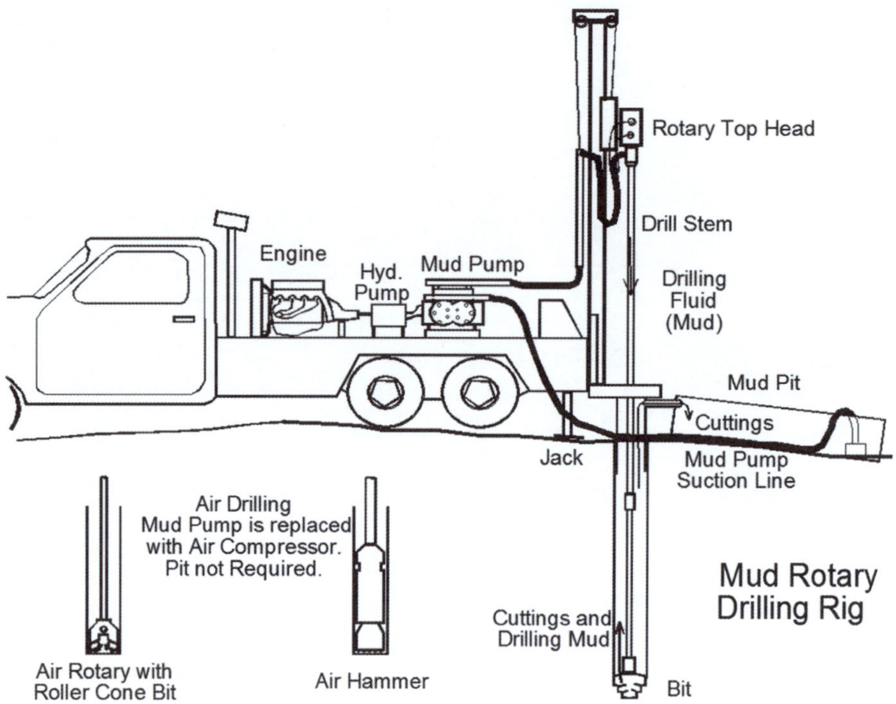

Figure D.1 Small Rotary Drilling Rig for Vertical-Loop Installation

D.2 VERTICAL-LOOP INSTALLATION

Figure D.2 presents the completed installation of U-tubes in two different formations. The left U-tube is shown installed in an unconsolidated formation, typically drilled with a mud rotary rig. The concern is often keeping the hole open enough to insert the U-tube. This may require some method of pushing the loop into the formation, which is often full of drilling fluid and some cuttings. Many installers tape a 3 to 5 ft (1 to 1.5 m) section of rebar or scrap metal to the side of the U-tube to keep the "curved" U-tube straight. Others may use some type of removable "pusher" bar in addition to the rebar.

The U-tube on the right of Figure D.2 is installed in a consolidated formation, typically drilled with an air rotary or air hammer drilling rig. The formation is relatively clear of cuttings, and the concern is the restriction of the water-filled U-tube when plated in the borehole. Many contractors have developed reels or carts to handle the insertion of the U-tubes in both consolidated and unconsolidated formations. One type is shown in Figure D.3. The cart assists in both the insertion and the handling of the U-tubes on the job site.

D.3 VERTICAL-LOOP BACKFILL AND GROUTING

The annular space between the tubing and the borehole wall must be filled to 1) ensure good heat transfer from the loop to the ground and 2) prevent flow of contaminated water from the surface (or from a contaminated aquifer) to the groundwater. Unfortunately, these goals are sometimes at cross purposes, because water movement is a primary mode of heat transfer. Traditional water well sodium bentonite grouts are poor

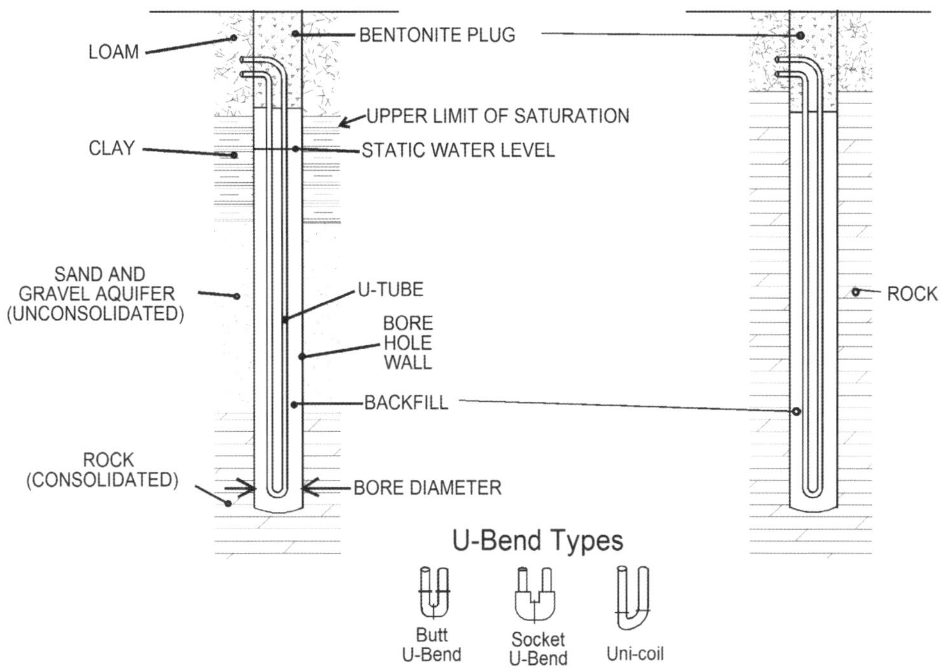

Figure D.2 Completed U-Tube Heat Exchangers

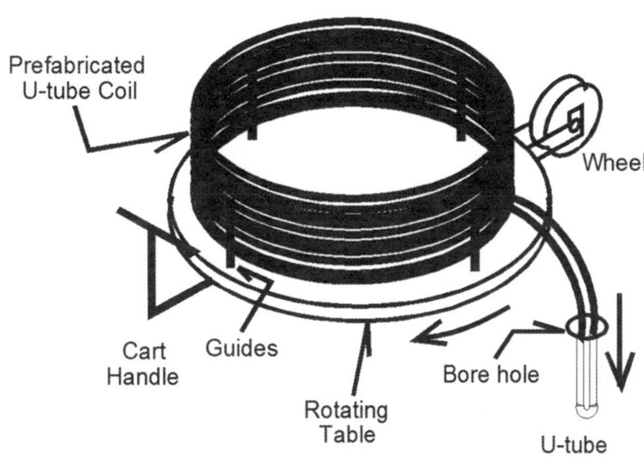

Figure D.3 "Lazy Susan" Cart for Handling U-Tube Coils at Construction Site

conductors of heat, and the bore annulus is in a critical high-heat-flux location. The placement of these grouts from the bottom to the top of the bore will result in poor system efficiency and/or much longer required loop lengths. Table 3.2 provides recipes for thermally enhanced grouts that provide positive seals with improved thermal performance. The traditional method of thermally enhancing grouts consists of one part sodium bentonite with two to eight parts of silica sand. More recently, combinations of sodium bentonite and graphite have provided equivalent thermal performance without the high weight and volume requirements of sand, which have been major complaints voiced by installers.

Figure D.4 shows a typical rig for grouting the boreholes. The pump shown is used to inject sand slurries, thermally enhanced grouts, and other backfills that are conducive to good heat transfer. Note that low-permeability grouts should always be placed in the

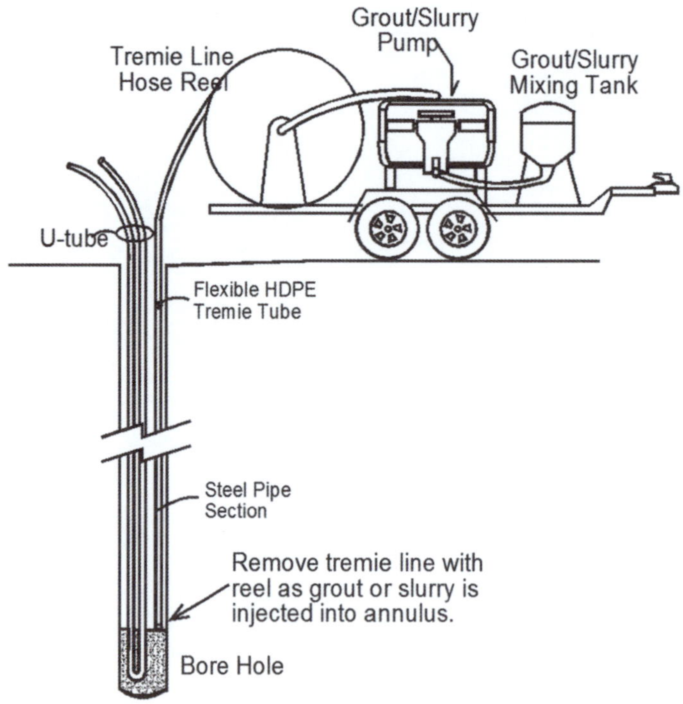

Figure D.4 Backfill/Grouting the Borehole Annulus

Figure D.5 Grout Mixer and Pump

upper portion of the borehole and above and below contaminated groundwater aquifers. Figure D.5 shows a grout mixer and pump.

Designers and regulators are encouraged to consult installation and grouting guidelines and standards published by the National Ground Water Association (NGWA 2010) and the International Ground Source Heat Pump Association (IGSPHA 2000). Table D.1 is provided to assist in the estimation of required grout/backfill volumes.

Table D.1 Grout Volumes Required to Fill U-Tube Bores (0% Waste)

U-Tube Nominal Diameter, in.	Gallons of Grout/Fill Required per 100 ft of Bore								
	Bore Diameter, in.								
	3	3.5	4	4.5	5	5.5	6	6.5	7
3/4	28	41	56	74	93	114	138	163	191
1		36	51	69	88	109	133	158	186
1 1/4			43	60	80	101	124	150	177
1 1/2				53	73	94	117	143	170

U-Tube Diameter, mm	Litres of Grout/Fill Required per 100 m of Bore								
	Bore Diameter, mm								
	80	90	100	110	120	130	140	150	160
25	404	538	687	852	1033	1229	1441	1669	1912
32		475	625	789	970	1166	1379	1606	1850
40			534	699	880	1076	1288	1516	1759
50					738	935	1147	1374	1618

D.4 REFERENCES

IGSHPA. 2000. *Grouting for Vertical GHP Systems*. International Ground Source Heat Pump Association. Stillwater, OK.

NGWA. 2010. *Guidelines for the Construction of Loop Wells for Vertical Closed Loop Ground Source Heat Pump Systems*. Columbus, OH: National Ground Water Association.

E Example of Field Study Results

E.1 COUNTY WATER AGENCY OPERATIONS AND MAINTENANCE OFFICE

A 20,600 ft^2 (1910 m^2) office building was retrofitted in 2010 with a 61 ton (215 kW) GSHP system. The 11 water-to-air heat pumps that serve the building are connected to a one-pipe interior piping distribution system. The ground heat exchanger consists of 32 400 ft (120 m) vertical bores with 1 1/4 in. (40 mm) HDPE U-tubes that are completed with thermally enhanced bentonite grout. The 32 ground heat exchangers are divided into four circuits, each with eight vertical bores and 2 in. (63 mm) supply and return headers. The circuit headers are routed into the mechanical room as shown in Figure E.1. Figure E.2 shows the two 1.5 hp (1.1 kW) pumps in a lead-lag control operation. A third primary 1/6 hp (0.12 kW) pump is used during unoccupied periods, primarily to meet the needs of a computer network that operates 24 hours per day. The control sequence calls for only one pump to operate. Ventilation air is provided via a 1500 cfm (2550 m^3/h) energy recovery ventilator (ERV).

In this piping system, a single pipe serves as the supply and return. A secondary circulator pump located inside the heat pump cabinet is activated when the unit calls for heating or cooling. After leaving the heat pump, the liquid is returned to the one-pipe loop downstream from the intake.

Figure E.3 demonstrates the ground-loop operating temperatures on a warm day late in the cooling season. The ground-loop leaving water temperature (LWT) (entering the heat pumps) was 76°F (24°C) in the morning and rose to 82°F (28°C) in the late afternoon. Although the system design conforms to recommendations in this book and the ground-loop temperatures were very good, the building did not attain ENERGY STAR designation. The design engineer of record investigated the operation of the system and discovered that the prescribed sequence of operation had been defeated and one of the larger primary pumps was operating during unoccupied periods. This is confirmed by noting the low differential temperatures in Figure E.3. A continuously operating primary pump consumes approximately 10,000 kWh per year. The sequence was returned to the design recommendation, and the ENERGY STAR rating will be recalculated after one year of operating results are obtained.

The following sections detail the specifics of the field study results.

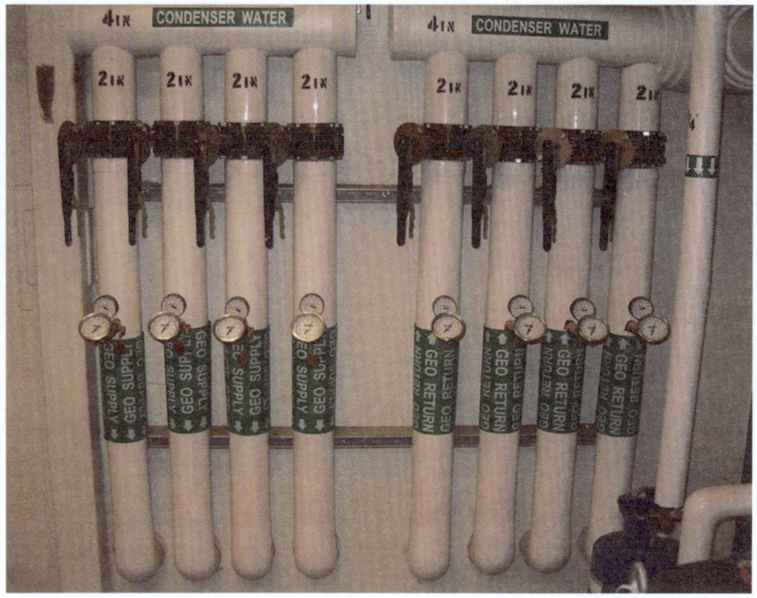

Figure E.1 Ground-Loop Headers

Figure E.2 Three Primary Pumps

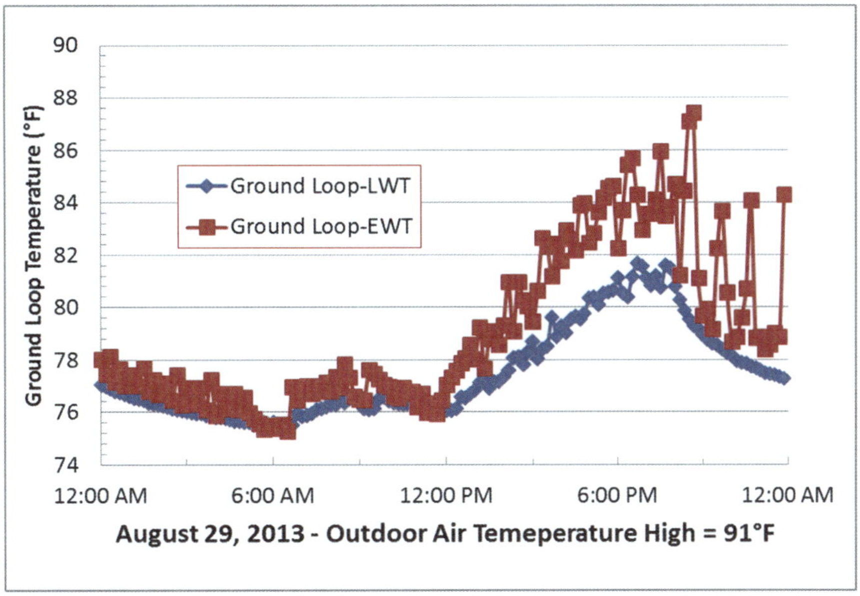

Figure E.3 Office Building Ground-Loop Temperatures on a Warm Day

E.1.1 Building and System Information

- 20,600 ft^2 (1910 m^2) office building
- Eleven heat pumps with a total capacity of 61 tons (215 kW)
- Three primary pumps (two at 1.5 hp [1.1 kW], one at 1/6 hp [0.12 kW], with on-off/lead lag control)
- Eleven secondary pumps: three at 1/6 hp (0.12 kW), seven at 1/12 hp (0.06 kW), and one at 1/25 hp (0.03 kW)
- Interior central/one-pipe loop
- Vertical ground loop: 32 400 ft (120 m) × 1 1/4 in. (40 mm) DR 11 HDPE U-tubes
- Bores on 20 ft (6 m) minimum spacing with thermally enhanced bentonite grout
- Outdoor air ventilation introduced through the wall into heat pump cabinet

E.1.2 System Metrics

- Loop = 210 ft/ton (55 W/m)
- Pump = 4.3 hp/100 tons (9.1 W_e / kW_t)
- Building area to cooling load = 338 ft^2/ton (8.9 m^2/kW)

E.1.3 Comfort and Satisfaction Survey Results

Surveys were provided to building occupants; they were completed by 11 people. Results on a scale of 1 (very dissatisfied) to 5 (very satisfied) are as follows:

- Indoor temperatures:
  - Cooling: 3.8/5
  - Heating: 3.4/5
- Air quality: 4.2/5
- Lighting: 3.8/5
- Acoustics: 3.7/5
- Maintenance responsiveness: 3.8/5
- Access to controls: 3.6/5

F Properties of Antifreeze Solutions

Table F.1 Properties of Antifreeze Solutions

Fluid	Solution Volume,	Freeze Point*		Viscosity (*cp*)			Density					
				32°F	59°F	86°F	lb/ft^3	kg/m^3	lb/ft^3	kg/m^3	lb/ft^3	kg/m^3
	%	°F	°C	0°C	15°C	30°C	32°F	0°C	59°F	15°C	86°F	30°C
Water	0	32	0	1.79	1.14	0.80	62.4	998	62.3	997	62.1	994
Ethanol	10	25	-4	3.00	1.67	1.09			61.4	982		
Ethanol	20	17	-8	4.62	2.32	1.42			60.7	971		
Ethylene glycol	10	25	-4	2.09	1.37	0.97	63.6	1018	63.4	1014	63.1	1010
Ethylene glycol	20	16	-9	3.03	1.89	1.31	64.7	1035	64.5	1032	64.1	1026
Ethylene glycol	30	3.5	-16	3.17	2.54	1.70	65.7	1051	65.4	1046	65.1	1042
Methanol	10	22	-6	2.44	1.48	0.99			61.4	982		
Methanol	20	11	-12	3.02	1.77	1.15			60.9	974		
Propylene glycol	10	26	-3	2.70	1.63	1.11	63.4	1014	63.1	1010	62.8	1005
Propylene glycol	20	19	-7	4.07	2.37	1.52	64.1	1026	63.8	1021	63.4	1014
Propylene glycol	30	10	-12	7.10	3.70	2.20	64.8	1037	64.4	1030	64.0	1024

**Freeze point values are for pure fluids and vary depending on inhibitor concentrations.*

G Volumes of Liquids in Pipe

Tables G.1a and G.1b provide volumes of liquid per length of pipe in I-P and SI units, respectively. They can be used to determine the volume of antifreeze solution required to obtain the various levels of freeze protection. Values can also be used to find the total volume of a liquid in a piping system. The thermal capacity of a system can be determined by multiplying the total volume by the liquid density and specific heat.

Table G.1a Gallons of Liquid per 100 Linear Feet of Pipe

Nominal Diameter, in.	Sch 40	Sch 80	DR 11	DR 13.5	DR 15.5	Copper, Type K	Copper, Type L	Cross-Linked Polyethylene (PEX) DR 9
3/4	2.8	2.2	3.0	3.3	3.4	3.1	2.3	2.5
1	4.5	4.9	4.7	5.1	5.4	5.2	4.0	4.3
1 1/4	7.8	6.7	7.5	8.2	8.5	7.7	6.3	6.5
1 1/2	10.6	9.2	9.9	10.7	11.2	10.8	8.9	9.2
2	17.4	15.3	15.4	16.7	17.5	18.4	15.7	16.1
2 1/2	25	22	NA	NA	NA	28	24	25
3	38	34	33	36	38	40	34	
4	66	60	55	60	63	69	61	
5	104	95	84	92	320	107	94	
6	150	135	120	130	136	153	134	
8	260	237	203	220	230	269	235	
10	410	373	316	342	358	418	364	
12	582	528	444	481	503	600	522	

Volume (gal/100 ft) = 4.08 × [ID (in.)]²

Table G.1b Litres of Liquid per 100 Linear Metres of Pipe

Nominal Diameter, mm	Outside Diameter, mm	Sch 10	Sch 40	Sch 80	Outside Diameter, mm	DR 9	DR 11	DR 13.5	DR 15.5
20	26.67	40	34	28	20	19	21	23	24
25	33.4	61	56	46	25	30	33	36	37
32	42.16	105	96	83	32	49	54	58	61
40	48.26	143	131	114	40	76	84	91	95
50	60.33	236	217	191	50	119	131	142	149
65	73.02	352	309	273	63	189	209	226	236
80	88.90	539	477	426	75	267	296	321	335
100	114.30	920	821	742	90	385	426	462	483
125	141.3	1436	1291	1174	110	575	636	690	721
150	168.27	2066	1864	1682	125	742	822	891	931
200	219.08	3515	3228	2946	160	1216	1346	1459	1525
250	273.05	5502	5087	4635	200	1900	2103	2280	2383
300	323.85	7779	7221	6557	250	2969	3286	3562	3724

Volume (L/100 m) = $(\pi/4) \times 100\ m \times 1000\ L/m \times [ID\ (mm) / 1000\ mm/m]^2 = 0.07854 \times [ID\ (mm)]^2$

H High-Density Polyethylene and Polypropylene Pipe Fusion Methods

This appendix presents the most successful pipe fusion methods for the most common GSHP pipe sizes and types.

Socket fusion is a useful method for 3/4 to 1 1/4 in. (25 to 40 mm) high-density polyethylene (HDPE) pipe since the equipment is small and manageable in confined spaces. Butt fusion is preferred for larger HDPE piping because the jig minimizes the handling effort and potential misalignment of larger-diameter pipe.

Socket fusion is the preferred method for 3/4 to 4 in. (25 to 110 mm) fiber-core polypropylene pipe. Butt fusion is used for larger polypropylene piping.

Electrofusion minimizes human error in the fusion process for both HDPE and fiber-core polypropylene piping. It is especially useful for repair, because the axial movement of pipe ends required by socket and butt fusion can be avoided. Fitting costs are significantly higher than socket and butt fusion fitting costs, however.

Figure H.1 shows socket fusion steps, Figure H.2 shows butt fusion steps for HDPE, and Figure H.3 shows electrofusion steps.

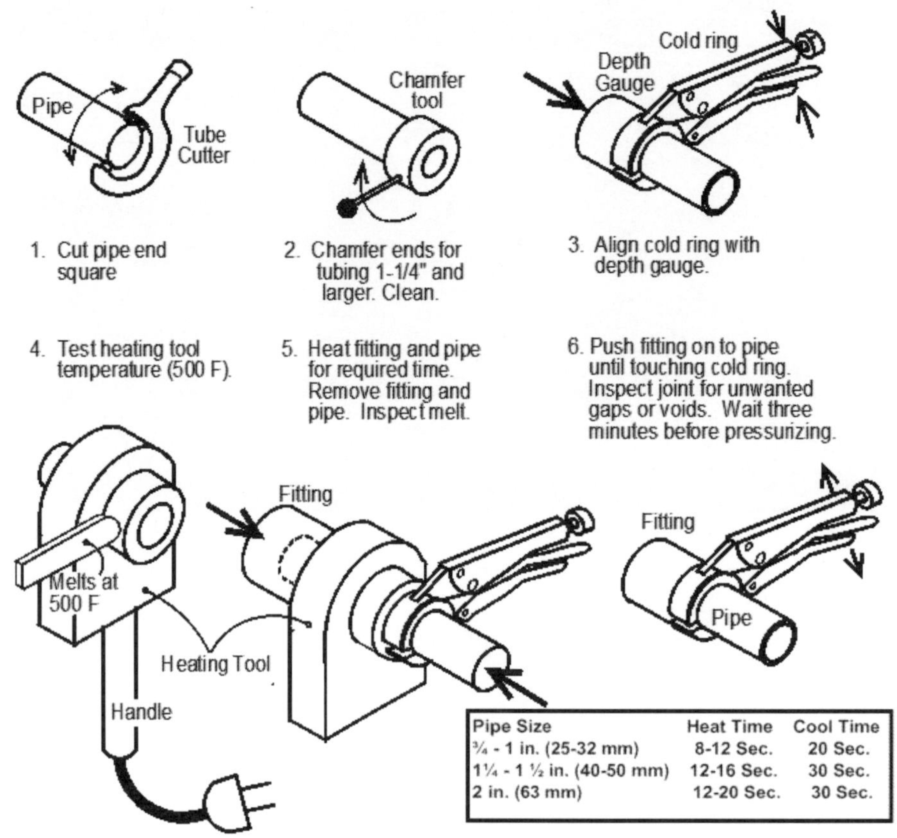

Pipe Size	Heat Time	Cool Time
¾ - 1 in. (25-32 mm)	8-12 Sec.	20 Sec.
1¼ - 1 ½ in. (40-50 mm)	12-16 Sec.	30 Sec.
2 in. (63 mm)	12-20 Sec.	30 Sec.

Figure H.1 Socket Fusion Procedure

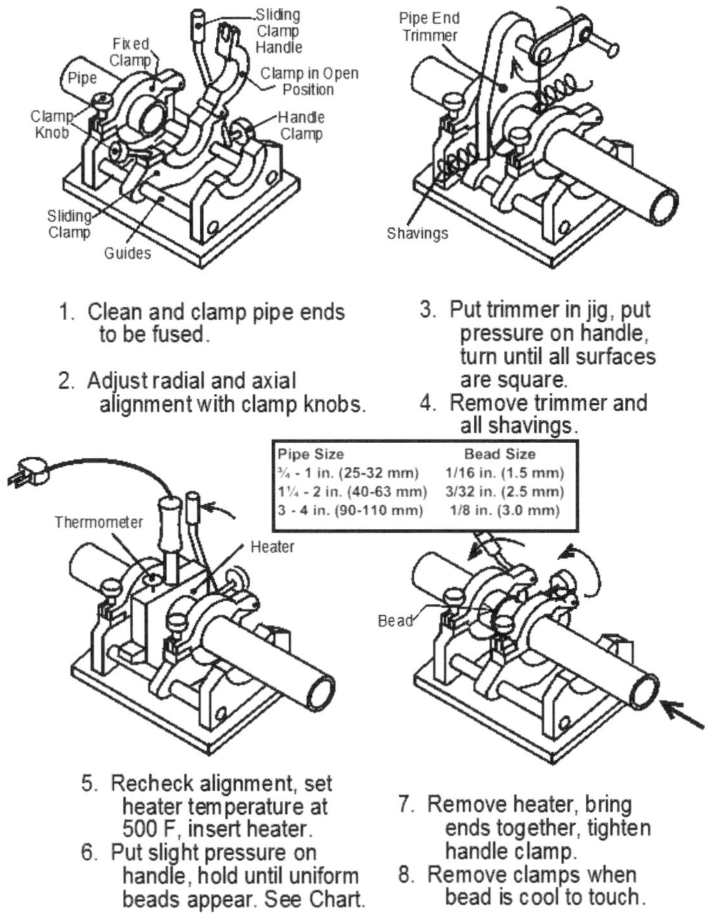

Pipe Size	Bead Size
¾ - 1 in. (25-32 mm)	1/16 in. (1.5 mm)
1¼ - 2 in. (40-63 mm)	3/32 in. (2.5 mm)
3 - 4 in. (90-110 mm)	1/8 in. (3.0 mm)

Figure H.2 HDPE Butt Fusion Procedure

1. Square ends of pipe, clean, remove burrs, and mark near required insertion depth.

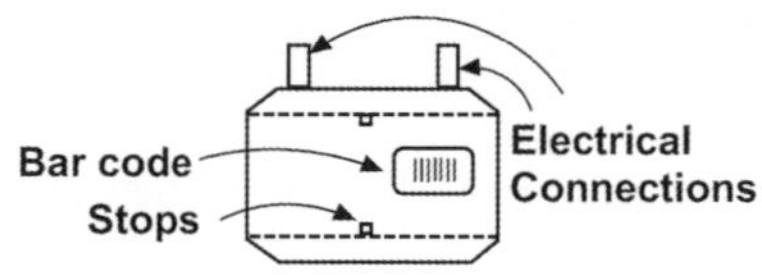

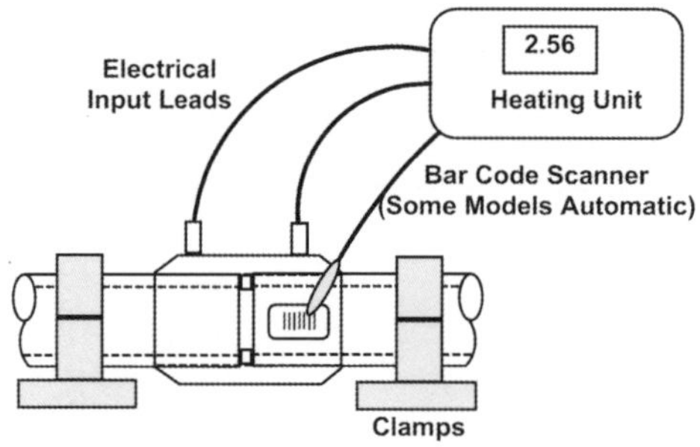

2. Insert pipe ends to stops, verify with marks, clamp pipes (if required), scan bar code (if required), activate heater, allow fitting to cool as recommended.

2a. When making repairs to existing pipes where axial movement of pipe is restricted, the stops can be removed so that fitting is moved over one pipe end, pipe ends aligned, and the fitting moved into place before fusing.

Figure H.3 Electrofusion Procedure

I Determination and Impact of Ground Coil Flow Imbalance

I.1 FLOW IMBALANCE IN CLOSED-LOOP GSHPs

The need for exact flow balancing in ground heat exchangers is less critical than in more compact heat exchangers such as fan-coils, tube-in-shell bundles, and heat pump units. It is typically unnecessary to resort to extreme measures such as specifying flow-balancing valves, circuit setters, or exact pipe lengths for individual U-tubes. These measures may make loop purging difficult and increase the potential for leaks and the cost of the ground loop.

The fundamental nature of ground-loop flow and heat transfer usually makes radical precautions unnecessary. Ground-loop heat transfer is not as dependent on fluid velocity and water-side temperature differences. The thermal resistance of the film at the water-to-pipe interface is only a small part of the overall thermal resistance of the ground loop, even when flow is laminar. The approach temperatures between the water and soil are very large. Thus, the water-side Δt, which changes with flow imbalances, does not greatly impact the log mean temperature difference (LMTD).

To demonstrate this concept, a detailed calculation was completed for the vertical ground heat exchangers shown in Figure I.1 in the predecessor to this book, *Ground-Source Heat Pumps: Design of Geothermal Systems for Commercial and Institutional Buildings* (Kavanaugh and Rafferty 1997). The loop consists of three circuits, each with 10 vertical U-tubes connected to a supply and return close-header. A total flow rate of 86 gpm (325 L/min) is specified for the system. Obviously there will be flow imbalances between the U-tubes near the close-headers compared to those that are more distant. Note the total length of tubing for the near U-tubes is 540 ft (2 × 260 ft vertical + 2 × 10 ft horizontal) (164 m [2 × 79 m vertical + 2 × 3 m horizontal]). The total length of tubing for the most distant U-tubes is 620 ft (2 × 260 ft vertical + 2 × 50 horizontal) (188 m [2 × 79 m vertical + 2 × 15 m horizontal]). The resulting flow rates for the near, middle, and most distant U-tubes are computed to be 2.6, 2.87, and 3.1 gpm (9.8, 10.8, and 11.7 L/min), respectively. This represents a flow rate variation of ±9% between the three vertical heat exchangers.

The computation of heat transfer rate in each of the three vertical heat exchangers was performed using the Number of Transfer Units (NTU) procedure (ASHRAE 2013). The resulting heat transfer rates for the near, middle, and most distant U-tubes are computed to be 20,870; 20,470; and 21,910 Btu/h (6.12, 6.29, and 6.42 kW). This represents a

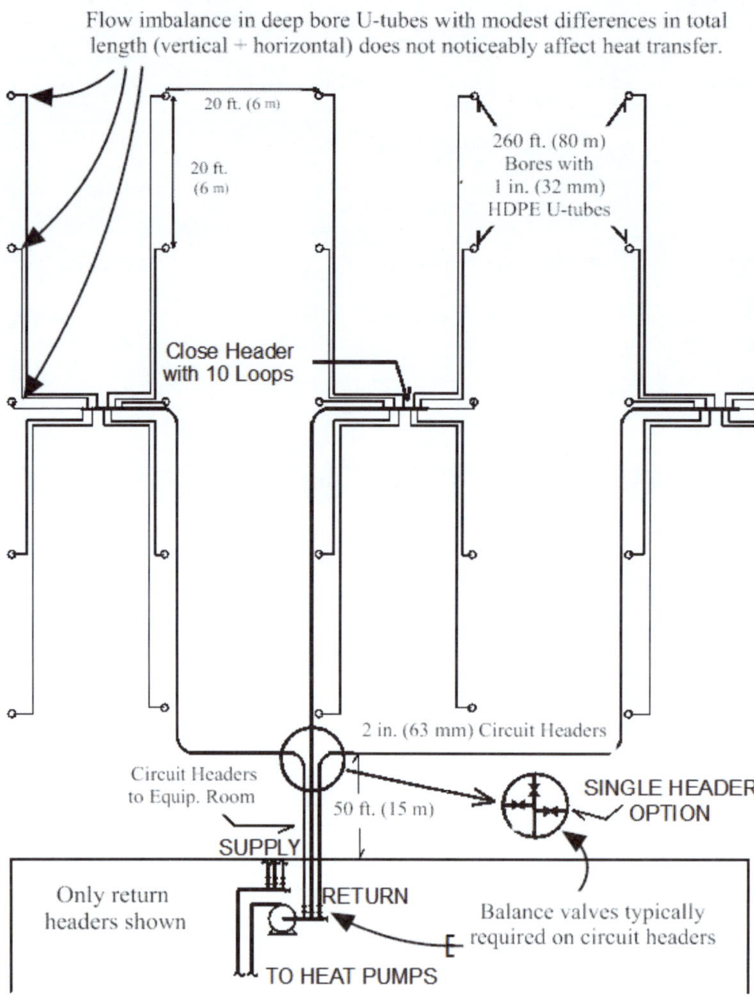

Figure I.1 Flow Imbalance and Heat Transfer Impact Example

heat transfer rate variation of less than ±3% between the three vertical heat exchangers. The overall heat transfer rate would be reduced by only 0.4% if all three U-tubes were balanced to 2.87 gpm (10.8 L/min).

It is therefore suggested that vertical heat exchanger liquid flow imbalances of up to ±15% can be tolerated with only a small impact on the overall heat transfer if the flow regime is nonlaminar.

Caution is advised against applying this concept to circuit flow imbalances as noted on Figure I.1. These sections of pipe tend to vary dramatically in overall length so imbalances are more pronounced. Thus, flow balancing is usually necessary on circuits, and it is recommended that the balancing devices have the capability of flow in both directions to allow more thorough purging.

I.2 REFERENCES

ASHRAE. 2013. *ASHRAE Handbook—Fundamentals*, “Heat Transfer,” pp. 4.21–4.23. Atlanta: ASHRAE.

Kavanaugh, S.P., and K. Rafferty. 1997. *Ground-Source Heat Pumps: Design of Geothermal Systems for Commercial and Institutional Buildings.* Atlanta: ASHRAE.

J Grain Size Classification

In the course of interpreting drilling completion reports and evaluating subsurface materials for hydrologic or thermal conductivity characteristics, terminology relating to particle size is often encountered. Table J.1 provides some of this terminology as well as commonly accepted size data for different materials.

Table J.1 Gran Size Classification

Component Description	Size, in.	Size, mm
Boulder	>10.08	>256
Cobble	2.52 to 10.08	64 to 256
Pebble	0.16 to 2.52	4 to 64
Very fine gravel	0.08 to 0.16	2 to 4
Very coarse sand	0.04 to 0.08	1 to 2
Coarse sand	0.02 to 0.04	0.5 to 1
Medium sand	0.01 to 0.02	0.25 to 0.5
Fine sand	0.005 to 0.01	0.125 to 0.25
Very fine sand	0.002 to 0.005	0.063 to 0.125
Silt	0.0002 to 0.002	0.004 to 0.063
Clay	<0.0002	<0.004
The United States Geological Survey subdivides the Pebble classification into the following:		
Very coarse gravel	1.26 to 2.52	32 to 64
Coarse gravel	0.63 to 1.26	16 to 32
Medium gravel	0.31 to 0.63	8 to 16
Fine gravel	0.16 to 0.31	4 to 8

K Well Drilling Methods

The type of well drilling method employed in a particular project is a function of a number of issues, but principal among them are the nature of the materials through which the well will be drilled, the diameter of the well, the presence or absence of water, and the depth at which the water is expected to be encountered. Within the context of GSHP systems, the materials through which the well or borehole will be drilled exerts the most impact. Though other methods are available for unusual conditions (most outside of what would typically be encountered in GSHP projects), four primary drilling methods are used in GSHP projects: cable tool, conventional rotary (also known as *mud rotary*), air rotary, and air hammer.

K.1 CABLE TOOL DRILLING

Cable tool drilling is the oldest method of well drilling, with examples of the technique present as early as 4000 years ago (Driscoll 1986). Unlike familiar rotary drilling methods, cable tool operations are reciprocating in nature. A heavy bit suspended on a cable is repeatedly raised and dropped on the subsurface materials to break up and loosen them. Cable rigs (Figure K.1) tend to be smaller than most rotary drilling rigs. They are typically operated by a deck engine turning a flywheel to which is attached a pittman arm. As the flywheel rotates, the pittman arm raises and lowers the spudder beam (the orange-colored portion of the rig in Figure K.1), to which the drilling line is attached via pulley. When a short interval of material (a few feet [metres]) is sufficiently crushed and loosened, the bit is removed from the hole and a device referred to as a *bailer* is inserted. The bailer is a 10 or 20 ft (3 to 6 m) section of pipe with a valve in the bottom. Different configurations of valve are available to suit different drilling conditions. As the bailer is lowered to the bottom of the hole, the loose cuttings enter the lower portion. Raising the bailer closes the valve in the bottom of the device, and the cuttings are brought to the surface and released from the bailer. The bit is then reinserted and drilling continues.

In unconsolidated formations, casing can be driven as the drilling operation progresses. Various methods are available for driving the casing into the hole. One approach involves using a weight (called a *casing clamp*) attached to the drill stem just above the bit. The weight is raised and dropped onto a fitting (drive head) attached to the top of the casing, and the casing is driven with a pile driver type action into the formation. A hardened "shoe" fitted to the bottom of the casing prevents deformation of the casing as it is driven into the formation. Typically the casing is driven a few feet at a time, either ahead

Figure K.1 Cable Tool Drilling Rig

of the bit or following the bit, depending on the formation. Drilling and bailing is then repeated. A second approach is to drive the casing with hydraulic jacks. As the ability to drive the casing against the resistance imposed by the borehole decreases, it is sometimes necessary to change to a smaller casing diameter as drilling progresses.

As this description suggests, the process required to remove the bit, bail, and drive casing then reinsert the bit and begin drilling again is cumbersome and time consuming. As a result cable, tool drilling proceeds much more slowly than most rotary drilling operations.

Despite this, cable tool methods offer a number of advantages that cause this type of rig to be used in specific applications. Among the more important of these are that this drilling method requires less water than other methods, it offers lower prospects for contamination of the formation since water is used for the drilling fluid, and it produces more accurate samples of each drilling interval and more accurate estimates of yield as drilling progresses. Both rig and labor costs are low compared to rotary drilling methods. This type of rig is unlikely to be used for the construction of closed-loop boreholes but remains an option for groundwater heat pump (GWHP) wells under certain circumstances.

K.2 CONVENTIONAL ROTARY DRILLING

Conventional or direct rotary drilling is often referred to as *mud rotary* due to the most common drilling fluid: a mix of water, bentonite clay, and additives. Figure K.2 presents the major components of a small rotary drilling rig. Rotary drilling equipment

Figure K.2 Rotary Drilling Rig
Printed with permission of Anderson Engineering & Surveying, Inc.

consists of four basic subsystems: motive power, rotating equipment, hoisting equipment, and circulating equipment.

The rig in Figure K.2 is powered by the truck's engine through a hydraulic system. Larger rigs sometimes have a separate deck engine for powering the various systems on the rig. In either case, the engine provides the power necessary to operate the drilling fluid pump, winches, and hydraulic pumps and to rotate the drill string, which is the term used to refer to the drill pipe, bit, and drill collars collectively.

Rotating equipment consists of the drill string and the device that imparts the rotary motion to the drill string. Two general types of drive arrangements are available for creating the rotating motion in the drill string. The original arrangement on rotary rigs was a rotating element known as a *table* located near the ground on the back of the rig at the base of the mast. The table is driven by the rig's engine and is equipped with a hole (square or circular with splines) in the center through which the kelly passes. The kelly, configured to mate to the shape of the hole in the table, is the element that connects to and turns the drill string. A second design, known as a *top head drive* eliminates the table and drives the drill string directly from an overhead hydraulic motor connected to the drill pipe. The advantage of the top head drive is faster handling of drill pipe; its limitation is less torque capability than the kelly drive arrangement (Driscoll 1986). Most wells drilled for GWHP systems use the top drive arrangement (Figure K.3).

The drilling fluid is circulated down the drill string and out through the bit. The drill string serves as both a conduit for the drilling fluid and the mechanical drive for the bit itself. As the bit turns, crushing, chipping, and loosening the subsurface materials, the

Figure K.3 Rotary Drilling Rig—Top Drive
Printed with permission of Anderson Engineering & Surveying, Inc.

fluid carries the cuttings to the surface as it passes up between the drill string and the borehole wall. The fluid also provides lubrication for the bit. At the surface, the drilling fluid is diverted to a tank or pit that serves several purposes—it separates out the cuttings from the fluid and it serves as a reserve drilling fluid reservoir and as a mixing vessel to accommodate additives to the drilling fluid. In many cases, such as that illustrated in Figure K.4, several vessels are used in series and in parallel to accommodate the functions of cuttings separation, mud storage, and mixing. The drilling fluid is drawn from the pit or tank by the mud pump and delivered to a device known as a *swivel* (on the top drive unit in Figure K.3) from which it reenters the drill string to begin the process again. In addition to the mechanical functions of cuttings removal and lubrication, the drilling fluid also supports the borehole by forming a "filter cake" on the surface of bore wall. This function, in unconsolidated formations, eliminates or greatly reduces the need for installing casing to support the hole as in the case of cable tool or air rotary drilling under similar conditions.

A variety of bits are available for different conditions, but in general drag or fishtail bits are used in unconsolidated materials and roller-cone bits are used for consolidated materials. Fishtail bits are flat steel designs similar in appearance to a large chisel, intended to break up softer formation materials. Roller-cone bits (Figure K.2) are equipped with hardened inserts on the rotating cutting surfaces and are intended to break up harder formations.

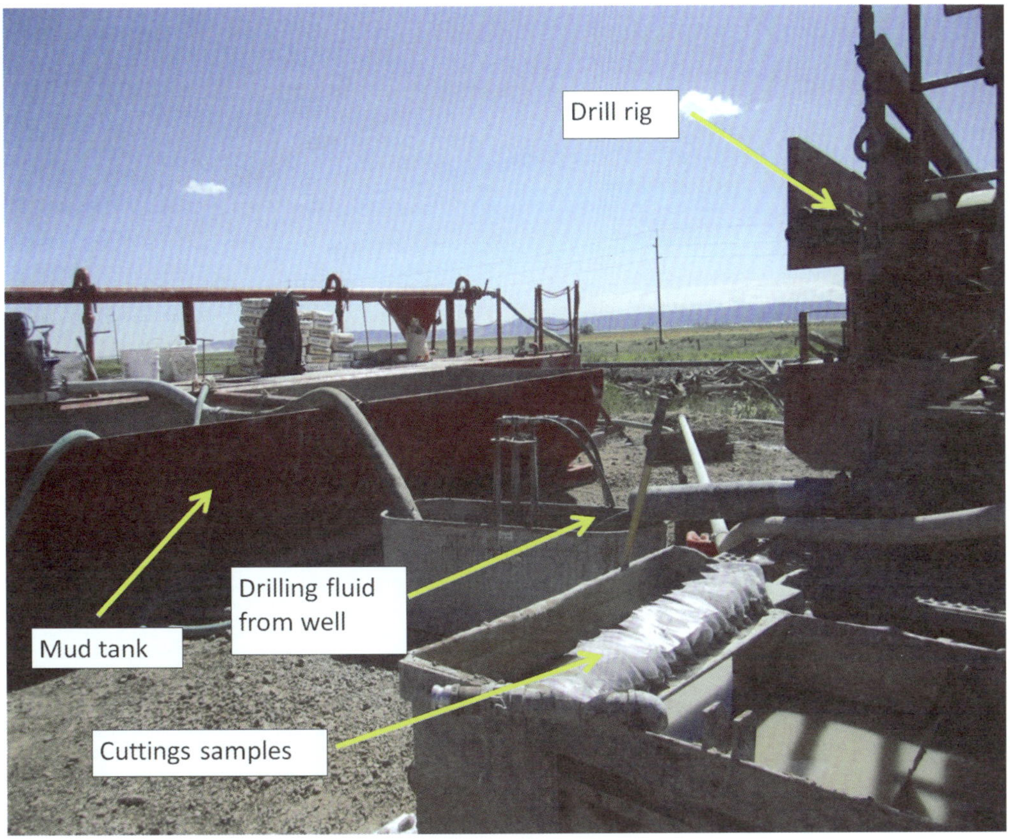

Figure K.4 Rotary Drilling Rig—Mud System
Printed with permission of Anderson Engineering & Surveying, Inc.

The two principal advantages of mud rotary drilling are the ability to drill through many unconsolidated materials without the need to advance casing as the drilling progresses and the speed of penetration achieved by this drilling method. On the negative side, drilling mud, if not carefully controlled, can invade water producing (or injection) zones and seriously damage permeability. The selection and control of the drilling fluid is a key aspect of successful rotary drilling. Because cuttings travelling up the annular space can mix with material eroded from the borehole, the quality of cuttings samples (in terms of the reliability as representative of a particular drilled interval) produced from direct rotary drilling can be lower than those of some other drilling methods.

K.3 AIR ROTARY DRILLING

A variation of direct rotary drilling uses air as the drilling fluid rather than mud. Referred to as *air rotary drilling*, it eliminates the prospect of mud contamination in production zones. It is not effective in unconsolidated settings unless casing is driven as drilling progresses. Many air rigs are equipped with a mud pump to allow drilling through surface unconsolidated materials. The operation can then be changed to air drilling when rock is encountered.

The basic operation of air drilling rigs as shown in Figures K.5 and K.6 are very similar to mud rotary rigs in terms of the components and the flow of the drilling fluid. The mud pump is replaced by an air compressor, and the need for pits or tanks for the drilling

Figure K.5 Air Rotary Drilling Rig—Side View
Printed with permission of Anderson Engineering & Surveying, Inc.

fluid is reduced. The key to effective air drilling is maintaining adequate airflow to lift the cuttings to the surface without excessive airflow that can cause erosion of the borehole and blowouts. The flow of air required is a function of the diameters of the hole and the drill pipe and the nature of the material penetrated. When drilling with air alone, water entering the borehole in small amounts can cause mud to form on borehole and drill string surfaces. Sufficient buildup of mud can cause problems with obstruction of airflow if not monitored closely. Additives can be injected into the air to facilitate drilling in various conditions. Controlled addition of water to the airstream (referred to as *air-mist drilling*) reduces dust and can help to control mud buildup caused by the uncontrolled entrance of water in the borehole. The cooling effect of the added water normally requires additional airflow to compensate for and maintain adequate uphole velocity. When air mist is used as the drilling fluid, only small amounts of water entering the borehole from the formation can be tolerated.

To accommodate larger volumes of water entering the borehole, water along with a surfactant is added to the airstream, resulting in the formation of a foam. Air foam drilling also reduces the airflow required due to the higher density of the foam drilling fluid. Uphole velocities with foam are in the range of 50 to 1000 ft/min (15 to 300 m/min), whereas velocities of 3000 to 5000 ft/min (900 to 1500 m/min) are required for air alone (Driscoll 1986). The reduced uphole velocity associated with foam allows less-consolidated formation to be drilled compared to air drilling. In addition, the foam adds some stability to the borehole walls.

Figure K.6 Air Rotary Rig—Rear View
Printed with permission of Anderson Engineering & Surveying, Inc.

A further increase in foam can be accomplished through the addition of small amounts of polymer or bentonite to the air/water/surfactant mixture. This increases the lifting capacity of the fluid beyond standard foam capabilities and is referred to as *stiff-foam drilling*.

K.4 AIR HAMMER DRILLING

Although air drilling performs well in consolidated rock settings for very hard rock (granite, basalt, and similar rocks), a pneumatic hammer-like device can be attached to the bottom of the drill string to increase the penetration rate in these settings. Referred to as *downhole hammer drilling*, it is more effective in hard-rock conditions than standard air drilling. The hammer operates from the compressed airstream supplied by the drill pipe and rotates slowly (typically < 30 rpm) while delivering blows to the rock. Carbide "buttons" on the cutting surface of the hammer serve to break up the rock, and the cuttings are carried to the surface by the air.

Table K.1 Drilling Method Performance Comparison (Adapted from Driscoll 1986)

Formation	Drilling Method				
	Cable Tool	Mud Rotary	Air Rotary	Air Hammer	Air Rotary w/ Casing Driver
Sand	difficult	fast	NR	NR	Very fast
Sand and gravel	difficult	fast	NR	NR	Very fast
Glacial drift	Slow to difficult	difficult	NR	NR	Very fast
Silt and clay	slow	fast	NR	NR	fast
Shale (firm)	fast	fast	NR	NR	fast
Shale (sticky)	slow	fast	NR	NR	fast
Shale (brittle)	fast	fast	NR	NR	fast
Sandstone (poorly cemented)	moderate	moderate	NR	NR	NR
Sandstone (well cemented)	moderate	slow	fast	NR	NR
Limestone	fast	fast	fast	Very fast	NR
Limestone (fractured)	fast	slow*	fast	Very fast	NR
Limestone (karst)	fast	slow to difficult*	difficult	Very fast	NR
Dolomite	fast	fast	fast	fast	NR
Basalt layers in sedimentary rock	fast	slow	fast	Very fast	NR
Basalt (massive)	slow	slow	moderate	fast	NR
Basalt (fractured)	slow	difficult*	slow	slow	NR
Marble, schist, quartzite, slate	slow	slow	moderate	fast	NR
Granite	slow	slow	Fast	fast	NR

**Not permitted in some regulatory jurisdictions*
NR = not recommended

K.5 DRILLING METHOD EFFECTIVENESS

Drilling methods vary in their effectiveness in different geologic conditions. Of the methods mentioned, mud rotary is most effective in unconsolidated conditions such as sand and gravel or clays or mixtures of these materials. Air or foam rotary works well in limestones, sandstones, and shales. It can also be effective in harder rocks if the bit is equipped with carbide inserts. For the hardest rocks such as basalt, chert, dolomite, and granite, the downhole hammer is the most effective strategy. Table K.1 provides a comparison of selected drilling methods in various geologic formations.

K.6 REFERENCE

Driscoll, F.G. 1986. *Groundwater and Wells*, 2d Ed. St. Paul, MN: Johnson Screens.

L Well Flow Test and Water Chemistry Analysis Specification

Carefully specifying the requirements of well flow testing and water chemistry analysis is the key to securing the necessary data for system design. The following text provides example specification language for the tests.

The contractor shall furnish, install, and remove the necessary measurement and pumping equipment capable of pumping to the point of discharge a minimum of _____ gpm (L/s), with a pumping level of ______ ft (m). Throttling or flow control means shall be provided such that the discharge may be reduced to ____ gpm (L/s). The pumping equipment shall be complete with ample power source, fuel, and controls to operate without interruption for a period of _____ hours.

Prior to starting the pump, the water level in the production well and any monitoring wells shall be measured and recorded hourly for a period of ____ hrs. Data shall be recorded and submitted in the same fashion as the flow test data.

The well shall be "step" tested at rates of approximately 25%, 50%, 75%, and 100% of the design capacity of ____ gpm (L/s). The contractor shall operate the pump in such a manner as to maintain the discharge rate within a range of ±5% at each step. Flow measurements shall be made with a device capable of producing an accuracy of ±___% over the entire range of expected flow rates. Water level measurements may be accomplished with an electric sonde equipped with calibrated wireline or a downhole pressure transducer. The reference point for water level measurements with an electric sonde shall be clearly recorded on the test data sheets.

Measurements of the pumping rate and well water level shall be made every 1 minute for the first 10 minutes of the test, every 2 minutes for the next 10 minutes, every 5 minutes for the next 40 minutes, every 15 minutes for the next hour and every 30 minutes thereafter. This sequence shall be repeated after each change in discharge rate. Discharge rate shall be changed after 3 consecutive equal water level readings are recorded or as directed by the owner. Recovery water level measurements, after the pump is stopped, shall be made at the same intervals until the water level reaches the pretest value.

In the event of a pump failure or other interruption of the test, the well water level shall be allowed to recover to the pretest value or until monitoring of well water level yields three consecutive readings of equal value prior to restart of the test.

Discharge Water

Discharge water shall be directed, through approved piping, to ____________________. The contractor shall allow no damage to occur to adjacent property as a result of flooding, leaking, or unauthorized discharge of water from the test.

Records

The contractor shall record complete records of the test results and furnish, within 24 hours of test completion, copies to the owner or his representative. The records shall also be available to the owner or his representative at any time during the test. For each well the report shall include well name or designation, depth, casing diameter, screen description, setting depth, length, a description of the reference point used for water level measurement, casing top elevation, description of instrumentation employed for water level, and water flow measurement. Records shall include the date of the test, time of pump start, elapsed time for each step, and time of each measurement, and each measurement shall include discharge rate, well water level, and any pertinent comments on test conditions (water turbidity, etc.). Frequency of measurements shall be as specified above.

Sand Content Testing

The sand content shall be determined by averaging the results of 5 samples collected at the following times during the pump test: (1) 15 minutes after the pump start, (2) after 25% of the test time has elapsed, (3) after 50% of the planned test time has elapsed, (4) after 75% of the planned test time has elapsed, and (5) near the end of the pump test.

The minimum volume (in gallons [litres]) of water sample collected for testing for sand content shall be no less than 5% of the pump test rate in gpm (L/s) or 50 gal (190 L) for test rates in excess of 1000 gpm (18 L/s).

In the event of the use of centrifugal sand testing equipment (Rossum Sand Tester), samples shall be taken at the same intervals specified above and the sand content reported in the same manner as specified below.

Sampling shall be done at an access port located along the centerline (3 o'clock or 9 o'clock position) of the discharge line or at the outfall and shall be collected at a discharge no less than 90% of the expected design flow rate. Samples shall be allowed to settle for a minimum of 10 minutes before the liquid is decanted. A data sheet recording the results of the test, including flow rate, test start time, time of sample collection, and volumes collected, and the well name or designation shall be furnished to the owner or his representative with 24 hours of completion of the test. The sand collected shall be carefully enclosed in sealed plastic bags and labeled with the date and time of collection. Sand samples shall be delivered to the owner of his representative with 24 hours of the test completion.

Sampling for Chemical and Biological Analysis

A sample of water at least 32 oz (1 L) shall be collected for microbiological analysis. A sterile sample bottle provided or approved by the testing laboratory shall be used. Nothing but the water to be analyzed shall be allowed to contact the bottle or cap, and the bottle shall not be rinsed prior to sampling. The water must not be allowed to contact the sampler's hands or other objects prior to entering the bottle. The sample shall be collected after the pump has been operated for a period of at least ____ minutes and delivered to the laboratory no more than 24 hours after the sample is collected. Analysis shall be conducted for the presence of iron bacteria, sulfate-reducing bacteria, and slime-forming bacteria species.

A sample of 64 oz (2 L) of water shall be collected for the purpose of chemical analysis. The sample shall be collected in a bottle approved or furnished by the laboratory performing the analysis. Nothing but the water to be analyzed shall be allowed to contact the bottle or cap, and the bottle shall not be rinsed prior to sampling. The water must not be allowed to contact the sampler's hands or other objects prior to entering the bottle. The sample shall be collected after the pump has been operated for a period of at least ____ minutes and delivered to the laboratory no more than 24 hours after the sample is collected. Results shall be reported for the following parameters:

pH	Chloride (Cl)
Total Dissolved Solids (TDS)	Carbonate (CO_3)
Iron (Fe)	Bicarbonate (HCO_3)
Total (M) Alkalinity	Hydrogen Sulfide (H_2S)
Phenolphthalein (P) Alkalinity	Carbon Dioxide (CO_2)
Sulfate (SO_4)	Oxygen (O)
Calcium (Ca)	Manganese (Mn)
Iron Bacteria	Total Hardness
Slime-Forming Bacteria	Sulfate-Reducing Bacteria
Langlier Saturation Index (LSI)	Ryznar Stability Index (RSI)

Stability and saturation index values shall be calculated at aquifer temperature (_____ °F [_____ °C]) and at ______°F (______°C) (maximum temperature of surfaces encountered—160°F [71°C] for direct use in heat pumps, 85°F [29°C] for systems with isolation heat exchangers).

M Example Well Chemical and Biological Analysis Results

This appendix provides an example of a water chemistry analysis and the accompanying report from the laboratory explaining and interpreting the data (Schnieders 2005). It is included here to provide the reader with a well-executed example of such material. The report is used with the permission of Michael J. Schnieders, PG, PH-GW, Water Systems Engineering, Inc. It has been edited to conform to the style of this book but otherwise includes all of the original data and text.

M.1 EXAMPLE

	Casing Sample, mg/L	Aquifer Sample, mg/L
Phenolphthalein Alkalinity*	0	0
Total Alkalinity*	212	212
Hydroxide Alkalinity	0	0
Carbonate Alkalinity	0	0
Bicarbonate Alkalinity	212	212
pH	8.5	8.5
Chlorides (as Cl)	11.2	10.8
Total Dissolved Solids (TDS)	389	429
Conductivity	505	557
Total Hardness*	192	244
Carbonate Hardness	192	212
Non-Carbonate Hardness	0	32
Calcium*	128	132
Magnesium*	64	112
Sodium (as Na)	26.2	27.1
Potassium (as K)	2.5	2.6
Phosphate (as PO_4)	0.4	0.6
Dissolved Iron (as Fe^{2+})	0	0
Suspended Iron (as Fe^{3+})	0.5	0.1
Total Iron (as Fe)	0.5	0.1
Iron (Resuspended)	0.8	0.0

Copper (as Cu)	0.2	0.1
Tannin/Lignin	1.8	0.3
Nitrate (Nitrogen)	0	0
Sulfate (as SO_4)	81	94
Silica (as SiO_2)	25.5	23.5
Manganese (as Mn)	0	0.2
Saturation Index	+0.7	+0.71
Chlorine (as Cl)	0	0
Total Organic Carbon	2.3	1.0
Oxygen Reduction Potential (ORP)	272 mV	245 mV
Biological Results		
Plate Count	>300	>300
Sulfate-Reducing Bacteria	Positive	Negative
Nitrate-Reducing Bacteria	Positive	Positive
Anaerobic Growth	50%	20%
Adenosine Triphosphate (ATP) (Cells per mL)	1,300,000	614,000
Total Coliform Bacteria	Negative	Negative
E. coli Coliform Bacteria	Negative	Negative
* *as $CaCO_3$*		

Microscopic Evaluation

Casing: Low visible bacterial activity, minor amount of crystalline debris, moderate iron oxide, no sheathed or stalked bacteria noted. Bacterial identification: *Acinetobacter calcoaceticus bv alc Comamonas testosterone Vagococcus fluvialis.*

Aquifer: Low visible bacterial activity, moderate amount of clay particulate matter, trace of crystalline debris and iron oxide, no sheathed or stalked bacteria noted. Bacterial identification: *Delftia acidovorans Pseudomonas pseudocaligenes ss alcaligenes.*

Observations

Chemical analysis of the submitted samples identified some variation in conditions observed in the two samples. Each sample exhibited elevated alkalinity, a slightly alkaline pH, moderate levels of total dissolved solids, and a calculated positive saturation index. The casing sample showed typical calcium to magnesium relationship with preference for calcium carbonate deposition. Iron levels were higher within the casing and at a level at which iron oxide deposition is expected. The tannin/lignin level and the total organic carbon content were both elevated in the casing sample, likely indicating the concentration of organic material due to biological activity. Within the aquifer sample, the magnesium and calcium concentrations were near equal, indicating the potential for calcium-magnesium carbonate (dolomite) deposition. As the profile shows, there appears to be a magnesium loss within the casing, which may indicate magnesium oxide or complex carbonate-hydroxide precipitation within the well. Manganese was high within the aquifer sample and at a sufficient level for the potential deposition of manganese dioxide. The phosphate concentration was elevated in both samples and may indicate residual drilling mud or developmental chemistry remaining downhole.

Biological analyses of the samples identified very high bacterial population levels of strong growth potential. Anaerobic among the bacteria was present in both samples and at considerably higher level in the casing sample. The casing sample tested positive for sulfate-reducing bacteria and nitrate-reducing bacteria. The sulfate-reducing bacteria were at a population level that would indicate considerable presence of this organism in the well sump and adjacent formation. The nitrate reducers are more indicative of activity within the aquifer. The lack of nitrates in either sample indicates an active nitrate reducer population, but more to the point it presents an active aerobic population in the aquifer, most probably sufficient to keep the aquifer fairly well cleaned through oxidation of any organics.

Bacterial identification of the two predominant species present within the casing sample identified two aerobic species not uncommon in well environments (*Acinetobacter* and *Comamonas*) The *Comanonas* species is known as a prolific slime (biofilm) producer. Also identified in the casing as a predominant species was the *Vagococcus* specie, a facultative anaerobe that is likely reflective of the high anaerobic growth observed in testing. Within the aquifer sample the two predominant species are common bacteria also known as prolific slime producers.

Microscopic evaluation sited low visible bacterial activity and no evidence of larger sheathed or stalked bacteria within the casing and aquifer samples. The casing sample did contain a minor amount of crystalline debris and moderate iron oxide accumulations. The aquifer sample contained a moderate amount of clay particulate and trace accumulations of crystalline debris and iron oxide.

Interpretation

The analyses show that the well has strong potential for fouling of several types, including mineral scale buildup, biofouling, and physical fouling from the accumulation of clay and crystalline particulate. Mineral accumulations expected within the well would most likely be carbonate in nature, with secondary composition of iron oxide and manganese dioxide. The heavy bacterial presence will likely enhance mineral scale development as well as contribute to the fouling process through biomass accumulation. The crystalline debris and clay particulate evident in both samples suggest that physical fouling of sediment fine components is likely within the borehole-aquifer interface zone.

The phosphate ion concentration may indicate that polymer-enhanced drilling mud and/or phosphate-based development fluids may remain downhole. Such phosphate-based products can remain within a borehole for many years, restricting flow and potentially stimulating biological activity. The concentrations as identified should be compared to regional aquifer background levels for further evaluation.

M.2 REFERENCE

Schnieders, M.J. 2005. Water Treatment Analysis and Control Report No. 16638, Water Systems Engineering Inc, Ottawa, KS, 1 Sept. Reprinted with permission.

N Well Problems and Strategies to Avoid Them

N.1 UNDERSTANDING WELL PROBLEMS

Wells, like any other system component, require careful design, diligent monitoring, and occasional maintenance if they are to provide reliable service. Unfortunately, one or more of these issues is often inadequately addressed. As with most problems, these issues are best dealt with in the design phase, or as early as possible if any are discovered after the system is in operation. Key to understanding and detecting well problems are monitoring and periodic data review. Operating wells should be monitored for production rate, drawdown, specific capacity, and sand content on a regular basis. The test should be conducted under similar conditions and the data recorded. Periodic review of trends in the data—reduced flow, increased drawdown, reduced specific capacity, and increasing sand content can indicate problems. A change of 20% in any of these parameters indicates action is required.

The most common water well problems include the following (Driscoll 1986):

- Incrustation and/or biofouling of screens
- Plugging of formation by fine components
- Sand pumping
- Casing or screen collapse
- Pump problems

Diagnosing many well problems is greatly assisted by performing a video examination of the well. In this procedure, a camera is lowered into the well and the resulting video provides very useful input to the determination of the problem as well as the most effective remedial approach. When problems are encountered involving scale or biological fouling, samples should be collected and the material analyzed by a lab experienced in groundwater issues.

Incrustation can arise from a number of factors, but precipitation of carbonate scale and iron on the screen are two of the most common occurrences. In both cases excessive drawdown can be a contributing factor in addition to a symptom. As the water level in a well is reduced, the pressure in the lower portion of the aquifer is reduced. For aquifers in which the water contains dissolved carbon dioxide (CO_2), the reduced pressure can lead to the evolution of the CO_2 from the water and a subsequent rise in the pH of the water. For water of a scaling character (see Chapter 7), this condition can result in precipitation of carbonate scale in the production zone and on the well screen.

Ferrous iron present in the water can also be influenced by drawdown. As the upper portion of the formation is dewatered in an unconfined aquifer, the water-filled pore spaces are replaced with air. During the pump off cycle, as the upper formation refills with water it is exposed to the air and ferrous iron present in the water is oxidized to the ferric state. Ferric iron has very low solubility in the water and is deposited on the surfaces of the formation materials and the well screen. If the well's cone of depression reaches a surface body (lake or river) and water saturated with oxygen is drawn into an aquifer with ferrous iron, fouling can also result.

Obviously operating with minimum drawdown reduces these problems. Minimizing drawdown can be accomplished with good well design practices (low water entrance velocity, careful gravel pack selection) and the use of minimum groundwater flow rates. This means selecting the minimum groundwater flow that will result in acceptable system performance (see Figure 8.6 and Table 8.16). Excessive groundwater flow rates produce higher drawdown, exacerbating the conditions discussed.

Treatment of incrustation is most often accomplished through the introduction of acid into the well water. Hydrochloric, sulfamic, and hydroxyacetic (glycolic) are the most common acids (Driscoll 1986). Hydrochloric acid offers low cost; sulfamic offers safe handling, as it is available in granular form; and hydrooxyacetic acid provides the additional benefit of serving as an effective biocide. Agitation or surging is recommended after the acid is introduced into the well. If iron or magnesium deposits are present in addition to the carbonate scale, the addition of a chelating agent is recommended if sulfamic or hydrochloric acid is used. Hydrooxyacetic acid provides a chelating function in addition to dissolving the scale (Driscoll 1986).

Biofouling of wells, particularly from iron bacteria, is a frequently misunderstood phenomenon and one that is often inadequately addressed in terms of treatment when an infestation occurs. *Iron bacteria* is a general term referring to a variety of organisms that metabolize ferrous iron and in the process secrete a gelatinous material that can severely clog wells and screens. Though often thought to "eat" iron alloy components in systems, iron bacteria do not directly attack these materials. They can create conditions, particularly under the thick gelatinous secretions, where accelerated corrosion can occur. The primary maintenance issue arising from an infestation of iron bacteria is obstruction of water flow entering the well—usually in the near-well zone or at the screen itself.

It is possible to identify iron bacteria by microscopic examination of a sample of material removed from a well. In terms of prediction of the likelihood of an iron bacteria infestation, one of the best strategies is to interview existing well owners in the area to determine their experience, if any, with biological problems. Water samples can be tested in the laboratory or field bacteriological activity reaction tests (BARTs) can be done. Because of the ubiquitous nature of bacteria in groundwater, it is not uncommon for tests to return positive results—but this does not necessarily mean that serious problems will arise. For this reason a combination of testing and existing experience with the aquifer is the preferred approach. One study of municipal well maintenance requirements determined that most biofouling problems occurred in unconsolidated materials and alluvial formations and were not a problem in wells completed in rock formations.

Treatment of an iron bacteria infestation can be accomplished with a variety of means, including biocides, heat, and ultraviolet light, but the most common is chlorine. For chlorine to be effective it must be applied in the correct dosage, at the correct pH range, for an adequate period of time along with necessary procedures on the well. The first step is to remove as much of the gelatinous mass associated with the bacteria as possible. This minimizes the biological material present in the well and reduces the chlorine dosage. In any disinfection use of chlorine, the concentration, both in terms of dosage and residual, is critical. Dosage is the amount of chlorine added to the water, and residual is the concentration left after a portion of the dosage is consumed by biological material in the well. In the case of iron bacteria, a sufficient dosage must be added to leave a residual of 50 to 200 ppm. Excessive concentrations of chlorine are not effective, as this tends to raise the pH into the range where the chlorine is not effective as a biocide. The form of the chlorine (hypochlorous acid) effective on the organisms is maximized at a pH of <8.0, preferably <7. Acid or proprietary chemical products are available for pH control. An exposure time of 18 to 24 h is necessary to completely kill the organisms in the treated area. Surging of the well either with the well pump or a well service rig is recommended to ensure that the treatment chemical reaches into the formation surrounding the well. Finally, the well should undergo redevelopment after the treatment, with brushing followed by jetting or surging. The nature of iron bacteria is to recolonize at some point, but the time between treatments is a direct function of the effectiveness of the treatment.

Plugging of a well with fine components or excessive sand production is often a natural consequence of the nature of the aquifer materials, and given sufficient time it will reduce well performance. It can also be a result of inadequate development, poor well design, or excessive flow rates. The importance of screen and gravel pack selection is discussed in Chapter 7, but it bears repeating as the selection of these materials is sometimes left to the contractor or vendor. Inaccurate matching of the well completion to the aquifer materials can result in poor performance. In addition, overpumping a well or operating it in excess of its intended yield also can result in excessive sand production. Inadequate development is also a cause of excessive sand production. If fine components in the near-bore zone are not fully removed in the development process, high suspended solids content in the production water can result. In aquifers prone to sand production, variable-speed operation of the well pump (instead of cycling a single-speed pump) can reduce the problem. Accumulation of fine components and some sand issues can be addressed through periodic redevelopment of the well. If sand production persists, surface separation using a strainer is strongly advised in applications where injection is the means of disposal. Centrifugal separators are not designed to achieve the levels of sand content necessary for injection. Ineffective performance can occur on start-up and if variable flow is used, as there may be insufficient velocity to achieve effective separation under peak flow conditions. Strainers are effective under all flow conditions. Depending on the size of the sand to be removed, it may be necessary to use multiple strainers in parallel to control pressure drop.

Casing and screen failure is often a result of poor material selection, inadequate wall thickness, lack of properly accounting for collapse strength (plastic casing and screens), or installation practices that are inappropriate for the screen type or casing. The material selection should be made with full understanding of the water chemistry present.

Though there is little in the way of regular maintenance that can be done on submersible well pumps, monitoring of flow and current draw is a useful strategy. Replacement is likely at some point in the life of the system, but for submersible pumps a service life of at least 10 years should be expected (15 to 20 years for lineshaft pumps) provided the pump is installed and operated according to the guidelines in Chapter 8. Many pump failures are attributable to cycling frequency above recommendations, motor overheating arising from inadequate water flow past the motor, and lightning strikes. Executing a design that includes protection from or avoidance of these issues is described Section 8.5.

The frequency of maintenance required on a well is a function of a variety of factors, including the geologic characteristics of the formation, the extent to which the well is used (continuous or intermittent pumping), the care with which it was designed and constructed, and the level of monitoring it receives. A study of municipal wells that were pumped continuously at maximum yield indicated that major maintenance intervals (defined as maintenance that amounted to a cost of approximately 10% of the well replacement cost) was a strong function of geology (Gass n.d.). Wells completed in metamorphic rock exhibited the longest service between major maintenance, 12 to 15 years. Wells completed in sandstone, limestone, and basalt showed a maintenance interval average of 6 to 12 years. Wells completed in interbedded sandstone and shale or a combination of consolidated and unconsolidated materials required major maintenance every 5 to 8 years. Wells completed in alluvium were the most maintenance intensive, requiring major maintenance on average every 2 to 5 years. The study noted that wells were not always constructed according to best practices. It is likely that the poorly constructed wells represent the shorter intervals between major service and the better-constructed wells the longer intervals between required service.

N.2 REFERENCES

Driscoll, F.G. 1986. *Groundwater and Wells*, 2d Ed. St. Paul, MN: Johnson Screens.

Gass, T.E., T.W. Bennett, J. Miller, R. Miller. n.d. *Manual of Water Well Maintenance and Rehabilitation Technology.* Reprinted by the National Water Well Association from the Robert S. Kerr Environmental Research Center, USPA, Ada, Oklahoma.

O Heat Exchanger Temperature Prediction Spreadsheet

O.1 SPREADSHEET TOOL

Heat Exchanger Temperature Prediction Spreadsheet, which is available with this book at www.ashrae.org/GSHP, is a convenient tool to estimate the surface area requirements for a heat exchanger and to predict the performance of a heat exchanger operating at flows and temperatures other than those for which it was designed. The equations on which the spreadsheet operates are based on the method described by Petrosky (n.d.).

In the context of groundwater heat pump (GWHP) design, this spreadsheet is most useful for predicting the heating-mode performance of a plate heat exchanger designed for the cooling mode.

Figures O.1 and O.2 are screenshots of the spreadsheet input and output sections. Table O.1 provides the equations available in the spreadsheet and their locations.

In using the spreadsheet, the first task is to identify the system of units desired by entering a 1 for Units (I-P or SI). The spreadsheet will operate in either mode, but all values must be entered in a consistent unit system and using the units identified to the right of the input fields. The overall U-factor for the exchanger is entered in input #1. Clean overall U-factors for water-to-water applications in plate exchangers generally result in values of between 700 and 1200 Btu/h·ft^2·°F (4000 to 6800 W/m^2·°C). Next a trial surface area is entered (input #2). This value will be adjusted later to meet the required load on the exchanger. The entering temperatures on the groundwater and building loop sides of the exchanger are entered in inputs #3 and #4. These values are available from the system performance calculation (Tables 8-16 and 8-17). Specific gravity and specific heat values for the groundwater and building loop fluid (usually plain water) are entered in input fields #7 through #10. The calculation is first run for the cooling-mode conditions. After the surface area requirement is established for the cooling mode, this value remains fixed and a trial heating-mode building loop entering temperature is made. The loop entering temperature is manipulated until the heat exchanger capacity matches the required heating load.

Using the design example from Section 8.7, the heat exchanger selection is based upon the cooling-mode duty as follows:

- Groundwater side = 141 gpm (8.89 L/s) entering at 54°F (12.2°C) and leaving at 72.6°F (25.6°C)

Units	IP	SI		
	1	0		
1. Overall "U"			900.00	Btu/h sq ft F or W/sq m C
2. Area			200.00	sq ft or sq m
3. Entering temp loop			76.60	F or C
4. Entering temp GW			54.00	F or C
5. Flow rate loop			248.00	gpm or lps
6. Flow rate GW			141.00	gpm or lps
7. Spec. gravity loop			1.00	
8. Spec. gravity GW			1.00	
9. Specific heat loop			1.00	Btu/lb F or W/kg C
10. Specific heat GW			1.00	Btu/lb F or W/kg C

Figure O.1 Heat Exchanger Temperature Prediction Spreadsheet Input Section

OUTPUT	in	out	
Hot Side	76.60	66.03	°F / °C
Cold Side	54.00	72.60	°F / °C
Capacity	1310463	Btu/hr	or Kw

Figure O.2 Heat Exchanger Temperature Prediction Spreadsheet Output Section

- Building loop side = 248 gpm (15.6 L/s), entering at 76°F (24.8°C) and leaving at 66°F (18.9°C)
- Load = 1,310,000 Btu/h (384 kW)

Assuming an overall U-factor of 900 Btu/h·ft^2·°F (5110 W/m^2·°C), a trial surface area of 250 ft^2 (23.2 m^2) is entered. The resulting capacity is 1,389,963 Btu/h (407 kW), which is much higher than the required load.

A new surface of 190 ft^2 (17.7 m^2) results in a capacity of 1,290,522 Btu/h (378 kW).

A new surface of 200 ft^2 (18.6 m^2) results in a capacity of 1,310,463 Btu/h (384 kW), which is sufficiently close to the required capacity of 1,310,000 Btu/h (384 kW).

For the heating-mode performance, the example system heating-mode calculation (Table 8-17) indicates a building loop return temperature of 42.7°F (5.9°C) the 141 gpm (8.89 L/s) groundwater flow rate. This suggests a heat pump COP of 3.92. Based on the space-heating load of 800000 Btu/h (234 kW), this requires a heat exchanger duty of approximately 596,000 Btu/h (175 kW). In evaluating the performance of the heat exchanger in the heating mode, it is necessary to reduce the overall U-factor to reflect the impact of heating-mode operating temperatures on the water viscosity. In this case a new U-factor of 825 Btu/h·ft^2·°F (4685 W/m^2·°C) is used. Entering the 42.7°F (5.9°C) value

Table O.1 Equations in Heat Exchanger Temperature Prediction Spreadsheet

Equation	Location
I-P	C4
SI	C5
Overall U-factor	E5
Specific Heat GW	E14
Building loop in =IF(D4=1,E7,K7)	C17
Building loop out =IF(D4=1,((I8-32)/1.8),I8)	D18
GW in =IF(D4=1,E8,K8)	C18
GW out =IF(D4=1,((I6-32)/1.8),I6)	D18
Capacity =IF(D4=1,I9/3412,I9)	C20
=K9*8.33*K11*K13*60	I4
=K10*8.33*K12*K14*60	I5
=((I4/I5)*(K7-I8))+K8	I6
=2.7128^((K5*K6)*((1/I4)-(1/I5)))	I7
=(((I5*K8)*(1-I7))-(K7*(I5-I4)))/(I4-(I5*I7))	I8
=K10*8.33*K12*K14*60*(I6-K8)	I9
=K10*8.33*K12*K14*60*(I6-K8)	I11
=IF(D4=1,E5/1.736,E5)	K5
=IF(D4=1,E6*3.28*3.28,E6)	K6
=IF(D4=1,(E7*1.8)+32,E7)	K7
=IF(D4=1,(E8*1.8)+32,E8)	K8
=IF(D4=1,E9/0.063,E9)	K10
=IF(D4=1,E10/0.063,E10)	K11
=E12	K12
=IF(D4=1,E13/4.186,E13)	K13
=IF(D4=1,E14/4.186,E14)	K14

as a starting point, the capacity of the heat exchanger is 638,143 Btu/h (187 kW)—more than the required load.

A new building loop entering temperature of 43.7°F (6.5°C) results in a capacity of 581,671 Btu/h (170 kW), which is slightly less than the requirement.

A new building loop entering temperature of 43.4°F (6.3°C) results in a capacity of 598,612 Btu/h (175 kW), which is close to the requirement.

This suggests that the heating-mode performance of the heat exchanger would actually be slightly better than that assumed in the Table 8.17 calculations of Chapter 8. The calculations assumed an approach of 3°F (1.7°C), and the estimated performance, based on the spreadsheet, is an approach of 45.5°F – 43.4°F (7.5°C – 6.3°C), or 2.1°F (1.2°C).

In evaluating heat exchangers for GWHP applications, the heating mode will frequently be the condition for which the heat exchanger performance is evaluated. The lower temperatures in the heating mode in addition to the potentially lower flow rates will have a negative impact on the overall U-factor compared to cooling-mode operation. To estimate the impact of these parameters on performance, the following may be used in the absence of more specific information on the heat exchanger:

- Reduced flow impact on overall U-factor:

$$\text{Heating-mode overall U-factor} = \left(\frac{\text{Groundwater heating-mode gpm}}{\text{Groundwater cooling-mode gpm}}\right)^{0.8} \times \text{Cooling-mode overall U-factor}$$

- Water viscosity correction to overall U-factor:

$$\text{Heating-mode overall U-factor} = \left[\frac{\left(\frac{1}{\mu_1}\right)^{0.8}(\mu_1)^{0.3}}{\left(\frac{1}{\mu_2}\right)^{0.8}(\mu_2)^{0.3}}\right] \times \text{Cooling-mode U-factor}$$

where

μ_1 = absolute viscosity of groundwater at heating-mode mean temperature

μ_2 = absolute viscosity of groundwater at cooling-mode mean temperature

The corrections for flow and viscosity can be multiplied to arrive at a total correction to cooling-mode U-factor for heating-mode U-factor.

O.2 REFERENCE

Petrosky, J.T. n.d. Direct calculation of heat exchanger exit temperatures, calculation and shortcut deskbook. *Chemical Engineering Magazine*, New York: McGraw-Hill.

Index

A

B

C

D

E

F

G

H

I

L

M

N

O

P

R

S

T

U

V

W